Arthritis Research

METHODS IN MOLECULAR MEDICINE™

John M. Walker, SERIES EDITOR

139. **Vascular Biology Protocols,** edited by *Nair Sreejayan and Jun Ren,* 2007
138. **Allergy Methods and Protocols,** edited by *Meinir G. Jones and Penny Lympany,* 2007
137. **Microtubule Protocols,** edited by Jun Zhou, 2007
136. **Arthritis Research:** *Methods and Protocols, Vol. 2,* edited by *Andrew P. Cope,* 2007
135. **Arthritis Research:** *Methods and Protocols, Vol. 1,* edited by *Andrew P. Cope,* 2007
134. **Bone Marrow and Stem Cell Transplantation,** edited by *Meral Beksac,* 2007
133. **Cancer Radiotherapy,** edited by *Robert A. Huddart and Vedang Murthy,* 2007
132. **Single Cell Diagnostics:** *Methods and Protocols,* edited by *Alan Thornhill,* 2007
131. **Adenovirus Methods and Protocols, Second Edition, Vol. 2:** *Ad Proteins, RNA, Lifecycle, Host Interactions, and Phylogenetics,* edited by *William S. M. Wold and Ann E. Tollefson,* 2007
130. **Adenovirus Methods and Protocols, Second Edition, Vol. 1:** *Adenoviruses, Ad Vectors, Quantitation, and Animal Models,* edited by *William S. M. Wold and Ann E. Tollefson,* 2007
129. **Cardiovascular Disease:** *Methods and Protocols, Volume 2: Molecular Medicine,* edited by *Qing K.Wang,* 2006
128. **Cardiovascular Disease:** *Methods and Protocols, Volume 1: Genetics,* edited by *Qing K. Wang,* 2006
127. **DNA Vaccines:** *Methods and Protocols, Second Edition,* edited by *Mark W. Saltzman, Hong Shen, and Janet L. Brandsma,* 2006
126. **Congenital Heart Disease:** *Molecular Diagnostics,* edited by *Mary Kearns-Jonker,* 2006
125. **Myeloid Leukemia:** *Methods and Protocols,* edited by *Harry Iland, Mark Hertzberg, and Paula Marlton,* 2006
124. **Magnetic Resonance Imaging:** *Methods and Biologic Applications,* edited by *Pottumarthi V. Prasad,* 2006
123. **Marijuana and Cannabinoid Research:** *Methods and Protocols,* edited by *Emmanuel S. Onaivi,* 2006
122. **Placenta Research Methods and Protocols:** *Volume 2,* edited by *Michael J. Soares and Joan S. Hunt,* 2006
121. **Placenta Research Methods and Protocols:** *Volume 1,* edited by *Michael J. Soares and Joan S. Hunt,* 2006
120. **Breast Cancer Research Protocols,** edited by *Susan A. Brooks and Adrian Harris,* 2006
119. **Human Papillomaviruses:** *Methods and Protocols,* edited by *Clare Davy and John Doorbar,* 2005
118. **Antifungal Agents:** *Methods and Protocols,* edited by *Erika J. Ernst and P. David Rogers,* 2005
117. **Fibrosis Research:** *Methods and Protocols,* edited by *John Varga, David A. Brenner, and Sem H. Phan,* 2005
116. **Inteferon Methods and Protocols,** edited by *Daniel J. J. Carr,* 2005
115. **Lymphoma:** *Methods and Protocols,* edited by *Timothy Illidge and Peter W. M. Johnson,* 2005
114. **Microarrays in Clinical Diagnostics,** edited by *Thomas O. Joos and Paolo Fortina,* 2005
113. **Multiple Myeloma:** *Methods and Protocols,* edited by *Ross D. Brown and P. Joy Ho,* 2005
112. **Molecular Cardiology:** *Methods and Protocols,* edited by *Zhongjie Sun,* 2005
111. **Chemosensitivity:** *Volume 2, In Vivo Models, Imaging, and Molecular Regulators,* edited by *Rosalyn D. Blumethal,* 2005
110. **Chemosensitivity:** *Volume 1, In Vitro Assays,* edited by *Rosalyn D. Blumethal,* 2005
109. **Adoptive Immunotherapy:** *Methods and Protocols,* edited by *Burkhard Ludewig and Matthias W. Hoffman,* 2005
108. **Hypertension:** *Methods and Protocols,* edited by *Jérôme P. Fennell and Andrew H. Baker,* 2005
107. **Human Cell Culture Protocols,** *Second Edition,* edited by *Joanna Picot,* 2005
106. **Antisense Therapeutics,** *Second Edition,* edited by *M. Ian Phillips,* 2005
105. **Developmental Hematopoiesis:** *Methods and Protocols,* edited by *Margaret H. Baron,* 2005
104. **Stroke Genomics:** *Methods and Reviews,* edited by *Simon J. Read and David Virley,* 2004
103. **Pancreatic Cancer:** *Methods and Protocols,* edited by *Gloria H. Su,* 2004
102. **Autoimmunity:** *Methods and Protocols,* edited by *Andras Perl,* 2004
101. **Cartilage and Osteoarthritis:** *Volume 2, Structure and In Vivo Analysis,* edited by *Frédéric De Ceuninck, Massimo Sabatini, and Philippe Pastoureau,* 2004
100. **Cartilage and Osteoarthritis:** *Volume 1, Cellular and Molecular Tools,* edited by *Massimo Sabatini, Philippe Pastoureau, and Frédéric De Ceuninck,* 2004
99. **Pain Research:** *Methods and Protocols,* edited by *David Z. Luo,* 2004
98. **Tumor Necrosis Factor:** *Methods and Protocols,* edited by *Angelo Corti and Pietro Ghezzi,* 2004
97. **Molecular Diagnosis of Cancer:** *Methods and Protocols, Second Edition,* edited by *Joseph E. Roulston and John M. S. Bartlett,* 2004
96. **Hepatitis B and D Protocols:** *Volume 2, Immunology, Model Systems, and Clinical Studies,* edited by *Robert K. Hamatake and Johnson Y. N. Lau,* 2004

METHODS IN MOLECULAR MEDICINE™

Arthritis Research

Methods and Protocols

Volume 2

Edited by

Andrew P. Cope

*The Kennedy Institute of Rheumatology
Imperial College London, London UK*

HUMANA PRESS ✸ TOTOWA, NEW JERSEY

© 2007 Humana Press Inc.
999 Riverview Drive, Suite 208
Totowa, New Jersey 07512

www.humanapress.com

All rights reserved. No part of this book may be reproduced, stored in a retrieval system, or transmitted in any form or by any means, electronic, mechanical, photocopying, microfilming, recording, or otherwise without written permission from the Publisher. Methods in Molecular Biology™ is a trademark of The Humana Press Inc.

All papers, comments, opinions, conclusions, or recommendations are those of the author(s), and do not necessarily reflect the views of the publisher.

This publication is printed on acid-free paper. ∞
ANSI Z39.48-1984 (American Standards Institute)
Permanence of Paper for Printed Library Materials.

Cover design by Nancy K. Fallatt.

Cover illustration: *(Background)* Figure 1 from Chapter 24 from volume 1. *(Inset)* Figure 2 from Chapter 15 from volume 1.

For additional copies, pricing for bulk purchases, and/or information about other Humana titles, contact Humana at the above address or at any of the following numbers: Tel.: 973-256-1699; Fax: 973-256-8341; E-mail: orders@humanapr.com; or visit our Website: www.humanapress.com

Photocopy Authorization Policy:
Authorization to photocopy items for internal or personal use, or the internal or personal use of specific clients, is granted by Humana Press Inc., provided that the base fee of US $30.00 per copy is paid directly to the Copyright Clearance Center at 222 Rosewood Drive, Danvers, MA 01923. For those organizations that have been granted a photocopy license from the CCC, a separate system of payment has been arranged and is acceptable to Humana Press Inc. The fee code for users of the Transactional Reporting Service is: [978-1-58829-918-5/07 $30.00].

Printed in the United States of America. 10 9 8 7 6 5 4 3 2 1

eISBN: 978-1-59745-402-5

Library of Congress Cataloging in Publication Data
Arthritis research : methods and protocols / edited by Andrew P. Cope.
    p. ; cm. — (Methods in molecular biology ; v. 135-136)
  Includes bibliographical references and index.
  ISBN 1-58829-344-0 (v. 1 : alk. paper) — ISBN 1-58829-918-X (v. 2 : alk. paper)
  1. Arthritis—Laboratory manuals. 2. Arthritis—Molecular aspects. I. Cope, Andrew P. II. Series: Methods in molecular biology (Clifton, N.J.) ; v. 135-136.
  [DNLM: 1. Arthritis—Laboratory Manuals. 2. Laboratory Techniques and Procedures—Laboratory Manuals. W1 ME9616J v.135-136 2007 / WE 25 A787 2007]
RC933.A665245 2007
616.7'220072—dc22
                                        2006019975

Preface

> "................ do not go where the path may lead,
> go instead where there is no path and leave a trail"
>
> Ralph Waldo Emerson

The postgenomic era is upon us and with it comes a growing need to understand the function of every gene and its contribution to physiological and pathological processes. Such advances will underpin our understanding of the molecular basis of common chronic inflammatory and degenerative diseases and inspire the development of targeted therapy. Any postgenomic approach for exploring gene function must necessarily address gene expression and regulation, localization of gene products in diseased tissue, manipulation of expression by transgenesis or knockdown technology, and combine these studies with appropriate manipulations in relevant in vivo models. To validate potential therapeutic targets in any depth requires a growing repertoire of assays and disease models that underpin key pathogenic pathways. The same repertoire of tools must be employed to rigorously evaluate process specific biomarkers, which may be of diagnostic and prognostic value. Indeed, measuring the impact of our interventions remains a major challenge for the future.

The rheumatic diseases encompass prototypic chronic inflammatory and degenerative diseases. It would be true to say that experimental procedures adapted for investigating the pathogenesis of diseases such as rheumatoid arthritis have contributed greatly to recent advances in biological therapy. *Arthritis Research: Methods and Protocols* seeks to crystallize methods and protocols that have contributed to such advances in molecular medicine. These volumes are timely because the tools are now accessible to most laboratories. Also included are newer technologies, some of which are still evolving and whose impact are yet to be realized. It is important to note that in these volumes there is something for everyone—basic scientists, clinician scientists, and clinicians alike—with contributions from leaders in their field covering imaging and immunobiology, animal models, and new technologies. Combine volumes 1 and 2 and the end product is a concise set of protocols condensing decades of experience and expertise. From the outset of this project it was always the intention that this compendium should provide a unique resource at the bench that would be used in ways that will facilitate the endeavors of clinicians at the bedside in the future.

Acknowledgments

I wish to thank many friends and colleagues for their enthusiasm, support, and invaluable contributions toward this project. I am also very grateful to Mandy Wilcox for her dedicated secretarial assistance in compiling the finished product. The research carried out by the Editor's laboratory at the Kennedy Institute is supported by grants from the Wellcome Trust and the Arthritis Research Campaign, UK.

Andrew P. Cope

Contents

Preface v
Ackowlegments vi
Contributors xiii
Color Plate xix
Contents for Volume 1 xxi

PART I IMMUNOBIOLOGY

1 Phenotypic Analysis of B-Cells and Plasma Cells
  *Henrik E. Mei, Taketoshi Yoshida, Gwendolin Muehlinghaus,
  Falk Hiepe, Thomas Dörner, Andreas Radbruch,
  and Bimba F. Hoyer* 3

2 Detection of Antigen Specific B-Cells in Tissues
  Marie Wahren-Herlenius and Stina Salomonsson 19

3 Single Cell Analysis of Synovial Tissue B-Cells
  Hye-Jung Kim and Claudia Berek 25

4 Tracking Antigen Specific CD4+ T-Cells with Soluble
  MHC Molecules
  John A. Gebe and William W. Kwok 39

5 Analysis of Antigen Reactive T-Cells
  *Clare Alexander, Richard C. Duggleby, Jane C. Goodall,
  Malgosia K. Matyszak, Natasha Telyatnikova,
  and J. S. Hill Gaston* 51

6 Identification and Manipulation of Antigen Specific T-Cells
  with Artificial Antigen Presenting Cells
  *Eva Koffeman, Elissa Keogh, Mark Klein, Berent Prakken,
  and Salvatore Albani* 69

7 Analysis of Th1/Th2 T-Cell Subsets
  Alla Skapenko and Hendrik Schulze-Koops 87

8 Analysis of the T-Cell Receptor Repertoire of Synovial T-Cells
  Lucy R. Wedderburn and Douglas J. King 97

9 The Assessment of T-Cell Apoptosis in Synovial Fluid
  *Karim Raza, Dagmar Scheel-Toellner, Janet M. Lord,
  Arne N. Akbar, Christopher D. Buckley,
  and Mike Salmon* 117

10 Assay of T-Cell Contact Dependent Monocyte-Macrophage Functions
Danielle Burger and Jean-Michel Dayer 139

11 Phenotypic and Functional Analysis of Synovial Natural Killer Cells
Nicola Dalbeth and Margaret F. C. Callan 149

12 Identification and Isolation and Synovial Dendritic Cells
Allison R. Pettit, Lois Cavanagh, Amanda Boyce, Jagadish Padmanabha, Judy Peng, and Ranjeny Thomas 165

PART II ANIMAL MODELS OF ARTHRITIS

13 The Use of Animal Models for Rheumatoid Arthritis
Rikard Holmdahl 185

14 Collagen-Induced Arthritis in Mice
Richard O. Williams 191

15 Collagen-Induced Arthritis in Rats
Marie M. Griffiths, Grant W. Cannon, Tim Corsi, Van Reese, and Kandie Kunzler 201

16 Collagen Antibody Induced Arthritis
Kutty Selva Nandakumar and Rikard Holmdahl 215

17 Arthritis Induced with Minor Cartilage Proteins
Stefan Carlsen, Shemin Lu, and Rikard Holmdahl 225

18 Murine Antigen-Induced Arthritis
Wim van den Berg, Leo A. B. Joosten, and Peter L. E. M. van Lent 243

19 Pristane-Induced Arthritis in the Rat
Peter Olofsson and Rikard Holmdahl 255

20 The K/BxN Mouse Model of Inflammatory Arthritis: Theory and Practice
Paul Monach, Kimie Hattori, Haochu Huang, Elzbieta Hyatt, Jody Morse, Linh Nguyen, Adriana Ortiz-Lopez, Hsin-Jung Wu, Diane Mathis, and Christophe Benoist 269

21 Analysis of Arthritic Lesions in the Del1 Mouse: A Model for Osteoarthritis
Anna-Marja Säämänen, Mika Hyttinen, and Eero Vuorio 283

PART III APPLICATION OF NEW TECHNOLOGIES TO DEFINE NOVEL THERAPEUTIC TARGETS

22 Gene Expression Profiling in Rheumatology
   Tineke C. T. M. van der Pouw Kraan, Lisa G. M. van Baarsen, François Rustenburg, Belinda Baltus, Mike Fero, and Cornelis L. Verweij 305

23 Differential Display Reverse Transcription-Polymerase Chain Reaction to Identify Novel Biomolecules in Arthritis Research
   Manir Ali and John D. Isaacs 329

24 Two-Dimensional Electrophoresis of Proteins Secreted from Articular Cartilage
   Monika Hermansson, Jeremy Saklatvala, and Robin Wait 349

25 Mapping Lymphocyte Plasma Membrane Proteins: A Proteomic Approach
   Matthew J. Peirce, Jeremy Saklatvala, Andrew P. Cope, and Robin Wait 361

26 In Vivo Phage Display Selection in the Human/SCID Mouse Chimera Model for Defining Synovial Specific Determinants
   Lewis Lee, Toby Garrood, and Costantino Pitzalis 369

27 Adenoviral Targeting of Signal Transduction Pathways in Synovial Cell Cultures
   Alison Davis, Corinne Taylor, Kate Willetts, Clive Smith, and Brian M. J. Foxwell 395

Index 421

Contributors

ARNE N. AKBAR • *Department of Immunology and Molecular Pathology, Royal Free and University College Medical School, London, UK*
SALVATORE ALBANI • *Departments of Medicine and Paediatrics, University of California San Diego, La Jolla, CA*
CLARE ALEXANDER • *University of Cambridge Addenbrooke's Hospital, Department of Medicine, Cambridge, UK*
MANIR ALI • *Leeds Institute of Molecular Medicine, University of Leeds, St. James's University Hospital, Leeds, UK*
BELINDA BALTUS • *Department of Molecular Cell Biology and Immunology, VU Medical Centre, Amsterdam, The Netherlands*
CHRISTOPHE BENOIST • *Section of Immunology and Immunogenetics, Joslin Diabetes Center, Boston, MA*
CLAUDIA BEREK • *Deutsches RheumaForschungszentrum, Berlin, Germany*
AMANDA BOYCE • *Centre for Immunology and Cancer Research, University of Queensland, Princess Alexandra Hospital, Brisbane Qld 4102, Australia*
CHRISTOPHER D. BUCKLEY • *MRC Centre for Immune Regulation, Institute of Biomedical Research Building, University of Birmingham, Birmingham, UK*
DANIELLE BURGER • *Clinical Immunology Unit, University Hospital, Geneva, Switzerland*
MARGARET F. C. CALLAN • *Department of Immunology, Imperial College London, Hammersmith Hospital Campus, London, UK*
GRANT W. CANNON • *Salt Lake City Veteran Affairs Health Care System Research Service and University of Utah, Division of Rheumatology, Salt Lake City, UT*
STEFAN CARLSEN • *Medical Inflammation Research, Lund University, Lund, Sweden*
LOIS LORNA CAVANAGH • *Centre for Immunology and Cancer Research, University of Queensland, Princess Alexandra Hospital, Brisbane Qld 4102, Australia*
ANDREW COPE • *The Kennedy Institute of Rheumatology Division, Imperial College London, London, UK*
TIM CORSI • *Salt Lake City Veteran Affairs Health Care System Research Service and University of Utah, Division of Rheumatology, Salt Lake City, UT*
NICOLA DALBETH • *Department of Immunology, Imperial College London, Hammersmith Hospital Campus, London, UK*

ALISON DAVIS • *Kennedy Institute of Rheumatology Division, Imperial College London, London, UK*
JOHN-MICHEL DAYER • *Clinical Immunology Unit (Hans Wilsdorf Laboratory), University Hospital, Geneva, Switzerland*
THOMAS DÖRNER • *Institute of Transfusion Medicine, Charité, Berlin, Germany*
RICHARD CHARLES DUGGLEBY • *University of Cambridge, Addenbrooke's Hospital, Department of Medicine, Cambridge, UK*
MIKE FERO • *Department of Functional Genomics, Stanford University School of Medicine, Stanford, CA*
BRIAN M. J. FOXWELL • *Kennedy Institute of Rheumatology Division, Imperial College London, London, UK*
TOBY GARROOD • *Department of Academic Rheumatology, GKT School of Medicine, King's College London, Guy's Hospital, London, UK*
J. S. HILL GASTON • *University of Cambridge, Addenbrooke's Hospital, Department of Medicine, Cambridge, UK*
JOHN A GEBE • *Benaroya Research Institute, Seattle, WA*
JANE CAROLINE GOODALL • *University of Cambridge, Addenbrooke's Hospital, Department of Medicine, Cambridge, UK*
MARIE M. GRIFFITHS • *University of Utah, Department of Medicine/Division of Rheumatology, School of Medicine, Salt Lake City, UT*
KIMIE HATTORI • *Section of Immunology and Immunogenetics, Joslin Diabetes Center, Boston, MA*
MONIKA HERMANSSON • *The Kennedy Institute of Rheumatology Division, Imperial College London, London, UK*
FALK HIEPE • *Clinic for Internal Medicine, Rheumatology and Clinical Immunology, Charité, Berlin, Germany*
RIKARD HOLMDAHL • *Medical Inflammation Research, Lund University, Lund, Sweden*
BIMBA F. HOYER • *Clinic for Internal Medicine, Rheumatology and Clinical Immunology, Charité, Berlin, Germany*
HAOCHU HUANG • *Section of Immunology and Immunogenetics, Joslin Diabetes Center, Boston, MA*
ELZBIETA HYATT • *Section of Immunology and Immunogenetics, Joslin Diabetes Center, Boston, MA*
MIKA HYTTINEN • *Departmnt of Anatomy, Institute of Biomedicine, University of Kuopio, Kuopio, Finland*
JOHN D. ISAACS • *Institute of Cellular Medicine, University of Newcastle Upon Tyne, Newcastle Upon Tyne, UK*
LEO A. B. JOOSTEN • *Rheumatology Research and Advanced Therapeutics, Radboud University Nijmegen Medical Centre, Nijmegen, The Netherlands*

Contributors

ELISSA KEOGH • *Departments of Medicine and Paediatrics, University of California San Diego, La Jolla, CA*
MARK KLEIN • *IACOPO Institute for Translational Medicine, La Jolla, CA*
EVA KOFFEMAN • *Departments of Medicine and Paediatrics, University of California San Diego, La Jolla, CA*
HYE-JUNG KIM • *Memorial Sloan Kettering Cancer Center, Department of Immunology, New York, NY*
DOUGLAS J. KING • *Department of Immunology and Molecular Pathology, Windeyer Institute, UCL, London, UK*
KANDIE KUNZLER • *Salt Lake City Veteran Affairs Health Care System Research Service and University of Utah, Division of Rheumatology, Salt Lake City, UT*
WILLIAM W. KWOK • *Benaroya Research Institute, Seattle, WA*
LEWIS LEE • *Department of Rheumatology, School of Medicine, King's College London, Guy's Hospital, London, UK*
JANET M. LORD • *MRC Centre for Immune Regulation, University of Birmingham, Birmingham, UK*
SHEMIN LU • *Medical Inflammation Research, Lund University, Lund, Sweden*
DIANE MATHIS • *Section of Immunology and Immunogenetics, Joslin Diabetes Center, Boston, MA*
MALGOSIA K. MATYSZAK • *University of Cambridge, Addenbrooke's Hospital, Department of Medicine, Cambridge, UK*
HENRIK E. MEI • *German Rheumatism Research Centre, Berlin, Germany*
PAUL MONACH • *Section of Immunology and Immunogenetics, Joslin Diabetes Center, Boston, MA*
JODY MORSE • *The Jackson Laboratory, Bar Harbor, ME*
GWENDOLIN MUEHLINGHAUS • *German Rheumatism Research Centre, Berlin, Germany*
KUTTY S. NANDAKUMAR • *Medical Inflammation Research, Lund University, Lund, Sweden*
LINH NGUYEN • *Section of Immunology and Immunogenetics, Joslin Diabetes Center, Boston, MA*
PETER OLOFSSON • *Medical Inflammation Research, Lund, Sweden*
ADRIANA ORTIZ-LOPEZ • *Section of Immunology and Immunogenetics, Joslin Diabetes Center, Boston, MA*
JAGADISH PADMANABHA • *Centre for Immunology and Cancer Research, University of Queensland, Princess Alexandra Hospital, Brisbane Qld 4102, Australia*
MATTHEW PEIRCE • *Kennedy Institute of Rheumatology Division, Imperial College London, London, UK*

JUDY PENG • *Centre for Immunology and Cancer Research, University of Queensland, Princess Alexandra Hospital, Brisbane Qld 4102, Australia*
ALISON R. PETTIT • *Centre for Immunology and Cancer Research, University of Queensland, Princess Alexandra Hospital, Brisbane Qld 4102, Australia*
COSTANTINO PITZALIS • *GKT School of Medicine, King's College London, Guy's Hospital, London, UK*
BERENT PRAKKEN • *IACOPO Institute for Translational Medicine, La Jolla, CA*
ANDREAS RADBRUCH • *German Rheumatism Research Centre, Berlin, Germany*
KARIM RAZA • *MRC Centre for Immune Regulation, Institute of Biomedical Research Building, University of Birmingham, Birmingham, UK*
VAN REESE • *Salt Lake City Veteran Affairs Health Care System, Research Service and University of Utah, Division of Rheumatology, Salt Lake City, UT*
FRANÇOIS RUSTENBURG • *Department of Molecular Cell Biology and Immunology, VU Medical Centre, Amsterdam, The Netherlands*
ANNA-MARJA SÄÄMÄNEN • *Department of Medical Biochemistry and Molecular Biology, University of Turku, Turku, Finland*
JEREMY SAKLATVALA • *The Kennedy Institute of Rheumatology Division, Imperial College London, London, UK*
MIKE SALMON • *MRC Centre for Immune Regulation, Institute of Biomedical Research Building, University of Birmingham, Birmingham, UK*
STINA SALOMONSSON • *Rheumatology Unit, Department of Medicine, Karolinska Hospital, Stockholm, Sweden*
DAGMAR SCHEEL-TOELLNER • *MRC Centre for Immune Regulation, University of Birmingham, Birmingham, UK*
HENDRIK SCHULZE-KOOPS • *Rheumatologie, Medizinische Poliklinik–Innenstadt, München, Germany*
ALLA SKAPENKO • *Nikolaus Fiebiger Center for Molecular Medicine, Clinical Research Group III, University of Erlangen-Nuremberg, Erlangen, Germany*
CLIVE SMITH • *Kennedy Institute of Rheumatology Division, Imperial College London, London, UK*
CORINNE TAYLOR • *Kennedy Institute of Rheumatology Division, Imperial College London, London, UK*
NATASHA TELYATNIKOVA • *University of Cambridge, Addenbrooke's Hospital, Department of Medicine, Cambridge, UK*
RANJENY THOMAS • *Centre for Immunology and Cancer Research, University of Queensland, Princess Alexandra Hospital, Brisbane Qld 4102, Australia*
LISA G. M. VAN BAARSEN • *Department of Molecular Cell Biology and Immunology, VU Medical Centre, Amsterdam, The Netherlands*
WIM VAN DEN BERG • *Radboud University Nijmegen Medical Centre, Nijmegen, The Netherlands*

TINEKE C. T. M. VAN DER POUW KRAAN • *Department of Molecular Cell Biology and Immunology, VU Medical Centre, Amsterdam, The Netherlands*

PETER L. E. M. VAN LENT • *Rheumatology Research and Advanced Therapeutics, Radboud University Nijmegen Medical Centre, Nijmegen, The Netherlands*

DOUGLAS J. VEALE • *St. Vincent's University Hospital, Dublin, Ireland*

CORNELIUS L. VERWEIJ • *VU Medical Centre, Department of Molecular Cell Biology and Immunology, Amsterdam, The Netherlands*

EERO VUORIO • *Department of Medical Biochemistry and Molecular Biology, University of Turku, Turku, Finland*

MARIE WAHREN-HERLENIUS • *Rheumatology Unit, Department of Medicine, Karolinska Hospital, Stockholm, Sweden*

ROBIN WAIT • *The Kennedy Institute of Rheumatology Division, Imperial College London, London, UK*

LUCY R. WEDDERBURN • *Rheumatology Unit, Institute of Child Health, UCL, London, UK*

KATE WILLETTS • *Kennedy Institute of Rheumatology Division, Imperial College London, London, UK*

RICHARD O. WILLIAMS • *Kennedy Institute of Rheumatology Division, Imperial College London, London, UK*

HSIN-JUNG WU • *Section of Immunology and Immunogenetics, Joslin Diabetes Center, Boston, MA*

TAKETOSHI YOSHIDA • *German Rheumatism Research Centre, Berlin, Germany*

Color Plate

The following color illustrations are printed in the insert that follows p. 138.

Chapter 12
    Fig. 3: Identification of synovial tissue DC by 2-color immunohistochemistry for RelB and HLA-DR.
    Fig. 4: CD11c and CD123 expression by DC in RA ST.

Chapter 27
    Fig. 2: Pictorial representation of the AdEasy system plasmid vectors.
    Fig. 3: Pictorial representation of the transfer vectors pAdKS17 and pAdSK16.
    Fig. 4: Modes of in vivo bacterial recombination of the AdEasy vectors.
    Fig. 6: Schematic representation of the AdEasy Vector System for the generation of recombinant adenoviruses.

Contents for Volume 1

Preface
Acknowledgment
Contributors
Color Plate

PART I SYNOVIAL JOINT MORPHOLOGY, HISTOPATHOLOGY, AND IMMUNOHISTOCHEMISTRY

1 Imaging Inflamed Synovial Joints
   Ai Lyn Tan, Helen I. Keen, Paul Emery, and Dennis McGonagle
2 Arthroscopy as a Research Tool: *A Review*
   Richard J. Reece
3 Immunohistochemistry of the Inflamed Synovium
   Martina Gogarty and Oliver FitzGerald
4 *In Situ* Hybridization of Synovial Tissue
   Stefan Kuchen, Christian A. See,ayer, Michel Neidhart, Renate E. Gay, and Steffen Gay
5 Subtractive Hybridization
   Jörg H. W. Distler, Oliver Distler, Michel Neidhart, and Steffen Gay
6 Laser Capture as a Tool for Analysis of Gene Expression in Inflamed Synovium
   Ulf Müller-Ladner, Martin Judex, Elena Neumann, and Steffen Gay
7 Preparation of Mononuclear Cells from Synovial Tissue
   Jonathan T. Beech and Fionula M. Brennan
8 Quantitative Image Analysis of Synovial Tissue
   Pascal O. van der Hall, Maarten C. Kraan, and Paul Peter Tak

PART II CARTILAGE MATRIX AND BONE BIOLOGY

9 Cartilage Histomorphometry
   Ernst B. Hunziker
10 Image Analysis of Aggrecan Degradation in Articular Cartilage with Formalin-Fixed Samples
   Barbara Osborn, Yun Bai, Anna H.K. Plaas, and John D. Sandy
11 *In Situ* Detection of Cell Death in Articular Cartilage
   Samantha N. Redman, Ilyas M. Khan, Simon R. Tew, and Charles W. Archer
12 Measurement of Glycosaminoglycan Release from Cartilage Explants
   John S. Mort and Peter J. Roughley

13 Assessment of Collagenase Activity in Cartilage
   Tim E. Cawston and Tanya G. Morgan
14 Assessment of Gelatinase Expression and Activity in Articular Cartilage
   Rosalind M. Hembry, Susan J. Atkinson, and Gillian Murphy
15 Analysis of MT1-MMP Activity
   Richard D. Evans and Yoshifumi Itoh
16 Analysis of TIMP Expression and Activity
   Linda Troeberg and Hideaki Nagase
17 Bone Histomorphology in Arthritis Models
   Georg Schett and Birgit Tuerk
18 Generation of Osteoclasts in vitro and Assay of Osteoclast Activity
   Naoyuki Takahashi, Nobuyuki Udagawa, Yasuhiro Kobayashi, and Tatsuo Suda

PART III CELL TRAFFICKING, MIGRATION, AND INVASION
19 Isolation and Analysis of Large and Small Vessel Endothelial Cells
   Justin C. Mason, Elaine A Lidington and Helen Yarwood
20 Analysis of Flow Based Adhesion In Vitro
   Oliver Florey and Dorian O. Haskard
21 Analysis of Leukocyte Recruitment in Synovial Microcirculation by Intravital Microscopy
   Gabriela Constantin
22 Angiogenesis in Arthritis: *Methodological and Analytical Details*
   Ursula Fearon and Douglas J. Veale
23 Analysis of Inflammatory Leukocyte and Endothelial Chemotactic Activity
   Zoltan Szekanecz and Alisa Koch
24 Acquisition, Culture and Phenotyping of Synovial Fibroblasts
   Sanna Rosengren, David L. Boyle, and Gary S. Firestein
25 Genotyping of Synovial Fibroblasts: cDNA Array in Combination with RAP-PCR in Arthritis
   Elena Neumann, Martin Judex, Steffen Gay, and Ulf Müller-Ladner
26 Gene Transfer to Synovial Fibroblast – Methods and Evaluation in the SCID Mouse Model
   Ingmar Meinecke, Edita Rutkauskaite, Antje Cinski, Ulf Müller-Ladner, Steffen Gay, and Thomas Pap
27 In Vitro Matrigel Fibroblast Invasion Assay
   Tanja C.A. Tolboom and Tom W.J. Huizinga
28 Culture and Analysis of Circulating Fibrocytes
   Timothy E. Quan and Richard Bucala

Index

I

IMMUNOBIOLOGY

1

Phenotypic Analysis of B-Cells and Plasma Cells

Henrik E. Mei, Taketoshi Yoshida, Gwendolin Muehlinghaus, Falk Hiepe, Thomas Dörner, Andreas Radbruch, and Bimba F. Hoyer

Abstract

B-cells and antibody-secreting plasma cells are key players in protective immunity, but also in autoimmune disease. To understand their various functions in the initiation and maintenance of autoimmune pathology, a detailed dissection of their functional diversity is mandatory. This requires a detailed phenotypic classification of the diversity of B-cells. Here, technologies of immunocytometry and ELISpot are described in detail, and their value for phenotypic characterization of cells of the B lineage, as well as for preparative cell sorting, to further characterize them functionally and on the molecular level are described.

Key Words: B-cells; plasma cells; flow cytometry; intracellular Ig; surface Ig; secreted Ig; ELISpot; antigen specific; memory B-cells; naïve B-cells; FACS; autoimmunity; cytometric secretion assay.

1. Introduction

With rising consciousness on the key relevance of autoreactive B-cells and autoantibodies in the pathogenesis of autoimmune disease, it becomes evident that a more detailed understanding of their function(s) will require more sophisticated methods for their characterization. B-cells may contribute in several ways to autoimmune pathogenesis: systemic and organ-specific pathology can be caused by autoantibodies, and complexes of autoantibodies and autoantigens. Immunological memory, as provided by memory B-cells and memory plasma cells can contribute to the persistence of inflammation. B-cells can control activation and differentiation of autoreactive T-cells through secretion of cytokines and presentation of antigens. Deciphering these various roles in the context of B-cell differentiation and immunopathology requires methods to address B-cells of all Whereas "antibodies" were discovered more than 100 yr ago as the principle of specific immunological protection provided by serum of immunized animals *(1)*, the cells making antibodies were discovered only in 1948 by A. Fagraeus *(2)*. She described the correlation of antibody secretion to numbers of "plasma cells," large cells with extended cytoplasm, endoplasmatic reticulum, and Golgi apparatus, and a peripheral nucleus. In 1959, Nossal provided direct evidence that plasma cells secrete antibodies *(3)*. At about the same time the precursors of plasma cells were discovered, the B-lymphocytes, as a result of the depletion of humoral immunity in bursectomized birds *(4,5)*.

B-lymphocytes look like other lymphocytes, with little cytoplasm and no distinct morphological features. Their character is reflected in their molecular signature. With the advent of monoclonal antibodies (mAbs) and their use for the immunofluorescent characterization of the immune system *(6)*, the molecular signatures of B-cell differentiation could be addressed by microscopy and the newly developed technology of flow cytometry *(7)*. More importantly, fluorescence-activated cell sorting (FACS) *(8)* and magnetic cell sorting (MACS) *(9)* provide efficient means to isolate B-cells of defined differentiation stages for more detailed molecular and functional analyses. It should be noted that B=lymphocytes provide the unique chance to reverse immunofluorescence, and use fluoresceinated antigen to stain B=cells expressing specific antibodies *(10)*.

B-lymphocytes develop from hematopoietic precursor cells in the embryonic yolk sac, fetal liver, and adult bone marrow *(11)*. B-cell differentiation is guided by the transcription factor Pax5 *(12,13)*. First steps of specific B-cell differentiation are the induction of rearrangement of gene segments for the antigen-binding domain of the antibody heavy chain in proB cells. As a "large" pre-B-cell, the cell will then proliferate and express an antibody-heavy chain, associated with commensal VpreB and lambda5 chains as "surrogate light chain." As a "small" pre-B-cell, the cell will then rearrange the antigen-binding domain gene segments of an antibody light chain. As an immature B-cell, the cell will leave the bone marrow and develop in the periphery through transitional stages into a mature B-cell of the B1 or B2 lineage *(14–16)*. Upon activation by antigen, B-cells can enter marginal zones or follicles of the spleen and lymph nodes, as marginal zone B-cells or follicular B-cells. Activated B-cells will develop into antibody-secreting plasmablasts and plasma cells, under the control of the transcription factors Blimp-1 and XBP-1 *(17,18)*. Their differentiation into memory B-cells and memory plasma cells is dependent on interaction with T-helper lymphocytes *(19)*. All these differentiation stages are characterized by the expression of distinct molecular signatures, which so far have not been fully elucidated *(20)*. In **Fig. 1**, representative molecules are listed, which allow classification of the most relevant B-cell differentiation stages, and are expressed on the cell surface, so that they can be used for the isolation of viable cells by fluorescence–activated or magnetic cell sorting.

Phenotypic Analysis of B and Plasma Cells

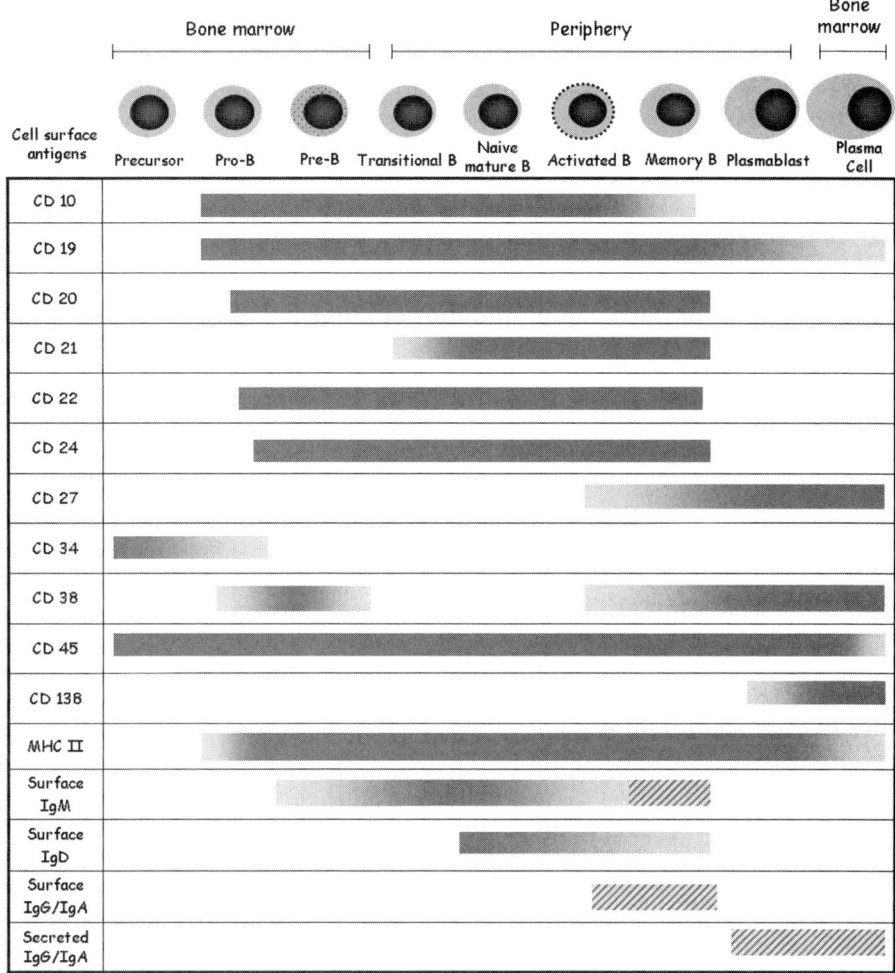

Fig. 1. Phenotypic changes during B-cell differentiation. B-cell differentiation stages are defined by phenotype. Representative stages of B-cell differentiation, their morphological features and their tissue localization are displayed. In order to distinguish between the subsets, time-frames for the expression of several markers are given (*11,15,31,33,34*). Shading intensity correlates to the level of expression, and hatched bars represent a heterogeneous behavior for a particular marker at a given maturational stage.

In principle, molecules characteristic for activation or differentiation stages of B-cells can be expressed in the cell, on its surface or can be secreted by the cell. Immunofluorescent detection of intracellular target molecules requires poration of the cell membrane, in order to get the rather large immunofluorescent staining reagents inside. This has two major disadvantages: first, the cell is killed by poration (and fixation) and can no longer be analyzed functionally, and second in that the conformation of the target molecule can be changed by fixation, decreasing staining intensity. To avoid this artefact, cells can be stained before fixation, but only with dyes that withstand fixation and do not lose fluorescence upon fixation, as do phycobiliproteins.

Immunofluorescent staining of surface molecules provides a detailed picture on the communication skills of the cell. It is technologically simple and has few limitations, (e.g., its sensitivity). In order to visualize expression of a given target molecule, a cell has to express more than 1000 copies of it. The use of magnetofluorescent liposomes allows detection of less than 100 copies/cell *(21)*. This approach has been used successfully to identify human pre-B-cells expressing the pre-B-cell receptor *(22)*.

A priori, immunofluorescence had not provided an option to characterize antibody-secreting cells according to their secreted antibodies. The original approach to characterize humoral immune responses on the single cell level was developed by Jerne and Nordin *(23)*, as the plaque-formation assay. Antibody-secreting cells embedded in agarose with antigen-coated sheep red blood cells, will appear as hemolytic plaques after addition of complement. In the meantime, this technology has been modified to directly visualize the secreted antibodies bound to an antigen-coated matrix. This solid-phase enyme-linked immunospot assay (ELISpot) technology was first described in 1983 *(24)*. Both the plaque-forming antibody-secreting cell assay and the ELISpot allow the precise enumeration of cells secreting antibodies of predefined characteristics (e.g. specificity and class of antibody), and to a certain degree also avidity, the latter by using different densities of antigen *(25)*. Any other information on the phenotype of the cell has to be circumstantial (e.g., based on sorting of the cells according to immunofluorescent phenotype prior to analysis of secreted antibodies) *(26)*. A second drawback of the plaque-forming cell and ELISpot assays is that cells cannot be sorted according to expression of secreted molecules.

In 1995, the immunofluorescent cytometric secretion assay was described for the cytometric identification and isolation of viable cells according to specific molecules they secrete *(27)*. Basically, an antibody, specific for the secreted molecule in question, is attached to the cell surface. The cell is then given a chance to secrete, and secreted molecules bound by the surface-bound antibody are detected by immunofluorescence.

2. Materials
2.1. Cell Preparation and Standard Reagents

1. Leucocyte separation medium (Ficoll-Paque PLUS, density 1.077 g/mL) (Amersham Pharmacia Biotech; Uppsala, Sweden).

2. Phosphate buffered saline (PBS): 137 mM NaCl, 2.7 mM KCl, 8.1 mM Na$_2$HPO$_4$,and 1.5 mM KH$_2$PO$_4$. Dissolve in distilled water and adjust pH to 7.2.
3. PBS/bovine serum albumin (BSA): PBS containing 0.5% BSA (Biomol, Hamburg, Germany).

2.2. Flow Cytometry

2.2.1. Detection of Surface-Bound Antigens

Monoclonal or polyclonal antibodies directed against particular antigens, conjugated to fluorescent dyes (various providers). Keep cold and protect from light.

2.2.2. Detection of Molecules of Intracellular Localization

1. 0.5 % and 0.1 % saponin (Sigma-Aldrich) in PBS/BSA.
2. 2% formaldehyde (Merck, Darmstadt, Germany) in distilled water.

2.2.3. Cytometric Antibody-Secretion Assay

1. NHS-L-biotin dissolved in 1 mg/mL PBS (Pierce; Rockford, IL).
2. RPMI 1640 medium/10% fetal calf derum (FCS) (Gibco Invitrogen; Carlsbad, CA).
3. Anti-human λ (clone HP6054; ATCC; Manassas, VA).
4. Pure anti-human λ (clone HP6054, ATCC) coupled to avidine.
5. Anti-human IgG, coupled to fluorochrome of choice (Becton Dickinson; San Jose, CA).
6. Anti-human IgG, coupled to digoxigenine (Becton Dickinson).
7. Anti-digoxigenine (anti-DIG) antibody coupled to a fluorochrome of choice.
8. Antibodies directed against particular surface molecules, coupled to different. fluorochromes other than the antibodies above and suitable for FACS.

2.3. ELISpot

2.3.1. Standard Reagents

1. Buffer 1: 15 g BSA (Biomol) in 500 mL PBS, filtrate using membranes with 0.22 λm pore size (Millipore, Bedford, MA). Store at 4°C.
2. Buffer 2: 500 mL buffer 1 and 250 µL Tween-20. Store at 4°C.
3. Roswell Park Memorial Institute Medium (RPMI), 1640 containing 10% FCS.
4. EIA/RIA flat-bottom, high-binding 96-well plates (Corning Costar; Corning, NY).
5. Streptavidin-alkaline phosphatase (SA-AP) (Hoffman-LaRoche; Indianapolis, IN).
6. 3% Agarose: 3 g agarose (low EEO) (Sigma-Aldrich), dissolve in 100 mL ddH2O. Store at 4°C.
7. AMP-buffer (stable at 4°C for at least 6 mo): 95 mL 2-amino-2-methyl-1-propanol (AMP), 0.1 mL Triton X-405, 150 mg/mL MgCl$_2$, and 900 mL H$_2$O. Adjust pH to 10.25 and store at 4°C.
8. Development buffer (prepare always fresh): 8 mg (BCIP) (Sigma-Aldrich), 8 mL AMP-buffer, incubation for 20 min at 65°C in a water bath, agitate gently, add

2 mL 3% agarose incubation for 10 min at 65°C. Best time for preparation is during the **step 3**.

2.3.2. Antibodies for Coating and Detection

1. Coating antibody (e.g., Anti-IgG, anti-IgM, anti-IgA).
2. Anti-immunoglobulin κ- or λ-light chain (Southern Biotech; Birmingham, AL) at a concentration of 1 to 2 µg/mL diluted in PBS.
3. Detection antibody (diverse biotinylated anti-Ig antibodies) (Sigma-Aldrich) at a concentration of 5 µg/mL diluted in buffer 2.

2.3.3. Reagents for Antigen-Specific Coating

1. Antigen for coating (concentration 5 µg/mL in PBS).
2. PBS/Tween-20: 100 mL PBS + 100 µL Tween-20.
3. Solution of methylated BSA in PBS at 0.01 mg/mL.

3. Methods

3.1. Isolation of Peripheral Blood Mononuclear Cells by Density Gradient Centrifugation

This protocol describes isolation of peripheral blood mononuclear cells (PBMCs) from whole blood using density centrifugation *(28)*.

1. Dilute whole blood with PBS in equal volumes and overlay carefully on 15 mL of separation media at room temperature (2-phase-system is forming) (*see* **Note 1**).
2. Centrifuge at 1140*g* for 20 min without brake (density gradient is generated) at room temperature (*see* **Note 2**).
3. Isolate the white layer of PBMCs (located between the 2 phases of separation medium and serum/PBS supernatant) and wash the cells at least 2 times in 50 mL of cold PBS/BSA (centrifuge 10 min at 320*g*) (*see* **Notes 3** and **4**).

3.2. Flow Cytometry

Flow cytometry allows high throughput analysis and sorting of single cells because of their phenotype. The phenotype is recognized by specific antibodies carrying different fluorescent molecules. In a cytometer, these fluorochromes are excited by laser light and subsequently emit light of various wavelengths, which is amplified and collected by detector units. Each event detected appears as a dot in evaluation software (e.g., CellQuest, BD, or FlowJo; TreeStar, Ashland, OR) as shown in **Fig. 2** (*see* **Notes 5–16**).

3.2.1. Surface Staining of Living Cells

1. Add staining antibody in appropriate dilution in PBS/BSA (use a volume of 50 to 200 µL for incubation, depending on sample size; consult manufacturer's data sheet for cell concentration).

Phenotypic Analysis of B and Plasma Cells

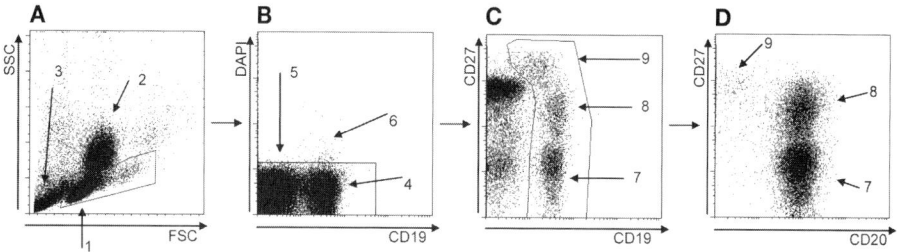

Fig. 2. Detection of circulating B-cell subsets by flow cytometry. The following anti-human surface markers were used: CD19-PerCp, CD27-Cy5, CD20-FITC and DAPI. This blood sample is taken from a healthy individual on day 7 after secondary tetanus immunization. At this time point, an increased plasma cell frequency can be observed *(31)*, which is beneficial for this demonstration. (**A**) In this picture, sideward light scatter (SSC) is plotted against forward light scatter (FSC). Whereas FSC correlates to the cell size, SSC is consistent with its granularity. The lymphocyte population (gated, no. 1) and other white blood cells (e.g., monocytes, no. 2) can be identified using these two optical parameters. (**B**) Gated cells from 2A are transferred to this plot. Dead cells (no. 6) are excluded on the *y*-axis (DAPI staining). DAPI detects dead and dying cells by their degraded membrane. Cells inside the gate shown here are alive. On the *x*-axis, CD19 expression is plotted. All B-cells in human blood express CD19. (**C**) Living lymphocytes are shown here (cells which are located in both gates in **A** and **B**, respectively) in a plot showing surface staining by mAbs directed against CD19 and CD27. B cells (gated cells) spread along the *y*-axis because of their different expression of CD27. On the basis of this plot, B-cell subsets can be identified: naïve B-cells (no. 7), expressing no or low CD27, memory B-cells (no. 8) expressing increased CD27 and plasmablasts/plasma cells (no. 9) showing extremely high expression of CD27 and a slight loss of CD19, resulting in a shift of this population to the left. (**D**) B-cells are shown in this plot (as gated in **C**), again demonstrating the different expression of CD27 and that CD20 differentiates between memory/naïve B-cells expressing CD20 and CD27^{high} plasmablasts/plasma cells not expressing this antigen.

Notes: No. 1, lymphocytes, small (left side population) and large (right side cells); no. 2, monocytes; no. 3, cell debris, erythrocytes and noncellular physical particles; no. 4, CD19 expressing, living lymphocytes (B-cells); no. 5, living non-B-cells, containing T-cells and NKcells); no. 6, dead cells; no. 7, naïve B-cells; no. 8, memory B-cells; no. 9, plasmablasts/plasma cells.

2. Incubation for 10 min at 4°C in the dark.
3. Wash cells at least once in a 10-fold PBS/BSA solution.
4. Use DAPI (22 μ*M*) or propidium iodide (0.6 ng/mL) for exclusion of dead cells. Add reagent directly prior to analysis if no intracellular staining is performed additionally.
5. Acquire cell sample using a cytometer (e.g., FACS Calibur).

3.2.2. Staining of Intracellular Epitopes

If surface and intracellular stainings are performed on the same sample, carry out the surface staining first and avoid using phycobiliprotein as conjugated dyes, because some epitopes and these dyes are not resistant to fixation.

1. Wash isolated PBMCs or cell suspension in PBS two times (removal of soluble protein).
2. Add formaldehyde (final concentration 2%).
3. Incubation 20 min at room temperature in the dark.
4. Wash 3 times in PBS (removal of formaldehyde).
5. Wash the cells in 1 mL 0.1% saponine solution.
6. Add staining antibody diluted in saponin 0.5%. Incubate for 15 min in the dark.
7. Wash at least once with 0.1% saponin.
8. Wash and resuspend in PBS/BSA before analysis (do not use DAPI or PI!).

3.2.3. Cytometric Antibody-Secretion Assay

The following protocol allows detection of cells secreting antibodies containing a λ-immunoglobulin light chain *(29)*. For other secreted products, such as κ-using antibodies or cytokines, the protocol and reagents have to be modified accordingly *(30)* (*see* **Notes 17–19**).

3.2.3.1. Biotinylation of Cells

1. Perform a surface staining as described in **Subheading 3.2.1.** using a fluorochrome-conjugated anti-human IgG antibody.
2. Wash the cells two times in cold PBS to remove the BSA.
3. Resuspend the cells in a 1 mL NHS-L-biotin solution prewarmed to 37°C using a 15-mL tube.
4. Incubate for 10 min at 37°C.
5. Add 1 mL RPMI 1640/10 % FCS and incubate another 10 min at 37°C.
6. Wash 3 times with 50 mL PBS/BSA.

3.2.3.2. Creation of a Cellular Affinity Matrix

1. Add 200 µL of 50 µg/mL anti human λ (clone HP6054) to the cells from the previous washing step and incubate for 5 min at 4°C (blocking of free λ-chains).
2. Add 50 µg/mL HP6054 coupled to avidin (catch-antibody) to the vial.
3. Incubate 10 min on ice.

3.2.3.3. Secretion and Detection

1. Add 2 mL RPMI 1640/10% FCS (37°C) to a final concentration of 3×10^6 cells/mL.
2. Incubate for 45 min at 37°C (Ig secretion phase, secreted λ + antibodies are attached to the matrix). Move the tube gently every 5 min.

Phenotypic Analysis of B and Plasma Cells

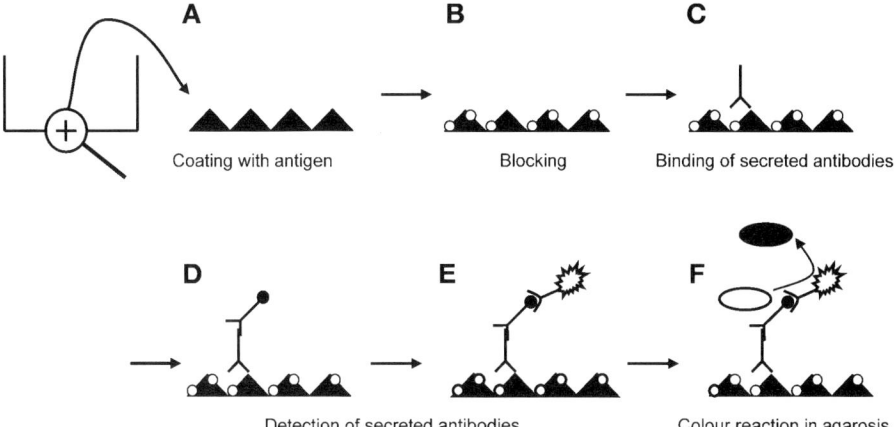

Fig. 3. Principles of ELISpot assay. The arrangement of coating and detection antibodies used in the protocol is depicted.

3. Cool the cells for 15 min on ice and gently agitate every 5 min.
4. Add digoxigenized anti-human λ antibody to a final concentration of 5 µg/mL and incubate for 10 min on ice.
5. Wash the cells in PBS/BSA.
6. Add anti-dig antibody conjugated to a fluorochrome (e.g., PE or FITC) and incubate for 10 min on ice.
7. Wash the cells in PBS/BSA .
8. Resuspend the cells and collect data using a cytometer.

3.3. ELISpot

The ELISpot allows specific and highly sensitive quantification of cells secreting defined molecules such as several antibodies or cytokines. In the protocol below, the procedure for enumeration of antibody-secreting cells is given (*see* **Notes 21–28**).

3.3.1. Enumeration of Antibody-Secreting Cells

3.3.1.1. COATING (SEE FIG. 3A,B)

1. Apply 50 µL coating antibody (5 µg/mL in PBS) to each well of a flat bottom, high-binding 96-well plate (*see* **Notes 20** and **21**).
2. Incubation for 1 h at room temperature.
3. Discard the solution from the plate.
4. Wash 1 time with buffer 1.
5. Add 100 µL of buffer 1 to each well and incubate for at least 30 min at room temperature (blocking free binding sites in order to reduce background).

3.3.1.2. Secretion Phase (SEE FIG. 3C)

1. Create a convenient dilution of the counted cells in RPMI 1640 medium/10% FCS, considering that the volume applied to one well will be 300 µL.
2. Wash the plate one time with PBS.
3. Add 300 µL of cell suspension to each first well of a column (*see* **Note 22**).
4. Fill the residual wells of the plate with 200 µL RPMI 1640 medium each.
5. Take out 100 µL from each well of the first line and create a dilution series downwards each column, skip the last row and discard the last 100 µL of cell suspension.
6. Fill 3 µL of serum into the skipped wells of the last line as a positive control.
7. Incubation for 3 h at 37°C and in a humid atmosphere containing 5 % CO_2.
8. Wash vigorously 5 to 6 times with buffer 2 (*see* **Note 23**).
9. Control by microscope that most of the cells have been washed away (*see* **Note 24**).

3.3.1.3. Detection Procedure (SEE FIG. 3D,E)

1. Add 100 µL of the corresponding biotinylated detection antibody diluted in buffer 2 to each well (*see* **Note 25**).
2. Incubation for 20 min at room temperature.
3. Wash 3 times with buffer 2.
4. Add 100 µL of SA-AP diluted in buffer 2 to each well.
5. Incubation for 20 min at room temperature.
6. Wash 3 times with buffer 2 (*see* **Note 26**).

3.3.1.4. Development of Spots (SEE FIG. 3F)

1. Add 100 µL of development buffer to each well while strictly avoiding bubble formation. Destroy eventual bubbles immediatly using a needle.
2. Move the plate as few as possible.
3. Leave the plates for solidification for 10 min at 4°C.
4. Tempering for 10 min at room temperature.
5. Spot development for 2 h at 37°C (*see* **Note 27** and **28**).

3.3.1.5. Counting of Spots

1. Count blue spots in each well using an inverted microscope.
2. Multiply with dilution factor to obtain total numbers of antibody secreting cells per seeded cell number.

3.3.2. Enumeration of Antigen-Specific Antibody-Secreting Cells

The detection of antigen-specific antibody-secreting cells becomes possible by coating with particular antigen instead of antibody. Some antigens (e.g., dsDNA) do not bind to the plates because of adverse distribution of electric charges within the molecule, therefore a precoating of the plates in order to enhance the binding is often mandatory.

3.3.2.1. PRECOATING

1. Add 100 µL of methylated BSA solution to each well.
2. Incubate for 2 h at 37°C.
3. Wash 3 times with PBS/Tween-20.

3.3.2.2. COATING

1. Add 100 µL of antigen (5 µg/mL) diluted in PBS to each well.
2. Incubation for 2 h at RT and then overnight at 4°C.
3. Discard coating solution.
4. Wash 1 time with buffer 1.
5. Add 100 µL of blocking buffer to each well.
6. Incubation for at least 30 min at room temperature.
7. Proceed with **Subheading 3.3.1.2., step 1** of the above ELISpot protocol.

4. Notes

1. Careful overlay. Try to avoid mixing of blood with the separation medium, separation will be of poorer quality (mainly impaired by erythrocyte contamination) and the recovery will drop.
2. No brakes. Using a brake in the separation centrifugation will destroy the gradient and prevent proper identification and isolation of the layer of PBMCs. Try to take out PBMCs in a small volume.
3. Keep cool. Keep the cells cool after isolation. After centrifugation, resuspend the cells by gentle pipetting to achieve a single cell suspension. Some proband's material is "stickier" than other's.
4. No clumps. Dissolving ethylene diamine tetracetic acid (EDTA) (2–5 mM) in the PBS/BSA buffer reduces cell clumping.
5. Work fast. Some cells taken ex vivo can rapidly (already within 30 min) change their expression profile and possibly therewith their phenotype by means like shedding or internalization of surface antigens. Being separated from their natural environment, many cell types are apoptosis-prone and will die within a short time.
6. Work cool. Cooling keeps your cells alive and freezes their metabolism, so that the cells' phenotype remains as stable as possible.
7. Stain in the dark. Light exposure results in degradation of fluorescent molecules and might affect you results.
8. Consider cell isolation procedures. Some stainings (e.g., of CD138) might be affected by the previous treatment of the cells (e.g., usage of collagenase for isolation of bone marrow cells).
9. Stain prior to fixation. Some epitopes are degraded by formalin treatment (e.g., CD20), so that they will not be recognized by specific antibodies anymore.
10. Reduce background. Some antibodies exhibit nonspecific binding, preferentially to monocytes. Adding human immunoglobulin (e.g., Beriglobin), results in blocking unspecific binding reactions. Do not use Beriglobin when targeting immuno-

globulin with your antibodies, these antibodies will be neutralized and cells will not be stained!
11. Titrate staining reagents. Create dilution series of every antibody before the first application and stain samples from the same origin, in order to determine the optimal antibody concentration ("titration"). Undersized concentration will lead to insufficient staining, an excess will result in nonspecific binding of the antibody. Vary the number of washing steps necessary in the staining procedure, especially if intracellular targets should be stained.
12. Open gates : Activated B-cells and plasma cells have higher FSC and SSC values than the majority of lymphocytes. They do not fit into the conventional "lymphocytes gate." Widen the scatter gate (as shown in **Fig. 2**) to include such cells.
13. Exclude doublets. Because some B-cell subsets are very small, even little contamination might change the results dramatically. One source of contaminants is doublet formation, especially after formaldehyde treatment, resulting in false positive events. Reduce doublet formation by adding EDTA to your buffer. Exclude doublets in your data sets by additionally plotting FSC-A vs FSC-H and gating on single cells.
14. Determine background. Isotype control antibodies are used to determine the degree of nonspecific staining caused by a fluorochrome-conjugated antibody. These control antibodies must match the actual antibodies in the following parameters: the species from which the antibody was derived, antibody class, the conjugated fluorochrome, and the conjugation procedure, and the concentration in the staining procedure.
15. Run controls. Many stainings are used frequently in many labs and are therefore rarely challenged. When using newly developed or rarely used antibodies or test new staining approaches, perform a check for contaminating events in your target population (e.g., try to stain for CD3, CD14, within the $CD19^+$ or $CD20^+$ B-cell population). Try to block your staining by adding an excess of uncoupled antibody to your sample prior to the actual staining procedure. This should mask the epitopes and the staining should be inhibited. When performing intracytoplasmatic stainings, the actual intracellular localization of the targeted antigen can be controlled by using PBS/BSA instead of saponine buffer, so that only a surface staining should be visible *(31)*.
16. Number of acquired cells. In order to allow a sufficient statistical analysis of cytometric data, the frequency of the cells of interest in the sample should be considered, estimating the number of events to be acquired. As a guideline, for "normal" analyses of total B-cells from a healthy donor's whole blood, 1×10^5 collected PBMCs should be enough. If antigen-specific cells are addressed or little subsets expressing an antigen of interest, the number of acquired events easily reaches approx 3×10^6. In addition to the hints given above for flow cytometry, the cytometric antibody-secretion assay desires some special considerations, because it is a simple, but laborious procedure requiring careful establishment.
17. Exclude background. Background staining results in false positive events mimicking antibody-secreting cells. Well-known contaminants are cells carrying sur-

face Ig, such as monocytes, natural killer (NK) cells and B-cells. Mostly, antibody-secreting cells are found in low frequencies, and therefore contaminations have a strong influence on the results. Therefore it is recommended to perform an enrichment or depletion procedure prior to the secretion assay. A suitable enrichment method is MACS.

18. Reduce background. Even after enrichment your sample likely contains residual cells generating a background by carrying surface bound λ-chains. Reduce the background by blocking free surface λ (*see* **Subheading 3.2.3.2., step 1**) in a quantitative manner.
19. Run controls. It is very important to control the specificity of the staining of secreted products running several controls (e.g., omitting the incubation with the catch-antibody should inhibit the staining completely). To do so, share the sample after **Subheading 3.2.3.2., step 2** and go on with two vials, with and without the catching antibody. Perform counterstainings for characterizing your antibody-secreting cell population for contaminants.
20. Titrate reagents. In order to obtain optimal spots clearly separated from each other, titrate the reagents and vary incubation times, if necessary.
21. Reduce background. In this ELISpot procedure, two kinds of background could appear. One is spots generated by particles mimicking antibody secretion by non-specific binding of one of the detection reagents. The rate of this background has to be determined in every single experiment. In other cases, already secreted antibodies in the cell suspension may cause a slight blue fog which sometimes becomes too intense to differentiate the spots.
22. Choose an appropriate sample size. Avoid giving too many cells in the wells as you will not be able to count them in the end. Try to design your experiment in a way, that a maximum of 150 spots/well is expected to appear in the first line. Apply more cells to the well, when the cells you are looking for are rare among the total cells (e.g., IgE-secreting cells or antigen-specific cells).
23. Tween kills. Do not use any Tween before adding the cells to the plates, because it kills the cells efficiently and inhibits antibody secretion. The only exception is the washing step following the preincubation prior to the coating with antigen.
24. Deplete cells after secretion. Be sure to eliminate the cells from the plate after the secretion phase (control by microscope), otherwise this will result in the appearance of background spots.
25. Control coating and detection. Some coating reagents do not stick to the plates readily. Run positive controls using serum reliably containing antibodies of question. Use a well without any cells or serum as a negative control. If available, use also a cell sample known to secrete the antibodies of question as a positive control. If precoating was necessary, run another control in order to determine the background caused by the precoating agent.
26. Avoid dry out. Overnight incubation and 37°C conditions may result in dehydration of the wells. Avoid this using a lid.
27. Storage. After the procedure has been finished, plates can be wrapped in cling film and stored for a maximum of 7 d at 4°C for latter counting.

28. Two color ELISpot. If necessary, two different products can be detected in the same assay. The two-color ELISpot *(32)* uses two different detection antibodies and color reactions resulting in spot formation of two different colors, each representing one product.

References

1. Behring, E. and Kitasato, S. (1980) Über das Zustandekommen der Diphterie-Immunität und der Tetanus-Immunität bei Tieren. *D. Medizinische Wochenschrift* **49**, 41–52.
2. Fagraeus, A. (1948) The plasma cellular reaction and its relation to the formation of antibodies in vitro. *J. Immunol.* **58**, 1–3.
3. Nossal, G.J. (2002) One cell, one antibody: prelude and aftermath. *Immunol. Rev.* **185**, 15–23.
4. Glick, B., Chang, T.S., Japp, R.G. (1956) The bursa of Fabricius and antibody production. *Poultry Sc* **35**, 224.
5. Cooper, M.D., Raymond, D.A., Peterson, R.D., South, M.A., and Good, R.A. (1966) The functions of the thymus system and the bursa system in the chicken. *J. Exp. Med.* **123**, 75–102.
6. Bernard, A.R., Boumsell, L., Dausset, J., Schlossman, S.F. (1984) Leucocyte Typing I., Berlin: Springer-Verlag.
7. Radbruch, A. (1999) Flow Cytometry and Cell Sorting. Berlin: Springer.
8. Herzenberg, L.A. and S.C. De Rosa (2000) Monoclonal antibodies and the FACS: complementary tools for immunobiology and medicine. *Immunol. Today* **21**, 383–90.
9. Miltenyi, S. et al (1990) High gradient magnetic cell separation with MACS. *Cytometry*, **11**, 231–238.
10. Leyendeckers, H., Odendahl, M., Lohndorf, A., et al. (1999) Correlation analysis between frequencies of circulating antigen-specific IgG-bearing memory B cells and serum titers of antigen-specific IgG. *Eur. J. Immunol.* **29**, 1406–1417.
11. Janeway, C.A.T., Walport, M., Shlomchik, M. (2001), *Immunobiology*. 5th ed. New York and London: Garland Publishing.
12. Rolink, A.G., Nutt, S.L., Melchers, F., and Busslinger, M. (1999) Long-term in vivo reconstitution of T-cell development by Pax5-deficient B-cell progenitors. *Nature* **401**, 603–606.
13. Nutt, S.L., Heavey, B., Rolin, A.G., and Busslinger, M. (1999) Commitment to the B-lymphoid lineage depends on the transcription factor Pax5. *Nature* **401**, 556–562.
14. Carsetti, R. (2004) Characterization of B-cell maturation in the peripheral immune system. *Methods Mol. Biol.* **271**, 25–35.
15. Carsetti, R., Rosado, M.M. and Wardmann, H. (2004) Peripheral development of B cells in mouse and man. *Immunol. Rev.* **197**, 179–191.
16. Hardy, R.R. and K. Hayakawa (2001) B cell development pathways. *Annu. Rev. Immunol.* **19**, 595–621.

17. Iwakoshi, N.N., A.H. Lee, and L.H. Glimcher (2003) The X-box binding protein-1 transcription factor is required for plasma cell differentiation and the unfolded protein response. *Immunol. Rev.* **194**, 29–38.
18. Johnson, K., Shapiro-Shelef, M., Tunyaplin, C., and Calame, K. (2005) Regulatory events in early and late B-cell differentiation. *Mol. Immunol.* **42**, 749–761.
19. Manz, R.A., Hauser, A.E., Hiepe, F., and Radbruch, A. (2005) Maintenance of serum antibody levels. *Annu. Rev. Immunol.* **23**, 367–386.
20. Klein, U., Tu, Y., Stolovitzky, G.A., et al. (2003) Transcriptional analysis of the B cell germinal center reaction. *Proc. Natl. Acad. Sci. USA* **100**, 2639–2644.
21. Scheffold, A., Miltenyi, S. and Radbruch, A. (1995) Magnetofluorescent liposomes for increased sensitivity of immunofluorescence. *Immunotechnology* **1**, 127–137.
22. Wang, Y.H., Stephan, R.P., Scheffold, A., et al. (2002) Differential surrogate light chain expression governs B-cell differentiation. *Blood* **99**, 2459–2467.
23. Jerne, N.K. and Nordin, A.A. (1963) Plaque formation in agar by single antibody-producing cells. *Science* **140**, 405.
24. Czerkinsky, C.C., Nilsson, L.A., Nygren, H., Ouchterlony, O., and Tarkowski, A. (1983) A solid-phase enzyme-linked immunospot (ELISPOT) assay for enumeration of specific antibody-secreting cells. *J. Immunol. Methods* **65**,109–121.
25. Blink, E.J., Light, A., Kallies, A., Nutt, S.L., Hodgkin, P.D., and Tarlinton, D.M. (2005) Early appearance of germinal center-derived memory B cells and plasma cells in blood after primary immunization. *J. Exp. Med.* **201**, 545–554.
26. Jacobi, A.M., Odendahl, M., Reiter, K., et al. (2003) Correlation between circulating CD27high plasma cells and disease activity in patients with systemic lupus erythematosus. *Arthritis Rheum.* **48**, 1332–1342.
27. Manz, R., Assenmacher, M., Pfluger, E., Miltenyi, S., and Radbruch, A. (1995) Analysis and sorting of live cells according to secreted molecules, relocated to a cell-surface affinity matrix. *Proc. Natl. Acad. Sci. USA* **92**,1921–1925.
28. Boyum, A. (1968) Isolation of mononuclear cells and granulocytes from human blood. Isolation of monuclear cells by one centrifugation, and of granulocytes by combining centrifugation and sedimentation at 1 g. *Scand. J. Clin. Lab. Invest.* **Suppl 97**, 77–89.
29. Arce, S., Luger, E., Muehlinghaus, G., et al. (2004) CD38 low IgG-secreting cells are precursors of various CD38 high-expressing plasma cell populations. *J. Leukoc. Biol.l* **75**, 1022–1028.
30. Assenmacher, M., Lohning, M., Scheffold, A., Manz, R.A., Schmitz, J., and Radbruch, A. (1998) Sequential production of IL-2, IFN-gamma and IL-10 by individual staphylococcal enterotoxin B-activated T helper lymphocytes. *Eur. J. Immunol.* **28**,1534–1543.
31. Odendahl, M., Mei, H., Hoyer, B.F., et al. (2005) Generation of migratory antigen-specific plasma blasts and mobilization of resident plasma cells in a secondary immune response. *Blood* **105**, 1614–1621.
32. Czerkinsky, C., Moldoveanu, Z., Mestechky, J., Nilsson, L.A., and Ouchterlony, O. (1988) A novel two colour ELISPOT assay. I. Simultaneous detection of distinct types of antibody-secreting cells. *J. Immunol. Methods* **115**, 31–37.

33. Tangye, S.G., Avery, D.T., Deenick, E.K., and Hodgkin, P.D. (2003) Intrinsic differences in the proliferation of Naive and Memory Human B cells as a Mechanism for Enhanced Secondary Immune Response. *J. Immunol.* **170**, 686–694.
34. Chung, J.B., Silverman, M., and Monroe, J.G. (2003) Transitional B cells: step by step towards immune competence. *Trends Immunol.* **24**, 343–349.

2

Detection of Antigen Specific B-Cells in Tissues

Marie Wahren-Herlenius and Stina Salomonsson

Abstract

The role of B-cells and autoantibodies in tissue destructive events of autoimmune diseases is emerging, and thereby increasing interest in identifying the presence and location of autoreactive B-cells and autoantibody secreting plasma cells. For visualization and analysis of the autoreactive B-cells, the antigen of interest is selected and produced and purified from native or recombinant sources. Biotinylation of the purified antigen and subsequent use in immunohistochemistry with sections from tissue under analysis permits detection of the autoreactive B-cells and plasma cells. Double staining with cell-specific markers or for the presence of intracellular Ig allows characterization of the cell or determination of Ig isotype of the autoantibody produced by the individual autoreactive B-cell, leading to better understanding of the role of the particular antibody in the inflammatory cascade of the organ. With this technique, identification as well as quantification of autoreactive cells within tissues may be performed; it is also possible to analyze the spatial relation to residual cells or other infiltrating cells of the target organ.

Key Words: Autoantigens; autoantibodies; B-cells; plasma cells; immunohistochemistry; immunofluorescence; antibody synthesis; antibody subclass; biotinylation.

1. Introduction

The role of B-cells and autoantibodies in tissue destructive events of autoimmune diseases is emerging *(1,2)*, and thereby increasing interest in identifying the presence and location of autoreactive B-cells and autoantibody secreting plasma cells with specificity for self-antigens in affected tissues. Labeling of autoantigens produced and purified from native or recombinant sources permits detection of the cells of interest *(3,4)*. Double staining with cell-specific markers or for the presence of intracellular Ig will allow determination of cell-subsets and the isotype or even subclass of the autoantibody produced by the individual autoimmune B-cell, leading to better understanding of the role of the particular antibody in the inflammatory cascade of the organ *(3,5,6)*. With this technique, identification as well as quantification of autoreactive cells within tissues may be performed; it is also possible to analyze the spatial relation to residual and other infiltrating cells of the target organ.

The antigen of interest may be simply ordered if commercially available, purified from tissue or cells or produced recombinantly in *Escherichia coli* or other expression systems. The antigen however needs to be pure and free from contaminants and other proteins before labeling. This protocol suggests labeling with biotin by covalent coupling, but labeling with other molecules is also feasible. This includes direct coupling of enzymes, fluorochromes, and iodine. The advantages of biotinylation are that the coupling is fast and easily performed, the conditions are mild, and the biotinylation normally does not have any adverse effects on the antigen. The high affinity for avidin and streptavidin (Ka 1014 mol–1) allows fast and specific binding (for illustrations of the first uses of biotin and avidin (*7–9*). The commercial availability of avidin/streptavidin coupled with different enzymes and fluorochromes permits great flexibility in use and detection systems. The assay can be rendered very sensitive by different amplification protocols (e.g., use of multimeric complexes). Both avidin and streptavidin are tetravalent and the additional binding sites can be used to increase the size of the detection complex and thereby the strength of the signal.

After coupling excess biotin is removed by dialysis, and following verification by immunoblotting of successful labeling the antigen is ready for use in immunohistochemistry to detect autoreactive B-cells and plasma cells. The immunohistochemistry protocol is adjusted to the tissue under investigation. It may also be modified to include double staining to specify subsets of cells or Ig isotype/subclass of produced antibodies. Analysis is performed under the light or fluorescence microscope.

2. Materials

1. Antigen solution. Purified antigen for which you wish to detect specific B-cells.
2. Tissue sample. Sample from tissue where you wish to investigate presence of antigen specific B-cells. Store at –70°C.
3. 0.1 M Sodium borate buffer (pH 8.8).
4. 10 mg/mL *N*-hydroxysuccinimide biotin (Sigma; St Louis, MO, USA), in dimethyl sulfoxide (DMSO).
5. 1 M Ammonium chloride.
6. Phosphate buffered saline (PBS).
7. Equipment for dialysis.
8. Equipment for sodium dodecyl sulfate-polyacrylamide gel electrophoresis (SDS-PAGE).
9. Equipment for Western blot.

Detection of Antigen Specific B-Cells 21

10. Sample buffer: 20 m *M* Tris-HCl (pH 8.8), 10% glycerol, 1% SDS, 5% β-merkaptoethanol, and 0.1% bromophenol blue.
11. Milk powder.
12. Bovine serum albumin (BSA).
13. Acetone.
14. 100 mL Tris-buffered saline (TBS) : 0.05 *M* Tris-HCl (pH 7.6), 900 mL distilled water, and 7.65 g NaCl.
15. Endogenous biotin blocking kit (Avidin/Biotin blocking kit, Vector laboratories; Burlingame, CA, USA)
16. ABComplex/HRP (DakoCytomation, Glostrup, Denmark).
17. AEC solution (2 mg/mL 3-amino-9-ethylcarbaxol in *n,n*-dimethyl formamide). Store in 0.5-mL aliquots at −20°C.
18. 0.02 *M* sodium acetate (NaAc) buffer (pH 5.2). Store at +4°C, check for bacterial growth before use.
19. H_2O_2 (30%, light sensitive, store at +4°C).
20. TPBS : 0.05% Tween-20 in PBS.
21. Meyer's hemalum (aqueous counterstain).
22. Aqueous mounting medium (Mount-Quick "Aqueous", Daido Sangyo, Japan).
23. Fluorescein isothiocyanate (FITC)-conjugated anti-IgG, anti-IgA, anti-IgM antibodies (DakoCytomation; Glostrup, Denmark).
24. Cy3-conjugated streptavidin (Jackson Immunoresearch, West Grove, PA, USA).
25. Anti-fading mounting medium (Vectashield mounting medium, Vector,).

3. Methods

3.1. Biotinylation of Antigen

1. Prepare a solution of your antigen at 1 to 3 mg/mL in sodium borate buffer (0.1 *M* [pH 8.8]). If the antigen is already dissolved in another buffer, change buffer by dialyzing against several changes of sodium borate buffer.
2. Prepare a solution of *N*-hydroxysuccinimide biotin in DMSO at 10 mg/mL.
3. Add the biotin ester-solution to the antigen at a ratio of 25 to 250 µg of ester/mg antigen. Mix and incubate for 4 h at room temperature (*see* **Note 1**).
4. Add 20 µL 1 *M* NH4Cl per 250 µg of ester. Incubate for 10 min at room temperature.
5. Dialyze the biotinylated antigen against PBS or buffer of choice to remove excess biotin (*see* **Note 2**).
6. Store biotinylated antigen in aliquots at −70°C. Avoid repeated freeze-thawing before use.

3.2. Check Biotinylation of the Antigen by Immunoblotting

1. Boil the protein sample with sample buffer for 5 min.
2. Separate the antigen on SDS-PAGE gels with acrylamide concentration adjusted to the molecular weight of the protein, usually between 7.5 and15% (*see* **Note 3**).

3. Transfer to nitrocellulose membrane by semidry or tank electroblotting.
4. Block nonspecific binding with 5% milk powder in TPBS for 30 min at room temperature with slight agitation.
5. Wash the membrane in PBS.
6. To detect biotinylated antigen, add ABComplex/HRP prepared according to the manufacturer's instructions. Incubate for 45 min at room temperature with slight agitation (*see* **Note 4**).
7. Wash the membrane in PBS.
8. Develop by adding AEC solution.
9. Stop reaction by rinsing with H2O.

3.3. Immunohistochemical Analyses Using Biotinylated Antigen

1. Cut tissue of interest into five 8 μm thick sections using a cryostat and collect on glass slides (*see* **Note 5**). Store at –70°C until use (*see* **Note 6**).
2. Taking the slides directly from the freezer and avoiding thawing, fix tissue sections in acetone at +4°C in 50% acetone/distilled water for 30 s and 100% acetone for 5 min.
3. Dry at room temperature. Between following steps, take care not to let the sections dry. To avoid evaporation, perform incubations below in a humidity chamber at room temperature.
4. Block endogenous peroxidase activity by adding freshly prepared hydrogen peroxidase (H2O2) diluted to 1% in methanol, incubate for 5 min (*see* **Note 7**).
5. Rinse sections in TBS; once briefly and once for 5 min. Take care to remove any remaining bubbles from the peroxidase quenching step by gentle rinsing.
6. Block endogenous biotin by adding avidin/biotin blocking solution for 15 + 15 min according to manufacturer's instructions.
7. Rinse sections in TBS; once briefly and once for 5 min.
8. Block nonspecific protein binding by incubating with 5% milk powder/4% bovine serum albumin in TBS for 15 min.
9. Incubate with biotinylated antigen diluted in TBS for 60 min. Optimal dilution of the biotinylated antigen must be, usually between 1:200 and 1:1000 from a 1 mg/mL solution (1–5 μg/mL).
10. Rinse sections in TBS; once briefly and once for 5 min.
11. Add ABComplex/HRP prepared according to manufacturer's instructions, incubate 45 min.
12. Rinse sections in TBS; once briefly and once for 5 min.
13. Prepare AEC chromogen solution by adding 0.5 mL of AEC solution (2 mg/mL) to 5 mL 0.02 M NaAc. Filter to remove precipitates and add 0.5 μL 30% H_2O_2.
14. Visualize by adding AEC substrate, incubate for 15 min.
15. Rinse sections in TBS; once briefly and once for 5 min.
16. Counterstain with Mayer's hemalum (*see* **Note 8**).
17. Wash sections with distilled water and air-dry before mounting in water based medium.

3.4. Double Staining for Intracellular Immunoglobulins

To explore the Ig isoform and confirm that cells stained with biotinylated antigen are Ig-producing B-cells/plasma cells double staining for intracellular Ig isoform production can be performed.

1. Repeat **steps 1–9** from **Subheading 3.3.**, but omit **step 4**. In **step 7**, substitute BSA for 10% serum from the animal species in which anti-IgG/IgA/IgM antibodies (below) were generated.
2. Incubate with Cy3-conjugated streptavidin diluted in TBS. From this step and onwards light should be eluded to preserve the fluorochromes (*see* **Note 9**).
3. Rinse sections in TBS; once briefly and once for 5 min.
4. Incubate with FITC-conjugated anti-IgG/IgA/IgM antibodies.
5. Rinse sections in TBS; once briefly and once for 5 min.
6. Mount in anti-fading medium.

3.5. Analysis of Slides

Computer assisted image analysis is suitable for determining degree of overlap when double staining of antigen specificity and isoform is performed. With the necessity for combining evaluation of staining intensity with morphologic assessment in terms of intracellular/ surface staining computer assisted analysis is of little extra help in evaluation of the single stainings.

4. Notes

1. High concentrations of the biotin ester will lead to multiple labeling of the antigen and increases the chances that every molecule is labeled. Lower ratios will keep biotinylation at a minimum and decrease the risk that epitopes are disrupted by modification. In our laboratory we have good experience from labeling with concentrations at the higher end of the proposed interval.
2. Despite its small size, biotin dialyzes slowly, so extensive dialysis is needed.
3. Intensive staining is often obtained in the blotting, and a dilution series of the antigen may be needed to evaluate the biotinylation.
4. The ABComplex is not a prerequisite, because enzyme-conjugated streptavidin alone will also yield a strong signal. Both here and when the biotinylated antigens are used in tissue analysis alternative substrates (e.g., DAB) or enzymes (e.g., alkaline phosphatase) with their respective substrates may be used according to preference.
5. Coated (chromalun, gelatin) or in other ways specially prepared glass slides might be needed for adherence depending on the tissue studied.
6. Storage of sections more than 2 wk before staining is not recommended, but its use after longer periods of time will depend the on tissue under study.
7. If problems with nonspecific tissue staining are encountered several different approaches may be tried. Further quenching of endogenous peroxidase activity

might be needed; different tissues contain different amounts of peroxidase. High levels are found in spleen and bone marrow. Between 0.3 and 2% can be used, whereas higher concentration may result in too vigorous bubble formation and destruction of the tissue. Instead, incubation time can be prolonged or 0.1% phenylhydrazine hydrocloride in PBS tried. Endogenous biotin blocking can also be modified by prolongation, and the protein source used for blocking of nonspecific protein-interactions changed to fetal calf serum (FCS).
8. If an alcohol soluble substrate is used a water based counterstain (e.g., Meyer's hemalum) should be used. Filter before use. If not regularly used, perform test stainings to determine the suitable incubation time.
9. Other fluorochrome combinations of preference can be used. Optimal dilution of the conjugated antibodies/streptavidin must be established locally, but manufacturers usually suggest starting ranges.

References

1. Chan, O. T., Madaio, M. P., and Shlomchik, M. J. (1999) The central and multiple roles of B cells in lupus pathogensis. *Immunol. Rev.* **169**, 107–121.
2. Davidson, A., and Diamond, B. (2001) Autoimmune diseases. *N. Engl. J. Med.* **345**, 340–350.
3. Tengnér, P., Halse, A.-K., Haga, H.-J., Jonsson, R., and Wahren-Herlenius, M. (1998) Detection of anti-Ro/SSA and anti-La/SSB autoantibody-producing cells in salivary glands from patients with Sjögren's syndrome. *Arthritis Rheum* **41**, 2238–2248.
4. Salomonsson, S., and Wahren-Herlenius, M. (2003) Local production of Ro/SSA and La/SSB autoantibodies in the target organ coincides with high levels of circulating antibodies in sera of patients with Sjogren's syndrome. *Scand J Rheumatol* **32(2)**, 79–82.
5. Salomonsson, S., Jonsson, M. V., Skarstein, K., et al. (2003) Cellular basis of ectopic germinal center formation and autoantibody production in the target organ of patients with Sjögren's syndrome. *Arthritis Rheum* **48**, 3187–3201.
6. Salomonsson, S., Larsson, P., Tengnér, P., et al. (2002) Expression of the B cell-attracting chemokine CXCL13 in the target organ and autoantibody production in ectopic lymphoid tissue in the chronic inflammatory disease Sjögren's syndrome. *Scand J. Immunol* **55**, 336–342.
7. Becker, J. M., and Wilchek, M. (1972) Inactivation by avidin of biotin-modified bacteriophage. *Biochim. Biophys. Acta* **264**, 165–170.
8. Heitzmann, H., and Richards, F. M. (1974) Use of the avidin-biotin complex for specific staining of biological membranes in electron microscopy. *Proc. Natl. Acad. Sci. USA* **71**, 3537–3541.
9. Heggeness, M. H., and Ash, J. F. (1977) Use of the avidin-biotin complex for the localization of actin and myosin with fluorescence microscopy. *J. Cell Biol.* **73**, 783–788.

3

Single Cell Analysis of Synovial Tissue B-Cells

Hye-Jung Kim and Claudia Berek

Abstract

Mononuclear cells often form highly organized lymphoid structures in the chronically inflamed synovial tissue of patients with rheumatoid arthritis (RA) within which B-cells are activated and may differentiate into effector plasma cells. The analysis of those activated B-cells and the determination of their specificity is of great importance for the understanding of the pathogenesis of RA. Here, we describe a technique that combines histological analysis of synovial tissue with a molecular analysis of the V-gene repertoire at the level of the single B-cell.

Immunohistochemical staining of tissue sections allows us to identify the activated B-cells. Those cells are then isolated using a micromanipulator and the rearranged immunoglobulin (Ig) genes amplified, cloned and sequenced. The combination of the V(D)J gene segments and the pattern of somatic mutations in the V-region genes, allows us to identify clonal relationships between the isolated B cells. Once Ig genes for a heavy and a light chain have been isolated from individual B-cells, they can be used to generate recombinant antibodies. These antibodies can be used to determine the antigens which support the activation of B-cells in the inflamed synovial tissue

Key Words: B-cells; plasma cells; immunoglobulin genes; rheumatoid arthritis; synovial tissue; germinal center; micromanipulation; single cell PCR; somatic hypermutation.

1. Introduction

Whereas the etiology of RA is not known, many studies have indicated that B-cells contribute to its pathogenesis by the production of autoantibodies *(1–6)*. Histological analyses of the affected joint tissue from rheumatoid arthritis (RA) patients have shown that, as disease progresses, massive infiltration of mononuclear cells occurs *(7,8)*. In contrast with many other chronic inflammatory diseases, the lymphoid infiltrates in the rheumatoid synovium often establish microstructures resembling the follicular structures in lymphoid organs *(9–11)*. Besides the cytokine and chemokine milieu in the synovium, self antigens might be of critical importance in triggering a local immune response and in inducing the formation of ectopic lymphoid structures in the inflamed synovial tissue.

In a T-cell dependent immune response the antigen activated B-cells normally migrate into the primary B-cell follicles of the secondary lymphoid organ. Here, a germinal center—a microenvironment within which affinity maturation of the immune response and the differentiation of B cells into memory and plasma cells takes place—may develop *(12,13)*. In the inflamed synovial tissue of about 30% of RA patients, T- and B-cells become organized into large follicle like structures. The analysis of B-cells revealed that these structures function as germinal centers and that antigen activated B-cells locally differentiate into effector cells *(14)*. Thus, the analysis of the local B cell response in the inflamed synovial tissue is of great importance for the understanding of the immune processes taking place in patients with RA.

During B-cell development in the bone marrow, each B-cell acquires a unique B-cell receptor by random rearrangement of VDJ gene segments for the heavy chain and VJ gene segments for the light chain *(15–19)*. Hence, the rearranged V-region genes can be used as clonal markers which allow us to follow synovial B-cell differentiation. In the course of an immune response the V-gene repertoire is further diversified by hypermutation. Single nucleotide exchanges are introduced into the rearranged V-region genes during B-cell proliferation in the germinal center. Competition for antigen presented by follicular dendritic cells (FDC) ensures that only those B-cells with high affinity receptors are selected to differentiate into plasma cells *(20)*.

Micromanipulation allows us to dissect the immune response taking place in the synovial tissue of RA patients. The analysis of single cells by this means is the technique of choice for the characterization of small defined B-cell subsets that are otherwise difficult to isolate *(21)*. The great advantage to other cell separation techniques is that the B-cells are analyzed within the histological context. Specific labeling allows us to differentiate between resting and activated B-cells and to distinguish between the different B-cell differentiation stages. With this method it becomes possible to examine the V-gene repertoire of the different synovial B-cell subsets. Furthermore, single cell polymerase chain reaction (PCR) analysis gives us the tools with which to express the rearranged V-region genes as a recombinant antibody. This offers the unique opportunity to determine the antigen specificity of synovial B-cells. It permits the definition of the antigens responsible for B-cell activation and differentiation in patients with RA.

Here, we describe a single cell PCR approach to study the V-gene repertoire of synovial B-cells. Methods to dissect B-cells directly from frozen tissue sections are described. Although those B-cells isolated from the germinal center structure are the best candidates to study the synovial immune responses, also plasma cells and various other B-cell subsets may also be analyzed. Further information on B-cell activation and differentiation taking place in the synovial tissue can be obtained using appropriate antibodies to stain the sections.

2. Materials
2.1. Preparation of Synovial Tissue
1. Fresh synovectomized tissue from RA patients.
2. Tissue-Tek OCT compound (Miles).
3. Plastic molds (10 mm × 10 mm × 5 mm) to freeze tissue samples (Miles).
4. Liquid nitrogen.

2.2. Immunohistochemistry
1. Microtome (Microm).
2. Microscope slides (Superfrost®/Plus, Fisher Scientific).
3. Cold acetone.
4. $CaCl_2$.
5. Phosphate buffered saline (PBS).
6. 3% BSA/PBS blocking solution: PBS supplemented with 3% *(w/v)* bovine serum albumin (BSA).
7. Hydrophobic pen (DAKO).
8. Primary Antibodies: for example CD20 (DAKO), CD38 (DAKO), Ki67 (DAKO), and mAb Wue-1 (kindly provided by A.Greiner *[10,22]*).
9. Secondary Antibodies: alkaline phosphatase anti-alkaline phosphatase (APAAP complex) DAKO.
10. Substrates: Fast-red (DAKO).
11. Mayer's Haematoxilin (Merck).
12. Dehydration solution (absolute ethanol; 90, 75, and 30% ethanol).

2.3. Microdissection
1. Inverse microscope equipped with a ×60 long distance lens (Nikon).
2. Digital camera and monitor.
3. 3D manual Micromanipulator (Nikon) connected to camera and monitor.
4. CellTram Oil (Eppendorf).
5. GD-1 Glass capillary (Narishinge).
6. Vertical pipette puller (Bachhofer).
7. Grinding machine (Bachhofer).
8. Sterile H_2O.

2.4. Preparation of Cells Prior to Polymerase Chain Reaction
2.4.1. Microdissected Cells
1. Proteinase K, PCR grade (Roche).

2.4.2. Sorted Cells

1. Lysis buffer: 200 mM KOH and 50 mM dithiothreitol (DTT).
2. Neutralization buffer: 900 mM Tris-HCl (pH9.0), 300 mM KCl, and 200 mM HCl.

2.4.3. Polymerase Chain Reaction

1. Thermocycler (Biometra).
2. AmpliTag Gold™ and 10X PCR buffer (Perkin Elmer).
3. Random primer (15mer) *(23)*.
4. dNTPs (Amersham Biosciences).
5. VH, Vκ and Vλ primers (**Table 1**) (*see* **Note 8**).
6. DNase free H$_2$O.
7. NuSieve GTG Agarose (Biozym).
8. Ethidium bromide (EB).
9. 123 bp DNA ladder (Invitrogen).
10. QIAquick® Gel Extraction Kit (Qiagen).

2.6. Cloning and Transformation

1. TA cloning kit (Invitrogen).
2. Agar plates: 3l capsule YT medium (Bio 101. Inc) + 15g Agar Select (Sigma).
3. LB medium for bacterial culture (formulation per 1l: 10 g bacto-tryptone, 5 g bacto-yeast extract, and 10 g NcCl).
4. X-Gal solution: 400 mg X-Gal (5-bromo-4-chloro-3-indolyl-β-D-galactopyranoside) + 10 mL dimethylformamide.
5. SOC medium: dissolve in 950 mL of deionized water 20 g of bacto tryptone, 5 g of bacto yeast extract, and 0,5 g MaCl. Add 10 mL 250 mM KCl, 5 mL 2M MgCl$_2$ 5 mL 2M MgSO$_4$ and adjust pH to 7.0 with 5N NaCl. Autoclave and allow it to cool to about 60°C and then add 20 mL of a 1M solution of glucose (sterile filtered). Adjust volume to 1 L.
6. Bacteria incubator.
7. Water bath.
8. Restriction enzyme EcoRI.

2.7. Sequencing

1. ABI PRISM DNA Sequence Analyser (Perkin Elmer).
2. BigDye Sequencing Kit (Perkin Elmer).
3. 3M NaAc (pH 4.6).
4. Absolute ethanol.
5. 70% Ethanol.
6. Sequencing primer (reverse and forward primer) to sequence the TA cloning vector.

2.8. Sequence Evaluation

1. VBASE data bank.

Table 1
Primer Sequences for Amplification of Ig V Genes

Heavy chain
	VH2	CAGATCACCTTGAAGGAGTCTGG
	VH3	GAGGTGCAGCTGGTGSAGTCTGG
	VH4	CAGGTGCAGCTGCAGGAGTCGGG
	VH6	TACAGCTGCAGCAGTCAGGTCCAGG
	VH4N	CAGGTGCAGCTACAGCAGTGGGG
	JHext	CTCACCTGAGGAGACGGTGACC
	JH1	TGAGGAGACGGTGACCAGGGTGCC
	JH3	TGAAGAGACGGTGACCATTGTCCC
	JH4	TGAGGAGACGGTGACCAGGGTTCC
	JH6	TGAGGAGACGGTGACCGTGGTCCC

Light chain
κ
	Vκ1	GACATCCAGWTGACCCAGTCTCC
	Vκ2b	GATGTTGTGATGACTCAGTCTCC
	Vκ2/4/6	GATATYGTGMTGACCCAGTCTCC
	Vκ3	GTCTKTGTCTCCAGGGGAAAGAG
	Vκ5	GAAACCACACTCACGCAGTCTCC
	Jκ1-4ext	TACTTACGTTTGATCTCCASCTTG
	Jκ5ext	GCTTACGTTTAATCTCCAGTCGTG
	Jκ1/4	ACGTTTGATYTCCACCTTGGTCCC
	Jκ2	ACGTTTGATCTCCAGCTTGGTCCC
	Jκ3	ACGTTTGATATCCACTTTGGTCCC
	Jκ5	ACGTTTAATCTCCAGTCGTGTCCC

λ
	Vλ1	CAGTCTGTGTTGACGCAGCCGCC
	Vλ2	CAGTCTGCCCTGACTCAGCCT
	Vλ3a	TCCTATGTGCTGACTCAGCCACC
	Vλ3b	TCTTCTGAGCTGACTCAGGACCC
	Vλ4b	CAGCTTGTGCTGACTCAATCGCC
	Vλ7/8	CAGACTGTGGTGACYCAGGAGCC
	Jλ1ext	GAGAGCCACTTACCTAGGACGG
	Jλ2/3ext	AGAAGAGACTCACCTAGGACGG
	Jλ7ext	CGGGGAGACTTACCGAGGACGG
	Jλ1	ACCTAGGACGGTGACCTTGGTCCC
	Jλ2/3/7	ACCKAGGACGGTCAGCTKGGTSCC

Abbr: R: A and G; W: A and T; Y: C and T; S: G and C.

2. DNA strider.
3. DNAplot.

3. Methods

3.1. Preparation of Synovial Tissue

1. Synovial tissue is taken up in PBS and cut in approximately 1 cm^3 size pieces.
2. Embed tissue in OCT compound.
3. Tissue is snap-frozen on the surface of liquid nitrogen.
4. Blocks can be stored in the −80° C freezer until ready for sectioning.

3.2. Immunohistochemistry

3.2.1. Preparation of Tissue Sections

1. Allow the temperature of the tissue block to equilibrate to the temperature of the cryostat (−20°C).
2. Tissues are cut into 6 µm sections to get a cell monolayer.
3. Sections are thaw-mounted onto adhesive slides and air-dried for 2 h.
4. Fix the tissue sections in ice cold acetone for 10 min and air dry
5. Frozen sections are now ready for staining (*see* **Subheading 3.2.2.**) (*see* **Note 1**). Alternatively, the frozen section can be stored for about 1 mo at −70°C.
6. To prevent rehydration of tissue sections during storage, put CaCl$_2$ wrapped in tissue paper into the box.

3.2.2. Immunohistochemistry

Immunohistochemistry permits the identification of cell types infiltrating the synovial tissue, and the analysis of their organization *in situ* (*see* **Note 2**).

1. Air-dry the tissue section for 5 min.
2. Circle the sections by hydrophobic slide marker.
3. To prevent nonspecific binding, sections should be incubate with 100 µL of blocking solution for 10 min at room temperature.
4. Incubate the sections with 75 µL of appropriately diluted primary Ab for 30 min at room temperature.
5. Wash twice by incubating in PBS for 5 min.
6. Sections are incubated with 75 µL of corresponding secondary antibody for 30 min at room temperature (*see* **Note 3**).
7. Wash twice by incubating in PBS for 5 min.
8. Prepare the substrates (Fast Red) according to the supplier's instructions. Color developing time will range from 5 to 15 min depending on the primary antibody used.
9. Wash twice with PBS.
10. Counterstain the sections with Haematoxilin
11. Cells of interest can now be isolated (*see* **Subheading 3.3.**).
12. Alternatively, the tissue sections can be dehydrated and stored at −70°C.

3.3. Microdissection

3.3.1. Preparation of Glass Capillaries

1. Using the vertical pipet puller glass capillaries are prepared.
2. Type 1 capillaries are prepared to cut out single cells from the tissue section. They can be used without polishing.
3. Type II capillaries are needed to take up the isolated cells. They have to be polished in order to make a tip of ca 15 µm in diameter.
4. Before use, autoclave both types of capillaries.

3.3.2. Microdissection (see Note 4)

1. The stained tissue sections are covered with PBS.
2. Cells of interest must be positioned in the center of the microscope field (*see* **Note 5**).
3. Under the lowest magnification glass capillary type 1 and 2 are brought onto the position from which cells should be isolated.
4. During the positioning of the capillaries the tissue should be kept in focus and capillaries are visible only as shadows
5. With the help of glass capillary 1, the cell of interest is cut out by scratching along the cell membrane. This is performed at ×600 magnification. The joystick allows precise three-dimentional movement of the glass capillary.
6. The cell is detached from the slide using the glass capillary 1.
7. Glass capillary 1 is used to move the isolated cell to the same height as glass capillary 2 (*see* **Fig. 1**).
8. The cell is positioned at the tip of capillary 2 and is absorbed with the help of CellTram.
9. Isolated cells are released into a PCR tube containing 20 µL of DNase free H_2O.
10. In order to make sure that the cell is transferred into the tube, the tip of capillary 2 may be broken into the tube and examined under the microscope afterwards.
11. PCR tubes are centrifuged shortly and are stored at –20°C.
12. Sections can be used for micromanipulation for up to 12 h without dramatic decrease of PCR efficiency (*see* **Note 6**).

3.4. Amplification of Rearranged V-Region Genes

3.4.1. Preparation of PCR Templates

1. To prepare DNA for PCR, add 1 µL of Proteinase K solution (16 mg/mL) to each of the tubes containing isolated cells.
2. Cells are inclubated for 1 h at 50°C.
3. Enzyme activity is stopped by a 10 min incubation at 95°C.

3.4.2. Preamplification of Genomic DNA Using Random Primers

1. Prepare 30 µL of reaction mixture (5U of Taq polymerase, 60 mM random primer and H_2O to a final volume of 30 µL).

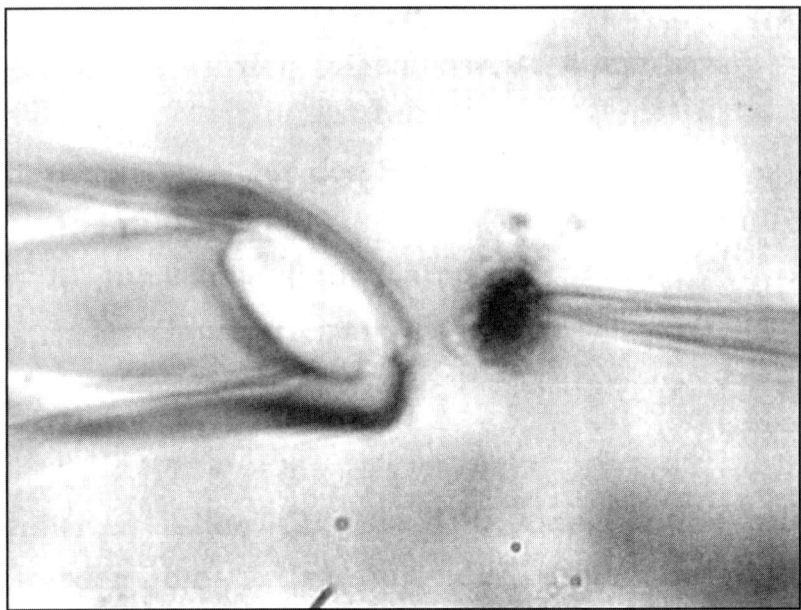

Fig. 1. Micro-dissection of a single cell with the help of glass capillaries.

2. Add to the 20 µL DNA solution prepared in **Subheading 3.3.2.**
3. Amplify genomic DNA to obtain sufficient timplate for the subsequent specific amplication *(23)*.
4. The initial denaturation step at 95°C for 10 min is followed by a 2 min annealing step at 37°C, a programmed ramping step of 10 s/deg to 55°C, and a 4 min incubation at 55°C for extension.
5. All subsequent denaturation steps are done for 1 min at 94°C.
6. Samples are stored at 4°C until further analysis.

3.4.3. Specific Amplification of V Genes (see Note 8)

Preamplified DNA is used as a template for the specific amplification of Ig heavy- and light-chain V genes. Two steps of amplification are performed to obtain the appropriate amount of PCR product. To improve the specificity of the second amplification step, semi-nested PCR reactions are conducted.

3.4.3.1. FIRST POLYMERASE CHAIN REACTION

1. Set up the PCR reaction by using: 2 U GoldTaq polymerase, 5 µL 10X PCR buffer, 10 pmol VH primer-mix, 10 pmol JHext primer, 10 pmol Vκ primer-mix, 10 pmol Jκext primer-mix, 10 pmol Vλ primer-mix, 10 pmol Jλ ext primer-mix, 200 µM dNTPs, 5 µL preamplification product, and H_2O to a final volume of 50 µL.

2. Aplify DNA by using the following PCR program: 95°C for 2 min, 40 cycles of (94°C 1 min, 65°C 1.5 min, 72°C 2 min), and 72°C for 15 min.
3. Run PCR product onn a Nusieve low melting agarose gel.
4. The region at approximately 350 bp is cut out and put into a 500 µL tube.
5. 200 µL H$_2$O is added to the tube.
6. The gel is melted by incubating the tubes at 65°C.

Second round PCR can be carried out as described in **Subheading 3.4.3.2.** Alternatively, the tubes can be stored at –20°C.

3.4.3.2. SECOND POLYMERASE CHAIN REACTION

1. Set up independent PCR reactions for Ig heavy, κ, and λ chains (see **Note 9**) using the following reaction mixture : 2 U GoldTaq polymerase, 5 µL 10X PCR buffer, 10 pmol VH (or Vκ- or Vλ-) primer-mix, 10 pmol JH (or Jκ - or Jλ) primer-mix, 200 µM dNTPs, 2 µL first PCR product (melted gel), and H$_2$O to a final volume of 50 µL.
2. Amplify the DNA using the following PCR program: 95°C for 2 min, 35 cycles of (94°C 1 min, 65°C 1.5 min, 72°C 2 min), and72°C for 15 min.
3. Analyze PCR products on a Nusieve low melting agarose gel.
4. Cut out DNA bands and extract using the Qiagen DNA extraction kit (see **Note 10**).

3.5. Cloning (see Note 11)

1. Amplified DNA is cloned into the TA cloning vector (Invitrogen).
2. Cloned are screened for the presence of inserts of 350 bp by performing restruction analysis using EcoRI.

3.6. Sequencing

1. Inserts are sequenced by using BigDye sequencing kit. The following reaction mixture is set up : 2 µL Big Dye reaction mix, 250 ng Plasmid DNA, and 10 pmol primer (+40 or –40 primer for TA cloning vector) in 10 µL.
2. Amplify DNA by using the following PCR Program: 35 cycles of 96° C, for 10 s, 45° C, for 5 s, and 60° C for 4 min.
3. Purify reaction products ethanol precipitation : add 90 µL H$_2$O, 10µL 3M NaAc (pH 4.6), and 250 µL 100% ethanol to the sequence reaction.
4. Centrifuge for 15 min at 20,000g.
5. Remove supernatant.
6. Wash pellet with 250 µL 70% ethanol.
7. Centrifuge for 5 min at 20,000g.
8. Air dry pellets and re-suspended in 25µl template suppression reagent (TSR) (BigDye sequencing kit).
9. Run sequence reaction products on the ABI 310 DNA SequenceAnalyzer
10. Sequence reactions can be stored at 4°C for up to 1 wk without a reduction in sequence quality.

3.7. Sequence Evaluation

1. V region sequences are identified using the IMGT database and DNAplot software (*http://www.dnaplot.de/input/human_v.html*).
2. The isolated sequence is compared to the most homologous germ line Ig sequence.
3. The number and type of somatic mutations can be determined.
4. The CDR3 region (VDJ joining region for heavy chain gene, VJ joining region for light chain gene) should be analyzed to confirm a functional rearrangement.
5. Overall V gene usage can be analyzed to determine if there is any preferential usage of certain genes or gene families, which might indicate an antigen driven B cell response.
6. The clonal relationship of identically rearranged Ig sequences isolated from different single cells can be determined and a phylogenetic tree be generated.

4. Notes

1. In only 30% of RA patients are organized follicles found in the synovial tissue. Therefore, a prescreening of tissue samples should be carried out. Once structures of interest are identified it is important to take serial sections if possible, as cells in the same cellular clusters localizing horizontally or vertically should be taken for single cell analysis. The number of each section should correspond to the real serial number of the section, i.e. number of sections lost in the presence of cutting should not be omitted, so that the size of the cell cluster can be determined.
2. Typically, anti-CD20 and anti-CD19 antibodies are used for B-cells, Wue-1, and anti-CD38 antibodies are for plasma cells, Wue-2 for FDCs, and anti-Ki67 for proliferating cells *(10,11, 22)*. T-cells can be stained using anti-CD4 antibody. Anti-CD79a antibody can be used to visualize proliferating germinal center B-cells which have down-regulated their B-cell receptor *(11)*. Staining with anti-CD38 antibody makes it possible to detect plasma cells and germinal center B cells at the same time, as CD38 is strongly expressed on the plasma cells and weakly expressed on the germinal center B-cells (*see* **Fig. 2**).
3. Five to ten micrograms per milliliter antibody concentration is used for staining, however, the appropriate concentration for each antibody has to be optimized by titration. To avoid nonspecific binding, antibodies should be diluted in 3% BSA/PBS solution. Every incubation step should be performed in a humid chamber to avoid drying out of tissue sections. Isotype control should be included for each staining experiment.
4. Single cell analyses can also be done after cells are prepared by laser microdissection. In this case, tissue sections are mounted on an ultra-thin transparent supporter membrane and cells are isolated together with the membrane (*see* **Chapter 6, Volume 1**).
5. The micro-dissction procedure can be observed and controlled on the monitor. To document the localization of the isolated cells in the histological context, photographs should be taken pre- and postisolation.

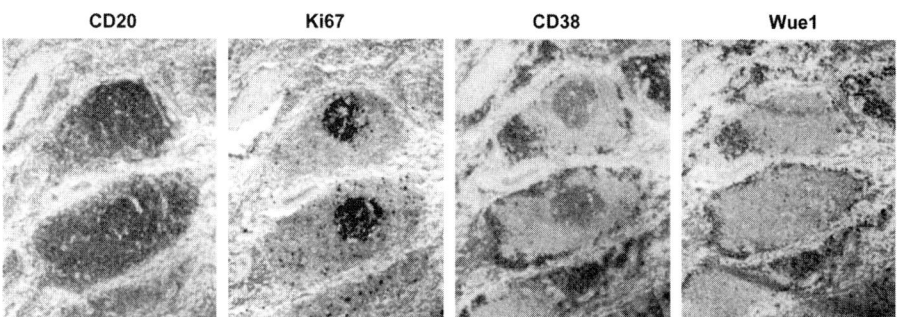

Fig. 2. Immunohistochemical analysis of synovial infiltrates. Consecutive sections are stained using anti-CD20 (for B-cells), anti-Ki67 (proliferating cells), anti-CD38 (germinal center cells and plasma cells) and, anti Wue-1 (plasma cells) antibodies.

6. It is possible to isolate more cells from the same section at a later time point, however, a second dehydration and storing of the section I necessary and that leads to a lower PCR efficiency. Therefore, to isolate additional cells it is advisable to use consecutive sections.
7. This protocol can be also used for performing single-cell PCR from sorted single cells *(23)*. Patient blood cells or cells prepared from synovial tissue by collagenase digestion are labeled for flow cytometry. Single cells are sorted into a 96-well plate containing cell lysis buffer (*see* **Subheading 2.4.2.**). Sorted cells are incubated for 10 min at 65°C in the lysis solution and neutralized This solution can be used as material for continuing with **Subheading 3.4.2.**
8. As more than 20 different primers are employed in the specific PCR for V genes, the purity of the primers are critically important for the high efficiency of PCR. We recommend that high-performance liquid chromatography (HPLC) purified primers are used. We routinely get a clear single bans using this type of primer while unpurified primers often result in a smear on the analyzing gel.
9. Each gene family can of course be separately amplified using the appropriate primers. While this increases the number of PCR reactions it allows sequencing directly after PCR. The same 5'-primer used for the PCR can be employed for the sequencing reaction.
10. PCR efficiency is variable depending on the tissue condition and on the efficiency of cell isolation. In our hands, the efficiency of single-cell amplification (defined as percentage of cells for which at least one PCR product was obtained) was in the range of 20 to 30%. When cells are isolated from the tissue section, many cells will be missing part of the nucleus. It is also conceivable that the primers employed in the experiment may fail to detect the V gene of a given cell, either because the V gene of that cell is not known yet or because a mutation occurred in the primer binding regions.

References

1. Wernick, R. M., Lipsky, P. E., Marban-Arcos, E., Maliakkal, J. J., Edelbaum, D., and Ziff, M. (1985) IgG and IgM rheumatoid factor synthesis in rheumatoid synovial membrane cell cultures. *Arthritis Rheum.* **28**, 742–752.
2. Natvig, J. B., Randen, I., Thompson, K., Forre, O., and Munthe, E. (1989) The B cell system in the rheumatoid inflammation. New insights into the pathogenesis of rheumatoid arthritis using synovial B cell hybridoma clones. *Springer Semin Immunopathol* **11**, 301–313.
3. Trentham, D. E., Townes, A. S., and Kang, A. H. (1977) Autoimmune to type II collagen an experimental model of arthritis. *J. Exp. Med.* **146**, 857–868.
4. Sebbag, M., Simon, M., Vincent, C., Masson-Bessiere, C., Girbal, E., Durieux, J. J., and Serre, G. (1995) The antiperinuclear factor and the co-called antikeratin antibodies are the same rheumatoid arthritis-specific autoantibodies. *J. Clin. Invest.* **95**, 2672–2679.
5. Schellekens, G. A., de Jong, B. A., van den Hoogen, F. H., van de Putte, L. B., and van Venrooij, W. J. (1998) Citrulline is an essential constituent of antigenic determinants recognised by rheumatoid arthritis-specific autoantibodies. *J. Clin. Invest.* **101**, 273–281.
6. Despres, N., Boire, G., Lopez-Longo, F. J., and Menard, H. A. (1994) The Sa system: a novel antigen-antibody system specific for rheumatoid arthritis. *J. Rheumatol.* **21**, 1027–1033.
7. Zvaifler, N. J. (1973) The immunopathology of joint inflammation in rheumatoid arthritis. *Adv. Immunol.* **16**, 265–336.
8. Takemura, S., Braun, A., Crowson, C., Kurtin, P. J., Cofield, R. H., O'Fallon, W. M., Goronzy, J. J., and Weyand, C. M. (2001) Lymphoid neogenesis in rheumatoid synovitis. *J. Immunol.* **167**, 1072–1080.
9. Randen, I., Mellbye, O. J., Forre, O., and Natvig, J. B. (1995) The identification of germinal centres and follicular dendritic cell networks in rheumatoid synovial tissue. *Scand. J. Immunol.* **41**, 481–486.
10. Schroder, A. E., Greiner, A., Seyfert, C., and Berek, C. (1996) Differentiation of B cells in the nonlymphoid tissue of the synovial membrane of patients with rheumatoid arthritis. *Proc. Natl. Acad. Sci. USA* **93**, 221–225.
11. Kim, H. J., and Berek, C. (2000) B cells in rheumatoid arthritis. *Arthritis Res.* **2**, 126–131.
12. Berek, C., Berger, A., and Apel, M. (1991) Maturation of the immune response in germinal centres. *Cell* **67**, 1121–1129.
13. Jacob, J., Kelsoe, G., Rajewsky, K., and Weiss, U. (1991) Intraclonal generation of antibody mutants in germinal centres. *Nature* **354**, 389–392.
14. Kim, H. J., Krenn, V., Steinhauser, G., and Berek, C. (1999) Plasma cell development in synovial germinal centers in patients with rheumatoid and reactive arthritis. *J. Immunol.* **162**, 3053–3062.
15. Matsuda, F., and Honjo, T. (1996) Organisation of the human immunoglobulin heavy-chain locus. *Adv. Immunol.* **62**, 1–29.

16. Cook, G. P., and Tomlinson, I. M. (1995) The human immunoglobulin VH repertoire. *Immunol. Today* **16**, 237–242.
17. Zachau. (1996) The Human Immunoglobulin k Genes. *The Immunologist* **4**, 49–54
18. Frippiat, J. P., Williams, S. C., Tomlinson, I. M., Cook, G. P., Cherif, D., Le Paslier, D., Collins, J. E., Dunham, I., Winter, G., and Lefranc, M. P. (1995) Organisation of the human immunoglobulin lambda light-chain locus on chromosome 22q11.2. *Hum. Mol. Genet.* **4**, 983–991.
19. Tonegawa, S. (1983) Somatic generation of antibody diversity *Nature* **302**, 575–581
20. Berek, C. and Ziegner, M. (1993) The maturation of the immune response. *Immunol. Today* **14**, 400–404.
21. Kuppers, R., Zhao, M., Hansmann, M. L., and Rajewsky, K. (1993) Tracing B cell development in human germinal centres by molecular analysis of single cells picked from histological sections. *EMBO J*, **12**, 4955–4967.
22. Greiner, A., Neumann, M., Stingl, S., Wassink, S., Marx, A., Riechert, F., and Muller-Hermelink, H. K. (2000) Characterisation of Wue-1, a novel monoclonal antibody that stimulates the growth of plasacytoma cells lines. *Virchows Arch.* **437**, 372–379.
23. Brezinschek, H. P., Brezinschek, R. I., and Lipsky, P. E. (1995) Analysis of the heavy chain repertoire of human peripheral B cells using single-cell polymerase chain reaction. *J. Immunol.* **155**, 190–202.

4

Tracking Antigen Specific CD4⁺ T-Cells With Soluble MHC Molecules

John A. Gebe and William W. Kwok

Abstract

The advent of soluble MHC multimer technology has allowed for the flow-cytometric direct identification of specific-MHC restricted antigen-specific T cells in mixed cell populations and also enabled the direct phenotyping and cloning of these cells at the same time. To date, MHC multimers have been used in characterizing the adaptive T cell repertoire under infectious, cancerous, and autoimmune states and has increased our understanding of the dynamics of T-cell immunity. Recombinant MHC multimers have been produced where MHC-binding peptide antigens are either covalently or non-covalently bound to the MHC, with the latter having the advantage of the ability to use a single recombinant MHC to investigate multiple MHC-binding peptides and their interacting T cells. In this method we describe how to generate recombinant non-covalently bound peptide MHC-multimers in insect cells. MHC multimers are generated as tetravalent complexes using a streptavidin scaffold.

Key Words: Tetramers; MHC; CD4⁺; T-cells; tracking.

1. Introduction

Antigen specific CD4⁺ T-cells reside in the peripheral immune system with measured precursor frequencies between 1 in 10^4 to 1 in 10^6. For T-cell mediated diseases the ability to track and characterize antigen specific T-cells holds great hope in understanding the pathologies of immunity and autoimmunity with the potential outcome of beneficial therapeutic interventions. Direct monitoring of antigen-specific T-cells within a mixed T-cell population is a formidable task, usually requiring antigen-driven short-term cultures for amplifying the low-precursor numbers to enable the measurement of functional properties, or the time consuming task of cloning and screening individual T-cells. Akin to the high specificities of antibodies as marker reagents for specific proteins or antigens, the advent of

From: *Methods in Molecular Medicine, vol. 136: Arthritis Research, Volume 2*
Edited by: A. P. Cope © Humana Press Inc., Totowa, NJ

MHC multimers with their property of binding antigen-specific T-cells within the enormously diverse T-cell repertoire has allowed one to study antigen-specific T-cells in the context of: (1) precursor frequencies in mixed cell populations *(1,2)*, (2) kinetics of in vivo antigen-mediated T-cell expansion *(3,4)*, (3) mapping of T-cell epitopes *(5)*, and (4) staging and characterization of disease prone individuals *(6)*. In whole protein responses where immune major histocompatibility complex (MHC)-dependent epitopes are many due in part to the many HLA molecules any one individual can have, the use of MHC-multimers can allow for T-cell specific therapeutic vaccinations that will target the greatest percentage of individuals. Recombinant MHC-multimers for tracking T-cells have been constructed as MHC bound to lipid micelles, MHC immunoglobulin dimers, chemically linked oligomers, and MHC streptavidin tetramers and have been made in prokaryotic and eukaryotic systems *(7–14)*. This particular protocol describes the production of class II MHC tetramers for the study of antigen-specific CD4$^+$ T-cells. This method produces tetramers containing noncovalently-linked MHC-binding peptides thus allowing the use of any MHC binding epitope.

2. Materials

2.1. Growing and Generating Recombinant Producing S-2 Cells

1. *Drosophila* expression vectors pRmHa-3 containing MHC class II α and β chain cDNAs, leucine-zipper, and biotinylation sequences (*see* **Fig. 1**).
2. pHSHNeo-2 (drug selection vector).
3. Schneider Cells S-2.
4. S-2 medium: Schneider's Drosophila medium 500 mL (Gibco; cat. no. 11720-034) containing 10% fetal bovine serum (FBS) (Gibco; cat. no. 10082-139), 1 m*M* sodium pyruvate (100 m*M* stock) (Gibco; cat. no. 11360-070) and penicillin (50 U/mL)/streptomycin (50 µg/mL) (Gibco; cat. no. 15070-063).
5. G418-sulfate (Gibco; cat. no. 11811-031).
6. 100 m*M* Cupric sulfate (Sigma; cat. no. C-8027).
7. *n*-Octyl β-D-glucopyranodise (OG) (Sigma; cat. no. O-8001).
8. Bellco spinner flask (Bellco; cat. no.1965-01000).

2.2. Purification and Biotinylation of Recombinant Class II

1. 2X coupling buffer: 0.2*M* NaHCO$_3$, 1.0*M* NaCl (pH 4.0) with HCl.
2. Blocking buffer: 1.5 g glycine (w/v) in 1X coupling buffer (pH 8.5).
3. Acetate buffer: 0.1*M* NaOAc, 0.5*M* NaCl (pH 4.0) with HCl.
4. Phosphate buffered saline (PBS)-N$_3$: PBS with 0.05% (w/v) NaN$_3$.
5. CNBr-sepharose 4B (Sigma).
6. OG.
7. 20 mg purified anti-MHC class II antibody (L243 hybridoma for MHC-DR purification, SPVL3 for MHC-DQ purification).

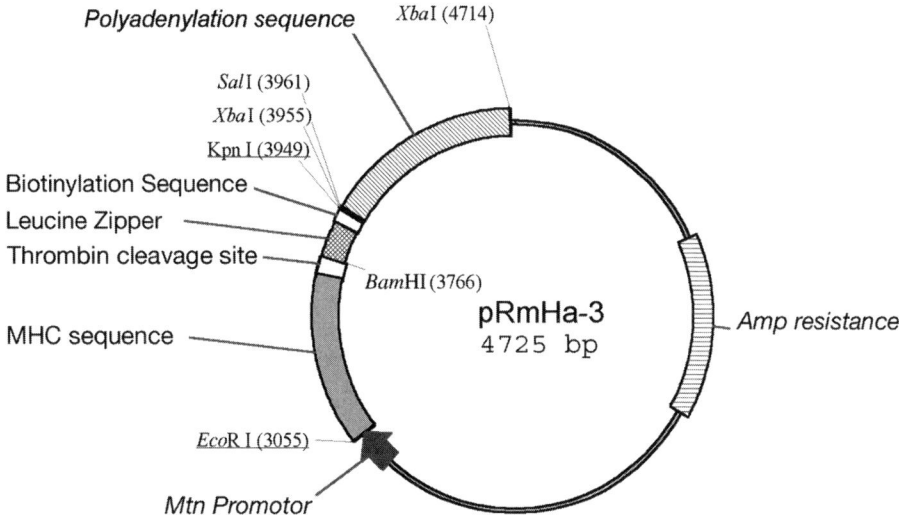

Fig. 1. pRmHa-3 copper-inducible promoter vector used to produce recombinant soluble MHC molecules. EcoR1 and BamHI restriction sites allow for subcloing extracellular regions of MHC molecules into pRmHa-3 vector.

8. Bio-Rad columns (20 cm × 1.5 cm).
9. Wash buffer 1: 0.02% *(w/v)* OG in 200 mL PBS (pH 7.5) Make fresh.
10. Wash buffer 2: 0.02% *(w/v)* OG in 50 mL of PBS containing 5.0M NaCl .Make fresh.
11. Wash buffer 3: 1.0M Tris-HCl (pH 7.5).
12. Elution buffer: 0.1M Tris-HCl and 0.5M NaCl (pH 11.5).
13. Biotinylation buffer: 20 mM Tris-HCl, 0.01M NaCl, and 7.5 mM MgCl$_2$ (pH 8.0).
14. Biotin Protein Ligase kit (Avidity; cat. no. BIRA).
15. Pefabloc SC: 100 mM in H$_2$O (Boehringer Mannheim; cat. no. 1585-916).
16. Pepstatin (1 mg/mL in absolute EtOH) (Boehringer Mannheim; cat. no. 1359-053).
17. Leupeptin (1 mg/mL in H$_2$O) (Boehringer Mannheim; cat. no. 1017-101).
18. Slide-A-lyzer dialysis cassette 10,000 MW cutoff (Pierce; cat. no. 664120).
19. Amicon Ultra-15 centrifugal filter unit 30,000 MW cutoff (Millipore; cat. no. UFC903096).

2.3 Peptide Loading and Chromophore-Labeled Tetramer Formation

1. Class II binding peptide at 50 mg/mL in dimethyl sulfoxide (DMSO) (approx 15 mer peptide).
2. 200 mM Na$_2$HPO$_4$ (pH 5.9).
3. Pefabloc SC: 100 mM in H$_2$O (Boehringer Mannheim; cat. no. 1585-916).

4. OG.
5. Phycoerythrin-labeled streptavidin (BioSource International; cat. no. SNN 1007).

3. Methods
3.1. Recombinant Soluble MHC Producing S-2 Cell
3.1.1. pRmHa-3 Construct

The copper-inducible metalothionine promoter vector pRmHa-3 is shown in **Fig. 1**. EcoRI and Bam HI provide unique restriction sites for subcloning of the extra cellular regions of MHC α and β regions. cDNA fragments encoding class II extra cellular region (signal sequence, α1 and α2 domains [β1 and β2 domains for β chain], plus the short stretch of amino acids just before the transmembrane region) are PCR amplified, sequenced, and subcloned into the pRmHa-3 (*see* **Note 1**).

3.1.2. Transfection, Selection, and Cloning of S-2 Cells

1. The following solutions are prepared prior to transfection:
   a. DNA solution: 10 μg MHC α pRmHa-3 vector, 10 μg MHC β pRmHa-3 vector, 250 ng pHSHNeo-2, 25 μL 2.0M CaCl$_2$, and sterile H$_2$0 to make 200 μL final volume.
   b. Stock solution: 100 μL 500 mM Hepes (pH 7.1), 125 μL 2.0M NaCl, 10 μL 150 mM Na$_2$HPO$_4$ (pH 7.0), 765 μL sterile H$_2$0.
2. Schneider (S2) cells are grown and maintained in S-2 (Schneider's) media with cell density between 0.5 and 2.0 × 10^6 cells/mL.
3. For transfection 3 to 4 million S-2 cells at greater than 90% viability are centrifuged at 220g (Beckman GS-6 centrifuge) for 8 min.
4. The cell pellet is resuspended and centrifuged a second time and resuspended in 4 mL S-2 media and transferred to a 6-well plate well.
5. DNA solution is added drop wise to 200 μL of stock solution. The DNA-stock solution should turn slightly cloudy as a reuslt of precipitation of the DNA.
6. DNA-stock solution is added drop wise to S-2 cells in a 6-well plate while shaking plate slightly.
7. Transfected cells are incubated at 28°C for 24 h.
8. At 24 h S-2 cells are transferred to a 15-mL conical tube along with 6 mL of S-2 media, spun at 220 g (Beckman GS-6 centrifuge) for 8 mins, aspirated, and resuspended in 10 mL fresh S-2 media.
9. Centrifugation is repeated again and pellet is resuspended in 6 mL S-2 media containing 2.0 mg/mL G418 and placed in a 6-well plate well and incubated at 28°C. S-2 media is replaced after 1 wk.
10. After 7 to 10 d, viable cells should begin to grow out. At this point transfected S-2 cells are single cell cloned.

3.1.3. Single Cell Cloning of Transfected S-2 Cells

1. Transfected S-2 cells from the 6-well plate well are replated into 96-flat bottom well plates at 1 to 2 viable cells/well and cultured in the presence of 2 mg/mL G418.
2. On day-5 onwards plates are viewed and scored for single cell colony growth.
3. Nearly confluent wells are tested for MHC class II protein production by dot western blots using anti-class II Ab (L243 Ab for DR detection and SPVL3 Ab for DQ detection).
4. Identified high producing clones are expanded in the presence of G418 and frozen down in 10% DMSO/FBS at 5×10^6 cells/vial aliquots and kept in liquid nitrogen freezers for future use.

3.2. Growth, Purification, and Biotinylation of Recombinant Class II

3.2.1. Growth of Transfected S-2 Cells

1. Frozen vials of MHC transfected S-2 cells are thawed and resuspended in 10 mL S-2 media, centrifuged, aspirated, and resuspended in S-2 media to a density of 0.5×10^6 cells/mL in T25 flask.
2. Transfected S-2 cells are expanded to 500 mL while maintaining cell density between 0.5 and 2×10^6 cells/mL and are then transferred to Bellco spinner flasks and grown to 4 to 6×10^6 cell/mL at which time cupric sulfate is added to a final concentration of 1.0 mM.
3. On day 5 of culture with cupric sulfate, cell cultures are centrifuged for 20 min at 2060g (Beckman GS-6 centrifuge) in 200-mL conical tubes and the supernatant is filtered through a 0.45 µm filter.
4. OG is added to filtered solution to a final concentration of 0.02% *(w/v)*. Filtered solution should be stored no more than 24 h at 4°C.

3.2.2. Purification of Recombinant Class II

3.2.2.1. COUPLING OF ANTI-CLASS II ANTIBODY TO SEPHAROSE-4B

1. Ab coupling to sepharose-4B is performed at room temperature. 1.5 g of sepharose-4B is suspended in 50 mL 1.0 mM HCl and allowed to sit for 30 min.
2. Swelled sepharose-4B is rinsed in a C-sinstered funnel (60 mL) with 200 mL of 1.0 mM HCl under light vacuum conditions so as to avoid drying sepharose.
3. Sepharose is then vacuum washed with 100 mL of 1X coupling buffer and ceased when 25 mL of coupling buffer remains.
4. Washed sepharose-4B in 25 mL 1X coupling buffer is transferred to a 50-mL conical tube, centrifuged at 60g (Beckman GS-6 centrifuge) for 5 min, aspirated and resuspended in 1X coupling buffer to a final volume of 10 mL.
5. 20 mg of purified anti-class II antibody is added to an equal volume of 2X coupling buffer and brought to a final volume of 25 mL with 1X coupling buffer and added to 50 mL conical tube.

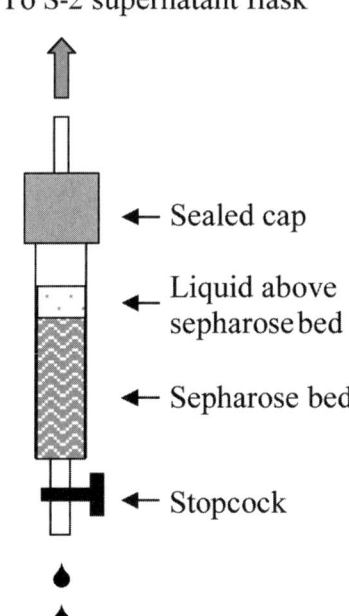

Fig. 2. Columns apparatus for purifying recombinant MHC from S-2 supernatant.

6. The 50-mL conical tube, containing sepharose-4B and antibody, is rotated for 2 h at room temperature.
7. Following the 2-h coupling period the 50-mL conical tube is centrifuged for 5 min at $60g$, supernatant aspirated, and pellet resuspended with 20 mL of 1X coupling buffer. Centrifugation and aspiration is repeated a second time.
8. Remaining sepharose-4B binding sites are blocked by resuspending pellet in 20 mL of blocking buffer and rotating for 2 h at room temperature.
9. 50-mL conical tube is centrifuged for 5 min at $60g$, aspirated, and pellet resuspended with 15 mL of 1X coupling buffer (see **Note 2**).
10. Antibody coupled sepharose-4B is poured into a Bio-Rad column (20 cm × 1.5 cm), allowed to settle, and washed with 50 mL acetate buffer followed by 50 mL 1X coupling buffer.
11. Column can be stored at this point by running PBS-N_3 through the column. Columns are stored such that the PBS-N_3 covers the top of the sepharose bed. Stored columns are used after rinsing with 50 mL of 1X coupling buffer.

3.2.2.2. PURIFICATION OF CLASS II FROM S-2 SUPERNATANT

1. Column loading and elutions are done at 4°C. Stopcock (see **Fig. 2**) on antibody-coupled sepharose-4B column is opened and liquid drained above the column so as to be even with the sepharose-4B bed.

2. Three to four milliliters of S-2 supernatant (containing your protein) is added slowly to the top of the sepharose bed and cap is attached to top of column.
3. After connecting the top of the column to the reservoir containing the S-2 supernatant (*see* **Note 3**) the stopcock is adjusted so that the flow rate is 1 drop/2 s (loading may take up to 48 h).
4. Flow-through is kept until protein is verified from the eluted column.
5. Once S-2 supernatant is loaded onto column, the column is washed with 200 mL of wash 1 buffer using a 1-drop/2 s flow rate (washed overnight).
6. Column is then washed with 50 mL of eash 2 buffer and then with 50 mL of wash 3 buffer in the same manner.
7. With wash 3 buffer in the column even with the sepharose bed and the stopcock closed, 5 mL of elution buffer is added to top of column.
8. Stopcock is opened and adjusted to a slow flow rate and 2 2-mL fractions are collected.
9. Column is equilibrated with elution buffer by letting column sit for 3 min, while connecting top of column to elution reservoir.
10. 14 × 2-mL fractions are collected and assayed for protein content using Bradford protein detection assay.
11. Pool 2-mL fractions with A_{570} values above 0.05 and dialyze at 4°C with biotinylation buffer changing buffer every 12 h over a 48-h period.
12. Concentrate dialyzed sample down to approx 2 mg/mL with 30,000 molecular weight cutoff centrifuge concentrators (Amicon Ultra-15).

3.2.3. Biotinylation of Purified Class II Protein

1. Biotinylation is carried out using Biotin Protein Ligase kit from Avidity (cat. no. BIRA).
2. To each 800 µL of purified class II (2 mg/mL) add the following: 100 µL Biomix A, 100 µL Biomix B, 100 µL Biotin, 10 µL BirA, pefablock 1:100 (stock 100 mM), pepstatin 1:1000 (stock 1 mg/mL), and Leupeptin 1:1000 (stock 1 mg/mL). Incubate overnight at 26°C.
3. Biotinylated class II is dialyzed at 4°C in phosphate buffer over a 48-h period with 4 1-L changes of phosphate buffer.
4. Protein concentration is quantitated again using the Bradford assay.

3.3. Peptide Loading and Chromophore-Labeled Tetramer Preparation

1. 2X peptide loading buffer is made fresh by adding OG (0.02% *[w/v]*) to 200 mM phosphate buffer (pH 5.9)w.
2. Purified and biotinylated MHC (α and β chain dimmer) at 1.0 mg/mL is added to an equal volume of peptide loading buffer in a 1.7 mL microcentrifuge tube.
3. Peptide (stock at 50 mg/mL in DMSO) is added to a final concentration of 500 µg/mL and pefabloc is added to a final concentration of 1.0 mM.
4. Peptide loading into MHC is complete following 72-h water bath incubation at 37°C.
5. For every 50 µL of peptide loaded MHC solution, 10 µL of PE labeled streptavidin (0.25 mg/mL) is added (*see* **Note 4**).

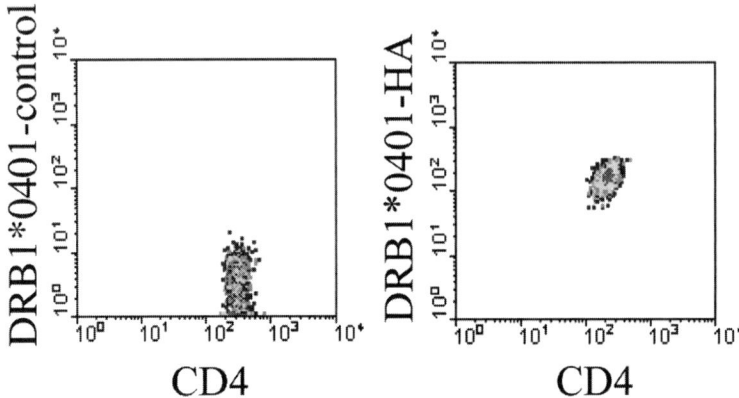

Fig. 3. Tetramer staining of a DRB1*0401 restricted HA (306–318) responsive clone. Clone was stained with tetramer for 15 min at 37°C followed by anti-CD4 Ab staining for 15 min on ice.

6. The streptavidin-chromophore MHC solution is incubated at room temperature for 8 h in complete darkness. It should be noted that adjustments to the volume of added chromophore-labeled streptavidin must be made if starting biotinylated MHC (α and β chain dimmer) is less than 1.0 mg/mL.

3.4. Staining of Cells

3.4.1. Clones

1. CD4+ T cell clones are stained in their resting phase following a 10 to 12 d in-vitro expansion.
2. Tetramer staining is carried out in T-cell growth media in a 37°C 10% CO_2 incubator using between 1 and 10 μg tetramer/mL in a volume of 50 to 100 μL for a period of 15 min to 3 h.
3. Amount of tetramer used and the length of time for staining are dependent on the avidity of the TcR-MHC interactions with high avidity interactions requiring less tetramer and/or a shorter binding time.
4. Surface antigen staining (e.g., CD antigens) are added directly to the sample following tetramer binding and incubated on ice for 15 min.
5. Samples are washed once and run on a flow cytometer. An example of DRB1*0401 tetramer staining of a DRB1*0401 restricted HA (306-318) responsive CD4+ T cells clone is shown in **Fig. 3**.

3.4.2. Peripheral Blood Mononuclear Cells

Direct ex vivo staining of antigen specific CD4+ T cells from peripheral blood mononuclear cells (PBMCs) or tissue is a difficult endeavor but has been

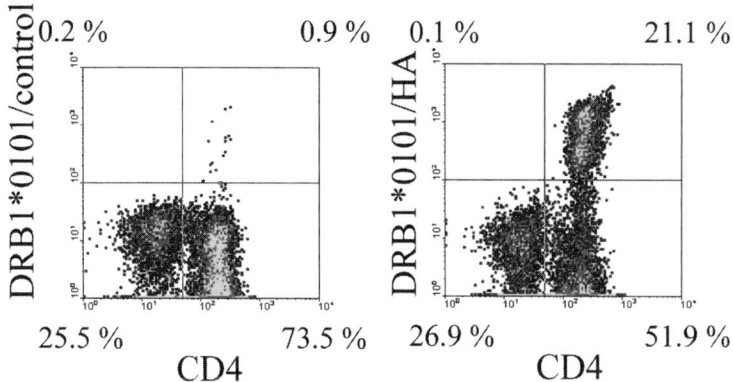

Fig. 4. Tetramer staining of 13-9d simulated PBMCs. PBMCs from a DRB1*0101 individual were stimulated with HA (306–318) peptide for 13 d and stained with DRB1*0101-control and DRB1*0101/HA(306-318) tetramers.

done *(1,3,15)*. Because of the low precursor frequencies of antigen-specific $CD4^+$ T-cell in the T-cell repertoire, antigen specific ex vivo expansion (7–14 d) is usually required. Whereas one can directly expand whole PBMCs for tetramer staining, better results are obtained when subpopulations are first purified (i.e,. $CD4^+$ T-cells) and then subsequently expanded by antigen with irradiated APC. This results in part from the fact that monocytes/macrophages can express the $CD4^+$ surface antigen complicating tetramer-binding analysis. We have also found that fresh as opposed to frozen PBMC gives better results. In addition to obtaining improved staining on purified cell populations, other flow cytometric reagents (e.g., antibody markers, surface cytokine staining kits, apoptotic markers, CFSE cell division marker) can be used for phenotyping tetramer positive cells. An example of whole PBMC from a DRA1*0101/B1*0101 individual that were stimulated with HA (306-318) peptide for 13 d and then stained with DRB1*0101-control and DRB1*0101-HA (306-318) tetramers is shown in **Fig. 4** (*see* **Note 5**).

3.5. Conclusion

Tracking of pathogenic T-cells with multimeric class II reagents in PBMCs of patients with rheumatoid arthritis and relapsing polychondritis has been attempted *(16,17)*. These experiments failed to detect collagen type II or gp39 responsive $CD4^+$ T-cells by using multiple class II tetramers. Failures in these experiments bring up future challenges and questions in using multimeric class II reagents in detecting pathogenic T-cells in rheumatic diseases.

What are the major auto-antigens and their MHC-specific binding epitopes involved in the pathogenesis in the various rheumatic diseases? Does posttranslational modification of these auto-antigens occur? Does the autoreactive T-cell repertoire consist of low avidity MHC-peptide-TcR interactions making monitoring of these T-cells a formidable task with MHC multimers? If so, what kind of experimental approaches can be utilized to detect these low avidity autoreactive T-cells?

Though class II multimers have been successfully used to detect antigenic T-cells in various infectious disease settings and autoimmune diseases such as diabetes, results from studies in rheumatic diseases are very limited. It is expected that the understanding of the molecular and cellular components in various rheumatic diseases in the coming decade will facilitate the development of multimeric class II reagents for the detection of rheumatic related T-cells.

4. Notes

1. When polymerase chain reaction (PCR) amplifying the extra cellular regions of the MHC α and β chains the demarcation between the transmembrane region and the extra cellular regions can unusually be identified from hydrophobicity plots.
2. Ab coupling efficiency can be estimated by measuring and comparing absorbance A_{280} values of sepharose-4B Ab supernatant (dilution corrected) to original A_{280} antibody concentration.
3. Column loading of S-2 supernatant can be done by gravity-syphon loading or by a peristaltic pumping apparatus.
4. APC (allophycocyanin), PE (phycoerythrin), CyChrome (PE-Cy5), Alexa-488, and Percp (peridinin chlorophyll protein) conjugated streptavidins have been used successfully as chromophoric markers in making flow cytometric MHC tetramers reagents. The above chromphores give nice tetramer staining with minimal background noise. FITC does not work well as a chromophore label for tetramers.
5. Whereas in vivo expansion of antigen specific CD4+ T-cells from PBMCs is variable depending on the antigen, the following method has been used successfully in our lab. PBMCs are culture with 0.1 to10 µg/mL peptide in RPMI-15 (RPMI containing 15% pooled human serum, Penn & Strep., and Na-pyruvate). Hemagen human IL-2 (Hemagen Diagnostics Inc) is added at 1/20 dilution between days 4 and 7, and culture media is split 1:2 with fresh IL-2 containing media every 2 to 3 d following initial IL-2 addition. Cells are split when necessary with IL-2 containing media. Tetramer staining of these cultures is done between 10 and 14 d post stimulation. It should be noted that we have at times failed to observe tetramer staining from in vivo stimulation of PBMCs, only to find that we can observe tetramer positive CD4+ T-cells in cultures of purified CD4+ T-cells with CD4-depleted adherent APCs.

References

1. Day, C.L., Seth, N.P., Lucas, M., et al. (2003) Ex vivo analysis of human memory CD4 T cells specific for hepatitis C virus using MHC class II tetramers. *J.Clin.Invest* **112**, 831–842.
2. Novak, E.J., Liu, A.W., Nepom, G.T., and Kwok, W.W. (1999) MHC class II tetramers identify peptide-specific human CD4(+) T cells proliferating in response to influenza A antigen. *J Clin Invest* **104**, R63–R67.
3. Danke, N.A. and Kwok, W.W. (2003) HLA class II-restricted CD4+ T cell responses directed against influenza viral antigens postinfluenza vaccination. *J.Immunol.* **171**, 3163–3169.
4. Bischof, F., Hofmann, M., Schumacher, T.N., et al. (2004) Analysis of autoreactive CD4 T cells in experimental autoimmune encephalomyelitis after primary and secondary challenge using MHC class II tetramers. *J.Immunol.* **172**, 2878–2884.
5. Novak, E.J., Liu, A.W., Gebe, J.A., et al. (2001) Tetramer-guided epitope mapping: rapid identification and characterization of immunodominant CD4+ T cell epitopes from complex antigens. *J.Immunol.* **166**, 6665–6670.
6. Reijonen, H., Novak, E.J., Kochik, S., et al. (2002) Detection of GAD65 specific T-cells by MHC class II multimers in type 1 diabetes patients and at-risk subjects. *Diabetes* **51**, 1375–1382.
7. Altman, J.D., Moss, P.A., Goulder, P.J., et al. (1996) Phenotypic analysis of antigen-specific T lymphocytes. *Science* **274**, 94–96.
8. Crawford, F., Kozono, H., White, J., Marrack, P., and Kappler, J. (1998) Detection of antigen-specific T cells with multivalent soluble class II MHC covalent peptide complexes. *Immunity* **8**, 675–682.
9. Appel, H., Gauthier, L., Pyrdol, J., and Wucherpfennig, K.W. (2000) Kinetics of T-cell receptor binding by bivalent HLA-DR.Peptide complexes that activate antigen-specific human T-cells. *J.Biol.Chem.* **275**, 312–321.
10. Gutgemann, I., Fahrer, A.M., Altman, J.D., Davis, M.M., and Chien, Y.H. (1998) Induction of rapid T cell activation and tolerance by systemic presentation of an orally administered antigen. *Immunity.* **8**, 667–673.
11. Mallet-Designe, V.I., Stratmann, T., Homann, D., Carbone, F., Oldstone, M.B., and Teyton, L. (2003) Detection of low-avidity CD4+ T cells using recombinant artificial APC: following the antiovalbumin immune response. *J.Immunol.* **170**, 123–131.
12. Cochran, J.R., Cameron, T.O., and Stern, L.J. (2000) The relationship of MHC-peptide binding and T cell activation probed using chemically defined MHC class II oligomers. *Immunity* **12**, 241–250.
13. Malherbe, L., Filippi, C., Julia, V., et al. (2000) Selective activation and expansion of high-affinity CD4+ T cells in resistant mice upon infection with Leishmania major. *Immunity.* **13**, 771–782.
14. Arnold, P.Y., Vignali, K.M., Miller, T.B., et al. (2002) Reliable generation and use of MHC class II:gamma2aFc multimers for the identification of antigen-specific CD4(+) T cells. *J.Immunol.Methods* **271**, 137–151.

15. Jang, M.H., Seth, N.P., and Wucherpfennig, K.W. (2003) Ex vivo analysis of thymic CD4 T cells in nonobese diabetic mice with tetramers generated from I-A(g7)/class II-associated invariant chain peptide precursors. *J.Immunol.* **171**, 4175–4186.
16. Buckner, J.H., Van, L.M., Kwok, W.W., and Tsarknaridis, L. (2002) Identification of type II collagen peptide 261-273-specific T cell clones in a patient with relapsing polychondritis. *Arthritis Rheum.* **46**, 238–244.
17. Kotzin, B.L., Falta, M.T., Crawford, F., et al. (2000) Use of soluble peptide-DR4 tetramers to detect synovial T cells specific for cartilage antigens in patients with rheumatoid arthritis. *Proc.Natl.Acad.Sci.USA* **97**, 291–296.

5

Analysis of Antigen Reactive T-Cells

Clare Alexander, Richard C. Duggleby, Jane C. Goodall,
Malgosia K. Matyszak, Natasha Telyatnikova, and J. S. Hill Gaston

Abstract

Because antigen-specific cells are the central coordinators of the immune response to infectious organisms, and the principal effector cells in autoimmune disease, there are many circumstances in which investigators may wish to examine the T-cell responses to particular antigens. This chapter outlines techniques for assessing the responses of polyclonal populations of T-lymphocytes by measuring a variety of outputs each of which gives different kinds of information about the response. The outputs discussed are proliferation and cytokine production, with methods for measuring cytokine secretion by the whole population together with techniques for making an estimate of the numbers of cells producing a cytokine in response to antigen, and examining the phenotype of the responsive cells.

In many cases detailed information about responses to particular antigens requires the isolation and characterization of antigen-responsive T-cell clones, and this is also described together with methods of identifying unknown antigens by screening recombinant expression libraries. Lastly, because the techniques differ in many respects, methods for isolating antigen-specific $CD8^+$ T-cells, particularly those which recognize bacteria, are also included.

Key Words: T-lymphocytes; proliferation assays; cytokines; ELISPOT; flow cytometry; T-cell cloning; $CD4^+$ T-cell clones; $CD8^+$ T-cell clones; recombinant antigens; expression libraries; dendritic cells.

1. Introduction

In principle the measurement of B-cell responses to antigens or other stimuli is straightforward, because antibody secretion is the major output of interest and is readily measured. In contrast, T-lymphocytes have several possible outputs in response to antigen (including help for B-cells), and this means that different outputs can be measured, depending on the information required about the T-cell response. This chapter deals with some of the techniques for measuring T-cell responses to antigen in vitro, including detailed characterization of the responses at a clonal level. Methods for generating bacteria-specific CD8$^+$ T-cells are also discussed.

2. Materials
2.1. Cell Separation

1. Preservative-free heparin.
2. Hyaluronidase.
3. Ficoll-Hypaque.
4. RPMI.
5. Hepes.
6. Nonessential amino acids.
7. Sodium pyruvate.
8. L-glutamine.
9. Penicillin/streptomycin.
10. Human AB- serum (heat inactivated).
11. Trypan blue.

2.2. T-Cell Outputs

1. 0.2 mL 96-well tissue culture plates (round or flat bottomed).
2. ^{3}H-thymidine.
3. Scintillation cocktail.
4. Cell harvester.
5. β-scintillation counter.
6. Cytokine enzyme-linked immunosorbent assay (ELISA) kits.
7. 0.1M NaHCO$_3$ buffer (pH 9.6).
8. Fetal calf serum (FCS).
9. Phosphate buffered saline (PBS).
10. Tween.
11. Strepavidin-alkaline phosphatase.
12. BCIP/NCT.
13. Monensin.
14. Brefeldin A.
15. Cytofix/Cytoperm™.
16. Perm/Wash™.
17. 5 mL polystyrene round bottom tubes.
18. MACS cytokine secretion assays.

2.3. T-Cell Cloning

1. Terasaki tissue culture plates.
2. Recombinant IL-2.

3. Phytohaemagglutinin (PHA).
4. 2 mL 12-well tissue culture plates.

2.4. Expression Library Screening

1. pTrc His A, B, and C expression vectors.
2. Sau 3AI.
3. *Escherichia coli* XL-2 blue.
4. YT growth medium.
5. Isopropyl-β-D-thiogalactopyranoside (IPTG).
6. Oligonucleotide primers.

2.5 CD8$^+$ T-Cells

1. Recombinant IL-7, IL-4, granulocyte macrophage-colony stimulating factor (GM-CSF).
2. Ethylene-diamine tetraacetic acid (EDTA).
3. Bovine serum albumin (BSA) (low LPS content).
4. Monoclonal antibodies (mAbs) (CD14, CD19, CD16, CD3, CD56, CD8; TCR γd, TCRαb).
5. Goat anti-mouse IgG antibodies conjugated to microbeads.
6. MACS selection columns (MS and LD).

2.6. Miscellaneous Reagents

1. Polymyxin B.
2. Limulus amebocyte lysate assay.
3. Nickel activated deleting sepharose columns.

3. Methods

This section will describe: (1) methods for preparing mononuclear cells for T-cell assays; (2) measuring T-cell outputs—proliferation and cytokine synthesis detected by assay of supernatants, ELISPOT assays, or flow cytometry; (3) production and characterization of T-cell clones; (4) screening expression libraries with T-cell clones; and (5) isolation and characterization of bacteria-specific CD8$^+$ T-cells.

3.1. Separating Mononuclear Cell Populations for Measuring Polyclonal T-Cell Responses

T-lymphocyte responses are often assessed within mononuclear cell (MC) populations in which they are found (e.g. peripheral blood mononuclear cells [PBMCs]). These, along with T-cells, contain the antigen presenting cells required for T-cell responses, and additional cell subsets (i.e., natural killer [NK]-cells, B-cells) which can influence the overall response. Similar populations may be obtained from sites of inflammation, including effusions and tissues;

thus in the case of joints one can examine mononuclear cells from synovial fluid (SFMC), or dissociated from synovial tissue. The precise composition of these mixtures will vary according to source, as well as from patient to patient.

3.1.1. Treatment of Samples

1. PB and SF require anticoagulation with preservative free heparin (10 U/mL).
2. Alternatively, for PB defibrinate by agitation for 15 to 20 min with glass beads.
3. SF requires treatment with hyaluonidase (10 U/mL) to prevent the formation of gels during centrifugation.

3.1.2. Separation

1. The MC population is obtained by carefully layering the PB (diluted 1:1 with PBS), or treated SF, over a separation medium such as Ficoll-Hypaque, chosen so that its specific gravity is similar to that of MC.
2. The MC float on the surface of the separation medium, whereas granulocytes and red cells are spun to the bottom (the latter because the separation medium makes them aggregate).
3. Centrifugation is carried out at $890g$ for 20 min and the cells at the interface carefully harvested using a Pasteur pipet. In addition to MC these cells will contain a certain number of platelets, but this contamination decreases, with subsequent washes since platelets do not pellet completely under the conditions used for spinning down MC.
4. The MC population is washed 3 times in PBS ($322g$ for 10 min) prior to resuspension in complete culture medium (RPMI supplemented 10 mM Hepes buffer, 1% nonessential amino-acids, 1% sodium pyruvate, 2 mM L-glutamine, penicillin/Streptomycin), and counting (*see* **Note 1**).

3.1.3. Assessment of Viability and Cell Numbers

1. Viability of the MC is assessed by their ability to exclude the dye trypan blue.
2. This can be conveniently assessed whilst counting cells in a haemocytometer (*see* **Note 2**).

3.2. Measuring T-Cell Outputs

This section is concerned with measuring T-cell responses to antigens, but a critical consideration is the quality of the antigen which is used, particularly with regard to recombinant proteins which are nowadays the most common source of pathogen-derived or autoantigens which are used in in vitro assays (*see* **Notes 3** and **4**).

3.2.1. Proliferation

When T-cells recognize antigens or other stimuli they divide, and proliferative responses have been the most widely used measurements of T-cell activation. Proliferation may be measured by the incorporation of radioactive thymidine into newly synthesized DNA by dividing cells. The measurement has in principle a high signal to noise ratio and detection of β-ray emission from tritiated thymidine ([^{3}H]-Thy) is very sensitive. Thus 10^5 resting T-cells in PBMCs might incorporate a few hundred counts per minute (cpm) whereas their response to a polyclonal stimulus such as PHA might be between 100,000 and 200,000 cpm.

1. PBMC or SFMC are cultured in sterile microtitre wells (0.2 μL) at 1 to 2×10^5 cells/well. For these cell numbers flat bottomed rather than U-shaped wells are preferable.
2. Cultures are performed at 37°C in an atmosphere of air and 5% CO_2 for the required number of days, according to the kinetics of the response being examined (typically 3 d for mitogens and 5–7 d for antigens).
3. 0.5 to 1μCi [^{3}H]-Thy/well is added 6 to 18 h before the end of the culture; the precise time is not critical since the experiment is internally controlled, but should be kept constant if results from different assays are to be combined for analysis.
4. At the end of the culture, the cells are washed out of the wells onto a glass-fiber filter mat using a cell harvester, and free unincorporated [^{3}H]-Thy is washed away.
5. The filter mats are dried, saturated in scintillation cocktail, and β particle emission from each well measured in a micro β-plate counter.
6. In addition to recording cpm for each well, and thence the mean and standard deviation of cpm for each stimulus tested, a stimulation index (SI) should also be calculated by dividing cpm recorded in response to a stimulus, by cpm recorded from cells cultured in medium alone.
7. There is inherent variability in assays of this kind, so responses should always be measured at least in triplicate or, if cell numbers permit, quadruplicate. A larger number of control (medium only) wells (e.g., 6) also permits more accurate estimates of SI's.

Alternative measurements of proliferation, which do not involve using radioactivity, have been devised. For example, viable cells can metabolize the dye MTT to an insoluble formazan product which, when solubilized, gives a purple color which can be quantified colorimetrically. However, such a technique does not have the sensitivity of [^{3}H]-Thy incorporation; cell doubling produces a theoretical SI of 2 for the MTT technique, since the undivided cells remain viable and able to metabolize the dye, whereas their incorporation of thymidine may effectively be zero (or very low) as compared with stimulated cells, giving rise to correspondingly high SI's. In practice, for the relatively low responses commonly evoked by antigens, the dye-based technique does not have the required sensitivity. This is particularly the case in inflammatory diseases where lymphocyte responses may be impaired, and where the spontaneous proliferation of cells may be substantial (particularly when testing SFMC, reflecting in vivo activation), so that only low SI's can be obtained when responses to antigen are assessed.

3.2.2. Measurement of Total Cytokine Production in Mononuclear Cell Cultures

Although measurement of proliferation has proved to be a very useful generic measurement of T-cell responses, it is also important to have information about the nature of the response, and in particular the cytokines which are produced. In these assays it is important to determine in advance the kinetics of secretion of the cytokine and the optimal time for measurements to be made.

1. Supernatant from PBMC/SFMC stimulated with antigen is removed and the cytokine(s) of interest measured using ELISA.
2. This can be done in conjunction with proliferation assays by removing up to 50 µL of supernatant prior to the addition of [^{3}H]-Thy.
3. Instructions supplied by the manufacturers of cytokine ELISA kits should be followed for assay of the cytokine of interest.

The introduction of techniques to measure multiple cytokines within the same small volume of supernatant (*1*) greatly expands the amount of information that can be obtained from single assays of T-cell responses but these techniques will not be dealt with in detail here. It has been particularly useful to assess the relative amounts of Th1 and Th2 cytokines (i.e., interferon [IFN]-γ, vs interleukin [IL]-4 and IL-5), or to measure cytokines associated with regulatory immune responses (IL-10, transforming growth factor [TGF]-β).

3.2.3. Measurement of Total Numbers of Cytokine Producing Cells by ELISPOT

As well as the total cytokine output in response to antigen, the numbers of cells in the culture making a particular cytokine can be estimated (this distinguishes between a small number of cells making a lot of cytokine, and a large number, each making a little). A general method used for detecting IFN-γ producing cells is described, but this may need to be modified for the detection of other cytokines. Likewise the description is for mixed populations of cells (PBMCs/SFMCs) but the same technique can be applied to purified populations of CD4$^+$ or CD8$^+$ cells provided with adequate antigen presenting cells. Required concentrations of antibodies are often recommended by manufacturers but require verification for each type of assay to be used (*see also* **Note 5**).

1. Day 0: The plate is coated with antibodies under sterile conditions; nitrocellulose membrane-base 96-well plates (sterile) are used; the capture mAb is diluted to

the appropriate concentration in filter-sterilized 0.1M NaHCO$_3$ buffer (pH 9.6), and dispensed in 100 µL into each well of test plate.
2. The plate is incubated overnight at 4°C on a level surface OR for 4 h at 37°C.
3. Day 1: Preparation of effectors: cells are retrieved from liquid nitrogen (*see* **Note 5**) and allowed to recover in T-cell culture medium at room temperature (RT) for 1 h.
4. Removal of mAb and blocking mAb binding sites on the plate; wells are washed five times with 180 µL PBS to remove nonbound coating mAb, using sterile filtered (0.2 µm) autoclaved PBS. 200 µL/well of PBS+5% FCS (blocking buffer) is added to the plate which is then incubated at 37°C for 30 min to 1 hr (maximum) (*see* **Note 5**).
5. Assay: Effector cells are counted after their 1-h recovery and added to the plate at 2×10^5 cells/well in 100 µL final volume.
6. The stimulus to be used (protein, peptide, mitogen) is diluted to the appropriate concentrations and added in 100µl volumes to the wells.
7. The plate is then incubated at 37°C, optimally for 20 h (16 h minimum).
8. Day 2: Detection of cytokine secreting cells; these and subsequent steps are performed under unsterile conditions. Cells are lysed and removed by washing and gentle flicking with PBST (PBS-Tween, 0.05% Tween-20), and the plate washed 3 times with PBST and a further 3 times with PBS (no Tween).
9. The plate is dried by gently blotting on a stack of tissues. It is important that the plate should not dry out at any stage, so wash solution should be left in the plate until the reagents for the next step have been prepared.
10. The diluted biotinylated detection mAb is dispensed in 100 µL volumes into each well and the plate incubated for 4 h at room temperature, wrapped in foil/in the dark, on a flat surface.
11. After incubation, washes are repeated as above.
12. Streptavidin-alkaline phosphatase (stored at 4°C) is diluted in filtered PBS to the recommended concentration (e.g., 1 µg/mL), and 100 µL added to each well, followed by incubation for 2 h at room temperature. The plate is then washed again as above.
13. The substrate is prepared by dissolving one BCIP/NBT tablet (−20°C) in 10 mL MilliQ H$_2$O. This is filtered and covered with foil, as it is photosensitive.
14. 100 µL/well of substrate solution is added to each well, the plate is wrapped in foil, and incubated at room temperature for 10 to 60 min for full development.
15. Color development is stopped by washing in tap water, repeated twice.
16. The plate is allowed to air dry overnight at room temperature in the dark.
17. Spots can now be counted manually under an inverted microscope, or using a calibrated ELISPOT counter.

3.2.4. Identifying the Phenotype of Cytokine Producing Cells by Flow Cytometry

This section describes two methods of assessing the polyclonal response that both rely on cytokine production by the antigen-specific T-cells. These are intracellular cytokine staining, and the use of capture antibodies to detect cytokine secretion. They are both considered to be reliable and sensitive approaches for monitoring antigen-specific T-cells ex vivo (i.e., with limited in vitro culture). PBMCs are purified as described above; best results are achieved by using freshly isolated PBMCs, but cryopreserved cells can also be used. For cytokine secretion assays special protocols are also available for detection of cytokine secreting cells in whole human blood.

3.2.4.1. ANTIGEN-SPECIFIC STIMULATION IN VITRO

1. For in vitro stimulation the cells are resuspended in culture medium at 2×10^6 to 10×10^6 PBMCs/2 mL in a 12-well plate.
2. If required, cells may be cultured in medium overnight prior to addition of antigen.
3. Stimulation with peptides requires 3 to 6 h at 37°C, 5% CO_2, and with proteins 6 to 16 h or overnight. Both require concentrations of approx 10 µg/mL; wells without antigen or peptide should be included as controls.
4. For intracellular staining of cytokines it is necessary to add a reagent which inhibits secrecetion of cytokines during the latter part of the culture. Either monensin ("Golgi stop") or brefeldin A ("Golgi block") can be used. One or other of these reagents may prove preferable for the detection of particular cytokines (e.g., monensin may give a better signal for detection of IFN-γ and IL-4), whereas brefeldin-A has been reported to give better results when detecting TNF-α.

3.2.4.2. INTRACELLULAR STAINING

1. Cells are collected from the plate by pipetting up and down, and the plate is rinsed twice with cold buffer to remove the remaining cells.
2. Cells are pelleted by centrifugation at 322g for 10 min, resuspended in buffer, and transferred at 100 µL/well into a 96-U well plate for staining.
3. Cells are then spun again at 300g for 5 min, resuspended in 100 µL of buffer containing the appropriate dilution of antibody(ies) for surface staining.
4. Cells are incubated on ice in the dark for 20 to 30 min. For best results it is important to work fast, keep the cells cold and use precooled solutions to prevent capping of antibodies on the cell surface and nonspecific labeling.
5. Cells are spun down at 322g for 5 min, washed once in buffer, and 100 µL Cytofix/Cytoperm™ is added neat to each well; the cells are then incubated on ice for 30 min.
6. Cells are then spun down at 322g for 5 min and washed once in Perm/Wash™ buffer, resuspended in 100 µL of Perm/Wash™ Buffer containing an appropriate dilution of antibody for intracellular staining, and incubated on ice in the dark for 30 min.
7. Cells are pelleted at 322g for 5 min and washed once in Perm/Wash Buffer, once in separation buffer, and transferred into the 5 mL polystyrene round-bottom tubes for flow cytometric analysis.

3.2.4.3. USE OF CAPTURE ANTIBODIES

Cytokine secretion assays rely on inducing in vitro cytokine secretion as described above. Full instructions are provided with commercially available MACS Cytokine Secretion Assays (Miltenyi Biotec).

1. The cells are labeled with a cytokine-specific "catch" reagent attached to the surface of all leukocytes.
2. The cells are then incubated for a short time at 37°C to allow cytokine secretion and binding of cytokine to the catch reagent on the secreting cells only.
3. The cells are subsequently labeled with a second cytokine-specific antibody for detection by flow cytometry.

3.3. Isolation and Characterization of Antigen-Specific T-Cells From Polyclonal Populations

If PBMCs from an individual mount a detectable proliferative response to an antigen of interest, then short-term antigen specific T-cell lines can be produced and used as a source of antigen specific T-cells for cloning. PBMCs/SFMCs are stimulated with either a complex mixture of antigens (e.g., whole bacteria, bacterial extracts, or cell extracts), purified antigen (natural or recombinant), or specific peptides, where responses to particular epitopes are being examined. Each of these has potential pit-falls; in the case of recombinant antigens, it is very important to ensure that the T-cells are not recognizing a contaminant protein derived from *E. coli*. This can be achieved by screening the T-cells using *E. coli* transformed with the expression construct which does not contain the recombinant antigen, or testing a nonrecombinant source of antigen. Note that *E. coli* lysates at high concentration can inhibit T-cell proliferation, so it is advisable to test equivalent amounts of *E. coli* lysate derived from both control and antigen-expressing bacteria. Preparation of synthetic peptides vary in their purity (this should be indicated by the manufacturer); T-cells which respond to synthetic peptides may fail to respond to intact protein antigens (i.e., the epitope is said to be hidden or "cryptic") and T-cells with this specificity will not normally be part of the physiologic T-cell response to the antigen. Therefore T-cells isolated by stimulation with peptides should always be checked for their ability to respond to the intact protein.

1. PBMCs are cultured with antigen, used at a previously determined optimal concentration, for about 6 d.
2. T-cell proliferation should be evident visually by observing T-cell blasts (elongated, enlarged cells) in wells containing antigen but not in those containing medium alone.
3. If a prominent response is evident, T-cell cloning can proceed immediately.
4. Otherwise 20 U/mL of recombinant human IL-2 can be added and cultures continued for a further 5 d before cloning.

5. Alternatively, a second cycle of stimulation can be initiated by addition of fresh irradiated autologous PBMCs plus antigen followed by cloning 7 to 10 d later. Note that repetitive stimulation can result in loss of the antigen-specific population which is out-grown by IL-2 responsive nonspecific cells.

3.3.1. T-Cell Cloning and Characterization of T-Cell Clones

1. T-cell blasts are cloned by limiting dilution in Terasaki cloning plates (20 µL/well); these require smaller numbers of autologous PBMCs/well as antigen presenting or accessory cells.
2. PBMCs are irradiated (3 Gy) and used at a concentration between 0.5 to 1×10^6/mL with antigen at an appropriate concentration and 100 U/mL IL-2.
3. T-cells harvested from antigen stimulated cell lines are carefully counted and diluted, typically in two stages (e.g., from 50×10^4 cells/mL to 50×10^2 cells/mL, and thence to 500 cell/mL).
4. This gives a suitable concentration for the addition of 18 cells (36 µL)/1.2 mL (enough T-cells for 60×20 µL wells at an average of 0.3 cells/well). In all cell dilutions thorough mixing is critical to get rid of any cell clumps because their presence will invalidate the assumptions of limiting dilution (i.e., dispersal of single T-cells to each well).
5. It is usually worthwhile setting up additional plates at higher cell numbers/well (10, 5, 1) because the precursor frequency of the antigen specific cells may be very low resulting in very few clones being obtained from wells seeded at 0.3 cells/well. Multiple seeding also allows assessment of the proportion of responding wells at each dilution and a calculation of the precursor frequency in the starting population by plotting cell seeding against the natural log of the proportion of negative wells. This should give a straight line plot with the precursor frequency corresponding to the dilution which results in 63% positive wells *(2,3)*.
6. The Terasaki plates are placed on damp tissue paper in a square Petri dish and left undisturbed for 10 to 12 d.
7. They are then assessed microscopically, and wells with visible T-cell growth transferred to 0.2-mL round bottomed wells and expanded by stimulation with 10^5 irradiated allogeneic PBMCs/well, 2 µg/mL PHA and IL-2 at 100 U/mL.
8. Growing wells are subcultured as required, with replacement of medium and IL-2 twice weekly.
9. If sufficient T cell proliferation has occurred over 10 d (e.g., expansion to 4 or more U wells) the specificity of the putative clones can be determined.
10. Screening should be conducted as soon as possible to avoid culturing large numbers of nonspecific T-cell clones. Typical screens can be carried out by harvesting the contents of a growing U well into 5 to 10 mL culture medium, washing (to remove IL-2) and redistributing the cells to wells containing medium alone, autologous APC and autologous APC + antigen.
11. Each of these can be measured in duplicate, and it is not necessary to count the T-cells in this initial screen since each well will receive the same number.

12. After 3 d proliferation is determined in the usual way, looking for clones which respond to APC + antigen and not to APC alone.
13. These clones are then expanded into 1 or 2 mL wells by nonspecific restimulation with mitogen every 3 to 6 wk; too frequent restimulation may give rise to activation-induced cell death and loss of the clone.
14. Avoid culturing T-cell clones at too low a cell density; 50 to 70% confluent cultures are ideal.
15. We recommend cryopreserving antigen-specific T-cell clones as soon as possible after verifying their specificity, as an insurance against losses resulting from infection, using a minimum of 1 million cells/ampoule.
16. T-cell clonality can be confirmed by polymerase chain reaction (PCR) amplification of TCR cDNA using TCR variable region specific oligonucleotide primers in combination with TCR constant region primers *(4)*. This will identify whether more than one alpha and beta chain sequence is present.
17. Sequencing the PCR product will confirm whether the sequences are in frame and confirm the clonality of the DNA. Note that T-cell clones not infrequently contain two rearranged TCR α genes, but usually only one rearranged β gene.
18. Staining with TCR variable gene specific mAbs will confirm surface expression of the TCR, but note that not all the products of TCR variable genes can be identified by Mab.

3.4. Identification of Specific Antigens Recognized by Specific T-Cell Clones: Screening of Expression Libraries

Identification of the immunogenic proteins that are recognized by microorganism-specific T-cell clones can be achieved by screening a *E. coli* expression library expressing fragments of DNA derived from the microorganism of interest. Here we describe a simple expression strategy that utilizes irradiated human PBMCs as a source of APC; these cells are able to process recombinant *E. coli* efficiently and stimulate the proliferation of bacteria-specific T-cell clones *(5)*.

3.4.1. Expression Library: Initial Considerations

The likelihood of finding a DNA sequence of interest in a random library can be estimated using a formula based on the Poisson distribution *(6)*.

$$N = \ln[1 - P]/\ln[1 - (I/G)] \qquad (1)$$

where N = number of clones to be screened, P = probability of finding a given insert (i.e., 99), I = insert size (i.e., 1 kB), and G = genome size (i.e., 1 Mb).

Therefore when screening a library from a organism with an approximate 1 Mb genome, the number of clones to be screened containing the 1kb insert is a minimum of 4602 to ensure a 99% chance of finding a given 1 kB insert. Because DNA can be inserted in either orientation and only 1 orientation will result in expression, a minimum of 9124 colonies should be screened. Due to small size of the DNA inserts we use a plasmid expression vector, PTrc His, enabling expression of inserted DNA as a polypeptide with an N-terminal epitope tag and polyhistidine sequence. The use of pTrc His A, B, or C expression vectors allows the expression of the inserted DNA in three reading frames. If stimulatory wells are not detected in the first library screen, then the other libraries may have to be addressed to ensure the identification of all potential antigen polypeptides.

3.4.2. Preparation of the Library

1. Genomic DNA from the microorganism of interest is isolated and partially digested with Sau 3A1, size fractionated on an agarose gel and 1-4 kB fragments purified and ligated to dephosphorylated, Bgl II cut pTrcHisC vector (Invitrogen).
2. The ligated plasmids are used to transform competent *E. coli* (Xl-2Blue, Stratagene).
3. The quality of the library should be assessed by analysis of at least 100 transformed bacteria using PCR and sequencing to ensure correct insert size, partial digestion and adequate % of recombinants.
4. Individual *E. coli* recombinants are picked and cultured in separate 96 U wells; pools of the individual recombinants are prepared and both the individual and pooled library stored in glycerol at –80°C.
5. The complexity of the pooled library is dependent on the sensitivity of the T-cell clone being used to detect a stimulatory recombinant; in practice we found that a *E. coli* recombinant expressing *Chlamydia trachomatis* (CT) hsp60, diluted 24-fold with an irrelevant *E. coli* recombinant, and cocultured with HLA matched PBMCs still provided sufficient signal to induce detectable proliferation of a T-cell clone that recognizes an epitope derived from CT hsp60.

3.4.3. Protocol for Screening an Expression Library

1. Using 24 recombinants/pool we have been able to screen 15,000 recombinants by screening a total of 625 pooled wells.
2. One screen requires a total of 62.5 million HLA class II- matched APC (10^5/well) and 6.25 million antigen specific T-cell clone (2×10^4/well).
3. PBMCs and T-cell clones are dispensed from sterile reservoir boats using a multichannel pipette. It is important the number of *E. coli* does not exceed 10^6 bacteria/ 96-U well containing APC and T-cell clone, because of the inhibitory effects of *E. coli* on proliferative responses: this will have to be approximated as pooled wells may grow at different rates and expression may inhibit growth significantly.
4. 200 µL of $2 \times$ YT growth medium in 96-U wells are inoculated with 10 µL from pooled library wells and grown overnight, diluted 1/10 and shaken at $37°C$ for 2 h followed by 4 h induction with IPTG.
5. The 96-U well plates are centrifuged and resuspended in 200 µL of RPMI and using a multi-channel pipet.

6. One microliter of each well is added to a separate well containing the irradiated PBMCs and T-cell clone.
7. The coculture is incubated for 2 d and pulsed overnight with tritrated thymdine.
8. Before harvesting, the cocultures are also analyzed microscopically for responding T-cells.
9. Following identification of a stimulatory well, the recombinants that constitute the pooled library are located from the original expression library and screened individually.
10. The bacterial DNA from the stimulatory *E. coli* recombinant is amplified by PCR using flanking pTrc His specific oligonucleotide primers, and the PCR product sequenced.
11. Sequences are analyzed using the search algorithm BLAST on DNA sequences deposited in GenBank.

3.5. Studies of $CD8^+$ T-Cell Responses to Bacterial Pathogens

In view of the differences in techniques which are required to study $CD8^+$ T-cell responses to intracellular pathogens, (such as the organisms associated with reactive arthritis) we include a short description of the techniques involved. Studies of $CD8^+$ T-cell responses to bacteria pathogens can be difficult using cells directly purified ex vivo because of the relatively low frequency of effector cells. It is necessary to expand these effectors in vitro, which means presenting antigen to them in an appropriate manner. This is best accomplished by using autologous dendritic cells (DC) infected with the bacterium of interest.

3.5.1. Preparation of DC From Human PBMCs

Sufficient numbers of DC can be differentiated in vitro from $CD14^+$ monocytes in the presence of GM-CSF and recombinant human IL-4. $CD14^+$ cells can be purified using one of the following three methods.

3.5.1.1. Purification of $CD14^+$ Cells With Positive Selection

1. PBMCs are purified on Ficoll gradient (Amersham Pharmacia Biotech AB).
2. Cells are collected from the interface and washed 3 times in sterile PBS and pelleted.
3. Pelleted cells are resuspended at 10^7 cells/ 200 µL of separation buffer (2 mM EDTA in PBS + 0.5% BSA [low LPS]).
4. Anti-CD14 antibody conjugated to microbeads (Miltenyi Biotec) is added to cell suspension at 20 µL for each 10^7 PBMCs.
5. Cells are incubated at 4°C for 15 to 20 min.
6. At the end of the incubation time, 1 to 3 mL of separation buffer is added to the sample (the volume of the separation buffer will depend on the number of cells in the sample. For samples containing less than 10^8 cells, use 1 mL of buffer).
7. The cells are spun at 322*g*.
8. The pellet is resuspended in fresh separation buffer and spun again.

9. Cells are resuspended in the separation buffer and purified on MACS columns.
10. Purification is carried out as recommended by the supplier. Positive selection columns (e.g., MS columns) are recommended for this purification.
11. CD14$^+$ cells eluted from the column are washed in sterile PBS and resuspended at 3×10^5 cells/mL in DC medium.
12. Plate in 6-well plates using 3mL/well.

3.5.1.2. PURIFICATION OF CD14$^+$ CELLS BY NEGATIVE SELECTION

This method is similar to that described below for CD8$^+$ T-cells.
1. PBMC are labeled with anti-CD19, CD16, CD3, and CD3 antibodies.
2. Cells are incubated with goat-anti mouse IgG MicroBeads and application to proprietary columns to deplete CD14 negative cells.

3.5.1.3. PURIFICATION OF CD14$^+$ CELLS USING PLASTIC ADHERENCE

1. PBMC are prepared as described above.
2. Cells are then resuspended at 5×10^6 cells/mL in complete medium supplemented with 5% FCS.
3. PBMC are plated in 6-well plates using 2 mL of cell suspension/well (using 6-well plates is recommended, although 24-well plates can also be used).
4. Plated cells are incubated for 45 min to 1 h at 37°C.
5. The nonadherent cells are washed in sterile PBS (you can use repeatable washes until only adherent monocytes remain on the plate).
6. DC medium (complete RPMI medium supplemented with 10% human AB serum, 50 ng/mL GM-CSF, and 1000 U/mL of IL-4 is added directly to the well (3 mL/well in a 6-well plate).
7. Cells are fed with fresh DC medium every 4 d. By days 7 to 8 of culture, DC are fully differentiated and ready to use.
8. DC are then infected with an appropriate pathogen.

3.5.2. Purification of CD8$^+$ T-Cells From PBMCs

CD8$^+$ T-cells should be purified using negative selection as follows:

1. PBMCs are incubated with a cocktail of antibodies (CD19, CD4, CD56, CD16, CD14, and anti-TCR γ/d). Unconjugated antibodies or antibodies directly conjugated to MicroBeads can be used to negatively select CD8$^+$ T-cells.
2. When using unconjugated antibodies, titration of the antibodies may be required to establish the optimal concentrations. Antibodies are made up in the separation buffer (as described for DC) and added to the cell suspension. Cells are incubated for 45 min at 4°C.
3. Cells are washed 2 times in the separation buffer and incubated with goat anti-mouse IgG, conjugated to 20 μL/10^7 PBMCs MicroBeads (Miltenyi Biotec), for 15 min at 4°C.
4. Cells are washed 2 times, resuspended in the separation buffer, and purified on a Macs column. The purification is carried out as recommended by the supplier.

Note that the fraction which contains CD8$^+$ T-cells is the fraction which has not bound to the column. Use columns for negative selection (e.g., LD columns).
5. When using primary antibodies directly conjugated to MicroBeads, use the concentration recommended by the supplier. Incubate the cells for 15 min and purify using LD columns.

3.5.3. Generation of Pathogen-Specific CD8$^+$ T-Cell Lines

1. DCs infected with an appropriate pathogen are seeded in 96-well plate at 10^4 cells/well.
2. Autologous purified CD8$^+$ T-cells are added to DC cultures at 5×10^4 cell/well and cultured for 2 wk in complete medium supplemented with 10% AB serum, 50 U/mL recombinant human IL-2, and 0.5 ng/mL IL-7 (BD-Pharmingen).
3. Cultures should be supplemented with rIL-2 and rIL-7 every 7 d.
4. After 2 wk of culture, the expanded CD8$^+$ T-cell line can be repurified by negative selection (as described above) to obtain a pure CD8$^+$ population.
5. Following two rounds of purification, the T-cell lines should contain >98% CD8$^+$ T-cells.
6. The purity of the line can be tested by flow cytometry for the surface expression of CD8 and TCRa/b.
7. The T-cell line can be tested for its ability to recognize pathogen-infected cells; this is best done not using cytotoxicity but by measuring IFN-γ production by ELISA, allowing a large number of clones to be screened rapidly.

4. Notes

1. Separation of mononuclear cells. In inflammatory diseases such as rheumatoid arthritis, larger numbers of platelets will be present because of the thrombocytosis associated with inflammation, and the PBMCs are also more likely to be contaminated with activated polymorphs and immature red cells than PBMC from healthy individuals.
2. Cell viability. Trypan blue exclusion is a very crude indication of cell viability, and cells which show extensive blebbing or convoluted morphology are likely not to be viable or able to contribute to T-cell responses, despite retaining an ability to exclude dye. Therefore a conservative policy when counting cells is recommended.
3. Quality control of antigen preparations used in testing immune responses. Many of the antigens and autoantigens used in arthritis research cannot readily be purified in sufficient quantity from tissues and therefore recombinant expression of these antigens is used. However, this approach does have inherent problems; first, the bacterial expression system is also a source of both contaminant proteins that are often antigenic, and entities such as LPS which can affect the immune response to the antigens by acting through pattern recognition receptors (e.g., Toll-like receptors [7]). Second, post-translational modifications do not occur in recombinant expression systems, and if the immune response is dependent on an epitope that requires post-translational modification (e.g., acetlyation, amidation [8,9]), this will not be detected when using recombinant antigens.

Purity of an antigen can be assessed by sodium dodecyl sulfate-polyacrylamide gel electrophoresis (SDS-PAGE) and immunoblotting both for the protein of interest or known possible contaminants, However, it should be noted that an apparent single band on a gel may conceal a contaminant of the same molecular mass, whilst the sensitivity of an antibody used for immunoblotting may be affected by the presence of an excess of a similar protein; thus we have found that the presence of large quantities of human hsp60 prevents detection of small quantities of *E. coli* hsp60 by immunoblotting even when using a monoclonal antibody which is completely specific for the *E. coli* protein.

Endotoxin contamination is also a potential problem with recombinant proteins *(10,11)* and can be tested for, using the limulus amebocyte lysate assay. LPS can also be removed by a variety of methods based on the use of polymyxin B (e.g., Polymyxin B-agarose, Pierce).

Even when all precautions have been taken, our experience has been that a totally pure preparation of an antigen is rarely obtained, and therefore the possibility that the activity recorded when using the antigen resulted from a contaminant rather than the antigen itself must constantly be borne in mind. For example, when using recombinant antigens, T-cell clones should routinely be tested for their ability to respond to a lysate of *E. coli* which does not express the antigen of interest. Ideally T-cell clones should also be tested against nonrecombinant antigens (e.g., purified from tissue, organisms). There are also instances where a clone may genuinely recognize an antigen and a homologue present in *E. coli*, in which case epitope mapping may be required to fully establish the specificity of the clone.

4. Methods of purification of antigens including recombinant proteins. As a general rule a robust purification method should separate antigen and contaminants by at least two different physical properties. Often the initial method of purification of a recombinant protein from a bacterial cell lysate will remove the vast majority of contaminants, but leave those few that share properties with the protein of interest. Hence a second method of purification, based on a different property of the protein to be purified, is required. Popular methods of purification include:

 a. Affinity chromatography: this method relies on the affinity of the protein of interest for a ligand which has been attached to a solid inert substance, usually sepharose beads. If a substrate and binding site (such as ATP and an ATP binding site) are not known, or their use is not practical, then the protein can be expressed as a fusion protein. A section of another protein such as glutathion-S-transferase (GST) or a short polyhistidine tag can be attached. These enable the protein to be separated by binding to a known ligand, glutathione-sepharose in the case of GST, and nickel activated chelating sepharose in the case of histidine tags. These fusion proteins and tags are often engineered so that they can be enzymatically removed if required. Nonspecific binding can be kept to a minimum but often cannot be eliminated completely.

b. Preparative electrophoresis and gel filtration both rely on separation by size. Preparative electrophoresis is essentially a protein purification method that utilizes SDS-PAGE to separate proteins. Gel filtration, a nondenaturing method of separation, relies on the speed at with proteins of different sizes move through a solvent permeated polymer. Unfortunately this method is not specific and the quality of the separation achieved can depend on a number of factors.

Whereas a number of standard purification methods exist, the purification strategy must be modified to take into account of the properties of any contaminants that are found to copurify with the protein of interest.
5. ELISPOT assays. In these assays both positive and negative controls are required; a nonspecific stimulus such as PHA or anti-CD3 serves as a positive control and unstimulated cells for the negative control. All solutions must be sterile filtered, otherwise fungal spores, bacteria, dust etc, will give high background readings.

For ELISPOT assays, cryopreserved PBMCs/SFMCs are optimal because red blood cells and neutrophils are lysed in this process and consistent use of cryopreserved samples allows comparisons between samples from different time points in the same assay.

When adding medium to wells, avoid touching the bottom of the plate with pipet tips, because this may damage the plate, by dispensing down the side of the wells.

Acknowledgments

The authors thank members of the EBV biology laboratory, Queensland Institute for Medical Research, Brisbane, Australia, for the ELISPOT assay protocol described in section 3.2.

References

1. Oliver, K. G., Kettman, J. R., and Fulton, R. J. (1998) Multiplexed analysis of human cytokines by use of the FlowMetrix system. *Clin. Chem.* **44,** 2057–2060.
2. Waldmann, H., Lefkovits, I., and Quintans, J. (1975) Limiting dilution analysis of helper T-cell function. *Immunology* **28,** 1135–1148.
3. Waldmann, H., Pope, H., and Lefkovits, I. (1976) Limiting dilution analysis of helper T-cell function. II. An approach to the study of the function of single helper T cells. *Immunology* **31**, 343-352.
4. Goodall, J. C., Bledsoe, P., and Gaston, J. S. H. (1999) Tracking antigen-specific human T lymphocytes in rheumatoid arthritis by T cell receptor analysis. *Hum. Immunol.* **60,** 798–805.
5. Goodall, J. C., Yeo, G., Huang, M., Raggiaschi, R., and Gaston, J. S. H. (2001) Identification of Chlamydia trachomatis antigens recognized by human CD4+ T lymphocytes by screening an expression library. *Eur. J. Immunol.* **31,** 1513–1522.
6. Clarke, L. and Carbon, J. (1976) A colony bank containing synthetic Col El hybrid plasmids representative of the entire E. coli genome. *Cell* **9,** 91–99.

7. Medzhitov, R. and Janeway, C. A. (2002) Decoding the patterns of self and nonself by the innate immune system. *Science* **296,** 298–300.
8. Smilek, D. E., Wraith, D. C., Hodgkinson, S., Dwivedy, S., Steinman, L., and McDevitt, H. O. (1991) A single amino acid change in a myelin basic protein peptide confers the capacity to prevent rather than induce experimental autoimmune encephalomyelitis. *Proc. Natl. Acad. Sci. USA* **88,** 9633–9637.
9. McAdam, S. N., Fleckenstein, B., Rasmussen, I. B., et al. (2001) T cell recognition of the dominant I-A(K)-restricted hen egg lysozyme epitope: Critical role for asparagine deamidation. *J. Exp. Med.* **193,** 1239–1246.
10. Ohashi, K., Burkart, V., Flohe, S. and Kolb, H. (2000) Cutting edge: heat shock protein 60 is a putative endogenous ligand of the toll-like receptor-4 complex [In Process Citation]. *J. Immunol.* **164,** 558–561.
11. Gao, B. C. and Tsan, M. F. (2003) Endotoxin contamination in recombinant human heat shock protein 70 (Hsp70) preparation is responsible for the induction of tumor necrosis factor a release by murine macrophages. *J. Biol. Chem.* **278,** 174–179.

6

Identification and Manipulation of Antigen Specific T-Cells with Artificial Antigen Presenting Cells

Eva Koffeman, Elissa Keogh, Mark Klein,
Berent Prakken, and Salvatore Albani

Abstract

T-cells specific for a particular antigen represent a small percentage of the overall T-cell population. Detecting the presence of antigen specific T-cells in patients, animal models or populations of cultured cells has presented a challenge to researchers. The T-cell capture method described here utilizes a truly artificial method of antigen presentation and requires only 50,000 cells for the detection of the major histomcompatibility complex (MHC) class II and antigen restricted T-cells. With this method, liposomes, prepared with readily available materials, are loaded with neutravidin "rafts" comprised of MHC/peptide complexes, anti-CD28, a costimulatory molecule, and anti-LFA-1, an adhesion molecule. These artificial APCs are easily manipulated to include any MHC, antibodies to cell surface markers and/or costimulatory signals of interest thereby enabling not only T-cell identification but also the manipulation of mechanisms of T-cell activation.

Key Words: Lipid lamellar sheets; liposomes; MHC; artificial antigen presenting cells; T-cell capture.

1. Introduction

In the adaptive immune response antigen specific T-cells play a key role. Recognition of a specific antigen by the T-cell receptor is essential in the defense against infections and prevention of tumor growth. Dysfunctional antigen specific T-cell reactions lead to immune deficiency and autoimmunity. In immunology research, identification of antigen specific T-cells and study of T-cell reactions are important to unravel the physiology of general immune mechanisms and the pathogenesis of diseases and to determine the efficacy of treatments. However, these cells are hard to detect because they are only a small fraction of the peripheral blood mononuclear cells. This chapter describes a method for identification and manipulation of rare antigen specific T-cells. In contrast to other techniques, that perform passive staining of the cells, this method implies an active process of binding the T-cell receptor (TCR). It is a truly artificial system that resembles the processes involved in the physiological situation of antigen recognition. We will discuss how this method can be used for manipulation and expansion of antigen specific T-cells for research and clinical purposes.

In the study of antigen specific T-cells, identification is the first step. Traditionally, the strategy used was antigen induced in vitro proliferation of peripheral blood mononuclear cells, followed by characterization of the reacting cells. Cells were characterized by subset (e.g., CD3, CD4), activation (e.g., CD69), the V-α and β patterns of their TCR *(1)* and the production of cytokines. These methods were based on passive staining and therefore did not isolate the T-cells based on their antigen specificity. Active binding of the TCR by the specific peptide, presented on a major histocompatibility complex (MHC) is needed to detect antigen specific cells exclusively. Therefore MHC-peptide complexes were developed, used soluble as monomers, dimers or tetramers, or bound on planar membranes *(2,3)*. Currently, the method of choice for detection of antigen specific T-cells is a soluble tetramer formation of MHC-peptide complexes *(4)*. Tetramers provide a tool to identify, isolate and characterize antigen specific T-cells. MHC class I tetramers work well *(4),* but the development of MHC class II tetramers has been more difficult *(5)*. Binding of $CD4^+$ cells with tetramers has been shown to be much less efficient *(6)*. The low numbers of $CD4^+$ cells and their low-affinity binding and high off-rates play a key role in this. *(5)* As will later be discussed in this introduction, $CD4^+$ cells need a more stable binding of the TCR to become activated.

To overcome the problem of the low numbers of antigen-specific cells in peripheral blood, methods to expand rare cells have been investigated. Expansion of cells is not only helpful in research situations, it can also be promising for clinical applications. Ex vivo activation and expansion of antigen specific T-cells can be used for the treatment of cancer and immunodeficiency states *(7,8)*. For expansion of cells, a number of methods have been developed based on natural APC *(7)*. For clinical purposes the use of autologous APC, usually dendritic cells (DCs), has been studied, but still poses a lot of challenges. Insect cells have been studied for research and clinical use. Cells derived from *Drosophila melanogaster* cells cannot load endogenous peptides on their class I MHC. They thus provide a well-controlled system to study exogenously loaded peptide epitopes, but the interaction with the T-cell is very unstable at body temperature. Mammalian cells have also been transfected to present exogenous peptide fragments, with mouse fibroblasts giving good results. However, the main problem of natural APCs for clinical purposes are the risks of xenoreaction and viral infection. The fact that natural cells express heterogeneous MHC and molecules implies a high cell-to-cell variety, which makes natural APCs difficult to control in research situations. This same problem applies for manipulation of antigen specific T-cells. When studying antigen specific cells, only identification or expansion of the cells does not yet provide insight in the physiology and pathophysiology of T-cell responses. To study the physiological process of antigen recognition, a tool to manipulate the reaction between the T-cell and the antigen is needed. An antigen presenting system to manipulate this process must resemble the physiological situation as closely as possible, but needs to be truly artificial as well in order to be fully flexible and controllable.

To further understand the difficulties in identification, manipulation and expansion of antigen specific cells, we will first give an outline of the physiological process of antigen recognition by T-cells. Before a T-cell becomes activated by an antigen presenting cell (APC), it has a series of transient interactions with different APCs *(9)*. When the specific APC is encountered, binding of TCRs to specific MHC-peptide complexes is of low affinity and has high "off-rates." The resulting short binding time of the individual TCRs, however, gives the opportunity for serial triggering of multiple TCRs on the membrane, which is necessary for a stable binding between the APC and the T cell *(10)*. The T-cell then undergoes an extensive membrane-cytoskeletal reorganization, a process called capping, resulting in the formation of the "immune synapse." The immune synapse consists of a central cluster of TCRs (cSMAC, central supramolecular activation cluster) and a peripheral ring of adhesion molecules (pSMAC, peripheral SMAC) *(9,11,12)*. The most important adhesion molecule on the APC is ICAM-1, that binds LFA-1 on the T-cell. The stability of the adhesion is dependent on binding by these adhesion molecules and by costimulatory molecules, most importantly B7-1 and B7-2 that respectively bind CTLA-4 and CD28 on the T-cell *(7,11)*. Without the involvement of these molecules, signaling through the TCR is not effective *(11)* and T-cells undergo anergy or apoptosis. *(13,14)*. Formation of the immune synapse triggers a cascade of signaling that ultimately leads to cell-cycle progression and cytokine production *(10)*. The process of "capping" has been shown to take about 5 min *(12)*. Then the immune synapse is maintained stable for several hours *(15)*. To prevent death of activated T cells through overstimulation, strong signaling over long times is limited by an adaptive control system in the immune synapse itself *(16)*.

Interestingly, $CD4^+$ T-cells need many TCRs (at least 8000) to be triggered before they respond by producing cytokines. $CD8^+$ cytotoxic T-cells, however, have a much lower threshold for inducing cytotoxicity. Possibly, they may be activated by the local triggering of only a few TCRs *(10)*.

This insight into the natural processes between T-cells and antigen presenting cells makes clear that the major problem in many techniques currently used for identification, manipulation and expansion of T-cells lie in the fact that the systems do not allow for immune synapse formation. Binding of low-affinity $CD4^+$ T-cells with tetramers has been shown to be difficult *(6)* because tetramers lack a membrane and the molecules needed for effective immune synapse formation and T-cell activation. Furthermore, the geometry of tetramers makes it very difficult for a cell to bind more then one peptide on one tetramer and the amount of MHC-peptide complexes on tetramers is too low to induce serial triggering above the threshold of $CD4^+$ T-cells *(6)*. For expansion and manipulation of T-cells, signaling through the TCR by immune synapse formation is essential. Natural APCs do this, but the major problem is that these cells cannot be controlled or manipulated. To combine the stable binding seen in nature with the flexibility of an artificial system, truly artificial APCs are needed. Magnetic beads have been developed, loaded with MHC-peptide complexes and costimulatory molecules *(7)*. Their limitation, however, is that the membrane is not fluid, which makes it impossible for ligands to cluster on the membrane of the APC.

We proposed a liposomal formulation as a carrier system for naturally derived MHC class II peptide complexes *(13,17)*. Liposomes have a lipid bilayer membrane that resembles the physiological situation. We showed for the first time that labeled MHC class II peptide liposomes can be used to specifically visualize T-cells *(13)*. Mallet-Designe et al. showed that liposomes with recombinant MHC formatted as monomers, trimers and tetramers within the membrane were much more sensitive in detecting $CD4^+$ T-cells than tetramers *(6)*. These data show that for detection of antigen specific T-cells, liposomes with MHC-peptide complexes are a sensitive system. In these liposomes and our first generation of liposomes, however, the MHC-peptide complexes were individually distributed in the membranes and the cells lacked costimulatory molecules *(17)*. To further simulate the actual structure of an APC, we changed our liposomes to artificial APCs (aAPCs) with "micro-domains" of MHC-peptide complexes and adhesion and costimulatory molecules on the membranes. A microdomain consists of a neutravidin molecule that has three sites available for binding the MHC-peptide complex and two molecules of choice *(18)*. We chose anti-LFA1 as an adhesion molecule and anti-CD28 as a costimulatory molecule *(7,9,11,15)*. This new generation of liposomal aAPCs efficiently stimulate the antigen-specific TCR, as was mirrored by PKC translocation from the cytosol. T-cell activation with these new aAPCs was of a significantly higher degree than with our first generation aAPCs *(18)*. Binding of antigen specific T-cells with the aAPC, or "T-cell capture," is made visible by flow cytometry. The population of interest is selected by staining for surface markers (e.g., CD4) and antigen specificity is ensured by loading a pure peptide on purified MHC (e.g., DR.*0401).

Our aAPC is currently the system that best resembles natural APCs in immune synapse formation and TCR signaling. The efficiency of TCR signaling provides sensitivity in detection of antigen specific cells and opens the way for expansion of rare antigen presenting cells. This may even have clinical applications, because these artificial APCs will not pose the risks of viral infection or xenoreaction. Because it is an artificial system, this system is very flexible and controllable. Changes in membrane molecules, MHC, and peptide types and densities can be made according to the need of each situation. The molarities of the different components are precisely defined by the user of the method. This flexibility allows for manipulation of physiological T-cell reactions.

The method has proven its value in immunology research. We used our first generation of aAPCs to identify antigen specific T cells *(17)* and to gain more insight into general mechanisms of T-cell responses *(17)* and specific T-cell responses in an autoimmunity model *(19)*. With the new generation of aAPCs, we studied T-cell reactions in juvenile dermatomyositis *(20)*, the role of regulatory T-cells in juvenile idiopathic arthritis *(21)*, and the efficacy of oral tolerance induction in a clinical trial in rheumatoid arthritis (RA) *(22)*. Besides autoimmunity research, we will now apply this technique for research on CD4$^+$ and CD8$^+$ T-cells in cancer research. We are exploring the use of the T-cell capture to expand numbers of rare antigen specific T-cells. Possible future applications of the method of T-cell capture are ex vivo enhancement of anti-inflammatory reactions in immunodeficiency, such as the post-transplantation state, or in cancer. To the contrary, downmodulation of (auto-) inflammation by ex vivo stimulation of regulatory T-cells is also an option. We continue optimizing this method. The protocol described here is based on our latest work on improvement of the sensitivity of the system. When new data on immune synapse formation will provide us with new adhesion and costimulatory molecules that are essential to T-cell signaling, we will investigate the possibilities of loading them on our aAPC. We believe that the close resemblance to the physiological situation and the flexibility of the protocol will remain essential characteristics, ensuring major opportunities for future research and therapy.

2. Materials

Please note that all buffers are prepared with distilled, deionized water unless otherwise noted.

2.1. Preparation of Lipid Lamellar Sheets

1. Pyrex glass tube (40 mL volume).

2. GM1 (monosialoganglioside, FW 1568.8) (Sigma; cat. no G-7641). Prepare a 10 mg/mL solution by dissolving in 1 mL of CHCl3. The solution can be stored at −20°C indefinitely.
3. Cholesterol (MW 386.7) (Sigma; cat. no. C-2044). Freshly prepare a 500 mg/mL solution in $CHCl_3$.
4. Phosphatidylcholine (FW 760) (Sigma; cat. no. P-2772). This is ready to use at 100 mg/mL.
5. Chloroform.
6. Buffer A: 140 mM Tris-NaOH, (pH 8.0) containing 100 mM NaCl. Prepare fresh and chill to 4°C. Dissolve 2.206 g Tris-HCl and 0.58 g NaCl in 100mL H_2O and titrate to pH 8.0 with 1M NaOH.
7. Deoxycholic acid (Sigma; cat. no. D-6750). Immediately prior to use, prepare 15 mL of 0.5% deoxycholic acid in buffer A. Vortex gently to mix and store on ice.
8. Nitrogen or argon gas.
9. Block heater or water bath set at 37°C.
10. Sonicator.
11. 1.5 mL microcentrifuge tubes.

2.2. Liposomes

1. 300 μL lipid lamellar sheets (*see* **Subheading 3.1.**).
2. Slide-A-Lyzer (0.1–0.5 mL; 10,000MWCO) (Pierce; cat. no. 66415 or 66383).
3. PBSO: 7.9 mM Na_2HPO_4, 1.5 mM KH_2PO_4, 2.7 mM KCl, and 140 mM NaCl (pH 7.1).
5. 1-mL syringes.
6. 22-Gage needles.
7. Blank calibration particles (6–6.4 μm, 10 × 106 particles/mL), (Becton Dickinson; cat. no. 556296) diluted 1:2 in phosphate buffered saline (PBS).
8. 1.2-mL Polypropylene microtubes (USA Scientific; cat. no. 1412-0000).
9. 5-mL Polystyrene round-bottom tube (Falcon; cat. no. 352052).

2.3. MHC Purification

1. Cell line(s): for example, Epstein-Barr virus (EBV)-transformed human B-cell lines expressing the class I or class II MHC of interest.
2. Complete medium: RPMI supplemented with 10% *(v/v)* heat inactivated fetal bovine serum (FBS) and 5 mL penicillin, streptomycin, glutamine mix (InVitrogen; cat. no. 10378-0161), or the defined medium for the cell line of interest.
3. 225 cm2 tissue culture flasks.
4. Trypan blue diluted 1:10 for determination of cell viability.
5. 100 mM Phenylmethylsulfonyl flouride (PMSF) in 2-propanol.
6. Ethylenediaminetriacetic acid (EDTA): 30 μg/mL.
7. Nonidet P40 (NP40).
8. Iodoacetamide.

9. 2X lysis buffer: 100 mL PBSO containing 2% NP40 *(v/v)* and 50 mM iodoacetamide. Immediately before use, add 230 µL PMSF (2 mM final) and 2 µL EDTA (0.03 µg/mL final)/10 mL of lysis buffer.
10. 0.45-mm syringe and bottle filters.
11. ImmunoPure Protein G IgG Plus Orientation Kit (Pierce 44990).
12. Sodium doedecyl sulfate (SDS).
13. Diethylamine (DEA_ (Merck; cat. no. 803010).
14. *n*-Octyl β-D-glucopyranodise(OG) (Sigma; cat. no. O-8001). Dissolve 25 g in 66 mL absolute ethanol by heating to 30°C. Measure final volume to determine concentration. Aliquot and store at –20°C. After thawing, OG is stable for 1 wk at 4°C. Do not refreeze.
15. Amicon Centricon-10 filter.
16. BCA protein assay kit (Pierce; cat. no. 23223/23224).
17. Column buffers:
    a. 1 L PBSO containing 1.0% NP40 and 0.1% sodium azide and filtered through a 0.45µm filter.
    b. 1 L PBSO containing 0.5% NP40, 0.1% sodium azide, and 0.1% SDS. Filter sterilize.
    c. 500 mL 0.05 M DEA buffer containing 0.15M NaCl (pH 11.0). Filter sterilize.
    d. 1 L PBSO containing 0.1% azide. Filter sterilize.
    e. DEA buffer (prepared in **step c**) containing 1% OG. Prepare 3 column volumes.
    f. PBSO/azide buffer (prepared in **step d**) containing 1% OG. Prepare 3 column volumes.
    g. Neutralization buffer: 2M Tris-HCl (pH 6.3) containing 0.1% azide. Prepare 100 mL and pass through a 0.45 µm filter.
18. Centrifugal filters (10,000 MWCO) (Millipore Amicon; cat. no. 4205).
19. Silver Xpress silver staining kit (Novex; cat. no. LC6100).
20. 0.1 M MES buffer (pH 5.5) (*N*-Morpholinoethane sulfonic acid) (Sigma; cat. no. M8250)
21. EZ-link™ Biotin-LC-PEO-Amine buffer (Pierce; cat. no. 21347). Prepare a 50 mM stock in MES buffer.
22. EDC (Pierce; cat. no. 22980). Immediately before use, dissolve 25 mg/mL MES buffer.
23. Protease inhibitor cocktail (Sigma; cat. no. P8340). The concentration of the cocktail is ×1000.

2.4. Artificial Antigen Presenting Cells (aAPCs)

1. Cholera toxin B subunit biotin Labeled, CTB biotin (Sigma; cat. no. C9972). Prepare a stock solution of 1 mg/mL ddH$_2$O and store at 4ºC. Just before use dilute 1:50 in PBSO for a 20 µg/mL (1.67 pmoles/µL) solution.
2. Biotinylated anti-CD28 (clone CD28.2) (BD Biosciences; cat. no. 555727) and anti-LFA1 (clone 38, Biodesign P01172B).

3. NeutrAvidin (NA) (Pierce; cat. no. 31000, 10 mg). Dilute 10 mg in 1 mL ddH$_2$O to prepare stock solution and store at 4°C. Just before use, prepare a 200 µg/mL (3.34 pmole/µL) solution by diluting 1:50 in PBSO.
4. Cholera toxin B subunit fluorescein isothiocyanate (FITC)-labeled (CTB FITC) (Sigma; cat. no. C9972). Prepare a stock solution of 1 mg/mL in H2O, stored at 4°C. Just before use, dilute 1:25 in PBSO for a 40 µg/mL (3.34 pmole/µl) solution.
5. MHC class II-peptide complex prepared in **Subheading 3.3.8.**
6. RPMI 1640 containing 10% human AB serum and 1% penicillin/strepamycin/glutamine mix.
7. Wash/stain/FACs (WSF) buffer: DPBS containing 2% heat inactivated FBS. Filter sterilize and store at 4°C.
8. PE-Cy5 (Cy-Chrome) anti-CD4 (clone RPA-T4, BD Biosciences 555348) and PE-Cy5-mouse IgG1 isotype control (clone MOPC-21) (BD Biosciences; cat. no. 555750).
9. FITC anti-CD3 (clone UCHT1) (BD Biosciences; cat. no. 555916) and FITC-mouse IgG1 isotype control (clone MOPC-21) (BD Biosciences; cat. no. 555748).
10. PBMCs: 50,000 per sample tube (*see* **Note 1**).

3. Methods

The methods described below outline the preparation of lipid lamellar sheets, liposomes, MHC purification, MHC/peptide complexes, artificial antigen presenting cells, and capture of antigen-specific T-cells.

3.1. Preparation of Lipid Lamellar Sheets

1. Bring the cholesterol, phosphatidylcholine and GM1 to room temperature.
2. Working in a fume hood, combine 1 mL of phosphatidylcholine and 0.029 mL cholesterol in a glass tube for a molar ratio of 7:2 (**Table 1**) *(23)*.
3. Add 0.5 mL of GM1.
4. Dry lipids under a stream of inert gas (nitrogen or argon) at 37°C for a minimum of 30 min.
5. Immediately add 11.45 mL buffer A containing deoxycholic acid to the lipids and vortex to resuspend.
6. Sonicate on ice for approx 2 to 5 min until solution becomes homogeneous and milky in appearance (*see* **Note 2**).
7. The concentration of lipids is calculated to be 10 mg/mL at this step (excluding GM1).
8. Proceed directly to **subheading 3.2.1., step 2** or aliquot 300 µL of the lamellar sheets into 1.5 mL microcentrifuge tubes and store at –20°C for up to 6 mo.

3.2. Liposomes

3.2.1. Preparation of Liposomes

1. If using frozen lamellar sheets, thaw 300 µL and sonicate on ice for 10 s before proceeding to **step 2**.

Table 1
The Molecular Weights and Concentrations of the Reagents Used in the Preparation of Liposomes

Reagent	Concentration of stock solution (mg/mL)	Formula weight	Volume used (concentration)	µmoles	Final molarity (mM)
GM-1	10	1568.8	500 µL (5.0 mg)	3.2	0.28
Cholesterol	500	386.7	29 µL (14.5mg)	37.5	3.27
Phophatidylcholine	100	760	1000 µL (100mg)	131.6	11.49

2. Prewet a Slide-a-Lyzer in PBSO and following the manufacturer's instructions, inject lipids into Slide-a-Lyzer.
3. Dialyze against 3 L PBS for 96 h at 4°C, changing PBSO every 24 h.
4. Remove liposomes from cassette and rinse the cassette twice with 100 µL PBS, adding rinses to liposomes.
5. Measure the final volume and calculate the liposome concentration. Adjust to 4 mg/mL with PBSO. The liposomes are used the same day they are prepared.

3.2.2. Sizing of Liposomes

The first time liposomes are prepared the preparation can be checked against calibration beads of a known size. The liposomes will be heterogeneous in size but should predominantly be in the 60 to 90 µm range (see **Fig. 1A**).

1. Dilute the liposomes (prepared in **Subheading 3.2.1.**) 1:10 in PBSO and transfer 200 µL to a 1.2-mL microtube.
2. Add 50 µL of calibration particles for a final volume of 250 µL and mix well.
3. Insert the microtube into a 5-mL tube for flow cytometry.
4. For sample acquisition, set forward and side scatter to linear scale. Adjust the cytometer settings so the bead population is in the upper right region of the dot plot (see **Fig. 1B**).
5. Acquire a total of 10,000 events.

3.3. MHC Purification

3.3.1. Generation of Cell Pellets

1. Cells expressing the MHC of interest are thawed and cultured following standard protocols.
2. The cells are grown to a density of approx 1×10^6 cells/mL in 225 cm² tissue culture flasks and are passaged until the desired cell number is achieved.
3. Collect the cells, centrifuge for 10 min at 500g and wash 1 time with 50 mL PBSO followed by 2 washes with 10 mL PBSO.

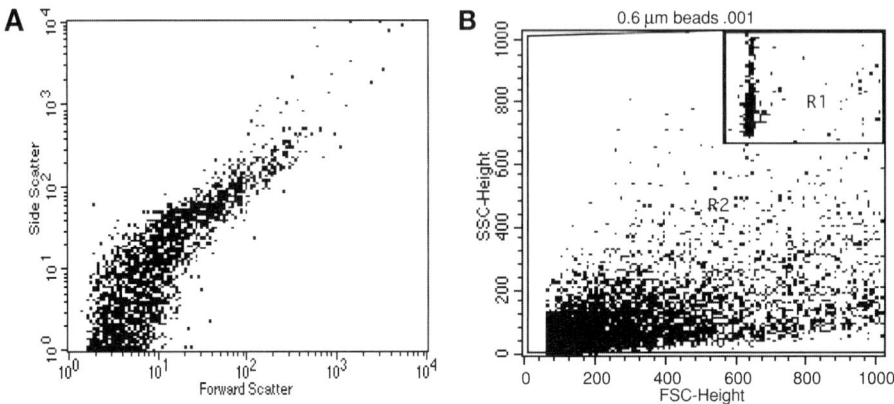

Fig. 1. Characterization of liposomes. (**A**) Liposomes alone. (**B**) Measurement of the relative size of liposomes compared to 6micron beads. The bead population is shown in the upper right of the dot plot (R2) and liposomes of varying sizes are seen in R1.

4. Count the cells with a hemacytometer and freeze 200 × 106 cells/pellet. The cell pellets may be stored indefinitely at –80°C.

3.3.2. Cell Lysates

1. Thaw the cell pellet on ice and resuspend in cold PBS0 at 2 × 108cells/mL.
2. Add an equal volume of 2X lysis buffer containing phenlymethylsulfonyfluoride (PMSF) and ethylene diamine tetraacetic acid (EDTA).
3. Incubate for 1 h at 4°C, mixing on a rotator.
4. Centrifuge at 3000g for 30 min at 4°C.
5. Filter lysate through a 0.45 µm syringe filter.

3.3.3. Antibody Coupled Affinity Column

The antibody linked column (ImmunoPure Protein G IgG Plus Orientation Kit) was prepared according to the Manufacturer's instructions.

3.3.4. Affinity Purification of MHC Molecules

1. Bring the column to room temperature and equilibrate with 3 column volumes of the 1% NP40 buffer by gravity flow.
2. Mix the lysate with the prepared antibody linked column resin and mix on a rotator for 2 h at room temperature.
3. To remove unbound material, wash the column with 3 column volumes of 1% NP40 buffer followed by 20 column volumes of 0.5% NP40 buffer, 3 column volumes of 0.05% NP40 buffer, and 3 column volumes of 1% OG buffer.
4. Elute the MHC with 3 column volumes of DEA buffer followed by 20 column volumes of PBS0/1% azide.

5. When the pH of the eluate is 11.0, begin to collect the eluate into a 50-mL tube containing 100 µL Tris/azide buffer (pH 6.3)/mL eluate (approx 600 µL total).
6. Collect the eluate until the pH changes to 7.0. Store the column at 4°C in PBS-azide buffer.

3.3.5. Concentration of MHC Eluate

1. Prerinse a sample reservoir once with distilled, deionized H2O and once with PBSO.
2. Fill the tube with the MHC eluate and centrifuge at 3000g at 4°C until the volume is reduced to between 300 and 500 µL.
3. Wash the membrane with 2 mL 1% OG buffer by spinning again until the desired volume is reached.
4. Place the sample cup on the collection tube and centrifuge for 5 min.
5. Remove the sample cup with the concentrated MHC and measure the volume.
6. Dialyze against 1 L of PBS-O overnight at 4°C using a 10,000 MWCO cassette.
7. The MHC preparation can be stored indefinitely at 4°C.

3.3.6. Determination of Quality and Concentration of the Purified MHC

1. Measure the protein concentration by BCA assay according to the manufacturer's instructions for a microassay in 96-well plates. Typically, 200×10^6 cells will yield between 20 and 200 µg of MHC (*see* **Note 3**).
2. Confirm the purity of the preparation with 2 12% sodium dodecyl sulfate-polyacrylamide gel electrophoresis (SDS-PAGE) gels run under reducing conditions.
3. Visualize the protein by silver staining one gel and perform a Western blot with the second gel (*see* **Note 4**).

3.3.7. MHC Biotinylation

1. Prerinse a Centricon-10 filter with MES buffer by adding 2 mL of MES buffer to the tube and centrifuging at 1000g until at least half the buffer passes into the filtrate vial.
2. Invert the tube and centrifuge at 300 to 1000g for 2 min.
3. Add the MHC to the reservoir and slowly add 1.0 mL MES buffer to the sample.
4. Spin at 1000 to 5000g until the sample volume has been concentrated to 150 µL.
5. Place the retentate vial over the sample reservoir, invert the unit and centrifuge at 300 to 1000g for 2 min to transfer the concentrate into the retentate vial.
6. Add 6 µL of the 50 mM EZ-linkTMBiotin-LC-PEO-Amine stock solution to 10,000 pmoles of MHC in MES buffer.
7. Add 6.4 µL of the freshly prepared EDC solution and incubate for 2 h at room temperature with stirring.
8. Transfer the sample to a Slide-A-Lyzer cassette and dialyze against PBS0 for 24 h at 4°C to remove any unreacted byproducts.
9. Remove the biotinylated MHC from the cassette and rinse the cassette twice with PBSO, pooling the concentrate and washes.

10. Spin the sample for 5 min (1000g at 4°C) to remove any aggregates.
11. Collect the supernatant and store in the dark at 4°C.
12. To calculate the moles of biotinylated MHC, assume 100% recovery of the starting material.

3.3.8. Preparation of Peptide-MHC Complexes

1. The T-cell capture requires 25 pmoles of MHC class II/5 × 10⁴ PBMCs.
2. Calculate the total amount of MHC required and mix a 20-fold molar excess of the peptide of interest with the MHC in PBS0. Add protease mix at a 1:1000 dilution (*see* **Note 5**).
3. Incubate for 48 h at room temperature.

3.4. Preparation of Artificial APCs (aAPCs)

The artificial antigen presenting cells are prepared fresh at room temperature (*see* **Note 5**) *(13)*, unless specified otherwise, while the cells are immunostained. All steps should be kept dark as much as possible. A schematic representation of an artificial APC is shown in **Fig. 2**.

1. For 50,000 cells, combine 5 pmoles CTB-biotin, 4.17 pmoles of each biotinylated antibody and 25 pmoles of the biotinylated MHC-peptide complex in a 1.5-mL eppendorf microfuge tube and incubate for 5 min while mixing on a rotator. **Table 2** shows the amount of each reagent required per assay tube whereas **Table 3** provides an example of an experimental set up (*see also* **Notes 6** and **7**).
2. Add 12.8 pmoles NA (*see* **Note 8**).
3. Incubate for an additional 15 min with mixing.
4. Add the liposomes (58.9 µg per 50,000 cells) and mix gently.
5. Cover the tube with foil and incubate for 90 min while mixing. While the liposomes are incubating, immunostain the cells (*see* **Subheading 3.5.1.**).
6. Add 62.5 pmoles of CTB-FITC and incubate for an additional 30 min on rotator.
7. Bring to 1.5 mL with PBS0 and centrifuge at 16,000g for 10 min. Carefully remove the supernatant with a pipet tip and discard. Wash twice more with 1.5 mL PBS0.
8. Resuspend the aAPCs in 100 µL RPMI containing 10% AB serum and penn/strep/glut mix with gentle pipetting and add to the well containing PBMCs (*see* **Subheadings 3.5.1.** and **3.5.2.**).

3.5. Preparation of Immunostained PBMCs

3.5.1. Immunostaining PBMCs for T-Cell Capture

1. Collect the PBMCs and wash twice with WSB Buffer.
2. Count the cells and stain 1 × 10⁶ cells with 20 µL PE-Cy5 (Cy-Chrome) conjugated anti-CD4 in a final volume of 250 µL of WSB buffer. Reserve an additional 1.0 × 10⁶ PBMCs for **Subheading 3.5.2**.
3. Incubate the cells at 4°C for 30 min.

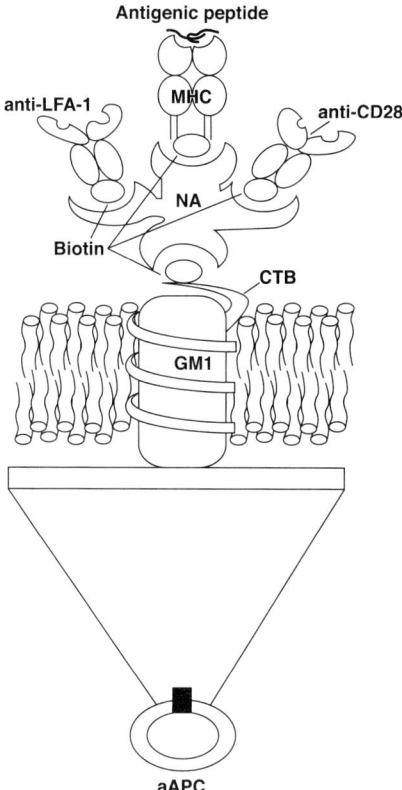

Fig. 2. Schematic of an artificial APC depicting a section of the lipid bilayer. GM-1 ganglioside incorporated into the bilayer binds one molecule of CTB biotin. In turn, one molecule of neutravidin anchors the biotinylated anti-CD28, the MHC-peptide complex and biotin-anti-LFA-1 to the aAPC through the CTB-biotin molecule. Different combinations and ratios of T cell ligands can be loaded on the aAPC as needed.

4. Wash twice with 2 mL WSB Buffer at 600g for 6 min.
5. Re-suspend the cells at 0.5×10^6 cells/mL of RPMI containing 10% human AB serum and penn/strep/glut mix.
6. Plate 100 µL/well of a 96-well U-bottom plate.

3.5.2. Preparation of Cytometer Set Up Controls

This step is started after the liposomes and PBMCs have been incubating for 1 h.

1. Stain 5 additional samples (0.2×10^6 cells/tube) for setting up the cytometer (using the concentration of antibody recommended by the manufacturer) (*see* **Note 8**):

Table 2
The Molecular Eights and Concentrations of the Reagents Used in the Preparation of TCC Rafts

Reagent	Formula weight	Concentration of stock solution (mg/mL)	Concentration of stock solution (µmoles/mL)	Dilution	Concentration of working solution (pmoles/µL)	Volume (µ:) /50,000 cells
Liposome-GM-1		4.0				14.7
CTB-Biotin	12,000	1	0.083	1:50	1.67	3.0
<LFA-1-Biotin	150,000	0.1	6.7×10^{-4}		0.67	6.2
<CD28-Biotin	150,000	0.0125	8.3×10^{-5}		0.083	50.2
MHC-peptide	60,000	variable			variable	
NeutrAvidin	60,000	10	0.167	1:50	3.34	3.8*
CTB-FITC	12,000	1	0.083	1:25	3.34	18.7

*The amount of NeutrAvidin needed will depend on the pmoles of biotinylated reagents used.

Table 3
Example of an Experimental Set-up for the Preparation of Artificial APCs

Tube no.	Sample	Liposomes (µL)	MHC-peptide (pmoles)	anti-Hu CD28 (µL)	anti-Hu LFA (µL)	CTB-biotin (µL)	NA (µL)	CTB-FITC (µL)
1	Empty Liposome Control	14.7	0	0	0	0	0	18.7
2	aAPC negative control (anti-CD28 + anti-LFA-1)	14.7	0	50	6.25	3.0	1.3	18.7
3	Irrelevant peptide control	14.7	25	50	6.25	3.0	3.8	18.7
4	T-cell capture	14.7	25	50	6.25	3.0	3.8	18.7

   a. Unstained.
   b. FITC isotype control.
   c. FITC anti-CD3 stained cells.
   d. PE-Cy5 (Cy-Chrome) isotype control.
   e. PE-Cy5 (Cy-Chrome) anti-CD4 stained cells.
2. Incubate the cells with the appropriate antibodies at 4°C for 30 min.
3. Wash twice with WSB Buffer and resuspend in 200 µL WSB Buffer.
4. Keep on ice until acquisition.

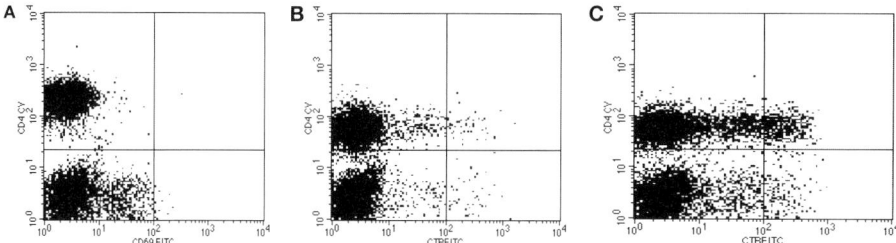

Fig. 3. Artificial APCs, loaded with HLA-DRB1*0401 bound to a peptide epitope, capture antigen specific T-cells as demonstrated by fluorescence activated cell sorting. For acquisition and analysis a gate was set around the lymphocyte population. (**A**) In this experiment, PBMCs from a patient with RA were stained with CyChrome anti-CD4 and FITC conjugated anti-CD69 (**B**) PBMCs stained with CyChrome conjugated anti-CD4 and incubated with aAPCs loaded with HLA-DRB1*0401 and an irrelevant peptide, anti-CD28 and anti-LFA-1. (**w**) PBMCs stained with CyChrome conjugated anti-CD4 and incubated with aAPCs loaded with HLA-DRB1*0401 and the bacterial heat shock protein derived dnaJP1 peptide, anti-CD28 and anti-LFA-1 bind antigen-specific T-cells from a patient with RA *(22)*. 10,000 gated events were acquired on a FACS Caliber.

3.6. T-Cell Capture Assay

1. The cells (from **Subheading 3.5.1.**) and the aAPCs (from **Subheading 3.4.1.**) are incubated together for 2.5 h at 37°C in the presence of 5% CO_2.
2. Wash cells twice with WSB Buffer by centrifuging at 400g for 5 min at room temperature.
3. Resuspend in 200 µL WSB buffer at room temperature.
4. Acquire immediately and collect as many events as possible (*see* **Fig. 3**) (*see* **Notes 9** and **10**).

4. Notes

1. When using frozen PBMCs, thaw and culture in complete media at 37°C in a 5% CO2 atmosphere for several hours to allow the cells to recover. Additionally, a larger number of cells may be useful to increase the sensitivity of the assay, particularly when the T-cell population of interest is less than 1% of the total population. This protocol has been tested with up to 2×10^5 cells/sample.
2. The sonicator used in this protocol is a Branson Sonifier set at an intensity level between 2 and 3. The sample is sonicated until the lipids are in suspension and evenly dispersed.
3. 200 ζy 10^6 cells will yield between 20 and 200 µg of purified MHC depending on the cell line and the expression levels of the particular MHC. Each T-cell capture tube requires 1.5 µg of MHC.

4. A general method for Western blots can be found at the BD Biosciences website *(http://www.bdbiosciences.com/pharmingen/protocols/)*.
5. The amount of peptide required, the pH of the MHC-peptide mix and the incubation time will depend upon the MHC and the binding affinity of the peptide for the MHC. Preliminary studies may be required to determine the optimal conditions for the peptide of interest.
6. Controls should also be included in the experimental set-up. Negative controls should include liposomes alone, aAPCs loaded with anti-CD28 and anti-LFA-1 without MHC-peptide and aAPCs loaded with MHC and a peptide with a lower binding affinity or one which is not recognized by the T-cells of interest. An additional negative control would be aAPCs loaded with anti-CD19, a B-cell marker, whereas a positive control could be aAPCs loaded with biotinylated anti-CD3.
7. If adding less than 3 biotinylated reagents, the amount of each biotinylated reagent used is increased proportionally. For example, when aAPCs are prepared with anti-CD28 and anti-LFA-1 but without MHC, 1.5 times the normal amount of antibodies are used.
8. The amount of Neutravidin required is determined by adding the amount of biotinylated moieties used and dividing by three.

    Neutravidin contains four biotin binding sites: in theory, one for the MHC-peptide complex, one for anti-LFA-1, one for anti-CD28, and one for CTB-biotin. The total pmoles are divided by 3 instead of the 4 biotinylated moieties to ensure that a slight excess of NA is added. For 50,000 cells, that includes 5 pmoles of CTB-biotin, 4.17 pmoles of anti-LFA-1, 4.17 pmoles of anti-CD28, and 25 pmoles of MHC-peptide for a total of 35.84 pmoles. 35.84/3= 12.8 pmoles NA.

    If the TCC rafts were to be modified to include less than 4 biotinylated proteins, the amounts of each would be increased proportionally to fill the available sites on the Neutravidin molecule.
9. Tubes a-e are used to establish the instrument settings and to set a gate on the $CD4^+$ population. If a PE conjugated antibody is used, cells from tubes c and e can be mixed together to set compensation.
10. It is important to keep the TCC samples at room temperature because the T-cell/aAPC complexes disengage more quickly at lower temperatures.

Acknowledgments

This work was supported by grants for NIAD and NIAMS of the National Institute of Health (NIH). Eva Koffeman was supported by grants from the Ter Meulen Fund of the Royal Netherlands Academy of Arts and Sciences, the Fulbright Fellow Scholarship, and the VSB Fund Scholarship. The authors wish to acknowledge Gisella Puga Yung for her help on the practical work. They thank Androclus Therapeutics and the Associazione Italiana per la ricerca contro l'artrite for their kind gifts.

References

1. Clark, D. M., Boylston, A. W., Hall, P. A., and Carrel, S. (1986) Antibodies to T cell antigen receptor beta chain families detect monoclonal T cell proliferation. *Lancet.* **2**, 835–837.
2. Albani, S., Keystone, E. C., Nelson, J. L., et al. (1995) Positive selection in autoimmunity: abnormal immune responses to a bacterial dnaJ antigenic determinant in patients with early rheumatoid arthritis. *Nat. Med.* **1**, 448–452.
3. Dustin, M. L., Miller, J. M., Ranganath, S., et al. (1996) TCR-mediated adhesion of T cell hybridomas to planar bilayers containing purified MHC class II/peptide complexes and receptor shedding during detachment. *Immunol.* **157**, 2014–2021.
4. Klenerman, P., Cerundolo, V., and Dunbar, P. R. (2002) Tracking T cells with tetramers: new tales from new tools. *Nat. Rev. Immunol.* **2**, 263–272.
5. Mallone, R. and Nepom, G.T. (2004) MHC Class II tetramers and the pursuit of antigen-specific T cells: define, deviate, delete. *Clin. Immunol.* **110**, 232–242.
6. Mallet-Designe, V. I, Stratmann, T., Homann, D., Carbone, F., Oldstone, M. B., Teyton, L. (2003) Detection of low-avidity CD4+ T cells using recombinant artificial APC: following the antiovalbumin immune response. *J. Immunol.* **170**, 123–131.
7. Kim, J. V., Latouche, J. B., Riviere, I,. and Sadelain, M. (2004) The ABCs of artificial antigen presentation. *Nat. Biotechnol.* **22**, 403–410.
8. Rosenberg S. A., Yang J. C., and Restifo N. P. (2004) Cancer immunotherapy: moving beyond current vaccines. *Nat. Med.* **10**, 909–915.
9. Dustin, M.L. (2004) Stop and go traffic to tune T cell responses. I*mmunity* **21**, 305–314.
10. Valitutti, S. and Lanzavecchia, A. (1997) Serial triggering of TCRs: a basis for the sensitivity and specificity of antigen recognition. *Immunol. Today* **18**, 299–304.
11. Pentcheva-Hoang, T., Egen, J. G., Wojnoonski, K., and Allison, J. P. (2004) B7-1 and B7-2 selectively recruit CTLA-4 and CD28 to the immunological synapse. *Immunity* **21**, 401–413.
12. Grakoui, A., Bromley, S. K., Sumen, C., et al. (1999) The immunological synapse: a molecular machine controlling T cell activation. *Science* **285**, 221–227.
13. van Rensen A. J., Wauben M. H., Grosfeld-Stulemeyer M. C., van Eden W., and Crommelin D. J. (1999) Liposomes with incorporated MHC class II/peptide complexes as antigen presenting vesicles for specific T cell activation. *Pharm. Res.* **16**, 198–204.
14. Jenkins, M. K., andSchwartz, R. H. (1987) Antigen presentation by chemically modified splenocytes induces antigen-specific T cell unresponsiveness in vitro and in vivo. *J. Exp. Med.* **165**, 302–319.
15. Krogsgaard, M., Huppa, J. B., Purbhoo, M. A., and Davis, M. M. (2003) Linking molecular and cellular events in T-cell activation and synapse formation. *Semin. Immunol.* **15**, 307–315.
16. Lee, K. H., Dinner, A. R., Tu, C et al. (2003) The immunological synapse balances T cell receptor signaling and degradation. *Science* **302**, 1218–1222.

17. Prakken, B., Wauben, M., Genini, D., et al. (2000) Artificial antigen-presenting cells as a tool to exploit the immune "synapse." *Nat. Med.* **6**, 1406–1410.
18. Giannoni, F., Barnett, J., Bi, K., et al. (2004) Clustering of T cell ligands on aAPC membranes influences T cell activation and PKC theta translocation to the T cell plasma membrane. *J. Immunology*, in press.
19. Bonnin, D., Prakken, B., Samodal, R., La Cava, A., Carson, D. A., and Albani, S. (1999) Ontogeny of synonymous T cell populations with specificity for a self MHC epitope mimicked by a bacterial homologoue: an antigen-specific T cell analysis in a nontransgenic system. *Eur. J. Immunol.* **29**, 3826–3836.
20. Massa, M., Costouros, N., Mazzoli, F., et al. (2002) Self epitopes shared between human skeletal myosin and Streptococcus pyogenes M5 protein are targets of immune responses in active juvenile dermatomyositis. A*rthritis Rheum.* **46**, 3015–3025.
21. de Kleer, I. M., Wedderburn, L. R., Taams, L. S., et al. (2004) CD4+CD25bright regulatory T cells actively regulate inflammation in the joints of patients with the remitting form of juvenile idiopathic arthritis. *J. Immunol.* **172**, 6435–6443.
22. Prakken, B. J., Samodal, R., Le, T. D., et al. (2004) Epitope-specific immunotherapy induces immune deviation of proinflammatory T cells in rheumatoid arthritis. *Proc.Natl.Acad.Sci. USA*. **101**, 4228–4233.
23. Brian, A. A. and McConnell, H. M. (1984) Allogeneic stimulation of cytotoxic T cells by supported planar membranes. *Proc.Natl.Acad.Sci. USA* **81**, 6159–6163.
24. Sidney, J., Southwood, S., Oseroff, C., del Guercio, M-F., Sette, A., and Grey, H. M. (1999) Ligand-Receptor Interactions in the Immune System. In: *Current Methods in Immunology*, Coligan, J. E.,Kruisbell, A. M., Margulies, D. H., Shevach, E. M., and Strober, W., eds., New York: New York, pp. 18.3.1–18.3.18.

7
Analysis of Th1/Th2 T-Cell Subsets

Alla Skapenko and Hendrik Schulze-Koops

Abstract

Specific immune responses are mediated by activated CD4$^+$ T-helper (Th) cells. Two major subsets, denoted Th1 and Th2, have been identified that are characterized by their distinctive cytokine secretion pattern and associated effector functions. The signature cytokines of Th1 and Th2 cells are interferon-γ and interleukin-4, respectively. Because of the dominant role of Th cells in directing specific immunity, the analysis of Th subsets by means of determining their signature cytokines has contributed greatly to the progress that has been made in recent years in the understanding of protective as well as pathogenic immune responses. Several methods, such as reverse transcriptase polymerase chain reaction, enzyme-linked immunosorbent assay, ELISpot, and intracellular flow cytometric analysis are used for the analysis of T-cell cytokines and, thus, of Th subsets. Here, we briefly discuss the advantages and disadvantages of these methods and describe in detail a standard protocol for the analysis of human Th subsets by means of detection of cytoplasmic cytokines by flow cytometry.

Key Words: T-cells; Th1/Th2; cytokines; flow cytometry.

1. Introduction

Based on their distinctive cytokine secretion pattern and effector functions, CD4$^+$ T-helper (Th) cells can be divided into at least three different subsets of which the two major are denoted Th1 and Th2. Upon activation, Th1 cells secrete the proinflammatory cytokines interferon (IFN)-γ, interleukin (IL)-2, and lymphotoxin. Th2 cells, in contrast, produce the anti-inflammatory cytokines IL-4, IL-5, IL-9, and IL-13 *(1,2)* (**Table 1**). Th2 cells can also secrete IL-6 and IL-10, however, in contrast to mice, those cytokines in humans are not confined to the Th2 subset but can also be produced by Th1 cells *(2,3)*. Recent evidence has suggested that Th1 and Th2 cell populations might display distinct arrays of chemokine receptors *(4)*. Yet, whereas several in vitro

Table 1
Characteristics of Th Cell Subsets

	Th1	Th2
Cytokines		
signature	IFN-γ	IL-4
other	IL-2, lymphotoxin	IL-5, IL-6*, IL-9, IL-10*, IL-13
Chemokine receptors	CXCR3, CCR5	CCR3, CCR4, CCR8
B-cell help		
human	IgG1	IgE, IgG2
mouse	IgG2a	IgE, IgG1

*Th2 cell cytokine in the mouse, but can be produced by Th1 and Th2 cells in humans.

studies and some animal experiments have indicated a preferential expression of chemokine receptors on activated effector cells from one or the other Th cell subset, a definite proof of concordant expression of a chemokine receptor with Th1 or Th2 cytokines on the single cell level is still missing. Moreover, analysis of individual CD4 memory T-cells from the rheumatoid synovium on the single cell level has failed to show a clear association of chemokine receptor expression and cytokine production. Thus, specific surface markers for a given Th population might not exist in an -situation in man (5), and their existence in animals needs to be confirmed. The gold standard to identify Th1 or Th2 cells, therefore, remains the detection of their respective signature cytokines, IFN-γ and IL-4.

Cytokine expression occurs generally within hours after stimulation of differentiated Th cells, but the precise kinetic of cytokine secretion depends on various factors, such as the mode of activation and the nature of the cells studied. It is important to understand that most applications for the analysis of Th cell subsets require in vitro activation of the T-cells. Thus, with the exception of reverse transcription-polymerase chain reaction (RT-PCR)-based approaches, the methods currently most widely employed for the analysis of Th cell subsets reflect the capability of the T-cells within the population of interest to produce cytokines after maximum stimulation, but do not necessarily mirror an in vivo situation.

Th cell cytokines can be detected in several ways. Cytokine mRNA levels, for example, can be assessed by different RT-PCR methods or by ribonuclease protection assays (RPA). These methods are very sensitive and may therefore not require in vitro activation, they may allow (semi)-quantification of messenger RNA transcripts, are relatively fast, simple, and reliable and do not rely on expensive nonstandard laboratory equipment. The need for radioactivity in some RPA-approaches might restrict its use. The main disadvantages of these assays, however, remain the rapid modulation of RNA levels in living cells (which is of particular importance for the analysis of samples that need to be transported to the laboratory, such as clinical or animal specimens) and the fact that cell populations are evaluated as a whole which excludes the determination of the frequencies of the individual Th cell subsets within the populations. Single cell RT-PCR, that would allow to address this issue, is technically demanding and a yet to be standardized procedure *(5)*.

Secreted cytokines from activated T-cells can be analyzed in fluids, such as cell culture supernatants or body fluids by enzyme-linked immunosorbent assays (ELISA). ELISA are commercially available for most cytokines, they are standardized and reliable, simple to perform, relatively cheap, allow quantification of the concentration of the molecules of interest, and they do not require expensive laboratory equipment. They are, however, time consuming. Moreover, similar to the PCR-based methods described above, determination of cytokines in fluids does not permit the analysis of cytokine profiles of individual Th cells, but rather of that of the cell population as a whole. Although in most (but, notably, not in all) systems cytokine concentrations in fluids roughly reflect the magnitude of the cell population secreting those particular cytokines, precise determinations of the frequencies of Th1 and Th2 cells within mixed populations cannot be performed by ELISA.

The enzyme-linked immunospot (ELISpot) assay is a single cell ELISA-based procedure that allows the analysis of the cytokine secretion profile of individual cells. The ELISpot assay detects locally secreted cytokines by capturing the cytokines onto a "culture plate" membrane coated with a cytokine-specific antibody. ELISpot assays are commercially available for numerous cytokines, they are simple to perform, quantitative, and do not depend on expensive laboratory equipment, except for a reverse microscope. Disadvantages are the length of the assays (similar to ELISA, they require 1–2 d before the results are available) and their sensitivity, which although tremendously improved, is still a problem for the detection of T-cell cytokines in some particular species. Moreover, detection of the cytokine together with the phenotype of the producing cells is almost impossible. Nevertheless, not least because of the possibility to detect more than one cytokine on the single cell level, ELISpot assays have become increasingly popular as a valid and useful approach to analyze cytokine secreting T cell subsets.

Analysis of cytoplasmic cytokines by flow cytometry (intracellular fluorescence activated cell sorting, icFACS) analysis is the most widely used method to determine cytokine secretion profiles of individual Th cells (**Note 1**) *(6–9)*. Anti-T-cell cytokine antibodies suitable for icFACS analysis are available for most species. IcFACS analysis is very reliable, technically undemanding, and fast. Because most antibodies are available conjugated to different fluorochromes, the simultaneous detection of numerous cytokines in individual cells is easy to accomplish *(10–13)*. For the simple analysis of IFN-γ and IL-4 producing T-cells, a single laser, four-parameter flow cytometer (forward scatter, sideward scatter, two channels for the two different fluorochromes labeled to the anticytokine antibodies) is sufficient. Cytometers that can detect more parameters enable the scientist to define the cells' entire spectrum of secreted cytokines. Moreover, simultaneous staining of cytoplasmic and extracellular targets can be performed and permits the precise determination of the cell surface phenotype of the cytokine producing cells. Thus, analysis of T-cell subsets by flowcytometry has tremendous advantages over the other assays described herein if a comprehensive characterization of T-cell cytokine profiles is required on the single cell level or if time is an issue.

In contrast to detection of secreted cytokines by ELISA or ELISpot assay, it is necessary for the detection of cytoplasmic cytokines by icFACS analysis to prevent the secretion of cytokines during the last few hours of stimulation, and, thus, increase the intracellular concentration of the targets (**Note 1**). Two different protein transport inhibitors are used, monensin and brefeldin A (**Note 4**). For best results, it is important to evaluate the efficacy as well as the best time of adding either of these substances in the specific assay systems employed. After stimulation, the cells need to be fixed to maintain their integrity during the permeabilization steps that are required for the fluorochrome-labeled anticytokine antibodies to reach their cytoplasmic targets (**Note 8**). It is essential that the fixative does not significantly alter the morphology of the cells while at the same time preventing the cytoplasmic targets from leaking out of the cells and preserving the accessibility of the epitopes for the antibodies. Crosslinking with paraformaldehyde has been shown to best meet these demands and is therefore most widely used for icFACS analysis of T-cell cytokines. Alternatively, glutaraldehyde may be used. Methanol, ethanol and acetone provide fixation by denaturing proteins, and they are therefore less favorable in icFACS analysis. Permeabilization can be accomplished by different chemicals, such as saponin, triton X-100, or *n*-octyl-β-d-glucopyranoside. Saponin, a glucoside with high affinity for cholesterol that creates membrane pores by replacing cholesterol, is most widely used. As the intercalation process by saponin into the cell membranes is reversible, it is important to have saponin present during all incubation and washing steps until final analysis *(9)*.

We have used the protocol presented here in greater detail for the detection of human T-cell cytokines for many years *(11,12 ,14–16)*. As the focus of the chapter is on the analysis of Th cell subsets, we will discuss the basic protocol for the analysis of the two signature cytokines of Th1 and Th2 cells, IFN-γ and IL-4, respectively. This protocol is very reliable and robust and can be easily adapted or expanded to specific, more demanding experiments necessary.

A final approach for analyzing Th cell subset activity should be mentioned. In vivo, Th1 cells promote the induction of complement fixing, opsonizing antibodies and of antibodies involved in antibody-dependent cell cytotoxicity (e.g., IgG1 in humans and IgG2a in mice). In contrast, Th2 cells provide potent help for B-cell activation and immunoglobulin class switching to IgE and subtypes of IgG that do not fix complement, e.g. IgG2 in humans and IgG1 in the mouse *(16,17)*. Consequently, serum IgG isotype levels have been used as indicators of in vivo Th subset activity. The advantages and problems of this indirect method to analyze Th subset functions are addressed elsewhere.

2. Materials

1. CD4 positive human T-cells.
2. Phosphate buffered saline (PBS).
3. 24-Well cell culture plate.
4. RPMI.
5. Serum: fetal calf serum (FCS) or normal human serum (NHS).
6. Phorbol 12-myristate 13-acetate, PMA (Sigma-Aldrich): 10 µg/mL stock solution: in dimethyl sulfoxide (DMSO). Store at –20°C (**Note 3**).
7. Ionomycin (Calbiochem): 0.5 mM stock solution in DMSO. Store at 4°C (**Note 3**).
8. Monensin: 10 mM stock solution in ethanol. Store at –20°C.
9. Staining vials (1.4-mL polypropylene reaction tubes) (Micronic BV, AP Lelystad).
10. Paraformaldehyde (Sigma-Aldrich). Fixation solution: 4% *(w/v)* paraformaldehyde in PBS (pH 7.5). Store at 4°C in the dark.
11. Sodium azide (NaN$_3$, Sigma Aldrich). Staining buffer: 2% FCS/PBS/0.01% *(w/v)* NaN$_3$. Store at 4°C.
12. Saponin. Stock solution: 5% *(w/v)* in PBS, 0.01% *(w/v)* NaN$_3$. Store at 4°C.
13. Permeabilization buffer: 0.1% *(w/v)* saponin in 2% FCS/PBS/0.01% NaN$_3$ (prepare by diluting saponin stock solution 1:50 with staining buffer).
14. Blocking buffer: 8% mouse serum/8% rat serum/permeabilization buffer.
15. Fluorescein isothiocyanate (FITC)-conjugated anti-human IFN-γ mAb (4S.B3), PE-conjugated anti-human IL-4 mAb (MP4-25D2), and PE-conjugated anti-human IL-2 mAb (MQ1-17H12) (BDPharMingen) (**Notes 6** and **7**).
16. Staining solution I: Optimal concentration of FITC-conjugated anti-IFN-γ antibody/optimal concentration of PE-conjugated anti-IL-4 antibody in 10 µL permeabilization buffer.
17. Staining solution II: Optimal concentration of FITC-conjugated anti-IFN-γ antibody/optimal concentration of PE-conjugated anti-IL-2 antibody in 10 µL permeabilization buffer.
18. Flow cytometer.

3. Methods
3.1. Activation of T-Cells

In order to analyze T-cell cytokines by icFACS analysis it is necessary to activate the T-cells in vitro (**Notes 1** and **2**). Different modes of activation can be employed that may have different effects depending on the cell population activated (e.g., peripheral blood mononuclear cells [PBMCs], isolated CD4+ T-cells, T-cell clones) and the stimulation chosen. For the purpose of simplicity, the protocol described here assumes a human CD4 T-cell population that can be isolated by various approaches (e.g., using magnetic beads, purification columns, or panning) from a mixed population of mononuclear cells to high purity (> 95%).

1. Wash cells 2 times with PBS at 4°C for 5 min at 380g. Resuspend in RPMI containing 10% FCS or NHS at a concentration of 1.5×10^6 cells/mL. Transfer 1 mL of the cell suspension into a 24-well plate.
2. Add 2 µL each from stock solutions of PMA (10 µg/mL in DMSO; store at –20°C) and ionomycin (0.5 mM in DMSO; store at +4°C) to stimulate the T-cells at final concentrations of 20 ng/mL PMA and 1 µM ionomycin (**Note 5**).
3. Dilute monensin from a stock solution (10 mM in ethanol, stored at –20°C) 1:50 in RPMI. Block cytoplasmic protein transport in the T-cells by adding 10 µL of the diluted monensin to the vials containing the activated T-cells (final concentration of monensin: 2 µM) (**Note 5**).
4. Incubate for 5 h at 37°C in a humidified atmosphere containing 5% CO_2.

3.2. Harvest, Fixation, and Permeabilization of the Cells

1. Harvest cells from the vials with a 1000 µL pipet by repeated aspiration of the cell suspension. Wash cells 2 times with PBS at 4°C for 5 min at 600g. Resuspend in 50 µL PBS.
2. Add 150 µL of a 4% PFA/PBS solution. Vortex well to avoid aggregation. Fix cells for 10 min at 37°C in the dark.
3. Wash cells twice with PBS at 4°C for 5 min at 3000g.
4. Wash cells once with permeabilization buffer at 4°C for 2 min at 3000g (**Note 4**). Resuspend in 50 µL permeabilization buffer.

3.3. Staining of Cytoplasmic Cytokines

1. Block nonspecific binding sites by adding 50 µL of blocking buffer (8% rat serum/8% mouse serum/permeabilization buffer). Vortex well. Incubate for 10 min on ice (**Note 9**).
2. Wash once with permeabilization buffer.
3. Resuspend in a total volume of 150 µL permeabilization buffer.
4. Transfer 50 µL each to new 1.4 mL polypropylene reaction tubes.
5. Prepare two staining solutions by adding the optimal concentrations of PE-labeled anti-IL-2 and FITC-labeled anti-IFN-γ, and the optimal concentrations of PE-

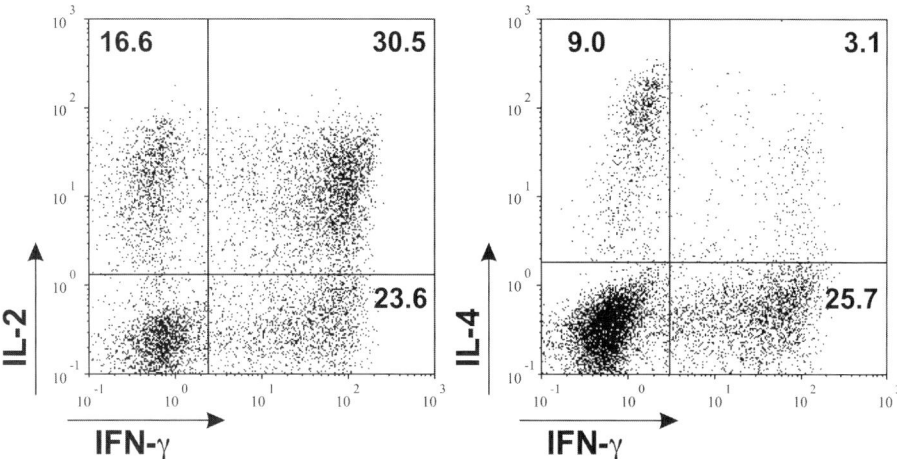

Fig. 1. Human CD45RO positive memory T-cells were isolated from the peripheral blood of a healthy individual to purity (>95% CD3/CD4, >90% CD45RO, viability >98%), primed for 5 d with mAbs to CD3 and CD28 (*left panel*) or with a mAb to CD28 in the absence of CD3 ligation (*right panel*), harvested and restimulated for 5 h with PMA and ionomycin, as described in the protocol. Cells were fixed with paraformaldehyde, permeabilized with saponin, and stained with monoclonal antibodies to IL-2 and IFN-γ (*left panel*) or IL-4 and IFN-γ (*right panel*). Analysis was performed on an Epics flow cytometer. The numbers indicate the percentage of cytokine-producing cells from the total population of gated viable lymphocytes.

labeled anti-IL-4 and FITC-labeled anti-IFN-γ each into permeabilization buffer to receive a final volume of 10 µL. Transfer one of the staining solutions each to the first two cell-containing tubes. The third tube serves as a control to determine background fluorescence.
6. Mix well; incubate for 25 min on ice in the dark.
7. Wash cells twice with permeabilization buffer at 4°C for 5 min at 3000g.
8. Wash cells with staining buffer at 4°C for 5 min at 3000g.
9. Analyze on a flow cytometer (*see* **Fig. 1**).

4. Notes
1. Cytokine levels in freshly isolated T-cells are usually very low, even if the T-cells are isolated from inflamed tissue. The concentration of cytoplasmic cytokines in T-cells that are to be analyzed for their cytokine expression in vivo might be increased if the freshly isolated cells are cultured for an additional 5 h in the presence of a protein transport blocking agent (e.g., monensin or brefeldin A).
2. Whereas T-cell clones, T-cell lines, and T-cell receptor transgenic T-cells can be activated by specific antigen (in the presence of appropriate APCs [APCs]),

mixed populations of T-cells require polyclonal activation (because of the low frequency of T-cells specific for any given antigen), which can be accomplished by different ways. The combination of PMA and ionomycin is the most widely used method to activate T-cells for icFACS analysis. Ligation of CD3 and/or CD28 by mAbs is a valid alternative for T-cell activation. Lectins (phytohemagglutinin or concavalin A) may be used if APCs are present. It is important to realize that the combination of PMA/ionomycin is not specific for T-cells but also activates almost all other cells, such as monocytes, natural killer (NK)-cells, and B-cells, which should be kept in mind if T-cell subsets are analyzed in cell mixtures. On the other hand, whereas stimulation via CD3 and CD28 is T-cell specific, it activates CD4 and CD8 T-cells alike. As B-cells, NK-cells and CD8 T-cells cannot be distinguished from CD4 T cells by scatter analysis, and IL-4 and IFN-γ can be expressed by CD4 T-cells, NKT-cells, basophils, eosinophils, and mast cells, and by CD4 and CD8 T-cells and NK-cells, respectively, it is essential to take into account the composition of the cell population and the effects of stimulation when interpreting the icFACS data. Moreover, a careful time kinetic is required for each stimulation to determine the optimal effect on the T-cell population in each particular assay.
3. Because of the hydrophobicity of PMA and ionomycin, avoid pipetting the compounds into serum free media or directly onto plastic. Store the stock solutions in glass vials at –20°C.
4. Monensin is a Na$^+$ ionophore that blocks glycoprotein secretion. Brefeldin A is a carboxylated ionophore that prevents protein transport from the Golgi apparatus to the endoplasmatic reticulum. Monensin is more toxic than brefeldin A and should not be used for prolonged periods of time. In fact, addition of monensin for more than 5 h to activated T-cells reduces the yield of cytokine expressing T-cells. In contrast, brefeldin A can be added to cell cultures for as long as 30 h if necessary.
5. Try to add as small volumes as possible of PMA, ionomycin, and monensin as all of these agents are solubilized in organic diluents that might influence the cells' viablility and functionality.
6. Prior to using antibodies, do a quick spin to recover the whole volume in small vials. For optimal performance, always store fluorochrome labeled antibodies at 4°C in the dark. Optimal concentrations of the fluorochrome-labeled antibodies have to be determined experimentally by titration.
7. FITC-conjugated antibodies should be used for the more abundant proteins, whereas the less prominent cytokines should be targeted by PE-labeled conjugates.
8. As the cells are fixed, staining can be performed at room temperature. Nevertheless, in our hands staining on ice improves the quality of the staining pattern.
9. The nonspecific background staining of fixed and permeabilized cells is generally higher than that of viable cells. In our experience, the addition of mouse and rat serum prior to staining substantially reduces nonspecific staining. Moreover, the autofluorescence of fixed, activated cells should be controlled, in particular when Th cell subsets are present at very low frequencies. Other controls that

need to be performed are staining of nonactivated, permeabilized cells, staining of activated, nonpermeabilized cells, and staining in the presence of excess amounts of unlabeled anticytokine antibodies or recombinant cytokines (to control specificity of the labeled antibodies).

Acknowledgments

The authors thank Daniela Thein for excellent technical assistance and Irina Prots, Francis Dodeller and Andreas Ramming for critical discussion. This work was supported by the Deutsche Forschungsgemeinschaft (Grants Schu 786/2-2, 2-3, and 2-4) and by the Interdisciplinary Center for Clinical Research (IZKF) at the University Hospital of the University of Erlangen-Nuremberg (Project B27).

References

1. Mosmann, T. R., Cherwinski, H., Bond, M. W., Giedlin, M. A. and Coffman, R. L. (1986) Two types of murine helper T cell clone. I. Definition according to profiles of lymphokine activities and secreted proteins. *J. Immunol.* **136**, 2348–2357.
2. Abbas, A. K., Murphy, K. M., and Sher, A. (1996) Functional diversity of helper T lymphocytes. *Nature* **383**, 787–793.
3. Yssel, H., de Waal Malefyt, R., Roncarolo, M. G., et al. (1992) IL-10 is produced by subsets of human CD4+ T cell clones and peripheral blood T cells. *J. Immunol.* **149**, 2378–2384.
4. Sallusto, F., Lanzavecchia, A., and Mackay. C. R. (1998) Chemokines and chemokine receptors in T-cell priming and Th1/Th2-mediated responses. *Immunol. Today* **19**, 568–574.
5. Nanki, T. and Lipsky, P. E. (2000) Cytokine, activation marker, and chemokine receptor expression by individual CD4(+) memory T cells in rheumatoid arthritis synovium. *Arthritis Res.* **2**, 415–423.
6. Andersson, U., Andersson, J., Lindfors, A., Wagner, K., Moller, G., and Heusser, C. H. (1990) Simultaneous production of interleukin 2, interleukin 4 and interferon-gamma by activated human blood lymphocytes. *Eur. J. Immunol* **20**, 1591–1596.
7. Jung, T., Schauer, U., Heusser, C., Neumann, C., and Rieger, C. (1993) Detection of intracellular cytokines by flow cytometry. *J. Immunol. Methods* **159**, 197–207.
8. Prussin, C. and Metcalfe, D. D. (1995) Detection of intracytoplasmic cytokine using flow cytometry and directly conjugated anti-cytokine antibodies. *J. Immunol. Methods* **188**, 117–128.
9. Sander, B., Andersson, J., and Andersson, U. (1991) Assessment of cytokines by immunofluorescence and the paraformaldehyde-saponin procedure. *Immunol. Rev.* **119**, 65–93.
10. Picker, L. J., Singh, M. K., Zdraveski, Z., et al. (1995) Direct demonstration of cytokine synthesis heterogeneity among human memory/effector T cells by flow cytometry. *Blood* **86**, :1408–1419.

11. Schulze-Koops, H., Lipsky, P. E., and Davis, L. S. (1998) Human memory T cell differentiation into Th2-like effector cells is dependent on IL-4 and CD28 and inhibited by TCR ligation. *Eur. J. Immunol.* **28**, 2517–2529.
12. Skapenko, A., Wendler, J., Lipsky, P. E., Kalden, J. R., and Schulze-Koops, H. (1999) Altered memory T cell differentiation in patients with early rheumatoid arthritis. *J. Immunol.* **163**, 491–499.
13. Maino, V. C. and Picker, L. J. (1998) Identification of functional subsets by flow cytometry: intracellular detection of cytokine expression. *Cytometry* **34**, 207–215.
14. Skapenko, A., Lipsky, P. E., Kraetsch, H. G., Kalden, J. R., and Schulze-Koops, H. (2001) Antigen-independent Th2 cell differentiation by stimulation of CD28: regulation via IL-4 gene expression and mitogen-activated protein kinase activation. *J. Immunol.* **166**, 4283–4292.
15. Dimitrova, P., Skapenko, A., Herrmann, M. L., Schleyerbach, R., Kalden, J. R., and Schulze-Koops, H. (2002) Restriction of de novo pyrimidine biosynthesis inhibits Th1 cell activation and promotes Th2 cell differentiation. *J. Immunol.* **169**, 3392–3399.
16. Skapenko, A., Leipe, J., Niesner, U., et al. (2004) GATA-3 in Human T Cell Helper Type 2 Development. *J. Exp. Med.* **199**, 423–428.
17. Coffman, R. L., Lebman, D. A., and Rothman, P. (1993) Mechanism and regulation of immunoglobulin isotype switching. *Adv. Immunol.* **54**, 229–270.

8

Analysis of the T-Cell Receptor Repertoire of Synovial T-Cells

Lucy R. Wedderburn and Douglas J. King

Abstract

The recognition of a wide diversity of antigens by lymphocytes is made possible by the expression of a large range of highly variable antigen specific receptors, coded for by tandem arrays of genes, which undergo rearrangement during T- and B-cell development. The study of T-cell receptor (TCR) diversity and clonal composition of mixed T-cell populations has taken advantage of the features of the TCR molecule in various ways. This chapter focuses on the study of T-cells obtained from the synovial fluid of patients with inflammatory arthritis. Methods to process and store the samples and to separate cell populations are described. Two alternative molecular methods to analyse TCR diversity, identify clonal expansions, and track specific T-cell populations over both time and location are also detailed.

Key Words: TCR diversity; repertoire; synovial fluid mononuclear cells (SFMC); clonal expansions; autoimmunity; CDR3; spectratyping; heteroduplex.

1. Introduction

Recognition of a vast potential array of foreign antigens by T-lymphocytes is achieved through the expression of a highly variable cell surface receptor for antigen, the T-cell receptor (TCR) *(1)*. In human peripheral blood approx 90% of T-cells bear the αβ heterodimer of TCR chains, while a minority of cells bear the γδ TCR. The diversity of the actual expressed TCR repertoire has been estimated to involve at least 2.5×10^7 and possibly up to 10^8 different TCR *(2,3)*, which are coded for by rearranging genes arranged in tandem arrays of V, D, and J genes, in a manner that is highly homologous to the B-cell receptor genes for immunoglobulin *(4)*. In contrast to B-cells, however, T-cells do not undergo class switching or affinity maturation and thus a TCR sequence for any particular clone of T-cells is both unique and unchanging for the lifespan of the clone (*see* **Note 1**). These features, combined with the fact that the highly variable sequence across the VDJ junction (also referred to as the complementarity determining region 3 or CDR3) of the TCR arises from a unique rearrangement event, make this sequence of DNA (of between approx 15–36 nucleotides long) a "molecular fingerprint" for a particular T-cell clone. The methods of analysis of TCR expression outlined in this chapter rely upon this small and specific sequence, and most focus on the TCR-β chain transcripts. The reason for the choice of the TCR-β chain is that allelic exclusion for TCR-β is tight, meaning that the majority of αβ+ T-cells express only one TCR-β protein. In contrast, in human T-cells, up to 20% of peripheral blood TCR-αβ+ T-cells may express two TCRα chains *(5)*, making analysis by molecular techniques which rely upon reverse transcriptase polymerase chain reaction (RT-PCR) difficult to interpret unless a single cell PCR approach is used. Studies using CDR3 analysis on either TCR-γ or TCR-δ chains have also been published but few in the field of arthritis research. Because the full sequencing of the human TCR genomic sequence and subsequent completion of the human genome sequencing, a new nomenclature has been proposed for TCR *(6)* (*see* http://imgt.cines.fr/), in which all T-cell receptor genes are designated TR with TRAV for TCR-α variable genes, TRBV for TCR-β variable genes (*see* **Note 2** for full discussion of nomenclature).

In this chapter, methods for the preparation of synovial fluid samples for TCR repertoire analysis will be covered, and then two molecular techniques to analyse this repertoire will be described. Both are based upon PCR amplification of cDNA prepared from T-cells, using a set of TRBV (V gene)- specific primers and a single TRBC primer. There are two constant region genes in the human TRBC locus, but the primer used is downstream to the polymorphisms between these two gene segments and should therefore amplify TCR rearrangements that use either TRBC gene segment. Heteroduplex (HD) analysis *(7)* separates TCR according to CDR3 sequence, whereas TCR spectratyping (ST), with or without TRBJ run off, is a CDR3 length based assay, sometimes also known as Immunoscope *(8)*. With either of these approaches it is advisable to separate CD4 and CD8$^+$ T-cells first, if possible, since clonal size and biology is very different between these two T-cell populations *(9)*. This is feasible with samples of 5×10^6 or more, such as samples of peripheral blood mononuclear cells (PBMCs) or synovial fluid mononuclear cells (SFMCs). Analysis from synovial biopsy tissue is not described here: however, these methods would in principle also be applicable to T-cells derived from biopsy material, when adequate cell numbers can be obtained. Although these molecular methods are highly sensitive, we advise aiming for a cell number of no less than 1×10^6 in each of the CD4 and CD8 populations in order to ensure a full representation of the repertoire (*see* **Note 3**).

For a rapid screen of TCR repertoire expression at the level of whole TCR-β families, monoclonal antibodies (mAbs) are commercially available which are specific for different TCR-β proteins, and which now cover over 70% of the TRBV repertoire (note that antibodies to the TCR molecules are still numbered according to the old nomenclature (*see* **Notes 2** and **4**). In our hands, three color analysis using antibodies to CD3, CD4, or CD8, and one TCR-β specific antibody per tube allows staining the whole repertoire of both CD4 and CD8⁺ T-cell populations on approx 1×10^6 cells in total *(10)*. Recently studies have been published using whole blood with very small cell numbers stained with premixed cocktails of anti-TCR-β mAbs *(11)*. However analysis by flow cytometry demonstrates only the overall expression of a TCR-β family (or group), and gives no information concerning the clonality of these cells and therefore the diversity of TCR expression. In contrast the molecular assays detailed here may provide information concerning TCR diversity, clonality, clone size and even TCR identity. In some instances absolute confirmation of clone identity may be needed, and if so, fast throughput DNA sequencing of PCR products is required.

2. Materials

2.1. Processing of Mononuclear Cells From Synovial Fluid Samples

All reagents are sterile.

1. Hyaluronidase (Sigma): Make 1000X stock in phosphate buffered saline (PBS) at 10,000 U/mL. Filter sterilize and freeze in aliquots at –20°C.
2. Preservative-free heparin (Monoparin, 1000 U/mL) (CP Pharmaceuticals).
3. Culture medium (CM): RPMI medium, supplied containing 2 m*M* L-glutamine and with 100U/mL penicillin, and 100 µg/mL streptomycin (Life Technologies).
4. Heat inactivated fetal calf serum (FCS).
5. Density centrifugation medium eg Lymphoprep (Axis Shield, Nycomed).
6. Dimethyl sulfoxide (DMSO) (Sigma).

2.2. Separation of Synovial T-Cells for TCR Analysis

1. Magnetic beads reagents for negative selection of CD4⁺ T-cells (Miltenyi CD4 T-cell isolation kit; cat. no. 130-091-155) or CD8⁺ T-cells (Miltenyi CD8 T-cell isolation kit; cat. no. 130-091-154). Store at 4°C.
2. Buffer for bead separation: PBS, 0.5% BSA, and 2 m*M* ethylene diamine tetraacetic acid (EDTA). Filter and degass 0.22 µm before use. Store at 4°C.
3. Columns for separation: for up to 10×10^6 cells, MS columns (Miltenyi; cat. no. 130-042-201); for more than 10×10^6 cells, LS columns, (Miltenyi; cat. no. 130-042-401).
4. Cell separator unit (Miltenyi; cat. no. 130-042-302) and stand (Miltenyi; cat. no. 130-042-303) for appropriate columns.
5. FACS buffer : PBS, 1% FCS, and 0.1% Na azide.

6. mAbs specific for human CD3, CD4, and CD8, each directly conjugated with different fluorochromes
7. RNAzol (Biogenesis): store at 4°C.
8. Chloroform (Sigma).
9. Isopropanol (Sigma)
10. Reagents for 1st strand cDNA synthesis, all made up in RNAse-free sterile dH_2O: oligo dT (Boehringer) stock at 1 mg/mL, dNTPs stock, each at 5 mM (Promega), and reverse transcriptase (Superscript) with buffer and DTT as supplied (Life Technologies).

2.3. Heteroduplex Analysis

The principle of heteroduplex (HD) analysis of TCR expression is to perform TRBV-specific PCR, and then for each reaction to denature PCR products at 95°C and allow them to reanneal at 50°C. Under these conditions, because the 5'- and 3'-ends of PCR products in each reaction are identical, but sequences across VDJ junctions differ for each unique TCR transcript, the products form a large number of heteroduplices in which the central sequences are non-complementary. This assay was originally described to analyze γδ T-cells *(12)*. The method was subsequently modified by Casorati et al. *(7,13)* by the use of a standard "carrier" DNA for each TRBV, which is produced separately by PCR and mixed at a standard concentration with the corresponding sample PCR product. Because the carrier DNA is in excess, the majority of heteroduplices then consist of one strand from the carrier DNA and one from a "sample" PCR product; the carrier also forms its own homoduplex. When run on a nondenaturing PAGE gel these heteroduplex molecules are differentially retarded according to sequence, not size. In a diverse, polyclonal T-cell population this produces a smear as each T-cell clone is at low frequency, with a dense band representing carrier homoduplex at the bottom of each lane whereas T-cell clones above a frequency of approximately 1 cell in 5000 *(14)* are detected as bands (*see* **Fig. 1**). The modification using a fixed carrier DNA includes the use of a specific 3'-TRBC primer for the carrier reactions (known

---

Fig. 1. (**A**) Schematic diagram of modified heteroduplex analysis. 26 TRBV- specific PCR reactions are carried out on sample cDNA and in parallel carrier TCR products are generated using the same BV primers but a down stream TRBC primer. For each TRBV samples are mixed, heated and annealed to allow heteroduplex formation and then run on a nondenaturing polyacrylamide gel. The gel is then probed with the downstream TRBC primer (shown in bold line). (**B**) Typical HD results of normal peripheral blood T-cells, from two healthy children (*left* and *middle panel*) and one healthy adult (*right panel*). Homoduplex carrier DNA, which is in excess, runs as a heavy band at the lower margin of the gel while clones are detected as bands within the

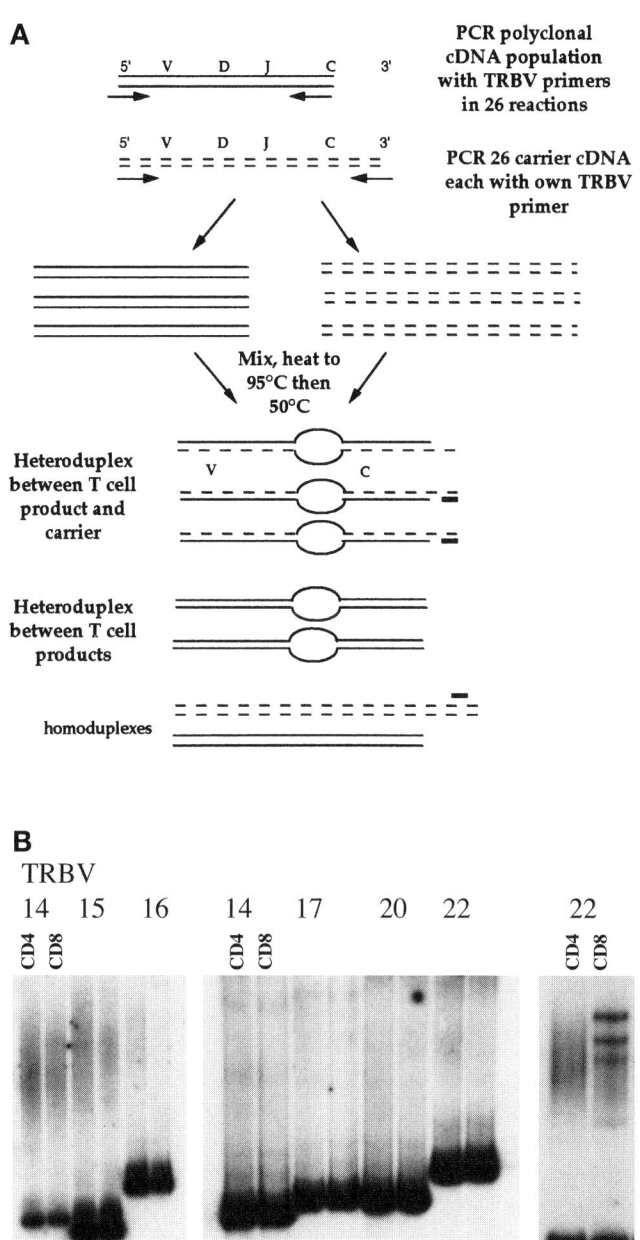

HD smear. For each sample T-cells have been separated into CD4[+] and CD8[+] cells and then used to make cDNA for TRBV- specific PCR. In the case of the adult PBMC (*right panel*) expanded clones are detected within CD8[+] T-cells expressing TRBV22. (TRBV designations are according to nomenclature in *ref. 16*).

as the external BC primer), which is downstream to the TRBC primer used in sample PCR; HD gels are then probed with this external TRBC primer allowing only detection of heteroduplices containing one strand of carrier and one of sample DNA. This strategy increases sensitivity of the HD assay and we have shown that this modified HD assay is more sensitive (though also more labor intensive) than spectratyping when using a TRBC run-off, for the detection of small clones in a mixed population *(14)*. Note that TRBJ run off increases the sensitivity of the spectratyping assay. A recent modification of the HD approach uses a fluorescently labeled TRBC primer to amplify carrier DNA *(15)*. This modification simplifies the assay since it removes the need to blot and probe gels, and analysis can be automated.

1. PCR primers: a set of 26 TRBV gene specific and two TRBC gene primers are used (high performance liquid chromatography [HPLC] purified). For sequences of primers *see* **Table 1**. These were designed using sequences and nomenclature in Arden et al. *(16)*. The equivalent TRBV gene group in the new nomenclature is also shown in **Table 1.** Keep primers at 1 mg/mL stock in dH$_2$O; make aliquots at 12.5 µM and freeze at –20°C.
2. Carriers : a set of 26 carrier DNA constructs (kindly provided by G. Casorati) (*see* **Note 5**), each containing a rearranged TCR with the relevant TRBV gene. Each carrier is prepared as plasmid DNA; carrier stock DNA are stored at 1 to 5 µg/ µL at –80°C and glycerol stocks are made and stored at –80°C.
3. Standard PCR reagents : mixed dNTPs each at 5 mM (Promega), stock MgCl$_2$ at 25 mM, Taq polymerase (Promega) with buffer as supplied. Where sequence will later be required from the same PCR products we advise the use of a high fidelity polymerase such as pfu (Stratagene). Thermofast skirted plates and Adhesive PCR film (both Abgene) for PCR.
4. Agarose, Tris, Borate, EDTA (TBE) and ethidium bromide for gel electrophoresis *(17)*.
5. Acrylamide 30%: Bisacrylamide 0.8% (Protogel, National Diagnostics), tetramethylethylenediamine (TEMED) (Sigma) and ammonium persulphate (APS)(Sigma). For 2 large gels make 80 mL of mix: 32 mL acrylamide (30/ 0.8%), 4 mL 10X TBE, 43.5 mL H$_2$O, 400 µL 10% APS, and 40 µL TEMED.
6. Large gel electrophoresis cells, tank, casting stand, clamps, combs, and cooling system to allow running of about 54 samples (i.e., two complete repertoire sets) or more, together (for example Biorad Protean ll vertical electrophoresis system), HD gel loading buffer: 40% sucrose, 0.25% xylene cyanol, and 0.25% bromophenol blue.
8. 20X saline sodium citrate (SSC): 175.3 g NaCl , 27.6 g NaH$_2$PO$_4$ · H$_2$O and 7.4 g EDTA in 800 mL H$_2$O. Adjust to pH 7.4 with NaOH (approx 6.5 m: of a 10N solution), and then adjust volume up to 1 L with dH$_2$O.
9. Standard reagents for blotting onto nitrocellulose and probing with oligonucleotide *(17)* using the TRBC external oligonucleotide, using either radioactively end-labeled probe or a biotinylated probe and commercially available detection system such as Bright Star (Ambion).

Table 1
Primer Sequences (5′–3′) for TRBV and TRBJ Gene PCR Used in HD and ST Analyses[a]

TRBV or TRBJ designation early nomenclature	Primer sequence for HD analysis	Primer sequence for ST analysis	TRBV subgroup or TRBJ gene in new nomenclature
1	GCACAACAGTTCCCTGACTTGCAC	CCGCACAACAGTTCCCTGACTTGC	9
2	TCATCAACCATGCAAGCCTGACCT	CACAACTATGTTTTGGTATCGTC	20
3	GTCTCTAGAGAGAAGAAGGAGCGC	CGCTTCTCCCTGATTCTGGAGTCC	28
4	ACATATGAGAGTGGATTTGTCATT	TTCCCATCAGCCGCCCAAACCTAA	29
5	5.1: ATACTTCAGTGAGACACAGAGAAAC 5.2: TTCCCTAACTATAGCTCTGAGCTG	GATCAAAAACGAGAGACAGC	5
6	AGGCCTGAGGGATCCGTCTC	6a: GATCCAATTTCAGGTCATACTG 6b: a 1:1:2 mixture of: CAGGGCCCAGAGTTTCTGAC CAGGGGCCAGAGTTTCTGAC and CAGCTGCGAGAGTTTCTGAC	7
7	CCTGAATGCCCCAACAGCTCTC	CCTGAATGCCCCAACAGCTCT	4
8	CCATGATGCGGGGACTGGAGTTGC	GGTACAGACAGACCATGATGC	12
9	CCTAAATCTCCAGACAAAGCT	TTCCCTGGAGCTTGGTGACTCTGC	3
10*	CTCCAAAAACTCATCCTGTACGTT	*	21
11	TCAACAGTCTCCAGAATAAGGACG	GTCAACAGTCTCCAGAATAAGG	25
12	GATACTGACAAAGGAGAAGTCTCAGAT	TCC(C/T)CCTCACTCTGGAGTC	10

(*continued*)

Table 1 *(continued)*

TRBV or TRBJ designation early nomenclature	Primer sequence for HD analysis	Primer sequence for ST analysis	TRBV subgroup or TRBJ gene in new nomenclature
13.1 (13a)	CAAGGAGAAGTCCCCAAT	13a :GGTATCGACAAGACCCAGGCA	6
13.2 (13b)	GGTGAGGGTACAACTGCC	13b: AGGCTCATCCATTATTCAAATAC	6
14	GTCTCTCGAAAAGAGAAGAGGAAT	GGGCTGGGCTTAAGGCAGATCTAC	27
15	AGTGTCTCTCGACAGGCACAG	CAGGCACAGGCTAAATTCTCCCTG	24
16	AAAGAGTCTAAACAGGATGAGT	GCCTGCAGAACTGGAGGATTCTGG	14
17	CAGATAGTAAATGACTTTCAG	TCCTCTCACTGTGACATCGGCCCA	19
18	GATGAGTCAGGAATGCCAAAGGAA	CTGCTGAATTTCCCAAAGAGGGCC	18
19*	CAATGCCCCAAGAACGCACCCTG	*	23
20	AGCTCTGAGGTGCCCCAGAAT	TGCCCCAGAATCTCTCAGCCTCCA	30
21	ATTCACAGTTGCCTAAGGATGA	GGAGTAGACTCCACTCTCAAG	11
22	GGGCAGAAAGTCGAGTTTCTGGTT	GATCCGGTCCACAAAGCTGG	2
23	TTTATGAAAAGATGCAGAGCGAT	ATTCTGAACTGAACATGAGCTCCT	13
24	AAGTCAAGTCAGGCCCCAAAGCT	GACATCCGCTCACCAGGCCTG	15
TRBC internal (for HD)	CACCCACGAGCTCAGCTCCACGTGGTC	N/A	
TRBC external (for HD)	TGCTGACCCCACTGTCGACCTCCTTCCCATT	N/A	

104

TRBC for first round PCR (for ST)	N/A	GGGTGTGGGAGATCTCTGC
Labeled TRBC run off primer (for ST)	N/A	ACACAGCAGCCTCGGGTGGG
TRBJ primers for ST run off:	N/A	
1.1	N/A	ACTGTGAGTCTGGTGCCTTGT	1.1
1.2	N/A	ACAACGGTTAACTTGGTCCCGAA	1.2
1.3	N/A	GGTCCTCTACAACAGTGAGCCAAC	1.3
1.4	N/A	AAGAGAGAGAGCTGGGTTCCACTG	1.4
1.5	N/A	GGAGAGTCGAGTTCCATCA	1.5
1.6	N/A	TGTCACAGTGAGCCTGGTCCCATT	1.6
2.1	N/A	CCTGGCCCGAAGAAACTGCTCA	2.1
2.2	N/A	GTCCTCCAGTACGGTCAGCCTAGA	2.2
2.3	N/A	TGCCTGGGCCAAAATACTGCG	2.3
2.4	N/A	TCCCCGCGCCGAAGTACTGAA	2.4
2.5	N/A	TCGAGCACCAGGAGCCGC	2.5
2.6	N/A	CTGCTGCCGCCCCGAAAGTC	2.6
2.7	N/A	TGACCGTGAGCCTGGTGCCCG	2.7

[a]Far left column shows designations of TRBV family/subfamily, on which these primers were based (16); far right column shows the TRBV subgroup designation of the same genes in the new nomenclature (6) (http://imgt.cines.fr/).
*See **Note 8**.

2.4. Spectratyping

CDR3 spectratyping (ST) allows a broad analysis of a T-cell compartment when a TRBC primer is used for the run-off labeling reaction, along with a far more sensitive analysis when TRBJ or even clonotypic primers are used for the run-off reaction. This allows a "focusing" of the assay, enabling full analysis of any CDR3 lengths of interest. First round PCRs each specific for a TRBV family are conducted. We have found it efficient to multiplex these into two families/reaction, cutting down on reagents and allowing sparse cDNA samples to be fully utilized. The amplicons from these first round PCRs are then subject to a single labeled primer linear PCR. This single labeled primer is specific to either the TRBC region or a TRBJ region, depending on the level of resolution required. The labeling system used depends upon the equipment available. Ideally, a sequencing automate such as one of the ABI Prism (Applera) sequencers or a MegaBace (Amersham) should be utilized, as these allow accurate sizing of the CDR3 chains. For these machines primers commercially produced with 5' additions of FAM, HEX, or NED are suitable. Radioactive labeling of the run-off primer and PAGE will still yield results though they will be less informative. Following electrophoresis through whichever equipment is selected, analysis of data can be conducted, intensity of signal at a given base pair length being relative to the number of transcripts within that family. Several software packages are available for this analysis, Immunoscope (Pasteur), Genetic Analyser (Amersham), GeneScan and Genotyper (Applera) being the most popular.

1. PCR Primers: a set of 25 TRBV-specific primers, first round TRBC primer and labeled run-off primers (TRBC or TRBJ), all HPLC purified (**Table 1**). Other primers can be used but we have found these to be reliable. They correspond to those utilized in Pannetier et al. *(8)*, and were derived according to nomenclature in Arden et al. *(16)*. Store at −20°C at a concentration of 50 pmoles/µL.
2. Standard PCR reagents (Promega). Taq DNA polymerase in Storage buffer B (M1665). 96-well Thermofast skirted plates and adhesive PCR film (both Abgene).
3. Sequencing automate for electrophoresis with Long read matrix and LPA (Amersham) in the case of the Megabace, or deionized Formamide and Performance Optimized Polymer 4 (POP4, Applera) in the case of an ABI Prism automate.
4. Size standard ET400-R (25-0205-01 Amersham): a ROX labeled size reference which works well for the sequencing automates.
5. If a polyacrylamide gel is to be cast for radiolabeled products then for 100 mL of a 5% gel the following is required: 42 g urea; 16.6 mL acrylamide 30%; bisacrylamide 0.8% premix (Protogel, National Diagnostics).

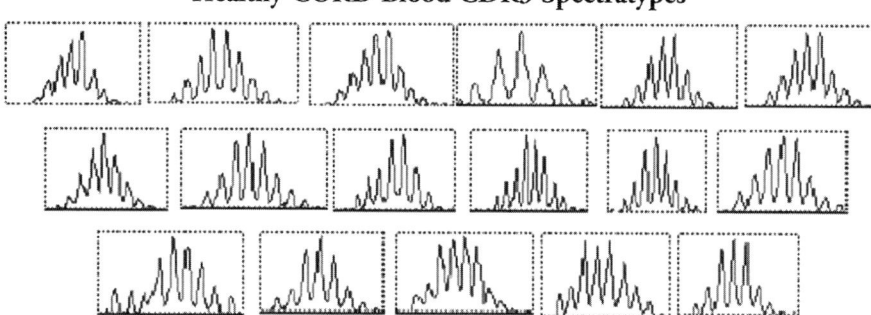

Fig. 2. Data from spectratyping (ST) analysis of TCR expression in human cord blood, showing typical Gaussian distribution of PCR products analysed by amplicon length.

3. Methods

The choice of method to analyse TCR expression and T-cell clonal diversity of clinical samples should be made according to the information required. CDR3 length based assays such as ST generate data on the frequency of TCR transcripts of specific lengths based upon differences in CDR3; in diverse and polyclonal T-cell populations these lengths are normally distributed around a modal length whereas in the presence of large clones this Gaussian distribution is distorted or even lost altogether (*see* **Fig. 2**). This strategy is sensitive to about 1 cell in 1000 but may tend to under represent clones of medium frequency whose CDR3 lengths are around the modal length for the TRBV family. ST can now be performed on fast throughput DNA-analysis equipment such as that used for genotyping. In contrast, HD provides information based upon CDR3 sequence and can be used to confirm identity of clones, for example, TCR expressed in samples which differ over time or in locality (*see* **Fig. 3**) *(18–20)*.

RT-PCR TRBV based assays assume that all T-cells contain an equal amount of TCR -specific mRNA: this is incorrect since activation and division of T-cells is associated with increased transcriptional activity of the TCR loci *(21)*. Therefore these assays may yield an overestimate of the size of individual T-cells that are activated or actively dividing, in the context of the whole popu-

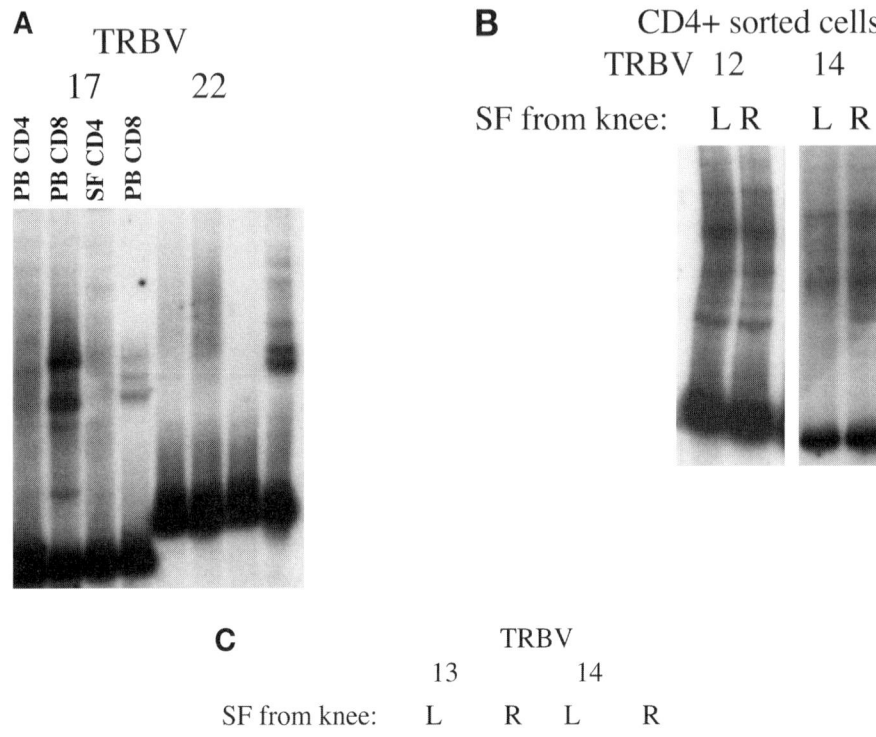

Fig. 3. Heteroduplex (HD) analysis of synovial fluid T-cells from three children with juvenile idiopathic arthritis (JIA). (TRBV designations are according to nomenclature *[16]*). (**A**) HD analysis of CD4 and CD8 cells from PBMCs and SFMCs for TRBV17 and 22. In both TRBV families, clonal expansions are apparent in the $CD8^+$ SFMC, and these are nonoverlapping with those in the PBMCs $CD8^+$ cells. (**B**) Clonal expansions detected in $CD4^+$ cells from SFMC are identical in two joints (right [R] and left [L] knees). (**C**) Clonal expansions in SFMC (unseparated) from two knees and at two time points show that some clones survive over 1 yr in vivo and that clonal hierarchy is identical between the two inflammatory sites. Reprinted with permission from **ref.** *(20)*.

lation. Some protocols include a control PCR reaction for the TRBC region for normalisation, to address this issue.

3.1. Processing and Preparation of Synovial Fluid Samples

1. Collect synovial fluid obtained fresh from joint aspiration into a sterile tube containing preservative free heparin (1 U/mL synovial fluid) and process within 2 h. Where possible paired venous blood should be collected and processed in parallel.
2. Dilute synovial fluid 1 in 2 in CM supplemented with 1 U/mL heparin, and incubate at 37°C for 30 min in the presence of 10 U/mL hyaluronidase. Venous blood does not require hyaluronidase treatment or additional heparin.
3. Prepare SFMC by standard density centrifugation, in parallel with the paired blood sample.
4. When cells are to be stored, wash and count SFMC and PBMC and resuspend at 5 to 10×10^6 cells /mL of FCS containing 10% DMSO for freezing.

3.2. Preparation of Synovial T-Cell Populations and cDNA for TCR Analysis

Stimulation of T-cells in vitro may alter levels of mRNA for TCR. Therefore for tracking and identification of T-cell clones ex vivo, minimal culture is advisable. A starting sample of 5×10^6 or more SFMC cells is required for TCR analysis. The final number of CD4 or $CD8^+$ T cells for the TRBV PCR-based assays need to be $>10^6$/sample (*see* **Note 3**), and samples need to be carefully matched for T-cell numbers. If large SFMC cell numbers are available separate samples can be processed by negative selection on magnetic beads to produce purified samples of $CD4^+$ and $CD8^+$ T-cells. This is ideal in that cells used to prepare cDNA have not received a signal through ligation by anti-TCR mAbs. However when SFMC numbers are limiting it is also possible to select $CD4^+$ T-cells by positive selection, and then use the remaining cells, which are enriched for $CD8^+$ T-cells (and negatively selected) to prepare cDNA for $CD8^+$ analysis (*see* **Note 6**). In either case FACS analysis before and after bead separation is required and samples are adjusted to equal number of T-cells (not total cells)/tube. The cDNA is then most useful specifically for amplification of T-cell-specific genes but not other cell types.

1. Cell counts should be recorded at each stage for calculation of yield.
2. When SFMCs are used in separation protocols they may be more difficult to handle than standard PBMCs (for which the columns are designed), as they have increased adherence and may clump easily. To avoid this, an extra wash in EDTA-containing buffer after thawing can be included. Large clumps should be removed through a wide pore filter before processing.
3. Before cell separation analyse a small aliquot of SFMCs by standard FACS analysis for CD3, 4, and 8^+ cells.

4. From two equal aliquots of SFMC (ideally each over 5×10^6 cells) prepare $CD4^+$ or $CD8^+$ T-cells by negative selection using Miltenyi Minimacs systems according to manufacturers instructions
5. Remove small aliquots (10^5) of the cells purified by negative selection to stain for CD3, 4, and 8 in each sample (in parallel with cells from **step 1**). Carry out standard 3 color FACS analysis and calculate the actual number of $CD3^+4^+$ or $CD3^+8^+$ cells in each tube
6. Aliquot samples containing 2×10^6 $CD4^+$ or $CD8^+$ purified T-cells into sterile Eppendorfs; spin and resuspend in 800 µL RNAzol; samples can be stored at –80°C if necessary at this point
7. Prepare total RNA according to the manufacturers instructions; if required remove $1/20^{th}$ of RNA for quality check, for example on Agilent analyzer.
8. Proceed to first strand synthesis of cDNA using the whole RNA sample, according to manufacturer's instructions
9. Resuspend final cDNA in 60 µL of sterile H_2O which allows two complete HD reactions to be carried out (*see* **Note 7**). Freeze in 30 µL aliquots at –80°C or proceed immediately.

3.3. Modified Heteroduplex Analysis of TCR Repertoire

3.3.1. Preparation of Carrier Plasmid Stocks

Carrier DNA stocks can be made once and stored such that they will provide ample material for many experiments.

1. Plate 26 carrier constructs from glycerol stocks onto L-amp agar plates and incubate overnight at 37°C.
2. For each carrier pick single colonies into L-amp containing broth, prepare standard mini-prep DNA for each carrier DNA, and verify correct insert by restriction digestion *(17)*.
3. Carry out 26 PCR reactions, one for each carrier, using approx 25 ng of plasmid DNA as template, TRBV primer (one per reaction) with TRBC external primer, as listed in **Table 1**. Concentrations of reagents for PCR are : 200 nM dNTPs, 1.5 mM Mg^+, and 250 nM of primer.
4. Adjust carrier PCR products to 200 ng/µL in H_2O. These stocks can be kept at 4°C for approx 2 wk, or aliquoted and frozen at –20°C for storage.

3.3.2. PCR Reactions of Sample cDNA for HD Analysis

Preparation of plates, master mix and PCR reactions are performed in a flow cabinet to reduce risk of contamination.

1. Prepare PCR plates with rows of TRBV specific primers, (**Table 1**) plating 5 µL of 2.5 µM primer per well (final concentration of 250 nM in 50 µL PCR reaction); for controls include one well with any TRBV primer and one well with H_2O alone. Because there are 26 TRBV reactions, and several controls, we routinely

set up 3 sets of PCR/plate in a fixed pattern. Primer plates can be prepared in batches, covered and frozen at –20°C.
2. Prepare PCR mix (buffer, dNTPs, MgCl$_2$, TRBC internal primer to give final primer concentrations of 250 nM, dNTPs of 200 nM, and Mg of 1.5 mM) and Taq (or high fidelity polymerase), enough for 30 reactions with final reaction volume of 50 µL.
3. Remove master mix into no-cDNA control well and then add 29 µL of sample cDNA to master mix.
4. Plate out master mix onto pre-prepared PCR plate containing TRBV primers taking great care to avoid contamination.
5. Perform TRBV-specific PCR in a 96-well format, including controls with no cDNA or no primer. PCR conditions: 95°C for 5 min, followed by 30 cycles of 95°C for 1 min, 57°C for 1 min, and 72°C for 1 min; with a final extension period at 72°C for 10 min.
6. Check 5 µl of products on a 1% agarose / TBE gel.
7. Products may be frozen at –20°C, or proceed straight to HD reaction.

3.3.3. Heteroduplex (HD) Reactions

1. Mix 20 µL of each PCR product with 200 ng (1 µL) appropriate HD carrier DNA. A 96-well plate format of same design as PCR is used.
2. HD reaction: 95°C for 5 min, 50°C for 60 min, and then keep at 4°C.
3. Mix each reaction with 4 µL of loading buffer and load onto a 12% acrylamide 0.5% TBE gel.
4. Wherever possible run all samples from one T-cell preparation together: for direct comparison of sampling over time or from different sites, load samples in adjacent wells (*see* **Fig. 3**).
5. Run at 4°C (if tank does not have cooler system built in, run in a cold room) overnight. Optimize to allow carrier homoduplex bands to reach lower edge of gel. Stain with ethidium bromide and visualize.
    Rinse in 2X SSC and then blot onto Hybond N+ in 20X SS.
    Fix in 0.4M NaOH. Rinse and dry filters.
    Membranes are then probed by standard methods (radioactive or using biotin labeled probe) generally using the TRBC external primer as the probe (**Table 1**). For radioactive labeling of TRBC external oligo we use kinase labeling of 400 ng probe for 2 large gels, and hybridize in 6X SSC/5X Denharts solution. Washes are then 5X SSC, 0.1% SDS followed by 1X SSC, and 0.1% SDS, each for 20 min. When a specific T-cell clone is under investigation, and if VDJ sequence is known, an N-region probe can also be used.

3.4. Spectratyping Analysis of TCR Repertoire

3.4.1. First Round PCR

1. Pipette TRBV primers (12.5 pmol/ primer) into 12 tubes/cDNA sample as per **Table 2**. Make volume of each well up to 15 µL with H$_2$O.

Table 2
Multiplex Array for PCR Reactions for First Round PCR in ST Analysis

Well number	TRBV* primer (expected size, bp of a 10 aa CDR3 length)
1	**9** (146), **18** (189)
2	**23** (165), **6b** (266)
3	**16** (146), **4** (192)
4	**3** (166), **13a** (282)
5	**22** (152), **11** (194)
6	**21** (172), **8** (285)
7	**24** (154), **20** (204)
8	**15** (178), **2** (302)
9	**17** (161), **13b** (253)
10	**7** (187), **6a** (317)
11	**12** (162), **14** (265)
12	**1** (189), **5** (350)

*Primer sequences as in Table 1 according to nomenclature in *ref. **16***.

2. Prepare master mix for each cDNA sample as follows. 34.1 μL 10X Taq buffer, 27.5 μL 25 m*M* MgCl$_2$, 27.5 μL of 2.5 m*M* each mixed dNTPs, 300 pmoles of TRBC primer, 3 μL of Taq polymerase 5 U/μL (Promega). Make up to a volume of 127.5 μL with cDNA and H$_2$O.
3. Pipette 10 μL of master mix into each of the wells containing TRBV primers.
4. Conduct PCR with the following program: 30 s at 95°C; followed by 40 cycles of 25 s at 95°C, 45 s at 60°C, and 45 s at 72°C; with a final extension of 72°C of 5 min.
5. Store plate at –20°C, or continue immediately to run-off labellng reaction (*see* **Subheadings 3.4.2. or 3.4.3.**).

3.4.2. Run-off Labeling Reaction Using TRBC Primer

1. Labeling master mix for global analysis of the T-cell repertoire by conducting the run-off reaction with a TRBC primer: 13.75 μL of 10X Taq buffer, 16.25 μL of 25 m*M* MgCl$_2$, 11 μL of 2.5 m*M* each mixed dNTPs, 13.75 pmoles of labeled TRBC primer, 0.5 μL of 5 U/μL Taq polymerase and make up to a volume of 111.9 μL with H$_2$O.
2. Dispense 8 μL of master mix into 12 wells as per first round PCR.
3. Transfer 2 μL from each first round PCR into a run-off labellng reaction.
4. Conduct PCR with the following programme: 30 s at 95°C; followed by 10 cycles of 25 s at 95°C, 45 s at 60°C, and 45 s at 72°C; with a final extension of 72°C for 5 min.
5. Store products at –20°C, or run immediately on a sequencing automate.

3.4.3. Run-Off Labeling Reaction Using TRBJ Primer

1. For visualization at the TRBJ level perform the following run-off labeling procedure. Make a master mix for each first round reaction, using 13.75 μL of 10X Taq

buffer, 16.25 µL of 25 m*M* MgCl$_2$, 11 µL of 2.5 m*M* each mixed dNTPs, 2 µL of first round PCR reaction, 0.5 µL of 5 U/µL Taq polymerase, and make up to a volume of 111.9 µL with H$_2$O.
2. Dispense 8 µL of each master mix into 13 wells.
3. To each of these wells add 1 pmole of labeled TRBJ primer, (**Table 2**) in 2 µL of H$_2$O to make a final volume of 10 µL/reaction.
4. Conduct PCR with the following programme: 30 s at 95°C, followed by 10 cycles of 25 s at 95°C, 45 s at 60°C, and 45 s at 72°C; with a final extension of 72°C for 5 min.
5. Store products at –20°C, or run immediately on a sequencing automate

3.4.4. Visualization of Spectratypes

1. 1 µL of each labeling reaction is loaded onto a sequencer, in 12 µL of formamide on an ABI Prism, with 0.5 µL of size standard or in 12 µL of 0.1% Tween-20, 0.35 µL of size standard on a Megabase. Alternatively, 2 µL of radiolabeled run-off is loaded onto a 5% denaturing manual gel.
2. CDR3 spectratypes from healthy cord blood are shown in **Fig. 2**. Each conforms closely to a Gaussian distribution as expected in a naïve T-cell population from a healthy individual. In contrast expanded clones, for example within SFMCs, distort this curve *(18)*. The reactions shown in **Fig. 2** have been conducted using a TRBC run-off; an example of TRBJ focussing can be seen in King et al. *(22)*.

4. Notes

1. Although it is generally true that T-cells do not alter their TCR DNA sequences after leaving the thymus, recent evidence has suggested that RAG genes can be re-expressed in peripheral T-cells in some conditions and that so-called receptor editing can occur *(23)*.
2. Nomenclature of TCR: the new proposed nomenclature has now been accepted by the HUGO Nomenclature committee and divides TCR-β chains into a total of 30 subgroups. Full details of this new nomenclature can be found at *http://imgt.cines.fr/*. In this system, all T-cell receptor genes start with the letters TR and genes coding for variable region of the TCR-β chain are TRBV. Note that subgroup designations are not based on the previous TRBV numbers at all (**Table 1**). As the international nomenclature becomes widely used, the ability to compare data using the old and new methods of TR gene classification will require careful reference to this site and related data bases.
3. 1×10^6 T-cells/tube are recommended for all TCR analysis because these methods rely on dividing the repertoire into at least 26 reactions for TRBV–specific PCR and may then require further division using TRJB run-off. Therefore each reaction may on average be from cDNA derived from approximately 35,000 T-cells: because the limit of detection of these methods is generally 1 cell in 5,000 to 1 in 10,000, starting with less than 30,000 cells/reaction could lead to PCR bias and false representation of the population.
4. The currently available mAbs specific for TCRβ chains are designated according to the old nomenclature (e.g., from Beckman Coulter, previously made by

Immunotech). These make it simple to compare to previously published repertoire studies, but may lead to confusion with studies using the IMGT nomenclature, if corresponding nomenclature is not used.
5. Carriers are in the form of a full length cDNA for a rearranged TRBV sequence, cloned into a Bluescript- based, or pUC- based construct: further information and requests can be made to Prof G. Casorati, Milan Italy at casorati.giulia@hsr.it
6. If cell numbers are very small and one population has to be used from the bead-ligated cells, it is recommended to use CD4 cells that are selected on anti-CD4 beads (Miltenyi, 130 045 101), whereas the $CD8^+$ beads are then enriched in the "flow through" cells. Samples are still normalised to equal numbers of T-cells/cDNA preparation. CD8 T-cells in general contain larger, expanded T-cell clones than $CD4^+$ cells and even a small number of $CD8^+$ cells in the $CD4^+$ sample may lead to the appearance of identical bands in both CD4 and CD8 lanes, resulting from contamination.
7. If cell numbers were low, cDNA is made to 30 µL volume and used for only one set of TRBV-PCR reactions.
8. Note that previous designations of TRBV10 (now 21) and TRBV19 (now 23) have subsequently been shown to be pseudo-genes. While rearranged transcripts can be detected which use these BV genes, these are excluded from some systems for this reason.

Acknowledgments

We thank Ms Alka Patel and Ms Hemlata Varsani for careful HD analysis, preparation of samples, and reading of the manuscript; staff of Great Ormond Street Hospital for help with sample collection, and the families and patients for generous donation of samples for this work. The work was supported by grants from SPARKS UK and the Cathal Hayes foundation.

References

1. Davis, M. M. and Bjorkman, P. J. (1988) T-cell antigen receptor genes and T-cell recognition. *Nature* **334,** 395–402.
2. Mason, D. (1998) A very high level of crossreactivity is an essential feature of the T-cell receptor. *Immunol Today* **19,** 395–404.
3. Arstila, T. P., Casrouge, A., Baron, V., Even, J., Kanellopoulos, J., and Kourilsky, P. (1999) A direct estimate of the human alphabeta T cell receptor diversity. *Science* **286,** 958–961.
4. Borst, J., Brouns, G. S., de Vries, E., Verschuren, M. C., Mason, D. Y., and van Dongen, J. J. (1993) Antigen receptors on T and B lymphocytes: parallels in organization and function. *Immunol. Rev.* **132,** 49–84.
5. Padovan, E., Casorati, G., Dellabona, P., Meyer, S., Brockhaus, M., and Lanzavecchia, A. (1993) Expression of two T cell receptor alpha chains: dual receptor T cells. *Science* **262,** 422–424.
6. Folch, G. and Lefranc, M. P. (2000) The human T cell receptor beta variable (TRBV) genes. *Exp. Clin. Immunogenet.* **17,** 42–54.

7. Wack, A., Montagna, D., Dellabona, P., and Casorati, G. (1996) An improved PCR-heteroduplex method permits high-sensitivity detection of clonal expansions in complex T cell populations. *J. Immunol. Methods* **196,** 181–192.
8. Pannetier, C., Even, J., and Kourilsky, P. (1995) T-cell repertoire diversity and clonal expansions in normal and clinical samples. *Immunol. Today* **16,** 176–181.
9. Maini, M., Casorati, G., Dellabonna, P., Wack, A., and Beverley, P. C. L. (1999) T-cell clonality in immune responses. *Immunol. Today* **20,** 262–266.
10. Wedderburn, L. R., Patel, A., Varsani, H., and Woo, P. (2001) The developing T cell receptor repertoire of children and young adults shows a wide discrepancy in the frequency of persistent oligoclonal T cell expansions. *Immunology* **102,** 301–309.
11. MacIsaac, C., Curtis, N., Cade, J., and Visvanathan, K. (2003) Rapid analysis of the Vbeta repertoire of CD4 and CD8 T lymphocytes in whole blood. *J. Immunol. Methods* **283,** 9–15.
12. Giachino, C., Granziero, L., Modena, Vet al. (1994) Clonal expansions of V delta 1+ and V delta 2+ cells increase with age and limit the repertoire of human gamma delta T cells. *Eur. J. Immunol.* **24,** 1914–1918.
13. Vavassori, M., Maccario, R., Moretta, A., et al. (1996) Restricted TCR repertoire and long-term persistence of donor-derived antigen-experienced CD4+ T cells in allogeneic bone marrow transplantation recipients. *J. Immunol.* **157,** 5739–5747.
14. Maini, M. K., Wedderburn, L. R., Hall, F., Wack, A., Casorati, G., and Beverley, P. C. L. (1998) A comparison of two techniques for the molecular tracking of specific T cell responses; CD4+ human T cell clones persist in a stable hierarchy but at a lower frequency than clones in the CD8+ population. *Immunology* **94,** 529–535.
15. Bernardin, F., Doukhan, L., Longone-Miller, A., Champagne, P., Sekaly, R., and Delwart, E. (2003) Estimate of the total number of CD8+ clonal expansions in healthy adults using a new DNA heteroduplex-tracking assay for CDR3 repertoire analysis. *J. Immunol. Methods* **274,** 159–175.
16. Arden, B., Clark, S. P., Kabelitz, D. and Mak, T. W. (1995) Human T-cell receptor variable gene segment families. *Immunogenetics* **42,** 455–500.
17. Sambrook, J., Russel, D.(2001) Molecular cloning: a laboratory manual, Third ed. ("Maniatis"). Cold Spring Harbour Lab press.
18. Wedderburn, L. R., Maini, M. K., Patel, A., Beverley, P. C. L. and Woo, P. (1999) Molecular fingerprinting reveals non-overlapping T cell oligoclonality between an inflamed site and peripheral blood. *Int. Immunol.* **11,** 535–543.
19. Wedderburn, L. R. (2000) Tracking T cells in arthritis. *Rheumatology (Oxford)* **39,** 458–462.
20. Wedderburn, L. R., Patel, A., Varsani, H. and Woo, P. (2001) Divergence in the degree of clonal expansions in inflammatory T cell sub-populations mirrors HLA-associated risk alleles in genetically and clinically distinct subtypes of childhood arthritis. *Int. Immunol.* **13,** 1541–1550.
21. Wotton, D., Ways, K., Parker, P. J., and Owen, M. J. (1993) Activity of both *raf* and *ras* is necessary for activation of transcription of the human T cell receptor β

gene by protein kinase C; *ras* plays multiple roles. *J. Biol. Chem.* **268,** 17975–17982.
22. King, D. J., Gotch, F. M., and Larsson-Sciard, E. L. (2001) T-cell re-population in HIV-infected children on highly active anti-retroviral therapy (HAART). *Clin. Exp. Immunol.* **125,** 447–454.
23. Ali, M., Weinreich, M., Balcaitis, S., Cooper, C. J., and Fink, P. J. (2003) Differential regulation of peripheral CD4+ T cell tolerance induced by deletion and TCR revision. *J. Immunol.* **171,** 6290–6296.

9

The Assessment of T-Cell Apoptosis in Synovial Fluid

Karim Raza, Dagmar Scheel-Toellner, Janet M. Lord, Arne N. Akbar, Christopher D. Buckley, and Mike Salmon

Abstract

T-cell apoptosis is central to the resolution of chronic inflammation. Inhibition of this process of programmed cell death contributes to disease persistence in conditions such as rheumatoid arthritis. An understanding of T-cell apoptosis and its regulation is clearly important for understanding the pathophysiology of inflammatory disease. This chapter describes a number of apoptosis assays that can be used to measure T-cell apoptosis in synovial fluid. The choice of assay depends, in part, on the phase of apoptosis under investigation and this review puts this into context by introducing these phases and their regulation.

Key Words: Apoptosis; T-cell; synovial fluid; rheumatoid arthritis.

1. Introduction

1.1. Apoptosis in Health and Disease

Apoptosis is a physiological mechanism for eliminating unwanted cells in an active, controlled and noninflammatory manner. It is essential for embryogenesis and the normal homeostatic control of cell populations in the adult *(1)*. In fact, all cells in multicellular organisms are programmed to die if they do not receive appropriate signals for survival *(2)*. Activated T-cells, for example, depend on the presence of specific cytokines including those that signal through the shared γ-chain of the interleukin (IL)-2 receptor (IL-2, -4, -7, -15, and -21) and type-1 interferons (IFN-α and -β) *(3–6)*. During the active phase of an immune response many such factors are produced, facilitating activated T-cell survival *(7)*. During resolution, levels of survival factors diminish and T-cell apoptosis restores a homeostatic balance *(8–10)*. These processes are normally closely regulated and coordinated to allow the inflammatory trigger to be dealt with appropriately and to prevent the inappropriate accumulation of leukocytes.

In patients with chronic inflammatory diseases, such as rheumatoid arthritis (RA), psoriatic arthritis, and inflammatory bowel disease, the normal mechanisms for resolution of inflammation are disrupted. In patients with RA for example, this is associated with the accumulation of large numbers of T-cells in the synovial compartment, which play an important role in disease persistence *(11)*. A phenotypic analysis of T-cells in the rheumatoid joint shows that they are almost exclusively primed and highly differentiated, with a CD45RB^{dull}RO^{bright} phenotype *(12)*. The progressive differentiation of T-cells, from a CD45RB^{bright}RO^{dull} to a CD45RB^{dull}RO^{bright} state is paralleled by loss of both Bcl-2 expression and the ability to synthesize IL-2. Associated with this, these cells have an increased sensitivity to apoptosis *(13)*. This, at first sight, presents a paradox as a high rate of T-cell apoptosis would be expected to contribute to an efficient removal of T-cells from the inflamed joint and consequent resolution of inflammation.

Apoptosis in the rheumatoid synovium was first studied by Firestein et al. *(14)*. Although apoptosis was seen in macrophages and other cells of the lining layer, a striking lack of apoptosis was reported amongst the CD45RO$^+$ T-cells in lymphoid aggregates. We subsequently reported that synovial T-cell apoptosis is profoundly inhibited in patients with RA, by fibroblast-derived mediators *(15)*; IFN-β was the key survival signal *(6)*. This led to the hypothesis that the expanded fibroblast network in the rheumatoid joint inhibited T-cell apoptosis locally and was the key determinant of disease persistence *(16–19)*. Over the last few years, T-cell apoptosis has been shown to be inhibited in a number of other chronic inflammatory diseases including Crohn's disease and eczema *(20,21)*. In addition to type-1 interferons, other mechanisms for rescue have been proposed. Exposure of CD4$^+$ T-cells to SDF-1α (produced by synovial fibroblasts) renders T-cells less susceptible to apoptosis induced by anti-CD3 stimulation *(22)* and cytokine deprivation *(23)*. In Crohn's disease and experimental colitis, a role has been suggested for IL-6 trans-signaling in mediating the inhibition of apoptosis of lamina propria T cells *(24)*. The relationship of such mechanisms to the type-1 interferon-mediated T-cell rescue that operates in the rheumatoid joint is, at present, unclear.

Understanding synovial T-cell apoptosis and its regulation in patients with inflammatory arthritis is central to an understanding of disease pathology. Furthermore, several groups have attempted to induce T-cell apoptosis as a therapeutic approach for inflammatory arthritis *(25,26)*. We suspect that this approach is flawed, because T-cell removal from the joint can be facilitated by conventional treatments, such as glucocorticoid therapy. The problem is that such modalities have no effect on the stromal population, so the T-cells return with subsequent immune challenges. The analysis of control mechanisms for synovial T-cell apoptosis is clearly of interest in many basic and clinically orientated investigations.

A wide range of assays is currently available to allow the detection of T-cell apoptosis. Many of these detect different stages in the process and the choice of assay is often influenced by the phase of apoptosis under investigation. A discussion of apoptosis assays is thus not possible without an introduction to the apoptotic pathway.

1.2. Mechanisms of Apoptosis

Apoptosis can be triggered either actively or through the withdrawal of essential survival factors. The initial signals used in cells undergoing apoptosis differ depending on the induction route, but the final execution pathways are common (*see* **Fig. 1**).

Caspase enzymes, or cysteine-dependent aspartic proteases are key effectors of apoptosis *(27,28)*. They are expressed as proenzymes, composed of an NH_2-terminal domain, and a large (approx 20–60 Kd) and small (approx 10 Kd) subunit domain. Activation involves cleavage between domains followed by association of the small and large subunits to form a heterodimer containing the catalytically active site. Caspases have an absolute requirement for cleavage after an aspartic acid residue. Recognition of at least four aminoacids NH_2-terminal to the cleavage site is also a necessary requirement for efficient catalysis. Significantly, the cleavage sites between the domains of caspase proenzymes are themselves caspase substrate site. Caspases are thus not only the key executioners of apoptosis, cleaving the structural and repair proteins whose degradation characterises the terminal stage of apoptosis, but are also central to the initiation of apoptosis, amplifying and consolidating apoptotic signals thorough autocatalysis and degradation of downstream effector caspases. Intriguingly, the link between apoptosis and inflammation is very close. Caspases originally evolved as death effectors. They fulfil purely this role in the well studied nematode worm *Caenorhabditis elegans*; however, in humans many of the caspase enzymes cleave pro-molecules crucial to inflammation. The best-known example of this is caspase-1, also known as IL-1β converting enzyme, which plays a crucial role in the generation of functional IL-1.

1.2.1.Initiation

1.2.1.1. Cell Surface Receptors Can Trigger the Apoptotic Program

Ligation of transmembrane cell surface receptors, such as Fas (CD95), leads to receptor clustering and the transduction of death signals *(29)*. This is medi-

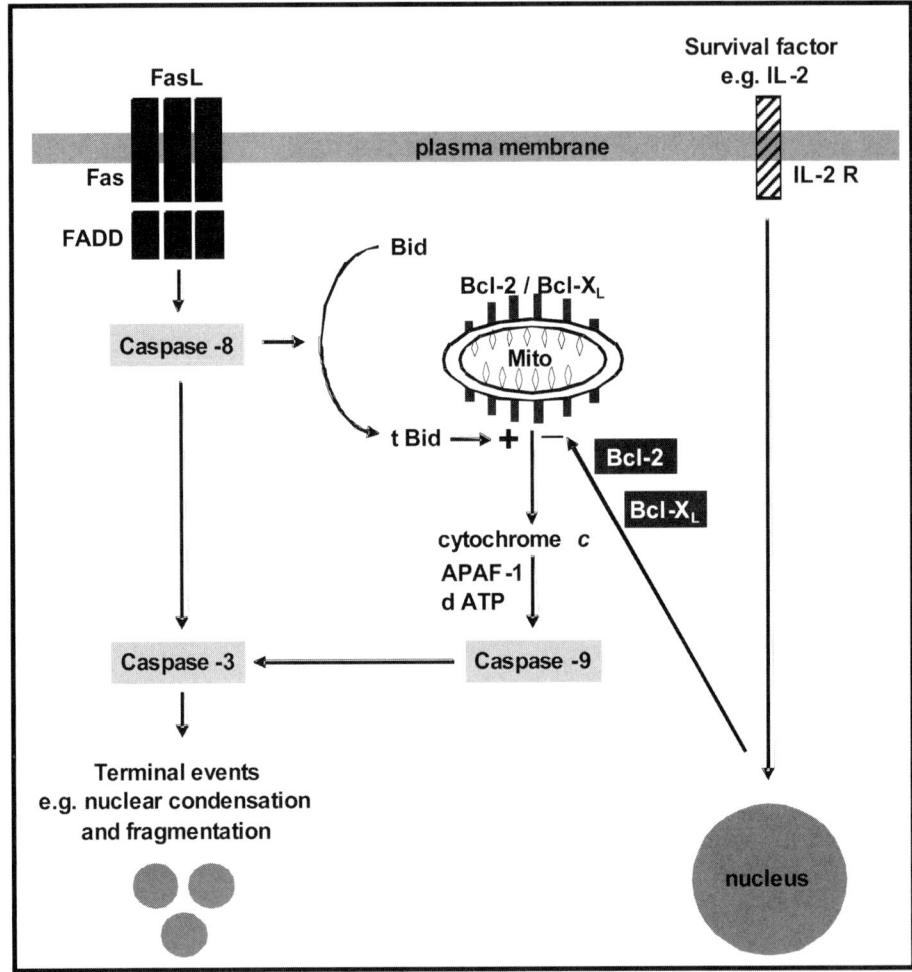

Fig. 1. The regulation of apoptosis in T-cells. Apoptosis may be initiated through the ligation of cell surface receptors (e.g., Fas) which lead to the activation of caspase-8 through the recruitment of the adapter protein FADD. Caspase-8 may directly activate caspase-3 or may operate via the mitochondrion through the cleavage of Bid leading to mitochondrial cytochrome c release. Cytochrome c complexes with APAF-1 and dATP to activate caspase-9 which in turn activates caspase-3. The activation of caspase-3 commits the cell to apoptosis leading to terminal apoptotic events such as nuclear condensation and fragmentation. Alternatively, a down-regulation of antiapoptotic Bcl-2 family members (induced for example by the withdrawal of IL-2) leads to mitochondrial cytochrome c release and the activation of caspase-3.

ated by the death-inducing signaling complex (DISC) of proteins which associates with the cytoplasmic tail of clustered Fas. The DISC complex brings pro-caspase-8 molecules into close proximity, leading to their activation *(30–32)*. In lymphocytes, Fas clustering and the formation of the DISC occurs in lipid rafts *(33)*. These processes occur very rapidly; active caspase-8 is released from the DISC into the cytosol within 10 seconds of receptor activation *(34)*.

Two pathways have been proposed whereby active caspase-8 can trigger apoptosis *(35)*. In "type I" cells, the activation of large amounts of caspase-8 at the DISC is followed by the rapid cleavage of pro-caspase-3. In contrast, in "type II" cells, little DISC is formed so the caspase cascade cannot be propagated directly, but is amplified via the mitochondrial route, through the cleavage and activation of Bid *(36–38)*.

1.2.1.2. Survival Factor Withdrawal Can Trigger the Apoptotic Program

T-cell death can also be triggered through the withdrawal of survival factors. Though very important for the homeostatic control of lymphocyte numbers, this pathway has been less well studied. Survival factors, such as IL-2 and IL-15, transcriptionally up-regulate the anti-apoptotic proteins Bcl-2 and Bcl-X_L *(3)* which insert into the outer mitochondrial membrane (*see* **Subheading 1.2.1.4.**).

1.2.1.3. The Mitochondrion and Cytochrome c Release

Mitochondria are central to the control of apoptosis. Many signals converge on this organelle which amplifies them by releasing proapoptotic molecules such as cytochrome c, a key step defining the commitment to apoptosis *(39)*. This phase of apoptosis is associated with a fall in mitochondrial transmembrane potential (ΔY) *(40)*. It is generally assumed that release of cytochrome c depends on the change in mitochondrial membrane potential, but actually these events occur synchronously in vivo *(41)*. A number of other proapoptotic proteins are released from depolarizing mitochondria. These include apoptosis inducing factor *(42)*, Smac/DIABLO *(43,44)* and pro-caspases-2 and -9. Cytochrome c, once released, binds Apaf-1 (apoptotic protease activating factor-1) *(45)*, pro-caspase-9, and dATP *(46)* to form the apoptosome. This leads to caspase-9 activation and the subsequent activation of caspase-3 *(46)*.

1.2.1.4. The Bcl-2 Family

The release of cytochrome c and other proapoptotic factors from mitochondria is controlled by members of the Bcl-2 family of molecules which localize to mitochondrial membranes *(47,48)*. The functions of Bcl-2 family members vary, with some proteins being anti-apoptotic (e.g., Bcl-2, Bcl-X_L) and others proapoptotic (e.g., Bax, Bak, Bid, Bad). Antiapoptotic Bcl-2 family members are integral membrane proteins found in the outer mitochondrial membrane, whereas many pro-apoptotic proteins localise to the cytosol or cytoskeleton prior to the induction of apoptosis. Following a death signal, these proapoptotic proteins undergo a conformational change that enables them to insert into the outer mitochondrial membrane. An important feature of Bcl-2 family proteins is their ability to homo- and heterodimerize. Heterodimerization between pro- and antiapoptotic proteins is believed to inhibit the activity of the partner protein.

1.2.2. Execution

In the execution phase of apoptosis many cellular processes are shut down in an orchestrated fashion. This stops processes such as proliferation and adhesion, and signals to neighboring cells that the apoptotic cell should be removed by phagocytosis. An impairment of this phagocytotic removal has serious consequences. Late apoptotic cells that enter secondary necrosis release neoantigens and induce a potent inflammatory signal. Uptake of apoptotic cells on the other hand delivers an anti-inflammatory signal. Caspase-3 is widely regarded as the key executioner of apoptosis with many of its target proteins being involved in processes such as DNA repair, cell proliferation, and adhesion *(49)*. Caspase-3 also initiates activation of the DNAses involved in DNA breakdown. Caspase-6, which is itself activated by caspase-3, cleaves Lamin B and initiates the dissolution of the nuclear envelope. Changes in intracellular ion composition and cleavage of cytoskeletal proteins lead to shrinkage and membrane blebbing.

2. Materials

2.1. Synovial Fluid Collection and Cytospin Preparation

1. Heparin sodium (CP Pharmaceuticals Ltd.).
2. Hyaluronidase (Sigma).
3. Phosphate buffered saline (PBS): 85g NaCl, 15g Na_2HPO_4, and 3g NaH_2PO_4 in 10 L distilled H_2O (pH 7.4).
4. Bovine serum albumin (Sigma).
5. Cytospin equipment (Shandon Inc).

2.2. Positive Selection of T-Cells From Synovial Fluid

1. Anti-CD4 and anti-CD8 Detach-a-Beads (Dynal Biotech).
2. Goat anti-mouse Detach-a-bead antibodies (Dynal Biotech).
3. RPMI 1640 (Sigma).

2.3. Early Stage of Apoptosis: Mitochondrial Depolarization

1. $DiOC_6(3)$ (Molecular Probes): 1 mg/mL in dimethyl sulfoxide (DMSO). Store in the dark at –20°.
2. JC-1 (Molecular Probes): 1 mg/mL in DMSO. Store in the dark at –20°C.

2.4. Early Stage of Apoptosis: Externalization of Phosphatidylserine

1. Annexin buffer: 10 mM Hepes (pH 7.4), 140 mM NaCl, 2.5 mM CaCl$_2$.
2. Annexin-V-PE (BD Pharmingen).
3. 7-AAD (Sigma, Poole, UK).

2.5. Execution Stage of Apoptosis: Caspase-3 Activation

1. Fix and Perm® cell permeabilization kit (Caltag Laboratories).
2. Heat inactivated fetal calf serum (FCS).
3. Rabbit anti-human active caspase-3 antibody (BD Pharmingen).
4. Normal rabbit immunoglobulin fraction (Dako).
5. Mouse anti-human CD3 antibody (Dako).
6. Fluorescein isothiocyanate (FITC) conjugated goat anti-rabbit IgG antibody (Southern).
7. Paraformaldehyde (Fisons Plc.): 10% (w/v) in distilled H$_2$O dissolved at 60°C in a fume cupboard. Ensure the glass container is covered with a loose fitting lid.
8. NP40 (BDH Chemicals).
9. Human immunoglobulin G (Institutio Grifols).
10. Goat serum (Dako).
11. Fish skin gelatine (Sigma).
12. Casein (Fisher Scientific).
13. Goat anti-mouse IgG Texas Red (Southern).
14. Streptavadin FITC (Dako).
15. DAPI (Sigma, Poole, UK). Stock solution made at 5 mg/mL in distilled H$_2$O. Store at –20°C.
16. AF-1 (Citifluor Ltd.).
17. PhiPhiLux G$_1$D$_2$ substrate (OncoImmunin Inc.).

2.6. Late Stage of Apoptosis

1. Lysis buffer: 10 mM ethylene diamine tetraacetic acid (EDTA), 50 mM Tris (pH 8.0), and 0.5% sodium lauryl sarkosinate.
2. Proteinase K: 1 mg/mL dissolved in 50 mM Tris-HCl (pH 8.0), and 1 mM CaCl$_2$.
3. RNAse A in lysis buffer: 10 mg/mL dissolved in 10 mM Tris-HCl (pH 7.5), and 15 mM NaCl, boiled for 15 min, cooled, and aliquoted. Store at –20°C.
4. Gel loading buffer: 0.25% bromophenol blue, 40% sucrose, 1% low-gelling-temperature agarose, and 10 mM EDTA in distilled H$_2$O.
5. TBE: 10.8g Tris, 5.5 g boric acid, 0.93 g Na$_2$ EDTA·H$_2$O in 1000 mL distilled H$_2$O (pH 8.3).
6. Ethidium bromide (EB).
7. TdT buffer (Gibco BRL).
8. Digoxigenin-UTP (11-2'-desoxy-uridin-5;-triphosphate) (Roche Diagnostics).
9. TdT (Roche Diagnostics).
10. ATP (Pharmacia).
11. Stop/wash buffer: 0.5 mM EDTA in PBS with 0.2% BSA.

12. Staining buffer: 50 µL 4X SSC, 5% nonfat milk powder (Marvel; Premier International Foods).
13. 20X SSC: 87.7 g NaCl and 44.1g sodium citrate dissolved in 400 mL distilled H_2O. Adjust pH to 7.2 with $10M$ NaOH and bring volume to 500 mL with distilled H_2O.
14. Anti-digoxigenin-FITC antibody.
15. Methanol.
16. Diff-Quik (Dade Behring AG, Switzerland).
17. DePeX mounting medium (BDH laboratory Supplies).

3. Methods

3.1. Synovial Fluid Collection and Initial Treatment

Synovial fluid can be aspirated from inflamed joints under palpation or ultrasound guidance *(50)*. For small joints such as those PIP or MCP joints where ultrasound examination demonstrates only very small amounts of fluid, the contents of the joint can be sampled using an ultrasound guided lavage technique *(51)*. For synovial fluid samples that will be processed immediately, the samples do not need to be collected in heparin. Otherwise the samples should be collected in a tube containing heparin sodium to prevent coagulation (synovial fluid requires approximately 5 iu/mL). We have not found that heparin used at this concentration has any effect on T-cell apoptosis.

Synovial fluid is often too viscous to process directly and requires incubation with hyaluronidase for 30 min at 37°C. We have not found that hyaluronidase used at this concentration has any effect on T-cell apoptosis.

3.2. Cytospin Preparation

1. Count synovial fluid cells using a haemocytometer and resuspend in PBS 2% BSA to give a final concentration of 0.5×10^6 cells/mL.
2. Load 100 µL of the cell suspension into a cytospin funnel and spin at 48*g* onto a glass slide.
3. Air dry the slides and either stain immediately, or wrap in aluminium foil and store at –20 or –80°C for subsequent staining (*see* **Note 1**).

3.3. Positive Selection of T-Cells From Synovial Fluid

For many apoptosis assays, T-cells do not have to be selected from synovial fluid and the apoptosis assay can be performed using the whole synovial fluid cell population, staining for CD3, to identify T-cells, as well as for the marker of apoptosis being assessed. For a few assays (e.g., detection of DNA laddering in synovial T-cells [*see* **Subheading 3.4.4.1.**]), synovial fluid T-cells should be positively selected from synovial fluid. This selection is performed with anti-CD4 and anti-CD8 coated beads rather than anti-CD3 coated beads as ligation of CD3 may activate the cells.

1. Incubate synovial fluid cells with CD4 and CD8 Detach-a-Beads (Dynal Biotech, Oslo, Norway) at an approximate bead to target cell ratio of 10:1.
2. Incubate on ice for 60 min with frequent gentle agitation.
3. Separate positively selected cells by magnetic attraction and wash extensively in ice cold RPMI 1640 to remove all nonselected cells.
4. Resuspend bead/cell rosettes in 100 µL RPMI containing 10% human serum and 20 µL goat anti-mouse Detach-a-Bead antibody/1.0×10^6 target cells (*see* **Note 2**).
5. Incubate for 60 min at room temperature on a rotary mixer.
6. Remove Dynabeads from unbound cells by magnetic separation and use these positively selected cells for analysis.

3.4. Apoptosis Assays

A very wide range of techniques can be used to detect apoptosis. Most of these have been developed over the last 15 yr. Ideally, the choice of which assay to use for a particular purpose is informed by the phase of apoptosis to be studied. More commonly, it is informed by fashion or habit. In this section we describe a number of the commonly used assays that we have applied to the analysis of T-cell apoptosis in the rheumatoid synovium and discuss their value. In addition to the phase of apoptosis, the number of cells available plays a key part in the choice of assay. It should be noted that even in tissues or fluids in which there is a high rate of T-cell apoptosis, the proportion of cells identified as apoptotic at any one time may be low as a consequence of the rapid removal of apoptotic cells by phagocytes. These issues will be discussed in more detail in relation to specific assays, but the following points are important for all apoptosis assays:

1. Ex vivo handling time should be minimized to reduce the possibility of artefactual results consequent upon ex vivo apoptosis.
2. T-cell apoptosis should be assessed in whole synovial fluid or in T-cells that have been positively selected on the basis of cell surface antigen expression (*see* **Subheading 3.3.**). The use of Ficoll or percol density centrifugation of synovial fluid is not recommended for the study of T-cell apoptosis, because apoptotic cells increase their density and consequently will be lost during this procedure.

3.4.1. Early Stage of Apoptosis: Mitochondrial Depolarization

Mitochondrial polarity can be measured using flow cytometric assays based upon the ability of intact mitochondria to take up and retain certain fluorescent dyes. These include $DiOC_6(3)$ (3,3'-dihexyloxacarbocyanine iodide) and JC-1 (5,5',6,6'-tetrachloro-1,1',3,3'- tetraethylbenzimidazolylcarbocyanine iodide) (*see* **Notes 3** and **4**).

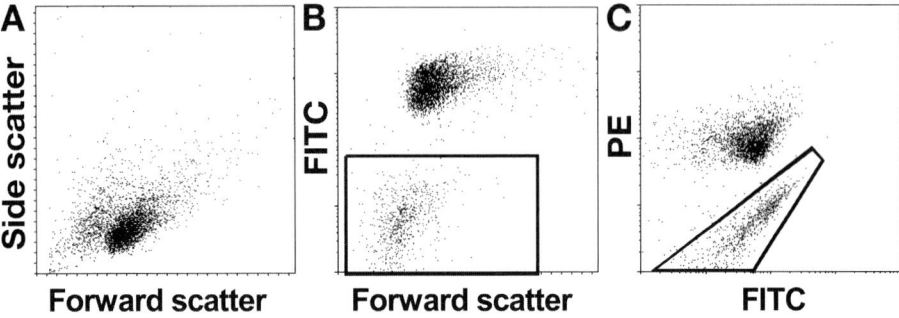

Fig. 2. T-cells analysed by flow cytometry. (**A**) Forward scatter/side scatter profile. (**B**) $DiOC_6(3)$ staining; cells with depolarised mitochondria are highlighted. (**C**) JC-1 staining; cells with depolarised mitochondria are highlighted.

3.4.1.1. $D_iOC6(3)$

1. Resuspend cells (0.1–1.0 × 10^6 cells) in 50 µL PBS and add 100 µL $DiOC_6(3)$ diluted from stock solution (*see* **Subheading 2.3.**) in PBS to give a final concentration of 20 ng/mL. Leave for 30 min at 37°C.
2. Wash cells in ice cold PBS.
3. Resuspended cells in ice cold PBS and analyze on a flow cytometer (*see* **Note 5** and **Fig. 2**).

Costaining for a cell surface antigen can be carried out before staining with $DiOC_6(3)$. Cells should be stained on ice with an antibody directly conjugated to a fluorochrome that emits light at wavelength longer than that of $DiOC_6(3)$.

3.4.1.2. JC-1

1. Resuspend cells (0.1–1.0 × 10^6 cells) in 50 µL PBS and then add 100 µL JC-1 diluted from stock solution (*see* **Subheading 2.3.**) in PBS to give a final concentration of 1.3 µg/mL. Leave for 30 min at 37°C.
2. Wash cells in ice cold PBS.
3. Resuspended cells in ice cold PBS and analyze on a flow cytometer (*see* **Note 6** and **Fig. 2**).

3.4.1.3. USE OF JC-1 FOR CELL RECOVERY ASSAYS

When T-cells are cultured it is often more useful to monitor the recovery of live cells from the system, rather than determine the proportion of apoptotic cells which may disintegrate during culture. For this we use in-culture labeling with JC-1 and quantify the number of cells with an intact mitochondrial membrane potential/unit volume.

1. Culture T-cells in 96-well plates (0.1×10^6 cells in 200 µL).
2. Add JC-1 stock solution to the cell culture to give a final concentration of 10 µM and incubate at 37°C for 30 min.
3. Resuspend cells and transfer the cell suspension directly into a flow cytometer-compatible plastic tube containing 300 µL of PBS.
4. If a Coulter EPICS XL flow cytometer is available, samples can be run using a fixed volume protocol that acquires data from 20 µL of sample (*see* **Note 7**). Cell survival is determined as a percentage of cells recovered compared to a similar test performed on a sample at the start of the culture experiment.

3.4.2. Early Stage of Apoptosis: Externalization of Phosphatidylserine

Quite early in apoptosis, phosphatidylserine (PS) is translocated from the inner to the outer layer of the plasma membrane, where it acts as a recognition marker for phagocytosis. The phospholipid-binding protein annexin V has a high affinity for PS and can be used to detect it on the outer leaflet of the membrane. This assay is very easy to perform, gives very clear results and, as a consequence, has been extremely popular. Unfortunately there are two distinct problems with PS exposure as a measure of apoptosis. First, in necrotic cells, which become permeable, annexin V can enter cells and bind to the vast quantities of PS on the inner leaflet of the plasma membrane. Costaining with propidium iodide (PI) or 7-amino actinomycin D (7-AAD), DNA stains that cannot enter intact cells, distinguishes apoptotic cells from necrotic cells. Apoptotic cells will stain annexin V positive, but PI negative. Necrotic cells will stain positive for both annexin V and PI because PI can cross the disrupted membrane. However, it is not possible with this method to tell whether the "necrotic" cells have entered secondary necrosis after the apoptotic program, or were simply damaged during isolation. The second problem with PS detection is more profound. A number of cell types, such as B-cells and neutrophils, have been reported to become transiently positive for Annexin V, then subsequently revert to a negative state. This appears to reflect a physiological role for PS in the outer membrane leaflet, which makes interpretation of this apoptosis assay difficult *(52,53)*.

1. Resuspend cells (0.1–1.0×10^6 cells) in 100 µL annexin buffer (*see* **Subheading 2.4.**).
2. Stain cells with 5 µL annexin-V-PE and 1 µL 7-AAD for 20 min in the dark.
3. Add a further 400 µL annexin buffer and analyze on flow cytometer.

3.4.3. Execution Stage of Apoptosis: Caspase-3 Activation

Activation of caspase-3 characterises the execution phase of apoptosis. This activation can be detected using a monoclonal antibody (mAb) specific for the active form of caspase-3 and with very low affinity for the proform of the enzyme (*see* **Subheadings 3.4.3.1.** and **3.4.3.2.**). Of all the assays developed in recent years, detection of active caspase-3 by this method is, in our opinion, by far the best. It is completely specific for apoptosis and relatively sensitive. It also produces attractive flow cytometric plots and rather pretty cytospins. The activation of caspase-3 can also be detected by using fluorogenic caspase substrates which have the amino acid sequence DEVDGI (*see* **Subheading 3.4.3.3.**). Cleavage of the substrate allows fluorochromes attached to the parent molecule to fluoresce when stimulated with light of an appropriate wavelength. This latter technique was briefly very popular, but was quickly found not to be specific for caspase-3. The substrate can just as readily be cleaved by caspase-7, which has identical specificity, and to a lesser extent by caspase-6. Consequently if a specific assessment of caspase-3 activation is required, this assay is not very good. On the other hand, a cell with sufficient active caspase-3, -6, or -7 to produce detectable fluorescence with this reagent is going to die, so as a simple apoptosis assay, it is perfectly adequate.

3.4.3.1. FLOW CYTOMETRIC ASSESSMENT USING AN ANTIACTIVE CASPASE-3 ANTIBODY

1. Fix and permeabilise 1.0×10^6 cells with Fix and Perm® cell permeabilization kit at room temperature.
2. Wash cells twice in PBS and resuspend in 50% heat inactivated FCS in PBS for 1 h to reduce nonspecific antibody binding.
3. Stain with affinity purified rabbit anti-human active caspase-3 mAb or normal rabbit immunoglobulin fraction as a negative control for 1 h. Cells can be costained for CD3.
4. Active caspase-3 can be detected with a FITC conjugated goat anti-rabbit IgG antibody. An appropriate antibody to detect the anti-human CD3 primary should be used.

3.4.3.2. IMMUNOFLUORESCENCE ASSESSMENT USING ANTIACTIVE CASPASE-3 ANTIBODY (*see* Note 8)

1. Draw 2 wax rings around cells cytospun onto a glass slide.
2. Fix cells by adding 100 µL 1% paraformaldehyde (diluted from 10% stock solution [*see* **Subheading 2.5.**] in PBS) for 30 min followed by 2 5-min washes in PBS (*see* **Note 9**).
3. Permeabilize cells with 0.1% NP40 for 30 min followed by 2 5-min washes in PBS.
4. Block nonspecific staining with 2.5% human immunoglobulin G (50% solution of Flebogamma® 5% in PBS) for 30 min followed by 20% goat serum, 0.2% fish skin gelatine, and 0.5% casein in PBS for 1 h (*see* **Note 10**).
5. Dilute the primary and secondary antibodies in 20% goat serum, 0.2% fish skin gelatin, and 0.5% casein in PBS (*see* **Note 10**).

6. Add 100 µL rabbit anti-human active caspase-3 mAb plus mouse anti-human CD3 at appropriate dilutions for 1 h at room temperature followed by 2 5-min washes in PBS.
7. Add 100 µL biotinylated goat anti-rabbit IgG plus goat anti-mouse IgG Texas red at appropriate dilutions for 30 min at room temperature followed by 2 5-min washes in PBS.
8. Add 100 µL streptavadin FITC diluted appropriately in PBS for 30 min followed by 2 5-min washes in PBS.
9. Counterstain nuclei with DAPI diluted from stock (*see* **Subheading 2.5.**) in PBS and used at a final concentration of 40 ng/mL. Wash DAPI off with PBS after 2 min. Mount in antifadant mounting medium (e.g., AF-1) and view under fluorescence microscope (*see* **Fig. 3**).

Flow cytometric assessment using PhiPhilux (*see* **Note 11**)

1. Wash cells (0.5–1.0×10^6 cells) in PBS.
2. Add 20 µL PhiPhiLux G_1D_2 substrate solution together with anti-human CD3-PE antibody and leave for 30 min at 37°C.
3. Harvest into 250 µL PBS and analyse immediately on flow cytometer (*see* **Note 12**).

3.4.4. Late Stage of Apoptosis: DNA Fragmentation

DNA is super-coiled around histone core proteins. Each histone has about 160 base-pairs (bp) of DNA wrapped around it. This structure is called a nucleosome. During the execution phase of apoptosis, DNA is cleaved by an endonuclease between the histones, resulting in fragments that are multiples of 160 bp. These nucleosomal units lead to the appearance of "DNA laddering" on agarose gels when DNA is extracted from apoptotic cells (*see* **Subheading 3.4.4.1.**). This technique is entirely specific for apoptosis, but its sensitivity can be limited when the frequency of apoptotic cells in a population is very low. In addition, this assay can only show that apoptotic cells are present, it cannot be used to quantify them. Alternatively, the TUNEL technique can be used to detect 3'-OH DNA ends generated by DNA fragmentation. Unfortunately this method cannot discriminate between the free ends of DNA released during apoptosis and shattered DNA resulting from necrosis. Consequently TUNEL may be more sensitive than an assessment of DNA laddering for the identification of fragmented DNA, but it is much less specific for apoptosis. Our own enthusiasm for this method was diminished some years ago, when we were studying the apoptosis of endothelial cells. As a rather crude negative control, we treated the endothelial cells with distilled water to induce necrosis. They became beautifully positive for TUNEL in less than 5 min. The third method in this category is simple morphological detection of nuclear condensation and cleavage. This is by far the oldest technique for apoptosis detection and it is still one of the best. Looking down a microscope, an apoptotic cell is

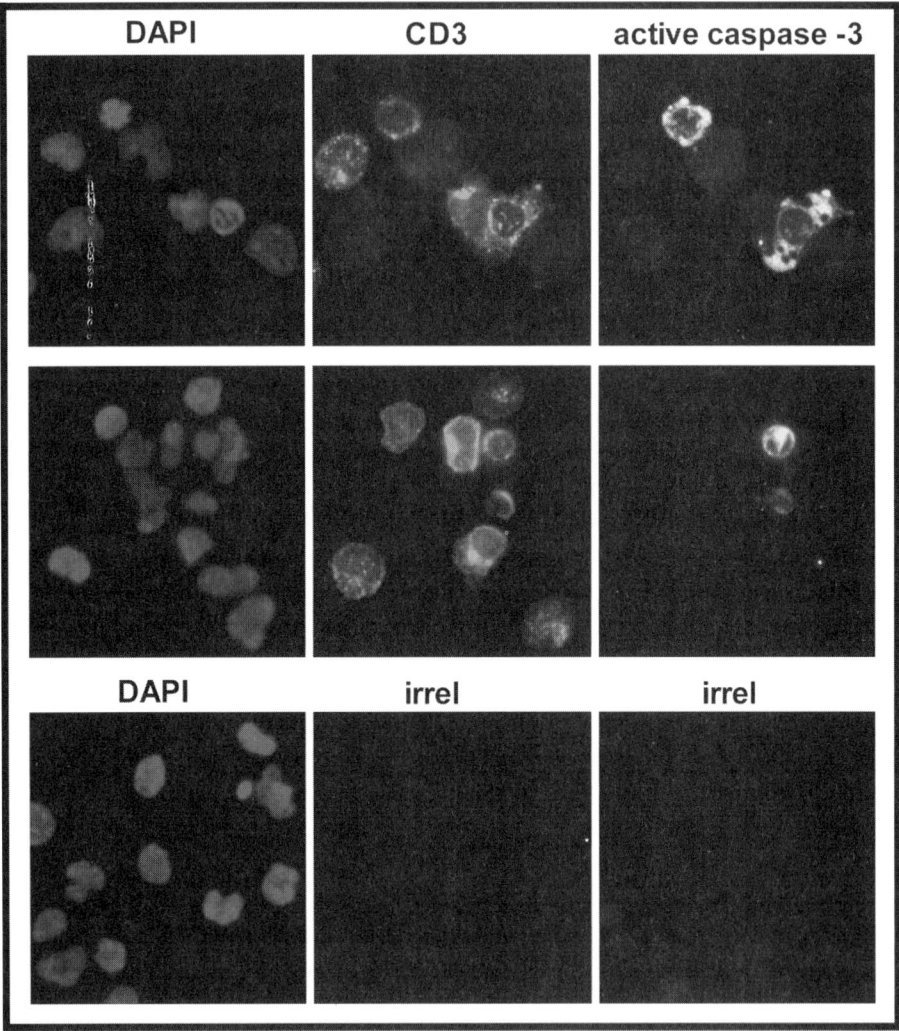

Fig. 3. Immunofluorescence staining for DNA, CD3, and active caspase-3 in synovial fluid cytospin preparations from patients with inflammatory arthritis as described in Subheading 3.4.3.2.

very difficult to confuse with anything else; the technique is cheap, requires no specialist equipment or experience and is both sensitive and specific.

3.4.4.1. DNA Laddering (see Note 13)

1. Pellet 1.0×10^6 cells in a microcentrifuge tube, remove supernatant, snap freeze in liquid nitrogen and store at –70°C until further use.

2. Defrost pellets on ice and resuspend in 20 µL ice cold lysis buffer and 20 µL proteinase K. Vortex cells.
3. Incubate at 50°C for 1 h in a preheated dry block.
4. Add 10 µL of 0.5 mg/mL RNAse A in lysis buffer and incubate at 50°C for 1 h.
5. Add 10 µL gel loading buffer and melt briefly at 70°C.
6. Load on 2% agarose gel in TBE with 0.1 µg/mL ethidium bromide.

3.4.4.2. TUNEL-STAINING OF CELLS FOR FLOW CYTOMETRY

In this technique residues of digoxigenin-nucleotide are catalytically added to the DNA break by terminal deoxynucleotidyl transferase (TdT). These are detected with a fluorescently labeled anti-digoxigenin antibody.

1. Resuspend cells ($0.1–1.0 \times 10^6$ cells) in 50 µL PBS.
2. Fix and permeabilize cells with Fix and Perm® cell permeabilisation kit.
3. Wash 3 times in PBS 2% BSA.
4. Wash cells in 25 µL equilibration buffer and 20 µL distilled H_2O.
5. Resuspend in 45 µL TdT mix 0.1 µL ATP prediluted 1:1000 from 100 mM stock solution, 39 µL distilled H_2O.
6. Incubate at 37°C for 30 min. Resuspend after 15 min.
7. Add 0.5 mL stop/wash buffer.
8. Spin down and remove supernatant.
9. Repeat **step 8**.
10. Resuspend in 50 µL staining buffer, 5% nonfat milk powder, with 0.125 µL anti-digoxigenin-FITC mAb.
11. Incubate at room temperature for 30 min. Resuspend after 15 min=.
12. Resuspend in 500 µL PBS 2% BSA. Spin down and remove supernatant.
13. Repeat **step 12**.
14. Run on flow cytometer.

A negative control should be run with all samples. Sham staining can be performed by substituting distilled H_2O for the TdT in **step 5**. A modification of this technique can also be applied to stain cells cytospun onto a slide.

3.4.5. Late Stage of Apoptosis: Morphological Assessment of Nuclear Condensation and Fragmentation

Synovial fluid cells can be stained with Romanowski stains.

1. Cytospin synovial fluid cells onto glass slide (*see* **Subheading 3.2.**).
2. Fix in methanol for 10 min.
3. Stain with Diff-Quik solution 1 (10 1-s dips) and Diff-Quik solution 2 (20 1-s dips)
4. Allow to air dry, mount in DePeX mounting medium and view under the light microscope. An apoptotic cell is defined on the basis of a condensed or fragmented nucleus. Apoptotic PMN and apoptotic lymphocytes are distinguished on the basis of acidophilic or basophilic cytoplasmic staining respectively (*see* **Fig. 4**).

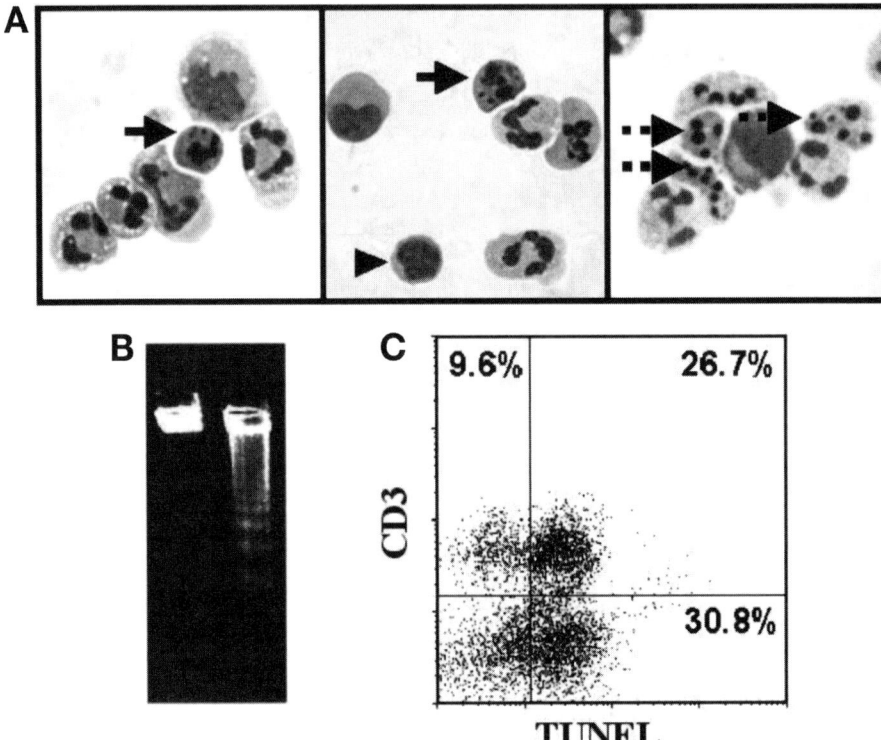

Fig. 4. (**A**) Apoptotic lymphocytes (*arrows*), healthy lymphocyte (*arrow head*) and apoptotic neutrophils (*dashed arrows*) in synovial fluid of patients with early inflammatory arthritis stained with Diff-Quik as described in Subheading 3.4.5. (**B**) Nucleosomal ladder of DNA (*see* Subheading 3.4.4.1.). (**C**) TUNEL technique on a sample of synovial fluid from a patient with gout (*see* Subheading 3.4.4.2.).

4. Notes

1. Thawing slides. Slides should be thawed at room temperature for 30 min before unwrapping aluminium foil.
2. Positive selection of T-cells. The goat anti-mouse antibody is idiotypic and interacts with the murine anti-human CD4 or CD8 antibody bound to the Detach-a-Beads, releasing the positively selected cells.
3. $DiOC_6(3)$ and JC-1. These assays depend upon the functional activity of mitochondria and so must be carried out at 37°C. Cells cannot be fixed or permeabilized before or after staining with these fluorochromes (*see* **Note 4**) and should be analyzed on the flow cytometer as soon as possible. The problem with mitochondrial polarization assays is that $DiOC_6(3)$ produces the most attractive results; JC-1 is less aesthetically pleasing and harder to interpret, but unfortu-

nately in some cells $DiOC_6(3)$ can also reflect changes in surface membrane polarity *(54)*. In T-cells we have not seen any evidence for this (by contrast, we find that JC-1 is a more specific marker of early mitochondrial potential changes in neutrophils) but critical reviewers might make publication of results with $DiOC_6(3)$ difficult.

4. Fixation resistant mitochondrial potential sensitive fluorochromes. Whereas cells stained with $DiOC_6(3)$ or JC-1 cannot be fixed or permeabilized before or after staining, cells can be stained with other mitochondrial potential sensitive fluorochormes. MitoTracker probes can be fixed and permeabilised after mitochondrial staining to allow co-staining for intracellular molecules. We have used this approach to stain T cells prior to cytospin preparation:
   a. Incubate 100 µL cell suspension (1×10^6 cells/mL) with 50 µL of Mitotracker red CMXRosbM-7512 (1 in 4000 dilution of stock solution), to give a final concentration of 50 nM, for 30 min at 37°C.
   b. Wash once in PBS 2%BSA.
   c. Resuspend in PBS 2% BSA and cytospin onto glass slide.
   d. Cells can be fixed/permeabilised with either 1% paraformaldehyde/ 0.1% NP40 for 30 min or acetone or 15 min.

5. Flow cytometric analysis of cells stained with $DiOC_6(3)$. The absorbance maximum of $DiOC_6(3)$ is 484 nm with a fluorescence emission maximum of 501nm. Mitochondria with a normal polarity actively take up $DiOC_6(3)$ and cells containing such mitochondria have a high fluorescence intensity in the FITC channel. By contrast cells with a low signal for $DiOC_6(3)$, indicative of a depolarised mitochondrial membrane potential, are taken as apoptotic. Cells undergoing apoptosis initially loose mitochondrial potential and then shrink. T-cells with an normal forward scatter (FSC) but low signal for $DiOC_6(3)$ on flow cytometry are thus early apoptotic cells whereas cells with a reduced FSC and a low signal for $DiOC_6(3)$ are late apoptotic.

6. Flow cytometric analysis of cells stained with JC-1. JC-1 exists as a monomer at low concentrations and yields green fluorescence (emission maximum of 525 nm). At higher concentrations, the dye forms J-aggregates with an emission maximum of 590 nm. The accumulation of this dye in mitochondria is dependent upon mitochondrial potential. In mitochondria with an intact membrane potential the dye accumulates at high concentration with the resultant formation of red aggregates. Following mitochondrial depolarization the concentration of JC-1 is lower and dye forms green fluorescent monomers. A shift from red to green fluorescence thus indicates depolarization.

7. Fixed volume counts. This facility was built into the Coulter EPICS XL flow cytometer to facilitate automatic sample loading in routine pathology laboratories, but it is an excellent way of counting cells. If a flow cytometer with fixed volume counting capacity is not available, Trucount™ tubes, supplied by Becton Dickinson, can be used. These tubes contain a fixed number of beads that can be used to assay absolute cell numbers in the sample.

8. Immunofluorescence staining for active-caspase-3. Relationship between caspase-activation and nuclear fragmentation. This assay is able to identify apoptotic cells which stain positively for active caspase-3 both before, and after, nuclear fragmentation. The ability to detect active caspase-3 in cells before nuclear fragmentation using this assay allows the detection of an earlier stage in the apoptotic program than is possible with an assessment of nuclear fragmentation by Diff-Quik staining (*see* **Subheading 3.4.5.**).
9. Fixation of synovial fluid cells on cytospin preparation prior to immunofluorescence staining. Inflammatory synovial fluid typically contains large numbers of neutrophils. When synovial fluid cells cytospun onto glass slides were fixed with either acetone or methanol and then taken through two or three layers of immunofluorescence staining neutrophil morphology was not preserved. If these cytospins are counterstained with a nuclear dye such as DAPI, streaks of DNA from neutrophils are seen across the slide. This problem is not observed when using 1% paraformaldehyde/0.1% NP40.
10. Blocking nonspecific neutrophil binding on synovial fluid cytospins. Activated neutrophils in synovial fluid actively bind the primary and secondary antibodies used in this cytospin staining protocol. To eliminate this nonspecific binding we developed the blocking regimen described. This is applied to the cytospun cells before the primary antibody is applied. In addition the primary and secondary antibodies are diluted in this blocking buffer. The streptavadin-FITC should clearly not be diluted in this buffer as biotin residues in the buffer will bind the streptavadin-FITC and prevent this from binding biotin conjugated to the secondary antibody.
11. PhiPhilux staining. Cells cannot be fixed or permeabilised after PhiPhilux staining.
12. Flow cytometric analysis of cells stained with PhiPhilux G_1D_2. Cleavage of the PhiPhilux G_1D_2 substrate yields a fluorophore with an absorbance peak at 505 nm and an emission peak at 530 nm.
13. Detection of DNA laddering in synovial T-cells. This approach can only be used on T-cells that have been selected from synovial fluid (*see* **Subheading 3.3.**).

Acknowledgments

Many of these techniques have been refined for use in synovial fluid by members of our group. In particular we wish to acknowledge the work of Lakhvir Assi, Emma Taylor and Darrell Pilling.

References

1. Kerr, J. F., Wyllie, A. H., and Currie, A. R. (1972) Apoptosis: a basic biological phenomenon with wide-ranging implications in tissue kinetics. *Br. J. Cancer* **26,** 239–257.
2. Raff, M. C. (1992) Social controls on cell survival and cell death. *Nature* **356,** 397–400.
3. Akbar, A. N., Borthwick, N. J., Wickremasinghe, R. G., et al. (1996) Interleukin-2 receptor common gamma-chain signaling cytokines regulate activated T cell

apoptosis in response to growth factor withdrawal: selective induction of anti-apoptotic (bcl-2, bcl-xL) but not pro- apoptotic (bax, bcl-xS) gene expression. *Eur. J. Immunol.* **26**, 294–299.
4. Vella, A. T., Dow, S., Potter, T. A., Kappler, J., and Marrack, P. (1998) Cytokine-induced survival of activated T cells in vitro and in vivo. *Proc. Natl. Acad. Sci. USA* **95**, 3810–3815.
5. Kaneko, S., Suzuki, N., Koizumi, H., Yamamoto, S., and Sakane, T. (1997) Rescue by cytokines of apoptotic cell death induced by IL-2 deprivation of human antigen-specific T cell clones. *Clin. Exp. Immunol.* **109**, 185–193.
6. Pilling, D., Akbar, A. N., Girdlestone, J., et al. (1999) Interferon-beta mediates stromal cell rescue of T cells from apoptosis. *Eur. J. Immunol.* **29**, 1041–1050.
7. Orteu, C. H., Poulter, L. W., Rustin, M. H., Sabin, C. A., Salmon, M., and Akbar, A. N. (1998) The role of apoptosis in the resolution of T cell-mediated cutaneous inflammation. *J. Immunol.* **161**, 1619–1629.
8. Akbar, A. N., Borthwick, N., Salmon, M., et al. (1993) The significance of low Bcl-2 expression by CD45RO T cells in normal individuals and patients with acute viral infections. The role of apoptosis in T cell memory. *J. Exp. Med.* **178**, 427–438.
9. Lenardo, M., Chan, K. M., Hornung, F., et al. (1999) Mature T lymphocyte apoptosis—immune regulation in a dynamic and unpredictable antigenic environment. *Annu. Rev. Immunol.* **17**, 221–253.
10. Plas, D. R., Rathmell, J. C., and Thompson, C. B. (2002) Homeostatic control of lymphocyte survival: potential origins and implications. *Nat. Immunol.* **3**, 515–521.
11. Salmon, M. and Gaston, J. S. (1995) The role of T-lymphocytes in rheumatoid arthritis. *Br. Med. Bull.* **51**, 332–345.
12. Matthews, N., Emery, P., Pilling, D., Akbar, A., and Salmon, M. (1993) Subpopulations of primed T helper cells in rheumatoid arthritis. *Arthritis Rheum.* **36**, 603–607.
13. Salmon, M., Pilling, D., Borthwick, N. J., et al. (1994) The progressive differentiation of primed T cells is associated with an increasing susceptibility to apoptosis. *Eur. J. Immunol.* **24**, 892–899.
14. Firestein, G. S., Yeo, M., and Zvaifler, N. J. (1995) Apoptosis in rheumatoid arthritis synovium. *J. Clin. Invest.* **96**, 1631–1638.
15. Salmon, M., Scheel-Toellner, D., Huissoon, A. P., et al. (1997) Inhibition of T cell apoptosis in the rheumatoid synovium. *J. Clin. Invest.* **99**, 439–446.
16. Salmon, M., Pilling, D., Borthwick, N. J., and Akbar, A. N. (1997) Inhibition of T cell apoptosis—a mechanism for persistence in chronic inflammation. *The Immunologist.* **5**, 87–92.
17. Salmon, M. and Akbar, A. N. (1999) The role of apoptosis in rheumatoid arthritis. In: *Challenges in rheumatoid arthritis*, Bird, H. A. and Snaith, M. L., eds., Blackwell Science, pp. 25–39.
18. Akbar, A. N. and Salmon, M. (1997) Cellular environments and apoptosis: tissue microenvironments control activated T-cell death. *Immunol. Today.* **18**, 72–76.
19. Buckley, C. D., Pilling, D., Lord, J. M., Akbar, A. N., Scheel-Toellner, D., and Salmon, M. (2001) Fibroblasts regulate the switch from acute resolving to chronic persistent inflammation. *Trends Immunol.* **22**, 199–204.

20. Ina, K., Itoh, J., Fukushima, K., et al. (1999) Resistance of Crohn's disease T cells to multiple apoptotic signals is associated with a Bcl-2/Bax mucosal imbalance. *J. Immunol.* **163,** 1081–1090.
21. Orteu, C. H., Rustin, M. H., O'Toole, E., et al. (2000) The inhibition of cutaneous T cell apoptosis may prevent resolution of inflammation in atopic eczema. *Clin. Exp. Immunol.* **122,** 150–156.
22. Nanki, T., Hayashida, K., El Gabalawy, H. S., et al. (2000) Stromal cell-derived factor-1-CXC chemokine receptor 4 interactions play a central role in CD4+ T cell accumulation in rheumatoid arthritis synovium. *J. Immunol.* **165,** 6590–6598.
23. Suzuki, Y., Rahman, M., and Mitsuya, H. (2001) Diverse transcriptional response of CD4(+) T cells to stromal cell-derived factor (SDF)-1: cell survival promotion and priming effects of SDF-1 on CD4(+) T cells. *J. Immunol.* **167,** 3064–3073.
24. Atreya, R., Mudter, J., Finotto, S., et al. (2000) Blockade of interleukin 6 trans signaling suppresses T-cell resistance against apoptosis in chronic intestinal inflammation: evidence in crohn disease and experimental colitis in vivo. *Nat. Med.* **6,** 583–588.
25. Liu, Z., Xu, X., Hsu, H. C., et al. (2003) CII-DC-AdTRAIL cell gene therapy inhibits infiltration of CII-reactive T cells and CII-induced arthritis. *J. Clin. Invest.* **112,** 1332–1341.
26. Ogawa, Y., Ohtsuki, M., Uzuki, M., et al. (2003) Suppression of osteoclastogenesis in rheumatoid arthritis by induction of apoptosis in activated CD4+ T cells. *Arthritis Rheum.* **48,** 3350–3358.
27. Thornberry, N. A. and Lazebnik, Y. (1998) Caspases: enemies within. *Science* **281,** 1312–1316.
28. Earnshaw, W. C., Martins, L. M., and Kaufmann, S. H. (1999) Mammalian caspases: structure, activation, substrates, and functions during apoptosis. *Annu. Rev. Biochem.* **68,** 383–424.
29. Krammer, P. H. (2000) CD95's deadly mission in the immune system. *Nature* **407,** 789–795.
30. Kischkel, F. C., Hellbardt, S., Behrmann, I., et al. (1995) Cytotoxicity-dependent APO-1 (Fas/CD95)-associated proteins form a death-inducing signaling complex (DISC) with the receptor. *EMBO J.* **14,** 5579–5588.
31. Muzio, M., Chinnaiyan, A. M., Kischkel, F. C., et al. (1996) FLICE, a novel FADD-homologous ICE/CED-3-like protease, is recruited to the CD95 (Fas/APO-1) death-inducing signaling complex. *Cell* **85,** 817–827.
32. Muzio, M., Stockwell, B. R., Stennicke, H. R., Salvesen, G. S., and Dixit, V. M. (1998) An induced proximity model for caspase-8 activation. *J. Biol. Chem.* **273,** 2926–2930.
33. Scheel-Toellner, D., Wang, K., Singh, R., et al. (2002) The death-inducing signalling complex is recruited to lipid rafts in Fas-induced apoptosis. *Biochem. Biophys. Res. Commun.* **297,** 876–879.
34. Scaffidi, C., Medema, J. P., Krammer, P. H., and Peter, M. E. (1997) FLICE is predominantly expressed as two functionally active isoforms, caspase-8/a and caspase-8/b. *J. Biol. Chem.* **272,** 26,953–26,958.

35. Scaffidi, C., Fulda, S., Srinivasan, A., et al. (1998) Two CD95 (APO-1/Fas) signaling pathways. *EMBO J.* 17, 1675–1687.
36. Yin, X. M., Wang, K., Gross, A., et al. (1999) Bid-deficient mice are resistant to Fas-induced hepatocellular apoptosis. *Nature* **400,** 886–891.
37. Li, H., Zhu, H., Xu, C. J., and Yuan, J. (1998) Cleavage of BID by caspase 8 mediates the mitochondrial damage in the Fas pathway of apoptosis. *Cell.* **94,** 491–501.
38. Gross, A., Yin, X. M., Wang, K., et al. (1999) Caspase cleaved BID targets mitochondria and is required for cytochrome c release, while BCL-XL prevents this release but not tumor necrosis factor-R1/Fas death. *J. Biol. Chem.* **274,** 1156–1163.
39. Li, F., Srinivasan, A., Wang, Y., Armstrong, R. C., Tomaselli, K. J., and Fritz, L. C. (1997) Cell-specific induction of apoptosis by microinjection of cytochrome c. Bcl-xL has activity independent of cytochrome c release. *J. Biol. Chem.* **272,** 30,299–30,305.
40. Zamzami, N., Marchetti, P., Castedo, M., et al. (1995) Reduction in mitochondrial potential constitutes an early irreversible step of programmed lymphocyte death in vivo. *J. Exp. Med.* **181,** 1661–1672.
41. Kim, J. M. and Weisman, M. H. (2000) When does rheumatoid arthritis begin and why do we need to know? *Arthritis Rheum.* **43,** 473–484.
42. Lorenzo, H. K., Susin, S. A., Penninger, J., and Kroemer, G. (1999) Apoptosis inducing factor (AIF): a phylogenetically old, caspase-independent effector of cell death. *Cell Death. Differ.* **6,** 516–524.
43. Du, C., Fang, M., Li, Y., Li, L., and Wang, X. (2000) Smac, a mitochondrial protein that promotes cytochrome c-dependent caspase activation by eliminating IAP inhibition. *Cell* **102,** 33–42.
44. Verhagen, A. M., Ekert, P. G., Pakusch, M., et al. (2000) Identification of DIABLO, a mammalian protein that promotes apoptosis by binding to and antagonizing IAP proteins. *Cell* **102,** 43–53.
45. Zou, H., Henzel, W. J., Liu, X., Lutschg, A., and Wang, X. (1997) Apaf-1, a human protein homologous to *C. elegans* CED-4, participates in cytochrome c-dependent activation of caspase-3. *Cell* **90,** 405–413.
46. Li, P., Nijhawan, D., Budihardjo, I., et al. (1997) Cytochrome c and dATP-dependent formation of Apaf-1/caspase-9 complex initiates an apoptotic protease cascade. *Cell* **91,** 479–489.
47. Kluck, R. M., Bossy-Wetzel, E., Green, D. R., and Newmeyer, D. D. (1997) The release of cytochrome c from mitochondria: a primary site for Bcl-2 regulation of apoptosis. *Science* **275,** 1132–1136.
48. Yang, J., Liu, X., Bhalla, K., et al. (1997) Prevention of apoptosis by Bcl-2: release of cytochrome *c* from mitochondria blocked. *Science* **275,** 1129–1132.
49. Zheng, T. S., Schlosser, S. F., Dao, T., et al. (1998) Caspase-3 controls both cytoplasmic and nuclear events associated with Fas-mediated apoptosis in vivo. *Proc. Natl. Acad. Sci. USA* **95,** 13,618–13,623.
50. Kane, D., Balint, P. V., and Sturrock, R. D. (2003) Ultrasonography is superior to clinical examination in the detection and localization of knee joint effusion in rheumatoid arthritis. *J. Rheumatol.* **30,** 966–971.

51. Raza, K., Lee, C. Y., Pilling, D., et al. (2003) Ultrasound guidance allows accurate needle placement and aspiration from small joints in patients with early inflammatory arthritis. *Rheumatology. (Oxford).* **42,** 976–979.
52. Frasch, S. C., Henson, P. M., Nagaosa, K., Fessler, M. B., Borregaard, N., and Bratton, D. L. (2004) Phospholipid flip-flop and phospholipid scramblase 1 (PLSCR1) co-localize to uropod rafts in formylated Met-Leu-Phe-stimulated neutrophils. *J. Biol. Chem.* **279,** 17,625–17,633.
53. Dillon, S. R., Constantinescu, A. and Schlissel, M. S. (2001) Annexin V binds to positively selected B cells. *J. Immunol.* **166,** 58–71.
54. Salvioli, S., Ardizzoni, A., Franceschi, C., and Cossarizza, A. (1997) JC-1, but not DiOC6(3) or rhodamine 123, is a reliable fluorescent probe to assess delta psi changes in intact cells: implications for studies on mitochondrial functionality during apoptosis. *FEBS Lett.* **411,** 77–82.

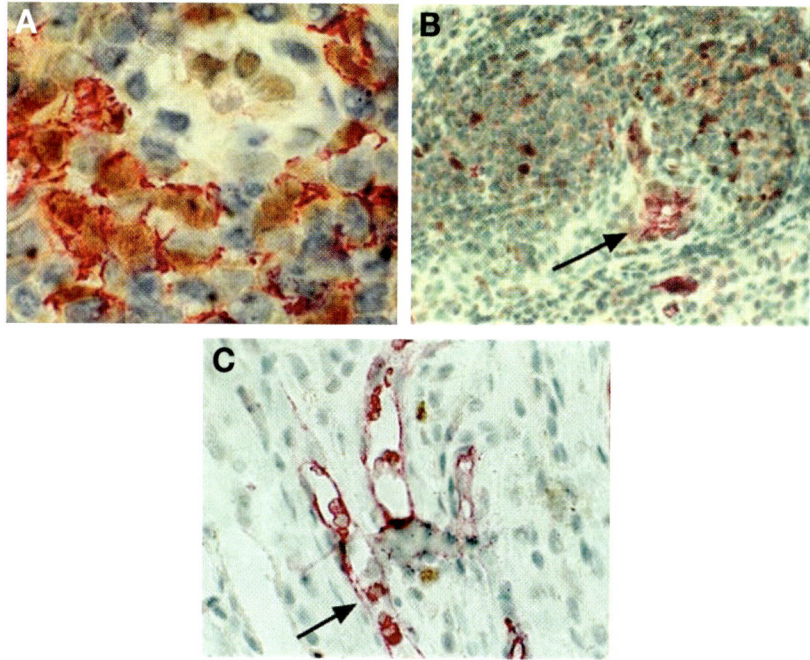

Fig. 3. Identification of synovial tissue DC by 2-color immunohistochemistry for Re1B and HLA-DR. (*See* discussion in Ch. 12 on p. 165 and complete caption on p. 176.)

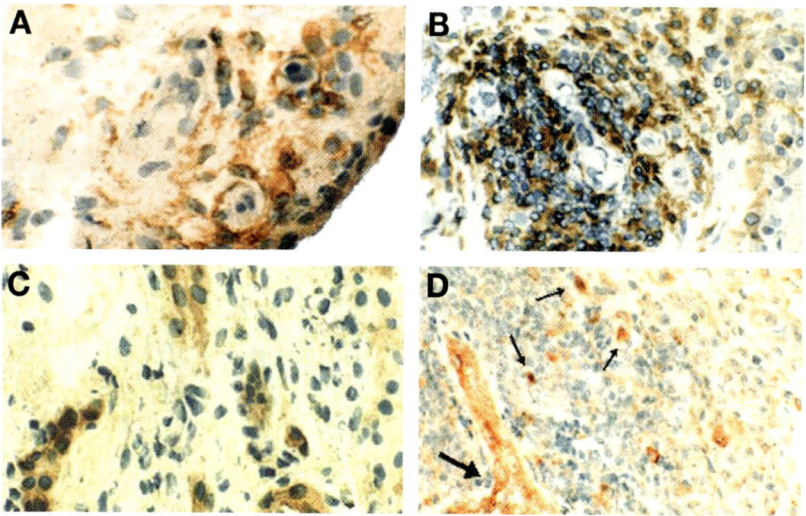

Fig. 4. CD11c and CD123 expression by DC in RA ST. (*See* discussion in Ch. 12 on p. 165 and complete caption on p. 178.)

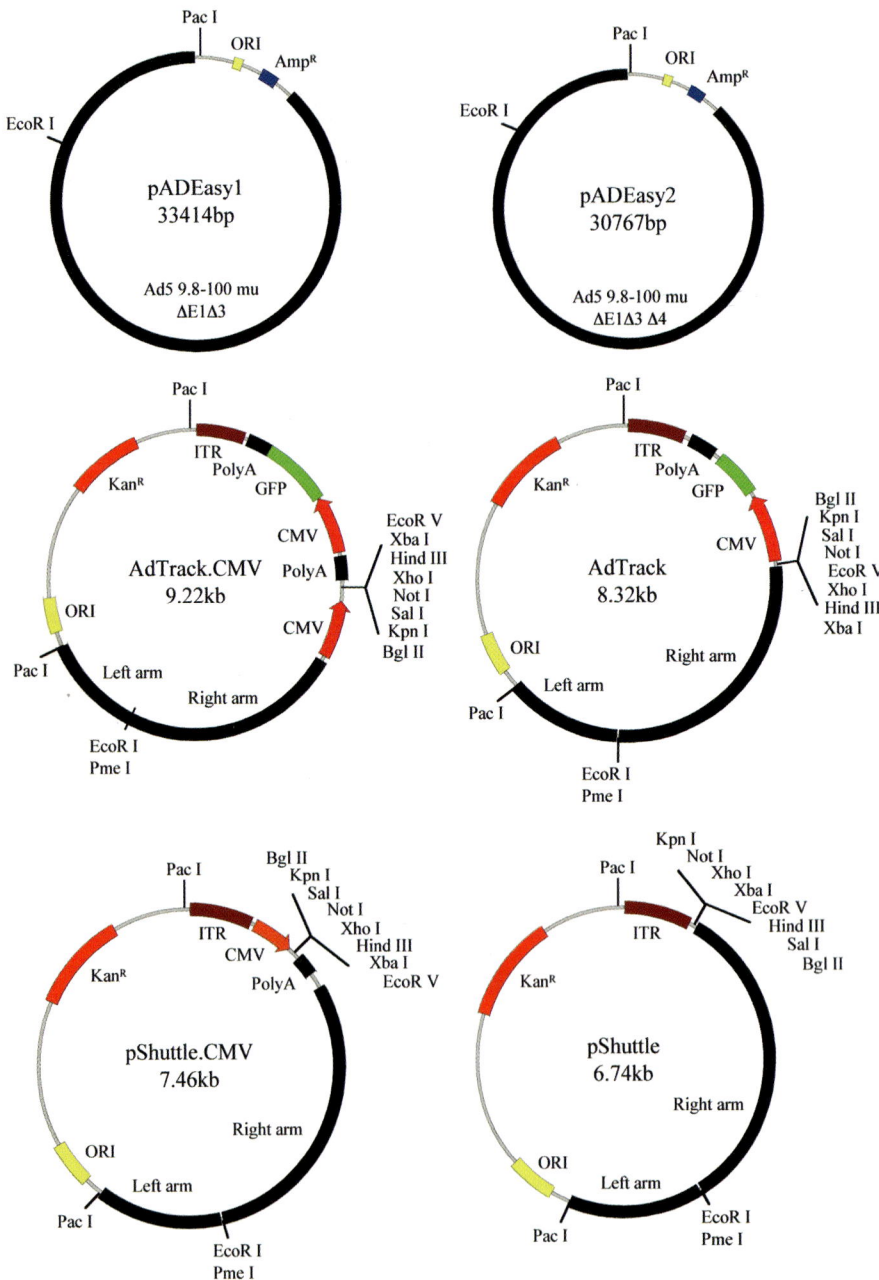

Fig. 2. Pictorial representation of the AdEasy system plasmid vectors. (*See* discussion in Ch. 27 on p. 395 and complete caption on p. 398.)

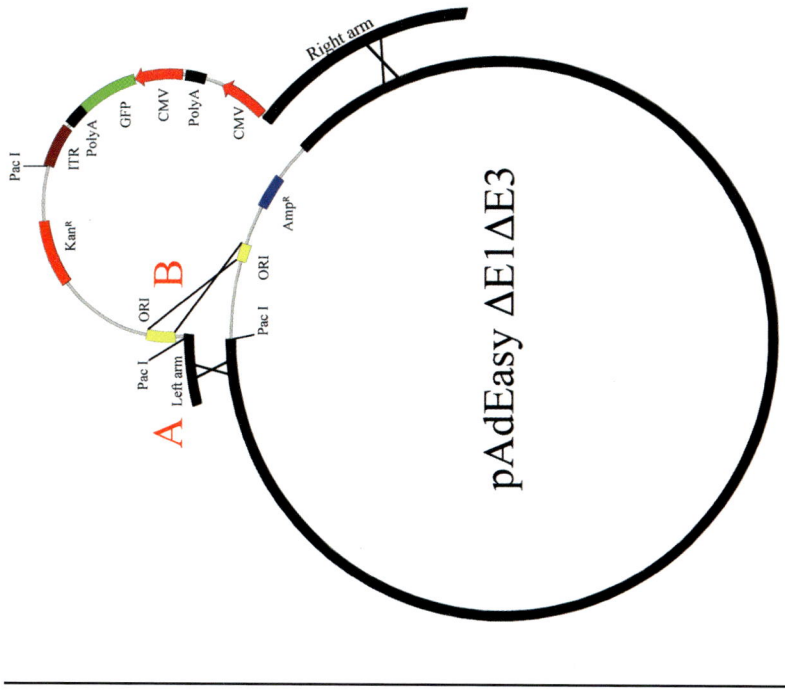

Fig. 4. Modes of in vivo bacterial recombination of the AdEasy vectors. (*See* discussion in Ch. 27 on p. 395 and complete caption on p. 400.)

Fig. 3. Pictorial representtion of the transfer vectors pAdKS17 and pAdSK16. (*See* discussion in Ch. 27 on p. 395 and complete caption on p. 399.)

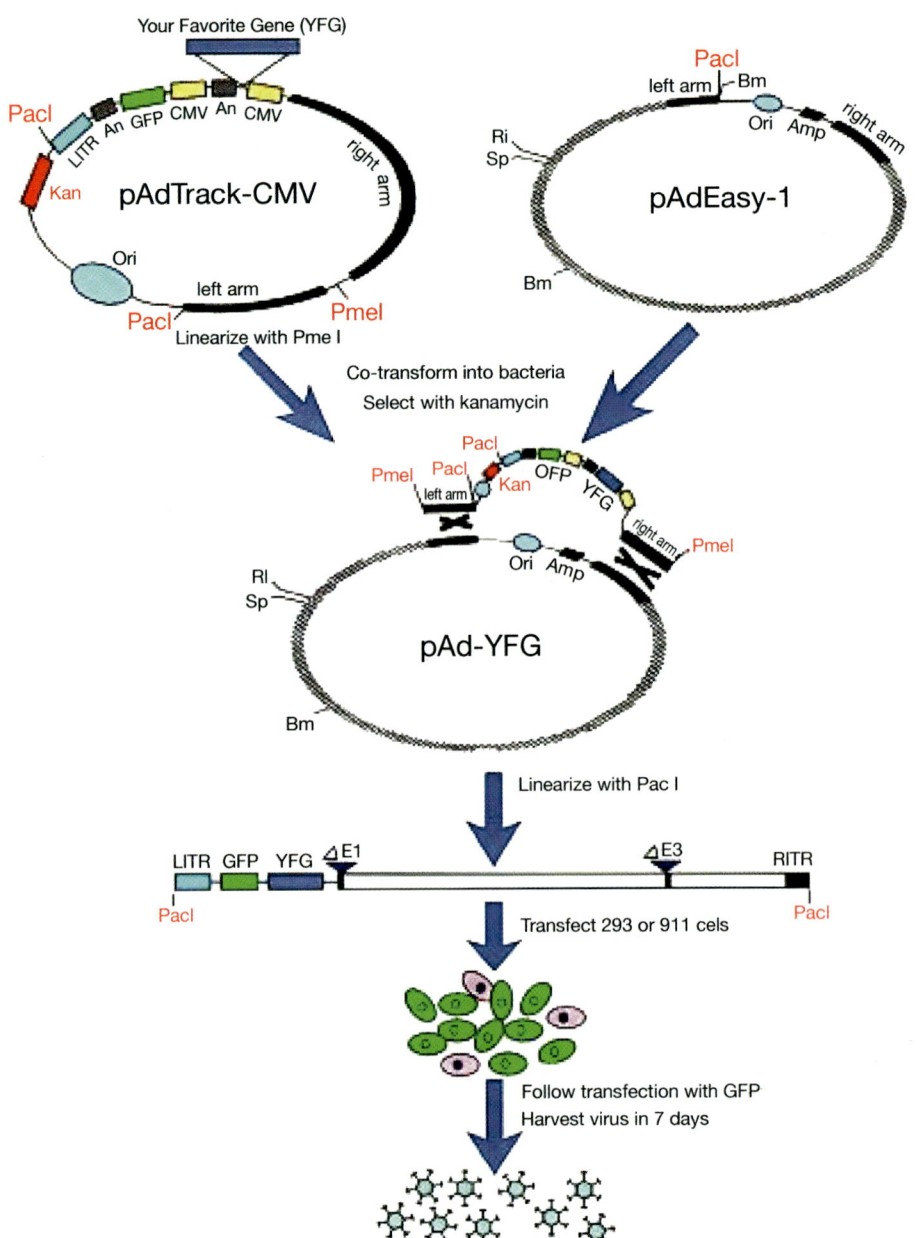

Fig. 6. Schematic representation of the AdEasy Vector System for the generation of recombinant adenoviruses. (*See* discussion in Ch. 27 on p. 395 and complete caption on p. 407.)

10

Assays of T-Cell Contact Dependent Monocyte-Macrophage Functions

Danielle Burger and Jean-Michel Dayer

Abstract

The production of cytokines and other inflammatory products in chronic/sterile inflammatory disorders such as rheumatoid arthritis might be induced in monocytes/macrophages by direct cellular contact with stimulated T-cells. Studies of cell–cell interactions are usually complicated by the simultaneous presence of at least two viable cell types. To obviate this problem, we developed strategies allowing only interactions between stimulated T-cell surface molecules and the monocytes/macrophages, and any products secreted into the medium or mRNA are necessarily of monocytic origin. This chapter aims at reviewing and describing the latter methods and strategies.

Key Words: Inflammation; cell–cell interactions; monocytes/macrophages; T-cells; cytokines; TNF; IL-1β; membrane isolation; membrane solubilization; cell isolation.

Like many chronic inflammatory diseases, rheumatoid arthritis (RA) is characterized by cell infiltration and proliferation, angiogenesis, inflammation, tissue destruction, and fibrosis. Joint destruction is partly the result of extracellular matrix degradation by proteolytic enzymes including matrix metalloproteinases and oxygen-free radicals, as well as the release of the mineral phase (Ca^{2+} release) by prostanoids such as prostaglandin E_2 (PGE_2). The production of metalloproteinases, O_2^- and PGE_2 is induced by the proinflammatory cytokines interleukin-1 (IL-1) and tumor necrosis factor (TNF) which play a major part in RA pathogenesis. IL-1 and TNF are mainly produced by monocyte-macrophages. In chronic inflammatory conditions, the production of IL-1 and TNF in monocyte-macrophages may arise through direct contact with stimulated T-cells at the inflammatory site *(1,2)*. This stimulation is independent of T-lymphocyte membrane-associated TNF or IL-1. Studies of cell–cell interactions are usually complicated by the simultaneous presence of at least two viable cell types. To obviate this problem, we developed strategies allowing only interactions between stimulated T-cell surface molecules and monocytes, such that any products secreted into the medium or induced mRNA are necessarily of monocytic origin. This chapter describes these methods and strategies.

2. Materials

1. Cell culture equipment, centrifuge, and Beckman rotors (e.g., SW60, SW28).
2. Eppendorf microcentrifuge and Falcon tubes (15 mL and 50 mL).
3. Sonicator with time control.
4. 10 mM ethylne diamine tetraaceti acid (EDTA) and 1 mM iodoacetamide.
5. Sonication buffer: ice-cold PBS containing 0.68M sucrose, 200 µM PMSF, 0.1 µg/mL pepstatin, and 5 mM EDTA.
6. T-cell membrane solubilisation buffer: ice-cold PBS containing 5 mM EDTA, 5 µM iodoacetamide, 1 µM leupeptin, 0.1 µg/mL pepstatin, and 200 µM PMSF.
7. 160 mM CHAPS (Uptima, cat.#no. UP333515) and PBS containing 8 mM CHAPS.
8. RPMI-1640 medium supplemented with 10% heat-inactivated fetal calf serum (FCS), 50 µg/mL streptomycin, 50 IU/mL penicillin and 2 mM L-glutamine (medium).
9. Paraformaldehyde powder, and 2% paraformaldehyde solution in PBS.
10. Polymyxin B sulfate (Sigma, catalogue #P1004)
11. Phytohemagglutinin (PHA) (E-Y Laboratories).
12. Phorbol-myristate-acetate (PMA).
13. Human myelomonocytic cell line THP-1 (ATCC) *(3)*.
14. Human cutaneous T-lymphoma HUT-78 cell line (ATCC) *(4)*.
15. Buffy coats (usually from Hospital or Red Cross blood centers).
16. Nylon wool columns: 10-mL syringe filled with nylon wool (Polysciences Inc.; cat. no. 18369) (*see* **Note 1**).
17. Sheep red blood cells (SRBC) (Biomérieux, 50% solution; cat. no. 72141).
18. 2-aminoethylisothiouronium bromide (AET) (Sigma; cat. no. A5879).
19. 4M NaOH
20. 22 µm Milipore filters.
21. Ficoll-Hypaque (Pharmacia Biotech) or Lymphoprep (Axis-Shield PoC AS)
22. Cytokine determination kits (Immunotechor Quantikine).
23. Bench top rolling machine (Spiramix 5, Denley Inc.).
24. UV lamp: Hg TUV 6WE (Cathodeon).
25. Unless stated otherwise, chemicals and reagents can be purchased from Sigma.

3. Methods

The methods described below outline the procedures for T-lymphocyte and monocyte isolation and the use of cell lines to study the effects of cell–cell contact between stimulated T-cells and monocytes (monocytic cells) on the production of various factors (mainly cytokines) by the latter cells.

3.1. T-Cells and T-Lymphocytes Preparation

To study contact-mediated activation of monocytes by T-cells, different T-cell types can be used: T-cell lines (e.g., HUT-78 cells and Jurkat cells) *(5)*, T-lymphocytes freshly isolated from peripheral blood (*see* **Subheading 3.1.1.**), or T-cell clones *(6,7)*.

3.1.1. Preparation of Peripheral Blood T-Lymphocytes

T-lymphocytes from human peripheral blood are prepared from buffy coats of healthy donors *(2)*.

1. Dilute the buffy coat with PBS to 100 mL, overlay 10 mL of Ficoll-Hypaque or Lymphoprep with 25 mL of diluted buffy coat, and centrifuge at 800g for 20 min at room temperature.
2. Collect peripheral blood mononuclear cells (PBMCs) at the interface and wash 3 times with PBS.
3. Resuspend the cells in 80 mL of RPMI-1640 medium, seed the cells into a large plastic plate (23 cm × 23 cm), and incubate for 1 h at 37°C. Meanwhile, equilibrate a nylon wool column (*see* **Subheading 3.1.1.1.**). Collect the supernatant, wash the plate thoroughly but not roughly with PBS, and centrifuge at 250g for 10 min. Discard the supernatant and resuspend the pellet in 1 to 2 mL of medium.

3.1.1.1. NYLON WOOL COLUMNS

1. Connect a nylon wool column (*see* **Note 1**) with a Connecta Plus 3 connector (Ohmeda). Equilibrate the column with 20 mL of medium, close the Connecta, and chase bubbles with a sterile pipet. Open the Connecta and wash the column with 20 mL of medium. Close the Connecta, overlay the nylon wool with 1 to 2 mL of medium, and close the top of the column with a cork (e.g., blue cork of 50-mL Falcon tube). Leave the column in a vertical position for at least 40 min at 37°C (cell incubator).
2. Warm 2 times 50 mL of medium up to 37°C for column elution.

3.1.1.2. DEPLETION OF B-LYMPHOCYTES AND REMAINING MONOCYTES ON TWO SUCCESSIVE NYLON WOOL COLUMNS

1. Open the Connecta, remove the excess of medium and close it. Load the cells (1–2 mL obtained as described in **Subheading 3.1.1., step 3**) onto the column, open the Connecta, and let the cells penetrate the wool. Close the Connecta. Add 1 mL of prewarmed (37°C) medium, and let penetrate the column). Add 2 to 3 mL of medium onto the top of the column to avoid drying. Close the top of the column with a cork, and incubate at 37°C for 45 min (incubator). Meanwhile, equilibrate a second nylon wool column.

2. Open the Connecta and collect the cells (drop by drop) in a 50-mL Falcon tube by washing the column with 20 to 30 mL of prewarmed medium. Centrifuge at 250g for 10 min. Resuspend the cells in 1 to 2 mL of medium and load on the second nylon wool column, incubate for 45 min at 37°C and elute as above. Resuspend the cells in the appropriate medium.

Usually, the T-cell population obtained consisted of >83% $CD3^+$ cells, <1% $CD14^+$ cells, <5% $CD19^+$ cells, and <10% $CD56^+$ cells.

3.1.2. Human T-Cell Lines

In addition to freshly isolated T-lymphocytes, the use of T-cell lines is an alternative to obviate experimental variability resulting from individual differences between blood donors. Good results can be obtained by using the HUT-78 human cutaneous T-lymphoma cell line *(4)* or the leukemic T-cell line Jurkat *(8)*.

3.2. T-Cell Activation

Although T-lymphocytes might be activated by different stimuli, activation by phytohemagglutinin (PHA) and phorbol-myristate-acetate (PMA) gives rise to high levels of activity. Optimal activation (maximal expression of activity) was determined for T-lymphocytes and HUT-78 cells.

3.2.1. T-Cell Stimulation With PHA/PMA

Each type of T-cell can be activated by PHA (1 µg/mL) and PMA (5 ng/mL) in medium, although stimulation time for optimal results differs depending on T-cell type.

1. T-lymphocytes (4×10^6 cells/mL) are stimulated for 48 h.
2. T-cell clones (4×10^6 cells/mL) are stimulated for 6 h.
3. HUT-78 cells (2×10^6 cells/mL) are stimulated for 6 h.

3.2.2. T-Cell Stimulation With Anti-CD3 Antibody (OKT3)

1. 24-well and 6-well plates are coated overnight at 4°C with 1 mL and 4 mL, respectively, of 1 µg/mL OKT3 in phosphate buffered saline (PBS), and washed 3 times with ice-cold PBS.
2. 1 mL and 4 mL of 2×10^6 T cells/mL in medium is added to each well of 24-well and 6-well plates, respectively, and cells are stimulated for 6 h.

3.3. T-Cell Fixation

T-cells are fixed in 1% paraformaldehyde. For optimal results, a 2% paraformaldehyde solution has to be prepared freshly on the day of utilization. Briefly, the required amount of PBS is heated to boiling point on a Bunsen flame. The Bunsen flame has to be stopped just before addition of paraformaldehyde powder, and the solution is chilled on ice immediately after paraformaldehyde dissolution.

1. Wash T cells 3 times in PBS and resuspend in 5 mL of PBS at 4°C (15 mL-Falcon tube). Add 5 mL of freshly prepared 2% paraformaldehyde solution in PBS (4°C), and incubate for 2 h at 4°C with gentle stirring on a "rolling machine."
2. After washing 3X with PBS (the cells are ready for use in contact activation of target cells. They can be used for 1 to 2 wk when stored in PBS at 4°C.

3.4. Preparation of T-Cell Membranes

1. Resuspend the cells (50×10^6 cells/mL maximum, $4–5 \times 10^6$ cells/mL minimum) in ice-cold sonication buffer. Depending on the volume, cells are placed in 1.5-mL Eppendorf tubes (1 mL/tube) or 50-mL Falcon tubes (10–25 mL/tubes). Cells can be frozen and kept at –20°C until used.
2. Keep the tubes on ice and sonicate for 5×5 s at 90 W in Eppendorf tubes, or 5×15 s in 50 mL tubes at 0°C. To avoid heating, leave at least 15 s intervals between each sonication step.
3. Centrifuge at 4000g for 15 min. and discard the pellet.
4. Centrifuge supernatants for 45 min at 100,000g in a swing-out rotor (i.e., 28,000 rpm in Beckman rotor SW60 or 24,000 rpm in Beckman rotor SW28).
5. Discard the supernatants and add 100 μL PBS or medium containing 20 μM EDTA (2 μL/mL, stock 10 mM), 5 μM iodoacetamide (5 μL/mL, stock 1 mM) to the pellet in SW60 tubes or 1 mL in SW28 tubes. Membrane pellets can be stored at –20°C.
6. To resuspend the pellets add the selected amount of buffer (usually 1 mL for small and 10 mL for large tubes) and sonicate 1×5 s (15 s for large tubes) as above. If some aggregates are still visible repeat sonication. Resuspended membranes can be stored at –20°C, but repeated cycles of freezing-thawing should be avoided. If aggregates are seen in thawed membrane preparations, sonicate as above.

3.5. Preparation of CHAPS Extracts From T-Cell Membranes

1. Resuspend the required amount of membrane pellets in 1 mL or 10 mL (depending on the tube) (i.e., 1.5-mL Eppendorf tube or 50-mL Falcon tube) of ice-cold T-cell membrane solubilization buffer.
2. Pool membrane solutions and add 10% of 160 mM CHAPS to obtain a final concentration of 16 mM CHAPS and a membrane concentration equivalent to 100×10^6 cells/mL.
3. Leave to solubilize for 1 h upon gentle agitation at 4°C, and centrifuge at 20,000 rpm (Beckman Rotor JA20).
4. Dialyze the supernatant for 24 h in PBS containing 8 mM CHAPS. The CHAPS extract (CE) can be kept at –20°C without significant loss of activity, although frequent freeze-thaw steps should be avoided.

3.6. Peripheral Blood Monocyte Preparation by Aggregation and Rosetting

To study contact-mediated activation of monocytes by T-cells, different types of monocytic cells can be used. We generally use THP-1 cells or freshly isolated blood monocytes *(9)* but MonoMac 6 and U937 can be used as alternatives. Finally, contact activation can be assessed in isolated human macrophages *(10)*. The preparation of monocytes from human blood is described in **Subheading 3.6.1.** and **3.6.2.**. This is an adaptation of the method described by Armant et al. *(11)*.

3.6.1. Treatment of Sheep Red Blood Cells (SRBC)

1. Isolate peripheral blood mononuclear cells (PBMC) by density gradient centrifugation (*see* **Subheading 3.1.1**).
2. Freshly prepare 2-aminoethylisothiouronium bromide (AET) solution (40.2 mg/mL). Add 10 mL H_2O to a 15 mL Falcon tube containing 0.603 g AET, adjust to pH 9.0 with 4M NaOH (about 23 drops) and adjust the volume to 15 mL. Sterilize by filtration using a 22-µm Milipore filter.
3. Wash SRBC 3X in PBS (5 min at 900g). Take 1 volume of the SRBC pellet, and add 4 volumes of AET solution.
4. Incubate for 15 min at 37°C and wash 3X with PBS (sometimes more washes are required to obtain a clear supernatant).
5. Resuspend the AET-SRBC pellet in heat-inactivated FCS (20% final cell concentration, (i.e., 500 µL cells and 2mL of heat-inactivated FCS).
6. Use at 1/10 dilution in medium (i.e., 2% final AET-SRBC concentration).
7. 20% stocks AET-SRBC can be kept at 4°C for at least 1 wk (check for hemolysis).

3.6.2. Monocyte Preparation

1. Wash PBMC 3X with PBS and resuspend the cells (50×10^6 cells/mL) in medium containing 1 µg/mL polymyxin B sulfate, and distribute 10 mL (max.) aliquots of cell suspension into 15-mL polypropylene tubes (Falcon; cat. no. 35.2097).
2. Incubate for 40 min at 4°C with gentle agitation (tubes in horizontal position) on a rolling machine (do not agitate the tubes upside down). Monocyte aggregates should be visible.
3. Leave the tubes in a vertical position on ice for 10 min in order to sediment aggregates.
4. With a long Pasteur pipette underlay 3 mL of cold, heat-inactivated FCS beneath aggregates (note, one full pipet is adequate).
5. Incubate on ice for 10 min. Monocytes penetrate the FCS layer when lymphocytes remain in the upper layer. Aspirate the supernatant and the very top of the FCS layer. Note that the upper layer can be used for the preparation of autologous T-lymphocytes.

6. Adjust the volume to 10 mL with complete medium containing 2 µg/mL polymyxin B sulfate and count the cells. Adjust cell concentration to 10×10^6 cells/mL with medium containing 2 µg/mL polymyxin B sulfate.
7. Mix 1 volume of monocyte solution with 1 volume of 2% AET-SRBC and distribute no more than 10 mL aliquots in 15 mL round bottom polypropylene tubes (Falcon; cat. no. 2059). Centrifuge for 5 min at $50g$ (500 rpm).
8. Incubate overnight at 4°C. If you are short of time, the rosetting can be carried out for 1.5 h on ice.
9. Gently resuspend the pellet by rolling the tubes slightly inclined between your hands.
10. Underlay 3 mL of Ficoll-Hypaque beneath the resuspended pellet with a long Pasteur pipet (nb one full pipet is adequate) and centrifuge at $900g$ for 20 min at room temperature.
11. Harvest the interphase that contains monocytes and wash the cells 3 times in PBS. If required, lyse the remaining SRBC (1 mL H_2O for 5 s followed by the addition of 10 mL of medium). Count the cells and resuspend at the required density. Keep at room temperature, because monocytes tend to aggregate in the cold.

3.7. Contact-Activation of Monocytes

Both freshly isolated monocytes and monocytic cell lines such as THP-1 cells (but also MonoMac 6 cells and U937 cells) can be used to assess the contact-activating capacity of stimulated T-cells. In order to avoid cross-stimulation by endotoxin, polymyxin B (2–5 µg/mL) should be added to culture medium. There are differences in sensitivity of monocytes/monocytic cells to contact-mediated activation, because freshly isolated monocytes are activated at a T-cell/monocyte ratio 10-times less that of monocytic cells (*9*). Thus, the amount of stimulus required to induce detectable production of cytokines has to be defined for each preparation of fixed T-cells, T-cell membranes, or CE by generating a dose response curve. Usually, with membranes or CE from PHA/PMA-stimulated HUT-78 cells, 5 to 7 µg/mL and 50 to 70 µg/mL proteins (final concentration) is appropriate for activating the production of cytokines and matrix metalloproteineases in freshly isolated monocytes and THP-1 cells, respectively. Fixed, PHA/PMA-stimulated T-cells have to be used at 1 to 4 and 4 to 16 T-cells by monocytes and THP-1 cells, respectively.

3.7.1. Cytokine Production by THP-1 Cells or Monocytes Upon Contact With Fixed, Stimulated T-Cells

1. Wash fixed, stimulated T-cells 3X with PBS and place them at the required density in 96-well plates in 150 µL medium. Sterilize the plates under ultraviolet (UV) irradiation for 5 min (UV lamp, *see* **Subheading 2., step 13**).
2. To each well add 50 µL of THP-1 cells or monocytes suspended (1×10^6 cells/mL) in medium containing 2 µg/mL polymyxin B. The final cell density is 50×10^3 cells/well/200 µL.

3. Culture at 37°C (cell incubator) and 5% CO_2 atmosphere for 24 h (monocytes) or 48 h (THP-1 cells).
4. Harvest culture supernatants and test for cytokine content using commercially available kits. Supernatants can be frozen before cytokine determination.

3.7.2. Cytokine Production by THP-1 Cells or Monocytes Activated by Membranes of Stimulated T-Cells

1. Add the required amount of isolated membranes to a total volume of 150 µL of medium in 96-well plates and sterilize the plates under UV irradiation for 5 min.
2. To each well add 50 µL of THP-1 cells or monocytes suspended (1×10^6 cells/mL) in medium containing 2 to 5 µg/mL polymyxin B. The final cell density is 50×10^3 cells/well/200 µL.
3. Culture as for monocyte and THP-1 cell activation with fixed T-cells.

3.7.3. Cytokine Production by THP-1 Cells or Monocytes Activated by CE of Stimulated T-Cell Membranes

1. Add 25 µL of the predetermined dilution of CE in PBS, containing 8 mM CHAPS, to a total volume of 125 µL of medium in 96-well plates and sterilize the plates under UV irradiation for 5 min. Alternatively, CE can be sterilized by filtration (cut-off: 22 µm) (*see* **Note 2**). To each well add 50 µL of THP-1 cells or monocytes suspended (1×10^6 cells/mL) in medium containing 2 to 5 µg/mL polymyxin B. The final cell density is 50×10^3 cells/well/200 µL.
2. Culture as for monocyte and THP-1 cell activation with fixed T-cells.

3.7.4. Monocyte and THP-1 Cell mRNA Analysis Upon Contact With T-Cells

Because fixed T-cells contain mRNA and can be solubilized by chaotropic agents, the induction of transcripts has to be analyzed by using either isolated membranes or CE.

1. 5×10^6 monocytes or THP-1 cells are stimulated in 24-well culture plates with the appropriate amount of membranes or CE. In the latter case, the final concentration of CHAPS should be 1 mM.
2. Depending on the read-out product, cells are cultured for between 1 and 24 h in medium containing 5 µM polymyxin (*see* **Note 3**).
3. Harvest cell mRNA using your usual method.
4. A broad analysis of cytokine mRNA can be carried out with RNase protection assay system kits (PharMingen). Usually 2 to 3 µg and 10 µg of total RNA from monocytes and monocytic cell line, respectively, are required to visualize the transcripts of around 10 cytokines (*12*).

4. Notes
1. In order to optimize T-lymphocyte purification, nylon wool columns should be packed with nylon wool that has been depleted of wool packs and aggregates.

This is achieved by using two pet brushes to card nylon wool. An optimal amount of 0.7 g of carded nylon wool.10 mL syringe should be used and the syringe piston should not be pushed beyond 5 mL. When the piston is removed, the wool level is around 9 to 10 mL. The column must be sterilized. If a sterilization center facility is not available, irradiate the columns at 3500 Rad.
2. To maintain optimal activity, a CHAPS concentration of 1 mM has to be maintained throughout the cell activation. This does not affect target cells (i.e., monocytes, macrophages, or monocytic cells) and keeps the stimulus soluble.
3. The incubation time has to be defined as a function of the read-out product. In freshly isolated monocytes, steady-state levels of IL-1β, IL-1Rα, and TNF-α mRNA are reached between 1 and 3 h. Metalloproteinase transcripts require 12 to 18 h to reach steady-state levels. In THP-1 cells, stimulation for longer time periods are required to observe steady-state levels of cytokines (3–6 h).

Acknowledgments

The authors are indebted to Mrs. R. Chicheportiche, L. Gruaz, and M.-T. Kaufmann for their skilful technical help in setting up the methods described in this chapter. Our research is supported by the Swiss National Science Foundation.

References

1. Burger, D. and Dayer, J. M. (2002) The role of human T lymphocyte-monocyte contact in inflammation and tissue destruction. *Arthritis Res.* **4 (3)**, S169–S176.
2. Vey, E., Zhang, J. H., and Dayer, J.-M. (1992) IFN-gamma and 1,25(OH)2D3 induce on THP-1 cells distinct patterns of cell surface antigen expression, cytokine production, and responsiveness to contact with activated T cells. *J. Immunol.* **149**, 2040–2046.
3. Tsuchiya, S., Yamabe, M., Yamaguchi, Y., Kobayashi, Y., Konno, T., and Tada, K. (1980) Establishment and characterization of a human acute monocyte leukemia cell line (THP-1). *Int. J. Cancer* **26**, 171–176.
4. Gazdar, A. F., Carney, D. N., Bunn, P. A., et al. (1980) Mitogen requirements for the in vitro propagation of cutaneous T-cell lymphomas. *Blood* **55**, 409–417.
5. Isler, P., Vey, E., Zhang, J. H., and Dayer, J. M. (1993) Cell surface glycoproteins expressed on activated human T-cells induce production of interleukin-1β by monocytic cells: a possible role of CD69. *Eur. Cytokine Netw.* **4**, 15–23.
6. Chizzolini, C., Chicheportiche, R., Burger, D., and Dayer, J. M. (1997) Human Th1 cells preferentially induce interleukin (IL)-1β while Th2 cells induce IL-1 receptor antagonist production upon cell/cell contact with monocytes. *Eur. J. Immunol.* **27**, 171–177.
7. Li, J. M., Isler, P., Dayer, J. M., and Burger, D. (1995) Contact-dependent stimulation of monocytic cells and neutrophils by stimulated human T-cell clones. *Immunology* **84**, 571–576.

8. Nagasawa, K., Howatson, A., and Mak, T. W. (1981) Induction of human malignant T-lymphoblastic cell lines MOLT-3 and Jurkat by 12-O-tetradecanoylphorbol-13-acetate: biochemical, physical, and morphological characterization. *J. Cell Physiol* **109**, 181–192.
9. Hyka, N., Dayer, J. M., Modoux, C., et al. (2001) Apolipoprotein A-I inhibits the production of interleukin-1beta and tumor necrosis factor-alpha by blocking contact-mediated activation of monocytes by T lymphocytes. *Blood* **97**, 2381–2389.
10. Ferrari-Lacraz, S., Nicod, L. P., Chicheportiche, R., Welgus, H. G., and Dayer, J. M. (2001) Human lung tissue macrophages, but not alveolar macrophages, express matrix metalloproteinases after direct contact with activated T lymphocytes. *Am. J. Respir. Cell Mol. Biol.* **24**, 442–451.
11. Armant, M., Rubio, M., Delespesse, G., and Sarfati, M. (1995) Soluble CD23 directly activates monocytes to contribute to the antigen-independent stimulation of resting T cells. *J. Immunol.* **155**, 4868–4875.
12. Molnarfi, N., Gruaz, L., Dayer, J. M., and Burger, D. (2004) Opposite effects of IFNβ on cytokine homeostasis in LPS- and T cell contact-activated human monocytes. *J. Neuroimmunol.* **146**, 76–83.

11

Phenotypic and Functional Analysis of Synovial Natural Killer Cells

Nicola Dalbeth and Margaret F. C. Callan

Abstract

Natural Killer (NK) cells are cells of the innate immune system with characteristic effector functions, including recognition and lysis of virus-infected or tumor cells and production of immunoregulatory cytokines, particularly interferon-gamma (IFN-γ). NK cells account for between 10 and 15% of peripheral blood lymphocytes and are also present in synovial fluid and tissue where they might potentially contribute to amplification of the inflammatory process through interactions with macrophages and dendritic cells. This chapter outlines methods of assessing the phenotype of NK cells through analysis of NK-cell markers and the function of NK cells through cytotoxicity assays and measurement of cytokine production.

Key Words: Natural Killer cells; phenotype; immunohistochemistry; cytotoxicity; cytokine; arthritis.

1. Introduction

Natural killer (NK) cells are an important component of the innate immune system. These cells account for between 10 and 15% of peripheral blood lymphocytes and are defined by expression of CD56 and lack of expression of CD3 *(1)*. Their activity is regulated by integration of both activatory and inhibitory signals from a wide range of cell surface receptors *(2)*. These include members of the C-type lectin family, such as CD94 and NKG2A, B, C, and D, the natural cytotoxicity receptors (NCRs), leukocyte immunoglobulin-like receptor-1 (LILR-1), and members of the immunoglobulin superfamily, such as the killer immunoglobulin-like receptors (KIRs). NK cells have a number of effector functions, including recognition and lysis of virus-infected or tumor cells and production of immunoregulatory cytokines, particularly interferon-gamma (IFN-γ).

Two subsets of NK cells in peripheral blood have been documented *(3)*. The majority (CD56^{dim} NK cells) express moderate levels of CD56 and high levels of CD16. The CD56^{dim} NK cells usually express KIRs and are heterogeneous with respect to expression of CD94 and NKG2A. The minor subset of NK cells (CD56^{bright} NK cells) accounts for only 10% of circulating NK cells. These cells express high levels of CD56, CD94, and NKG2A and tend to lack expression of CD16 and the KIRs *(4)*. The two subsets of NK cells differ also in terms of chemokine receptor *(5)* and adhesion molecule expression, suggesting that they have different homing properties *(6)*. Lastly they show important functional differences; the CD56^{dim} subset has superior cytotoxic capacity *(4)*, whereas the CD56^{bright} subset has greater ability to produce proinflammatory cytokines on exposure to low concentrations of monokines *(3)*. We have demonstrated that the CD56^{bright} subset of NK cells is expanded in inflamed synovial fluid *(7)*. These cells express CD69, a lymphocyte marker of recent activation, and respond to low doses of monokines by producing interferon (IFN)-γ (**Table 1**).

This chapter describes protocols to analyze the phenotype of NK cells in synovial fluid and tissue. Methods to analyse the functional capacity of NK cells, namely cytotoxicity and cytokine production have also been included.

2. Materials
2.1. Phenotypic Analysis of Synovial NK Cells
2.1.1. Isolation of Mononuclear Cells from Blood, Synovial Fluid, and Synovial Tissue

1. Freshly obtained samples from patients: paired blood and synovial fluid collected in 50-mL Falcon tubes with 100 U unfractionated heparin. Synovial tissue samples collected in 50-mL Falcon tubes containing 20 mL of complete medium (*see* **Subheading 2.1.1., item 4**)
2. RPMI 1640 (Gibco).
3. Lymphoprep (Nycomed Pharma) (protect from light and store at room temperature).
4. Complete medium: RPMI 1640 supplemented with 10% fetal calf serum (FCS), 2 m*M* glutamine, 100 U/mL penicillin and 100 µg/mL streptomycin.
5. Collagenase Type IV (Sigma).
6. Cell strainer (70 µ*M*) (BD Biosciences).

2.1.2. Analysis of NK Cell Surface Markers

1. Mononuclear cells obtained from blood, synovial fluid, and synovial tissue (*see* **Subheadings 2.1.1.** and **3.1.1.**).

Table 1
NK Cell Markers in Blood and Synovial Fluid

NK cell marker	Ligand	$CD56^{dim}$ NK cell dominates in peripheral blood	$CD56^{bright}$ NK cell dominates in synovial fluid/tissue
CD56 (NCAM)	Unknown	Mid expression	High expression
CD16 (FcγRIII)	IgG	++	– (or low-frequency +)
CD94	HLA-E	±	++
NKG2A	HLA-E	±	++
KIR	HLA Class I	±	– (or low-frequency +)
CD62L	PEN5	±	++
CCR7	ELC, SLC	–	+
CD69	unknown	–	– (peripheral blood) ± (synovial fluid)
Perforin	–	++	–
Granzyme A	–	++	±

2. FACS buffer: phosphate buffered saline (PBS) supplemented with 0.16% bovine serum albumin (BSA) and 0.1% sodium azide.
3. Primary monoclonal antibodies (mAbs): anti-CD3-PE (BD Biosciences), anti-CD56-PC5 (Immunotech), anti-CD16-FITC (Dako), anti-CD94-FITC (BD Biosciences), anti-CD69-FITC (BD Biosciences), anti-CD158a (for KIR2DL1 and KIR2DS1) (Immunotech), anti-CD158b (for KIR2DL3, KIR2DS2, and KIR2DL2) (Immunotech), anti-KIR3DL1 (clone DX9, R&D), anti-CD62L-FITC (BD Biosciences), anti-CCR7 (BD Biosciences), anti-NKG2A (clone Z199, Immunotech), and isotype control IgG-FITC (BD Biosciences).
4. Secondary mAbs: rabbit F(ab')$_2$ anti-mouse IgG-FITC (Dako).
5. Human serum (Sigma).
6. Mouse serum (Sigma).
7. FACS FIX: 1% FCS and 1% formaldehyde in PBS. Store at 4°C and protect from light.
8. Flow cytometer and software. All flow cytometry protocols described have been optimized for the FACSCalibur 4-color flow cytometer and analysis may be performed using CellQuest software.

2.1.3. Analysis of Perforin and Granzyme Expression by NK Cells

1. Mononuclear cells obtained from blood, synovial fluid, and synovial tissue (*see* **Subheadings 2.1.1.** and **3.1.1.**).
2. FACS buffer (*see* **Subheading 2.1.2., item 2**).
3. Cytofix/cytoperm kit (BD Biosciences): Cytofix contains 4% formaldehyde. Cytoperm buffer (1X) contains PBS with 0.1% saponin and 1% FCS.
4. Primary antibodies: anti-CD3-APC (BD Biosciences), anti-CD56-PC5 (Immunotech), anti-granzyme A-FITC (BD Biosciences), anti-perforin-PE (BD Biosciences), and isotype controls IgG-FITC (BD Biosciences) and IgG-PE (BD Biosciences).
5. FACS FIX (*see* **Subheading 2.1.2., item 7**).
6. Flow cytometry machine and software (*see* **Subheding 2.1.2., item 8**).

2.1.4. Detection of CD56+ Cells in Fixed Synovial Tissue

1. Samples: slides of formaldehyde-fixed, paraffin-embedded synovial tissue and small cell lung cancer (positive control) (*see* **Note 1**).
2. Citroclear (HD Supplies).
3. 100% ethanol.
4. Retrieval buffer (pH 6.0) (Dako).
5. Goat serum (Sigma).
6. Antibodies: anti-CD56 (Zymed) and isotype control (mouse IgG1).
7. EnVision+ Kit (Dako): containing peroxidase block (0.03% hydrogen peroxide containing sodium azide), peroxidase labeled polymer conjugated goat anti-mouse IgG, buffered substrate and DAB+ chromogen.
8. Haematoxylin (Dako).
9. Aquamount (BHD) and coverslips.
10. Light microscope ($\times 20x–\times 800$).

2.2. Isolation of NK Cells and NK Cell Subsets

2.2.1. Isolation of NK Cells by Magnetic Selection

1. Freshly isolated mononuclear cells from blood, synovial fluid or synovial tissue (*see* **Subheadings 2.1.1.** and **3.1.1.**).
2. Buffer: PBS with 2 m*M* ethylene diamine tetraacetic acid (EDTA) and 0.5% BSA. Degas and filter with 0.20 µm filter before use. Store at 4°C.
3. Anti-CD56 MACS beads (Miltenyi).
4. LS MACS column (Miltenyi).
5. MACS magnet and stand (Miltenyi).
6. Anti-CD3 Dynabeads (Dynal).
7. Dynal MPC magnet (Dynal).
8. Complete medium (*see* **Subheading 2.1.1., item 4**).

2.2.2. Isolation of Blood NK Cell Subsets by FACS Sorting

1. Freshly isolated blood mononuclear cells (*see* **Subheadings 2.1.1.** and **3.1.1.**).

2. FACS sort buffer: PBS with 2 m*M* EDTA and 0.5% BSA. Degas and filter with 0.20 µm filter before use. store at 4°C.
3. Antibodies: anti-CD56-PE (BD Biosciences) and anti-CD3-PE-Cy5 (BD Biosciences).
4. Cell strainer (70 µm) (BD Biosciences).
5. FACS sorting apparatus.

2.3. Functional Analysis of NK Cells

2.3.1. Cytokine Analysis by Intracellular Staining

1. Freshly isolated mononuclear cells from blood, synovial fluid and synovial tissue (*see* **Subheadings 2.1.1.** and **3.1.1.**).
2. Complete medium (*see* **Subheading 2.1.1., item 4**).
3. Cytokines: IL-12, IL-15, IL-18 (all from R&D).
4. Brefeldin A (Sigma) resuspended to a stock solution of 5 mg/mL in 100% ethanol.
5. FACS buffer (*see* **Subheading 2.1.2., item 2**).
6. Cytofix/cytoperm kit (BD Biosciences) (*see* **Subheading 2.1.3., item 3**).
7. mAbs: anti-CD3-PE, anti-IFNγ-APC, anti-tumor necrosis factor (TNF)-γ-APC, IgG-APC (all from BD Biosciences) and anti-CD56-PC5 (Immunotech).
8. FACS FIX (*see* **Subheading 2.1.2., item 7**).
9. Flow cytometry machine and software (*see* **Subheading 2.1.2., item 8**).

2.3.2. Cytotoxicity Assays

1. Effector cells: blood and synovial fluid NK cells isolated as per **Subheading 3.2.1.** or NK cell subsets isolates as per **Subheading 3.2.2.**
2. Target cells: K562 cell line (ATCC).
3. ^{51}Chromium (Amersham Pharmacia Biotech).
4. RPMI 1640 (Gibco).
5. Complete medium (*see* **Subheading 2.1.1., item 4**).
6. Triton-X (Sigma).
7. Appropriate safety equipment including ^{51}Chromium radioactivity monitors and lead containers for transporting/handling tubes.
8. 96-well & counter plates with sealing tape,
9. Scintillation liquid (Supermix, note light sensitive).
10. Micro & plate reader.
11. Minishaker.

3. Methods

These protocols require the use of human biological samples with the associated risk of transmission of fluid and tissue-borne pathogens. All work should be undertaken in class II containment (or class III if there is a significant risk of infection with ACDP risk group 3 pathogens).

3.1. Phenotypic Analysis of Blood and Synovial NK Cells

3.1.1. Isolation of Mononuclear Cells From Blood, Synovial Fluid, and Synovial Tissue

3.1.1.1. Digestion of Fresh Synovial Tissue Specimens

1. Collect and transport fresh synovial tissue in 50-mL Falcon tubes with complete medium.
2. Cut synovial tissue samples into small pieces using a sterile scalpel.
3. Digest tissue with 1 mg/mL collagenase Type IV in complete medium for 1 h at 37°C in the presence of 5% CO_2 (see **Note 2**).
4. Pass the resulting suspension through a sterile cell strainer.
5. Wash three times in complete medium and resuspend in RPMI 1640.

3.1.1.2. Isolation of Mononuclear Cells From Blood, Synovial Fluid, and Synovial Tissue

1. Use samples of blood or synovial fluid in 50-mL Falcon tubes with 100 U unfractionated heparin or cell suspension from digested synovial tissue.
2. To each sample add RPMI 1640 (at 37°C) at 1:1 ratio.
3. Layer 12 mL of sample/RPMI 1640 over 6 mL Lymphoprep in a 25-mL universal tube and centrifuge at 280g (see **Note 3**) for 30 min at room temperature with the centrifuge brake off.
4. Gently remove the mononuclear layer found just above the Lymphoprep and transfer into a new universal tube using a pasteur pipette and add RPMI to a total volume of 25 mL (see **Note 4**).
5. Centrifuge at 775g for 15 min at room temperature. The centrifuge brake may be left on.
6. Pour off supernatant and resuspend cells in 10 mL of RPMI.
7. Centrifuge at 280g for 10 min at room temperature. The centrifuge brake may be left on.
8. Discard supernatant and resuspend cells in complete medium

3.1.2. Analysis of NK Cell Surface Markers

There are significant differences between blood and synovial NK cell surface and intracellular markers. These differences are summarized in **Table 1**. In this protocol, all samples are labeled for CD3 and CD56. Other individual NK cell markers are detected by fluorescein isothiocyanate (FITC)-labeled mAbs. An example of surface staining of some of these markers on blood and synovial fluid NK cells is shown in **Fig. 1**.

1. Aliquot 10^6 mononuclear cells (from blood, SF or ST) into each FACS tube, add 1 mL FACS buffer, centrifuge at 280g (see **Note 3**) for 5 min at 4°C and decant the supernatant. The addition of FACS buffer, centrifugation and decanting of supernatant will subsequently be termed "washing."

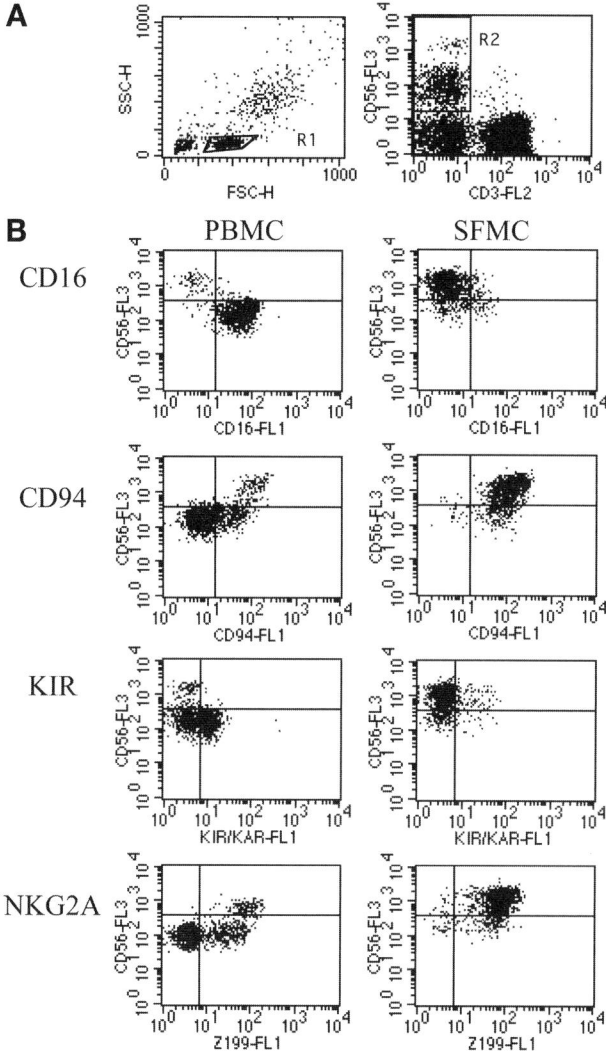

Fig. 1. Surface staining for NK cell phenotypic markers. Peripheral blood and synovial fluid mononuclear cells (PBMC and SFMC) from a patient with rheumatoid arthritis were stained for NK cell surface markers (*see* **Subheading 3.1.2.**). (**A**) Gating of NK cells: this is a typical dot plot for PBMC following staining for CD3 and CD56. NK cells lie within the lymphocyte gate (R1) and the CD3$^-$/CD56$^+$ gate (R2). (**B**) Blood NK cells are predominantly of the CD56^{dim} phenotype with medium expression of CD56, high expression of CD16 and variable expression of CD94, KIRs, and NKG2A. In a minority of blood NK cells and the majority of SF NK cells, there is high expression of CD56 with low expression of CD16 and the KIRs, and high expression of CD94 and NKG2A.

2. Add one of the following primary antibodies to individual FACS tubes and incubate for 30 min on ice (protect from light): anti-CD16-FITC, anti-CD94-FITC, anti-CD62L-FITC, anti-CD69-FITC, anti-CCR7, anti-NKG2A, and isotype control IgG-FITC. For all FITC-conjugated and unconjugated antibodies use 5 μL antibody solution/10^6 cells. The three mAb specific for the killer immunoglobulin receptors may be added separately or in combination to the same sample: for 10^6 cells, 5 μL 58.1/5 μL 58.2/ 50 μL DX9 (*see* **Note 5**).
3. Wash with FACS buffer.
4. For antibodies that are not flurochrome conjugated, add 50 μL 1:50 rabbit F(ab')$_2$ anti-mouse IgG-FITC in FACS buffer supplemented by 1.5% human serum and incubate for 30 minon ice (protect from light).
5. Wash with FACS buffer.
6. To all samples add anti-CD3-PE (1 μL/10^6 cells) and anti-CD56-PC5 (2.5 μL/10^6 cells) in 20 μL FACS buffer supplemented by 1.5% human serum and 1% mouse serum (*see* **Note 6**). Incubate for 30 min on ice (protect from light).
7. Wash with and resuspend in FACS buffer. Alternatively, if there is to be a delay prior to analysis resuspend in FACS FIX and keep in the dark at 4°C.
8. Acquire data on flow cytometer that allows for 4-color analysis, such as a FACS Calibur and analyze using an appropriate software programme such as CellQuest.

3.1.3. Analysis of NK Cell Intracellular Granules

An example of intracellular perforin and granzyme A staining of blood and synovial fluid NK cells is shown in **Fig. 2**.

1. Add 10^6 mononuclear cells (from blood, SF or ST) to FACS tubes and wash with FACS buffer (*see* **Subheading 3.1.2., step 1**).
2. To each tube, add 2.5 μL anti-CD3-APC and 2.5 μL anti-CD56-PC5 and incubate for 30 min on ice (protect from light).
3. Wash with FACS buffer.
4. Fix cells with 250 μL Cytofix buffer for 20 min on ice (vortex every 10 min).
5. Centrifuge cells at 435*g* for 5 min at 4°C and discard supernatant.
6. Incubate with 1 mL 1X Cytoperm buffer for 10 min.
7. Centrifuge cells at 435*g* for 5 min at 4°C and discard supernatant.
8. Add 5 μL anti-granzyme A-FITC and 5 μL anti-perforin-PE in 90 μL 1X Cytoperm buffer (*see* **Note 7**), or 5 μL IgG -FITC and 5 μL IgG-PE in 90 μL 1X Cytoperm buffer, and incubate for 30 min on ice (protect from light).
9. Wash with and re-suspend in FACS buffer. Alternatively, if there is to be a delay prior to analysis resuspend in FACS FIX and keep in the dark at 4°C.
10. Acquire and analyse data (*see* **Subheading 3.1.2., step 8**).

3.1.4. Detection of NK Cells in Fixed Synovial Tissue

3.1.4.1. Dewaxing and Rehydrating of Paraffin Sections

Add slides to solutions in the following order: Citroclear, 6 min; Citroclear, 6 min; 100% alcohol, 5 min; 100% alcohol, 5 min; 50% alcohol, 5 min; and H$_2$O, 5 min.

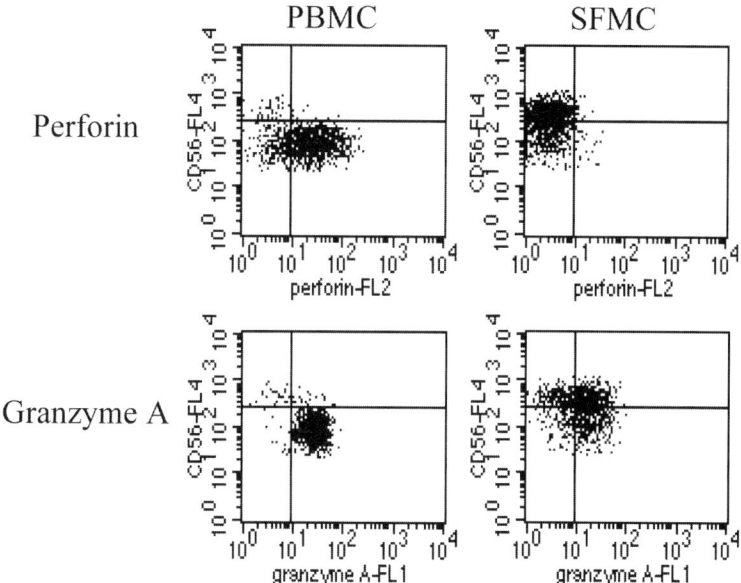

Fig. 2. Intracellular granule staining of NK cells. Peripheral blood and synovial fluid mononuclear cells (PBMC and SFMC) from a patient with rheumatoid arthritis were stained for intracellular granules as outlined in **Subheading 3.1.3.** The majority of CD56^{dim} NK cells in blood are positive for intracellular perforin and granzyme A. CD56^{bright} NK cells in blood and SF NK cells are perforin negative and have reduced granzyme A staining.

3.1.4.2. Antigen Retrieval by Microwave

1. Preheat retrieval buffer (4–5 min at 700W).
2. Transfer slides to a glass tray and place in the buffer.
3. Incubate for 10 min at 95 to 100°C (80W).
4. Remove slides and submerge in PBS for 5 min.

3.1.4.3. Staining Procedure (All Reagents and Incubations at Room Temperature)

1. Block slides with 10% goat serum in PBS for at least 5 min.
2. Add peroxidase block for at least 5 min, then wash with PBS.
3. Incubate slides with 1/50 anti-CD56 or 1/50 mouse IgG in 10% goat serum in PBS for 60 min, then wash with PBS.
4. Incubate with peroxidase labeled polymer conjugated goat anti-mouse IgG for 30 min, then wash with PBS.
5. Prepare DAB solution, and apply to specimens for 5 min, then wash with distilled water.

6. Counterstain with haematoxylin and wash with water after 2 to 3 s, transfer to water bath, then back to PBS (until tissue color changes to blue).
7. Mount slides with Aquamount onto coverslip.
8. Acquire images with light microscope.

3.2. Isolation of NK Cell and NK Cell Subsets

Isolation of NK cells or NK cell subsets may be required for functional assessment of NK cells (e.g., cytotoxicity assays), some cytokine assays (where cytokine production is not measured on a single cell level), and experiments analyzing the interaction of NK cells with other cells.

3.2.1. Isolation of NK Cells by Magnetic Selection (see **Note 8**)

3.2.1.1. POSITIVE SELECTION OF $CD56^+$ CELLS

1. Resuspend mononuclear cells in 80 µL buffer/10^7 total cells (if less than 10^7 total cells, add 80 µL buffer), then add 20 µL anti-CD56 beads/ 10^7 total cells (if less than 10^7 total cells, add 20 µL beads). Mix well and incubate for 15 min at 6 to 12°C.
2. Wash cells by adding 1 mL buffer /10^7 total cells, centrifuge at $193g$ (*see* **Note 3**) for 10 min at 4°C, discard supernatant and resuspend in 1 mL buffer.
3. Set up LS column in magnet and flush with 3 mL buffer.
4. Place fresh tube under column and add cell suspension to top, collecting $CD56^-$ cells.
5. Wash through column with 3×3 mL buffer, collecting the remaining CD56 negative cells.
6. Remove column from magnet and place over fresh tube. Add 5 mL buffer to top and use plunger to force buffer through column, collecting $CD56^+$ cells.
7. Centrifuge the cell suspension at $280g$ for 5 min at room temperature.

3.2.1.2. DYNABEAD NEGATIVE SELECTION OF $CD3^+$ CELLS

1. Resuspend $CD56^+$ cells at $0.25/2.5 \times 10^6$ cells/mL buffer.
2. Wash anti-CD3 labeled dynabeads with buffer.
3. Add the sample of $CD56^+$ cells to the washed dynabeads (use at least 4 dynabeads/target cells and $>2 \times 10^7$ beads/sample).
4. Incubate sample for 20 min at 2 to 8°C (rotating).
5. Place the sample in the Dynal magnet and leave for at least 2 to 3 min.
6. Transfer the supernatant to a fresh tube and wash NK cells in complete medium.
7. Retain small aliquot for purity analysis (by surface staining using anti-CD56 and anti-CD3 mAbs).

3.2.2. Isolation of NK Cell Subsets by FACS Sorting

1. Obtain blood mononuclear cells
2. Resuspend cells in sterile FACS sort buffer and stain with anti-CD56-PE and anti-CD3-PE-Cy5 for 30 min on ice (protect from light).

Synovial Natural Killer Cells

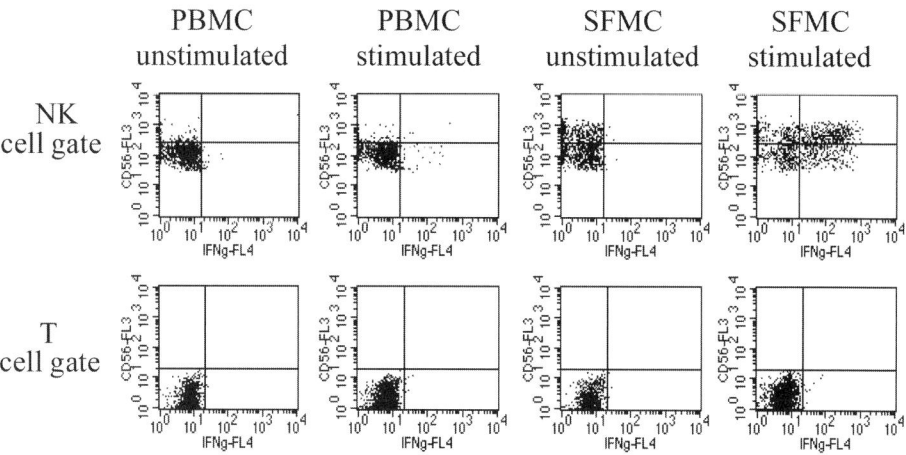

Fig. 3. Analysis of cytokine production on a single cell level. Peripheral blood and synovial fluid mononuclear cells (PBMC and SFMC) from a patient with rheumatoid arthritis were stimulated and stained for intracellular IFN-γ as outlined in **Subheading 3.3.1.** Following stimulation, SF NK cells have a greater capacity to produce IFN-γ compared with blood NK cells, or with T-cells (CD3$^+$ lymphocyte gate) from blood or synovial fluid.

3. Wash cells with PBS and pass through cell strainer.
4. Sort NK cells based on presence within lymphocyte gate, expression of CD56 and absence of CD3. NK-cell subsets may be sorted based on intensity of CD56 and other cell surface markers such as CD16 expression.
5. Wash cells with complete medium.
6. Retain small aliquot of each subset for purity analysis.

3.3. Functional Analysis of NK Cells

3.3.1. Cytokine Analysis by Intracellular Staining (see **Note 9**)

An example of cytokine analysis of blood and synovial fluid NK cells is shown in **Fig. 3**.

1. Isolate mononuclear cells from blood, synovial fluid and synovial tissue.
2. Resuspend cells at 2×10^6/mL and incubate in culture medium alone or culture medium supplemented with IL-12 (5 ng/mL) and IL-18 (10 ng/mL) in 24-well plates for 18 h at 37°C in the presence of 5% CO_2.

3. Add Brefeldin A after 2 h of incubation to a final concentration of 5 µg/mL.
4. After 18 h, wash cells in FACS buffer and surface stain for anti-CD3-PE (1 µg/10^6 cells), anti-CD56-PC5 (2.5 µg/10^6 cells).
5. Fix and permeabilise cells as outlined in **Subheading 3.1.3**.
6. Add anti-IFN-γ-APC (1/200), anti-TNF-γ-APC (1/100), or IgG-APC (1/100) in 100 µl 1x Cytoperm buffer (*see* **Note 7**) and incubate for 60 min on ice (protect from light).
7. Wash with and resuspend in FACS buffer. Alternatively, if there is to be a delay prior to analysis resuspend in FACS FIX and keep in the dark at 4°C.
8. Acquire and analyse data (*see* **Subheading 3.1.2., item 8**).

3.3.2. Cytotoxicity Assays (see **Note 10**)

This protocol involves the use of ionising radiation. Users should be trained in the handling of radioactive material before use. Monitoring equipment and appropriate safety measures must be used.

1. On the day prior to the assay, resuspend K562 cells at 2×10^5 cells/mL in complete medium.
2. On the day of assay, obtain samples and separate mononuclear cells.
3. Separate NK cells or NK cell subsets as outlined in **Subheading 3.2**.
4. Resuspend K562 cells to 5×10^6 cells in 100 µL complete medium. Label K562 cells with ^{51}Chromium by adding 200 µCi and incubating for 1 h at 37°C.
5. Wash K562 cells with RPMI 1640 3 times and perform viable cell count.
6. Resuspend K562 cells in complete medium at 5×10^4 cells/mL and use 100 µL of cell suspension/well (5000 cells).
7. Resuspend NK cells in complete medium and use 100 µL of cell suspension/well.
8. Add cells to 96-well U bottomed plates in triplicate: K562 cells alone, NK:K562 at ratios 20:1, 10:1, 5:1, 1:1 and K562 cells in 10% Triton-X (*see* **Note 11**). Incubate for 4 h at 37°C in the presence of 5% CO_2.
9. Add 100 µL scintillation liquid into each well of a 96-well & counter plate. Add 35 µL supernatant from the sample plate to the scintillation liquid. Seal plate using sealing tape and mix with minishaker.
10. Obtain data on Beta plate reader.
11. Calculate % specific lysis = 100 × (cpm experimental wells – cpm spontaneous release)/(cpm maximum release–cpm spontaneous release). Maximum release = K562 cells in Triton X, and spontaneous release = K562 in complete medium

4. Notes

1. The majority of small cell lung cancers (SCLC) express CD56 (NCAM) (*8*). Therefore, SCLC tissue is a useful positive control for immunohistochemistry.
2. Collagenase treatment of synovial tissue may lead to cleavage of CD56 from the cell surface (*see* **Fig. 4**). This effect may result from protease contamination of collagenase preparations (*9*) and can be reduced by adding additional serum to the culture, reducing the concentration and duration of collagenase exposure, and

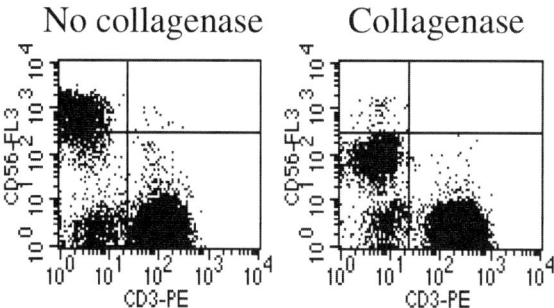

Fig. 4. Collagenase preparations reduce the florescence intensity of CD56. Synovial fluid mononuclear cells from a patient with rheumatoid arthritis were cultured in complete medium with and without 5 mg/mL collagenase type IV (Sigma) for 1 h, then washed and stained for surface expression of CD3 and CD56.

culturing collagenase-exposed cells in complete medium for 24 h before surface staining.
3. This speed is used on a centrifuge with a centrifugal radius (r) of 173 mm. For use of centrifuges with other r values, the equivalent revolutions/min (rpm) may be calculated using the formula:

$$\text{Relative centrifugal force (RCF)} = (\text{RPM}/1000)^2 \times r \times 1.12$$

4. When removing the mononuclear cells from Lymphoprep, take care not to remove Lymphoprep with the cell layer as the presence of Lymphoprep will prevent formation of the lymphocyte pellet in the next spin.
5. Different individuals have different KIR repertoires, and single cells may express more than one KIR *(10)*. Therefore, pooling of the available mAbs to the various KIRs allows an estimation of the percentage of cells expressing the KIRs to be tested.
6. The addition of mouse serum at this stage is critical if a rabbit anti-mouse antibody (e.g., FITC-conjugated) has been added previously. The mouse serum should bind to any excess rabbit anti-mouse antibody and prevent antibody from binding to antibodies that are added subsequently. Failure to add the mouse serum will result in the subsequently added murine antibodies fluorescing on the FITC axis as well as on the appropriate axis for their individual fluorochromes.
7. Antibodies used for intracellular staining must be added in the presence of Cytoperm buffer to ensure that cells remain permeabilized throughout the incubation.
8. NK cells can be separated using a number of methods *(11–13)*. In our hands, purities using negative selection of CD3 (T-cells), CD14 (monocytes), and CD19 (B-cells) were inconsistent and often poor. Purities were significantly and consistently improved by positive selection of CD56 cells, followed by negative selection of CD3 cells (*see* **Fig. 5**).

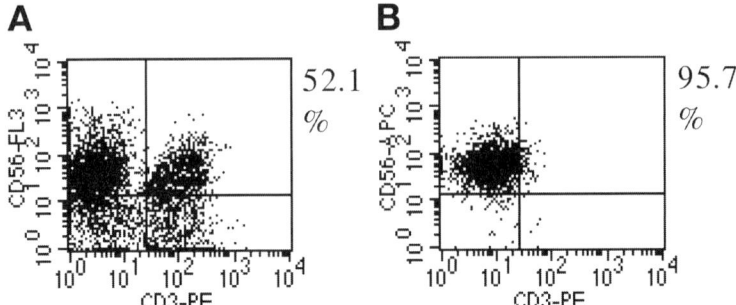

Fig. 5. Positive selection of NK cells. (**A**) Purity analysis of blood NK cells following negative selection of CD3, CD14, and CD19 positive cells using pan-mouse IgG Dynabeads. (**B**) Purity analysis of blood NK cells following CD56$^+$ selection by MACS followed by negative selection using anti-CD3 Dynabeads. Values indicate the percentage of NK cells (CD3$^-$, CD56$^+$) in the live cell gate.

9. A number of methods for detection of cytokine production by NK cells have been described. Intracellular staining following incubation with Brefeldin A offers the advantage of detecting cytokine production at a single cell level (*14*), and allows comparison of cytokine production between NK cells and other cell types and also between different NK cell subsets without the need for separation of NK cells (*see* **Fig. 3**).
10. This protocol describes analysis of NK cell cytotoxicity using freshly isolated ex vivo NK cells. The lymphokine IL-2 activates NK cells and may be added to NK cell cultures to enhance cytotoxicity (*15*).
11. Triton-X wells should be separated from other wells to avoid contamination of wells containing NK cells.

References

1. Cooper, M. A., Fehniger, T. A., and Caligiuri, M. A. (2001) The biology of human natural killer-cell subsets. *Trends. Immunol.* **22**, 633–640.
2. McQueen, K. L. and Parham, P. (2002) Variable receptors controlling activation and inhibition of NK cells. *Curr. Opin. Immunol.* **14**, 615–621.
3. Cooper, M. A., Fehniger, T. A., Turner, S. C., et al. (2001) Human natural killer cells: a unique innate immunoregulatory role for the CD56(bright) subset. *Blood* **97**, 3146–3151.
4. Jacobs, R., Hintzen, G., Kemper, A., et al. (2001) CD56bright cells differ in their KIR repertoire and cytotoxic features from CD56dim NK cells. *Eur. J. Immunol.* **31**, 3121–3127.
5. Campbell, J. J., Qin, S., Unutmaz, D., et al. (2001). Unique subpopulations of CD56+ NK and NK-T peripheral blood lymphocytes identified by chemokine receptor expression repertoire. *J. Immunol.* **166**, 6477–6482.

6. Fehniger, T. A., Cooper, M. A., Nuovo, G. J., et al. (2003) CD56bright natural killer cells are present in human lymph nodes and are activated by T cell-derived IL-2: a potential new link between adaptive and innate immunity. *Blood* **101**, 3052–3057.
7. Dalbeth, N. and Callan, M. F. (2002) A subset of natural killer cells is greatly expanded within inflamed joints. *Arthritis Rheum.* **46**, 1763–1772.
8. Kaufmann, O., Georgi, T., and Dietel, M. (1997). Utility of 123C3 monoclonal antibody against CD56 (NCAM) for the diagnosis of small cell carcinomas on paraffin sections. *Hum. Pathol.* **28**, 1373–1378.
9. Mulder, W. M., Koenen, H., van de Muysenberg, A. J., Bloemena, E., Wagstaff, J., and Scheper, R. J. (1994) Reduced expression of distinct T-cell CD molecules by collagenase/DNase treatment. *Cancer Immunol. Immunother.* **38**, 253–258.
10. Uhrberg, M., Valiante, N. M., Shum, B. P., et al. (1997) Human diversity in killer cell inhibitory receptor genes. *Immunity* **7**, 753–763.
11. Ferlazzo, G., Tsang, M. L., Moretta, L., Melioli, G., Steinman, R. M., and Munz, C. (2002) Human dendritic cells activate resting natural killer (NK) cells and are recognized via the NKp30 receptor by activated NK cells. *J. Exp. Med.* **195**, 343–351.
12. Gerosa, F., Baldani-Guerra, B., Nisii, C., Marchesini, V., Carra, G., and Trinchieri, G. (2002) Reciprocal activating interaction between natural killer cells and dendritic cells. *J. Exp. Med.* **195**, 327–333.
13. Piccioli, D., Sbrana, S., Melandri, E., and Valiante, N. M. (2002) Contact-dependent stimulation and inhibition of dendritic cells by natural killer cells. *J. Exp. Med.* **195**, 335–341.
14. Pala, P., Hussell, T., and Openshaw, P. J. (2000) Flow cytometric measurement of intracellular cytokines. *J. Immunol. Methods* **243**, 107–124.
15. Domzig, W., Stadler, B. M., and Herberman, R. B. (1983) Interleukin 2 dependence of human natural killer (NK) cell activity. *J. Immunol.* **130**, 1970–1973.

12

Identification and Isolation of Synovial Dendritic Cells

Allison R. Pettit, Lois Cavanagh, Amanda Boyce, Jagadish Padmanabha, Judy Peng, and Ranjeny Thomas

Abstract

In rheumatoid arthritis patients, three compartments need to be considered: peripheral blood, synovial fluid, and synovial tissue. Dendritic cells characterized from each compartment have different properties. The methods given are based on cell sorting for isolation of cells, and flow cytometry and immunohistochemical staining for analysis of cells in these compartments.

Key Words: Dendritic cells; synovial tissue; synovial fluid; peripheral blood; rheumatoid arthritis; flow cytometry; immunohistochemistry.

1. Introduction

In rheumatoid arthritis patients, three compartments need to be considered: peripheral blood (PB), synovial fluid (SF), and synovial tissue (ST). Dendritic cells (DCs) characterized from each compartment have different properties. The methods given are based on cell sorting for isolation of cells, and flow cytometry and immunohistochemical staining for analysis of cells. Myeloid non-T-cells are first enriched by density gradient centrifugation, sheep erythrocyte rosetting, and, in some cases, magnetic immunodepletion. By flow cytometry, DCs can then be analyzed or sorted based on 2- or 3-color immunofluorescence. Some variations on this basic theme are also outlined. The basic protocol for 1- and 2-color immunohistochemistry of formalin-fixed synovial tissue are then given. The 2-color protocol is based on the localization of the NF-κB family member, RelB, to the nucleus of differentiated DCs, and exclusion of B-cells, macrophages and follicular DCs by double staining. Single color protocols examine various DC populations based on several recent publications. Some variations, that are particularly useful in frozen sections, follow.

2. Materials
2.1. Solutions and Reagents

1. Phosphate Buffered Saline (PBS): 8 g NaCl, 0.2 g KCl, 0.92 g Na_2HPO_4 (anhydrous), and g KH_2PO_4. Add H_2O to final volume of 1 L. Filter or autoclave sterilize and adjust to pH 7.4. Store at room temperature or at 4°C for up to 6 mo for use in cell preparation.
2. Hanks Balanced Salt Solution (HBSS): HBSS without sodium bicarbonate is purchased from Sigma (St. Louis, MO). Bicarbonate is added on preparation of the solution: 5.4 mM KCl, 0.3 mM Na_2HPO_4, 0.4 mM KH_2PO_4, 4.2 mM $NaHCO_3$, 1.3 nM $CaCl_2$, 0.5 mM $MgCl_2$, 0.6 mM $MgSO_4$, 137 mM NaCl, and 0.02% phenol red (optional). Add H_2O to final volume of 1 L and adjust to pH 7.4. Filter sterilize. Store at 4°C for up to 6 mo.
3. RPMI supplemented information: Penicillin G (200 U/mL), gentamicin (10 mg/mL), and L-glutamine (0.3 mg/mL). Store at 4°C for up to 3 mo after the addition of l-glutamine, and check manufacturer's specifications beyond this.
4. 10X NH_4Cl: 82.9 g NH_4Cl, 10 g $KHCO_3$, 327 mg ethylene diamine tetraacetic acid (EDTA). Make up to 1 L with millipore water. Dilute 1:10 for 1X NH_4Cl and filter.
5. 1% and 4% Paraformaldehyde: 1 mL 1M NaOH. Heat at 65°C until dissolved. Add 10mL 10X PBS. Cool to room temperature. Adjust to pH 7.4. Adjust volume to 100mL (4%) with H_2O. Filter sterilize. Dilute 1:4 with H_2O for 1% PFA. Store at 4°C in the dark. Wear gloves and avoid breathing vapors.
6. 10 mM EDTA: Add 37 mg of disodium salt EDTA to 800 mL of H_2O and adjust to pH 7.5. Note that the pH is very unstable and should be checked just before use. Store at room temperature for up to 6 mo.
7. 10 mM citrate buffer: Add 8.4 g citric acid to 4 L H_2O and adjust to pH 5.0 with 2M NaOH. Check pH before use. Store at room temperature for up to 6 mo.
8. Tris buffered saline (TBS): 7.88 g Tris-HCl and 8.8 g NaCl. Add H_2O to 1 L and adjust to pH 7.6. Store at room temperature for up to 6 mo.
9. Serum Block—10% fetal calf serum (FCS) 10% swine serum in TBS: 1 mL of swine serum and 0.01% azide. Add TBS to 10 mL. Store at 4°C for up to 2 wk.
10. 3% Milk block: 3 g skim milk powder. Add TBS to 100 mL and store at 4°C for up to 5 d.
11. Peroxidase Block: 0.5% H_2O_2 in TBS. Stable for 1 d at 4°C, light sensitive.
12. Collagenase: 100 mg collagenase (Type XI, Sigma), 5 mL Hepes buffer, 1 mL heat inactivated FCS, and 94 mL HBSS. Stir for 3 min on a magnetic stirrer, filter, and aliquot.
13. Strep-ABC-alkaline phosphatase (DAKO): Follow kit instructions. Add 5mL Tris buffer to 1 drop each of vial A and vial B. Make fresh each time required.
14. Diaminobenzidine (DAB) (DAKO): Add 1 drop DAB to 1mL TBS. Make fresh each time required.
15. Fast Red c hromogen (DAKO): Kit buffer is stored at 4°C and fast red tablets at –18 to –20°C. Add 1 tablet to 2 mL buffer, vortex to dissolve. Use filter dropper provided with the kit to add to slide. Solution is stable for 3 d.

16. Neuraminidase (GIBCO-BRL).
17. Nylon mesh filters (Becton Dickenson).
18. Ficoll-Paque, Ficoll-diatrizoate, research grade (Pharmacia Biotech).
19. MACS immunomagnetic beads (Miltenyi Biotech).
20. MACS reagents and columns (BD Pharmingen).
21. Recombinant granulocyte macrophage-colony stimulating factor (GM-CSF) (Schering-Plough) and IL-4 (Peprotech, Rocky Hill).
22. Streptavidin conjugated horseradish peroxidase (HRP) ().
23. Vectorbond-coated slides (Vector Laboratories).
24. Cytospin centrifuge (Shandon).

2.2. Antibodies

1. Anti-CD16, clone Leu-11b for NK cells (BD Pharmingen)
2. Anti-CD19, clone Leu-12 for B-cells (BD Pharmingen).
3. Anti-CD56, clone Leu-19 for NK cells (BD Pharmingen).
4. Anti-CD3, clone OKT3 for T-cells (American Type Culture Collection).
5. Anti-CD33, clone Leu-M9 for myeloid enrichment (BD Pharmingen).
6. Anti-CD14, clone Leu-M3 for myeloid enrichment (BD Pharmingen).
7. CMRF-44 (BD Pharmingen).
8. CMRF-56 (BD Pharmingen).
9. Mouse and Rabbit Ig (DAKO).
10. Biotinylated anti-rabbit and anti mouse Ig (DAKO).
11. Anti-DR, clone TAL IB5 (DAKO).
12. Anti-CD20, clone LZ6 (DAKO).
13. Anti-CD23, clone NCL-CD23-1612 (Novacastra Laboratories).
14. Anti-CD68, clone PG-M1 (DAKO).
15. Anti-CD11c (BD Pharmingen).
16. Anti CD123, clone 9F5 (BD Pharmingen).

3. Methods

3.1. Analysis of Cells in Suspension by Flow Cytometry and Isolation by Flow Cytometric Sorting

3.1.1. Preparation of Synovial Tissue Cell Suspension

Collect fresh ST into sterile medium. Gently tease the tissue into small pieces with scissors and forceps and add 1 mL collagenase. Digest at 37°C for 1 to 2 h, mixing every 20 min, until the pieces have disintegrated. Filter the suspension through 70 µm nylon mesh to remove tissue debris. Wash in a 50 mL volume of PBS by centrifuging for 10 min at 410g at 10°C. Resuspend in 20 mL 10% FCS/RPMI.

3.1.2. Preparation of N-SRBC

1. Add 10 mL of sheep's blood in Alsever's solution to a 50-mL tube and add saline to 50 mL to wash.

2. Centrifuge for 5 min at 910g at 10°C.
3. Aspirate the supernatant carefully as pellet is soft. Add 40 µL of neuraminidase. Add HBSS to 40 mL and mix.
4. Incubate at 37°C in the water bath for 1 h and mix every 10 min. Alternatively, use a shaker-water bath set for gentle agitation.
5. Fill to 50 mL with saline to wash.
6. Centrifuge for 5 min at 910g at 10°C.
7. Aspirate supernatant, top with 50 mL HBSS and centrifuge as in **step 6**. Repeat this wash step 3 times.
8. After last wash, resuspend in 50 mL of RPMI +PGG. Store at 4°C for up to 1 wk.

3.1.3. Preparation of Dendritic Cells from Peripheral Blood, Synovial Fluid, or Synovial Tissue Cell Suspension

3.1.3.1. PREPARATION OF MONONUCLEAR CELLS AND ERYTHROCYTE ROSETTE FRACTIONS

1. When preparing cells from a fresh SF it is necessary to prevent clumping of cells and consequent cell loss. Dilute the SF 1:2 with saline, and filter the suspension through 70 µm nylon mesh filters into 50-mL tubes. Aliquot 40 mL/tube. If extensive cell clumping occurs at any stage during the cell purification procedure, repeat this filtration step. Aliquot 20 mL PB or ST cell suspension into 50-mL tubes and dilute 1:2 with saline.
2. Underlay with 10 mL Ficoll-Paque.
3. Centrifuge for 45 min at 180g at room temperature with no brake (density gradient centrifugation).
4. Carefully aspirate 20 mL of supernatant from the top of the solution and discard.
5. Collect monolayer containing mononuclear cells into 50-mL tubes by gently aspirating just above the monolayer. Remove as much supernatant as possible without disturbing the red blood cell (RBC) and granulocyte pellet using this technique. Discard RBC pellet.
6. Wash cells in a 50 mL volume with saline by centrifuging for 10 min at 410g at 10°C. Resuspend pellet in 1 mL saline using a 1-mL pipet, disrupting any cell clumps.
7. Repeat wash step 2 times. Pool the resuspended cells each time so that by the final wash the cells are in a single tube. Count the cells after the second wash and resuspend after the final wash at 10^7 cells/mL.
8. To rosette the T-cells, aliquot 5 mL of the cell suspension into 50-mL tubes and to each tube aliquot 2.5 mL of both FCS and N-SRBC and mix. Incubate for 10 min in a 37°C water bath.
9. Centrifuge for 10 min at 170g at 10°C. Incubate, without resuspending the pellet, for 1 h at 4°C. Gently resuspend pellet using a 10-mL pipet and underlay with 10 mL Ficoll-Paque as in **step 2**.
10. Centrifuge for 35 min at 180g at room temperature with no brake. Harvest the monolayer as in **step 5**. The monolayer is the erythrocyte-negative non-T-cell

Dendritic Cells

fraction and contains DCs as well as other cells, and the pellet is the erythrocyte-positive T-cell fraction.

11. Discard the pellet if T-cells are not required. Wash the non-T-cells in a 50 mL volume of PBS by centrifuging for 10 min at 410g at 10°C. Add 10 mL NH$_4$Cl to each T-cell pellet and invert several times for approx 3 to 5 min. This will lyse the contaminating N-SRBC.
12. Pool cell suspensions. If RBC contaminate the non-T-cell pellet, lyse with 10 mL NH$_4$Cl. Make up the volume to 50 mL with PBS and centrifuge for 6 min at 410g at 10°C.
13. Wash the T- and non-T-cells again in a 50 mL volume of PBS by centrifuging for 6 min at 410g at 10°C.
14. Count the cells. For yield and purity *see* **Note 1**.

3.1.3.2. Myeloid Enrichment of Non-T-Cells

Non-T-cells are enriched for myeloid cells by negative selection. Cells are first labeled with anti-CD16, anti-CD19, anti-CD56, and OKT3 then depleted with immunomagnetic beads, using the MACS system.

1. Resuspend non-T-cells (*see* **Subheading 3.1.3.1.**) in 1 mL PBS containing 1% FCS and add 5 µg/mL of each antibody: anti-CD16, –CD19, –CD56, and CD3.
2. Incubate at room temperature for 10 min, or on ice for 1 h.
3. Wash cells once in a 50 mL volume of PBS by centrifuging at 410g at 10°C for 6 min. Resuspend pellet in 0.5 mL of 1% FCS PBS.
4. Add 1 µL of goat anti-mouse IgG for every 10^6 cells that are expected to be depleted (i.e., expect to deplete 1/3 of the cells) and gently shake. Do not vigorously mix the cell-microbead mixture or the weak microbead-antibody bond will be disrupted.
5. Incubate at 4°C for 15 min.
6. During the incubation, set up the MACS separator. Choose a depletion column for the appropriate number of cells being selected. For example, the AS column is suitable for up to 3×10^7 positive cells (*see* **Note 2**). Place the column in the magnetic field of the MACS separator with a 3-way-stopcock. Add enough 1% FCS/PBS to fill the column and attach a flow resistor to the 3-way-stopcock (25-gage needle, supplied).
7. Mix cell suspension gently and, using a pasteur pipet, apply the cell suspension to the depletion column. Allow the cells to run through but do not let the column run dry. Pass the cells through the column approximately 3 times. Rinse with 3 column volumes of 1% FCS/PBS (6 mL). Change flow resistor to a higher flow rate (23-gage needle) and rinse column with approximately 20 mL of 1% FCS/PBS. The myeloid-enriched cell fraction is contained in the effluent.
8. Wash in a 50 mL volume by centrifugation at 410g for 6 min at 10°C.
9. Resuspend in 1% FCS/PBS for staining.

3.1.3.4. FLOW CYTOMETRIC ANALYSIS OF DENDRITIC CELLS

1. To the myeloid-enriched non-T-cell suspension add 5 μg/mL of phycoerythrin (PE)-conjugated anti-CD33 and FITC-conjugated anti-CD14 and incubate for 30 min on ice or for 10 min at room temperature. Control tubes will contain:
   a. Negative 5 μg/mL of PE-conjugated IgG and FITC-conjugated IgG.
   b. 5 μg/mL of PE-CD33 alone.
   c. 5 μg/mL of fluorescein isothiocyanate (FITC)-CD14 alone.
2. Wash 3 times in PBS.
3. To fix, add 100 μL of 1% paraformaldehyde for 10 min at room temperature. Filter through 70 μm nylon mesh.
4. Set the negative gates using the control sample.
5. Set the compensation for PE and FITC by optimizing each of the single positive samples.
6. Run the 2-color sample. Typical plots are shown in **Fig. 1**. DCs are gated as $CD33^+CD14^{dim}$ cells and monocytes as $CD33^+CD14^{bright}$ cells.

3.1.3.5. FLOW CYTOMETRIC CELL SORTING FOR DENDRITIC CELLS

1. Resuspend the myeloid enriched non-T-cell suspension in 0.5 mL 1% FCS/PBS. Add 5 μg/mL anti-CD33-PE and anti-CD14-FITC and incubate for 30 min on ice or for 10 min at room temperature.
2. Wash in a 50 mL volume with PBS by centrifugation at 410g at 10°C for 6 min.
3. Resuspend cells in 10% FCS/RPMI +PGG at 10^7/mL and filter them as above in preparation for sorting.
4. Cell sorting is carried out according to the manufacturers' instructions. DCa have been successfully sorted using Becton Dickinson, Coulter, and Cytomation instruments.
5. After gating, cells are sorted and collected into 15-mL tubes containing 1 mL 10% FCS/RPMI + PGG.
6. Once collected, wash cells in a 15 mL volume of PBS by centrifugation at 410g for 6 min at 10°C.
7. Count the cells. For yields and purity *see* **Note 3**, and for tips and techniques *see* **Note 4**.

3.2. Alternative Protocols

3.2.1. 3- or 4-Color Analysis and Sorting of Dendritic Cells

Using an argon laser, myeloid-enriched cells or peripheral blood mononuclear cells (PBMCs) can be analyzed using fluorochromes of four different absorption/emission spectra (e.g., FITC, PE, and PE/Cyanin-5 and PE/Texas Red [ECD, Coulter] or allophycocyanin [APC]). For greater flexibility, biotinylated monoclonal antibody (mAb) can be used and then labeled with streptavidin-conjugated fluorochromes. For low-abundance antigens, such as cytokine receptors, detection with PE or APC is more sensitive than with FITC. In this way, the phenotype of gated DC and monocytes and their subpopulations can be assessed without sorting.

Dendritic Cells

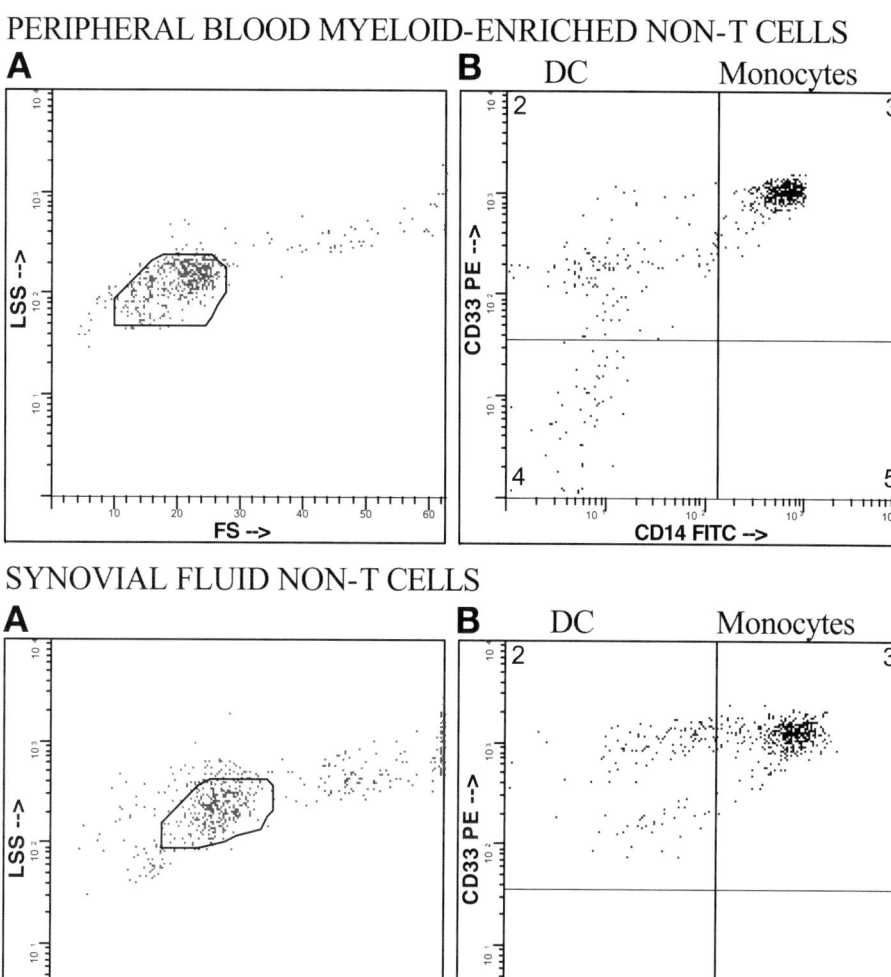

Fig. 1. Expression of myeloid antigens by PB (A by PB and SF DCs). Myeloid-enriched RA PB non-T-cells or SF non-T-cells were and stained with anti-CD33-PE and anti-CD14-FITC. (**A**) Live cell gate, (**B**) DC and monocyte gates.

3.2.2. Alternative Method for Dendritic Cell Analysis or Sorting

PB or SF MNC can be stained with a combination of fluorochromes to analyze CD14/CD19/CD16/CD56/CD3 ("lineage markers"), HLA-DR, CD11c, and CD123. Lineage$^-$HLA-DR$^+$ CD11c$^+$ (myeloid) or CD123$^+$ (plasmacytoid) DCs can be analyzed in this way *(1)* (**Fig. 2**). Alternatively, using a similar staining approach, the BDCA mAb can be used to identify DC subsets *(2)*. In blood, BDCA-2 and BDCA-4 are expressed by CD11c$^-$ CD123$^+$ plasmacytoid DCs, whereas BDCA-3 is expressed by a small population of CD11c$^+$ CD123$^-$ DCs. None of the three are detectable on CD1c$^+$ CD33^{bright} CD11c$^+$ CD123^{dim} DCs, or on any other cells in blood. BDCA-4 is also expressed by monocyte-derived DCs. Using a similar staining approach, Hart and others have developed methods for counting DCs by flow cytometry, in body fluids such as PB *(3)*.

The mAb CMRF-44 and CMRF-56 recognize cell surface antigens expressed by relatively differentiated DC in PB, SF, and ST. Between 20 and 50% of SF DCs express CMRF-44. As in PB, these DC are CD33^{bright} *(4,5)*. DC enrichment can also be carried out using CMRF-44 or CMRF-56 to select PB DCs that express these antigens after a short overnight incubation *(6)*. CD83 is a member of the Ig superfamily, and is expressed by differentiated DC cultured from PB, and by skin Langerhans cells *(7–9)*. It is expressed by neither freshly-isolated PB nor SF DC. DC freshly-isolated from PB or SF are round. PB and SF DC can be induced to differentiate by overnight incubation in medium in vitro, at which time they will express typical markers of differentiation and bear the characteristic dendritic morphology.

3.3. Preparation of Monocyte-Derived Dendritic Cells From Blood or Synovial Fluid

Monocytes are prepared after PBMCs have been obtained (*see* **Subheading 3.1.3.1., step 7**), by positive immunomagnetic selection. Monocytes are then cultured in the presence of GM-CSF and interleukin (IL)-4 to generate DCs.

3.3.1. Positive Selection of Monocytes From PB or SF Using MACS

1. The amount of CD14-conjugated microbeads and buffer depends on the number of cells required; for effective binding, the recommended ratio is 1:4 (beads: cells + buffer). Rule of thumb: use a minimum of 50 µL microbeads; the number of positively selected cells is equivalent to the amount of beads used (e.g., 50 µL beads should select for around 50 million cells). Suspend PBMCs in 400 µL PBS + 1% bovine serum albumin (BSA) buffer and add beads. Mix gently then incubate for 15 min at 4°C.
2. Wash cells in 50 mL volume in PBS and resuspend in buffer to 2×10^8/mL.
3. While cells are incubating, prepare column; add stop-cock, and 18-gage needle and charge 5-mL syringe with buffer. Run about 6 mL of buffer through column. Adjust stop-cock so that the speed is approx 2 drops/s.
4. Run cells through column 3 times then collect the negative fraction in a tube, for T-cell purification if required, (discard if not required). Continue to flush MACS column with buffer until the eluent starts to look clear.
5. Remove the column and attachments from the magnet. Knock column with syringe to disturb the cells in the column (this would dislodge any negative cells

Dendritic Cells 173

that may have been retained in the column). Replace column on magnet, adjust stop-cock to full-speed and flush column again. Repeat this procedure 3 to 4 times
6. Place column in the top stop cock position. Place a 10-mL syringe in the side position of the stop cock. Fill column with buffer and pull syringe plunger to collect CD14$^+$ cells. Continue until all cells are collected, and solution becomes clear.
7. Take 10 µL of cells for a cell count
8. Wash cells in PBS at 1500 rpm for 5 min, tip off supernatant and re-suspend cells in 10%FBS/RPMI at 1×10^6 /mL.
9. Add GM-CSF 800 U/mL and IL-4 400 U/mL, incubate at 37ºC, 5% CO2 for 2 to 7 d *(10)*. Feed cells every second day with fresh medium and cytokines. *See* **Note 5** for yield and purity.

3.4. Identification of Dendritic Cells in RA ST-Immunohistochemical Staining of Formalin-Fixed Biopsies

3.4.1. Dewaxing and Retrieval of Antigens

1. Place mounted slides at 37°C in a drying oven on a flat slide tray for a minimum of 2 h.
2. Transfer to a 60°C oven for 1hr.
3. Transfer slides to a glass slide rack and place in a research grade xylene bath for 5 min.
4. Repeat **step 3** in a fresh xylene bath.
5. Rehydration: transfer slides to a 100% ethanol bath for 5 min.
6. Repeat **step 5** in a fresh ethanol bath 2 times (i.e., total of 3 changes of 100% ethanol)
7. Transfer slides into a 95% ethanol bath for 5 min.
8. Transfer slides into a 70% ethanol bath for 5 min.
9. Transfer into autoclavable pots containing 10 m*M* EDTA (pH 7.5) or 10 m*M* sodium citrate pH 5.0 for 1 min. Tip this off. Replace with fresh EDTA or citrate buffer. Loosely secure pot lids.
10. Autoclave for 10 min at 121°C.
11. Allow autoclave to cool for a least 2 h, but preferably overnight before opening. Do not air vent autoclave.
12. Transfer slides into a TBS bath for a least 5 min. For trouble shooting and tips, *see* **Note 6**.

3.4.2. Immunohistochemical Staining

1. Remove antigen-retrieved slides from TBS bath and dry the slide area surrounding the tissue section to remove the excess liquid.
2. Immerse slides in a staining jar containing 3% milk block, or using a pipet, cover the tissue section with serum block. Usually, 100 µL is sufficient when adding any reagent at any step during the staining protocol. Incubate at room temperature in a moist environment for 20 min.

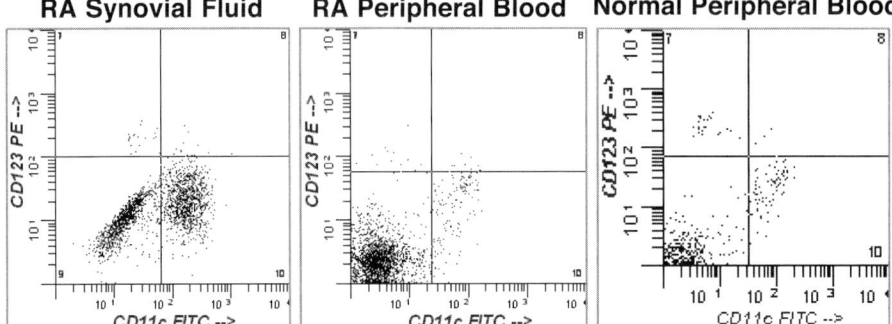

Fig. 2. CD123 and CD11c expression by peripheral blood and synovial fluid DCs. After purification of mononuclear cells from either normal or RA PB or RA SF, cells were stained with CD123-PE, CD11c FITC, CD14-Pe/Cy5, and HLA-DR-APC. Live HLA-DR+CD14- mononuclear cells were gated; polymorphonuclear cells were excluded on the basis of forward and side light scatter. Typical histograms depicting expression of CD123 and CD11c by the gated cells are shown.

3. Remove blocking solution by gently tapping slide onto absorbent paper OR, in case of milk, pour off the milk and wash once with TBS for 30 s.
4. Add the first primary antibody, as per dilutions given in **Table 1** and **Note 7** to the sections, and the relevant control antibody to negative tissue sections, and incubate at room temperature for 60 min.
5. Wash in a TBS bath for 2 times for 3 min each on a shaker. Dry the slide area surrounding the tissue section to remove the excess liquid.
6. Add the peroxidase block solution and incubate at room temperature for 30 to 45 min. If milk block is used, then this step is omitted.—proceed from **step 5** to **step 8**.
7. Wash in a TBS bath 2 times for 3 min each on shaker. Dry the slide area surrounding the tissue section to remove the excess liquid.
8. Add the appropriate biotinylated secondary antibody and incubate at room temperature for 30 min. Important: While planning to do a tertiary step make sure that the secondary antibody is whole and not $F(ab)_2'$.
9. Wash in TBS bath 2 times for 3 min each on shaker. Dry the slide area surrounding the tissue section to remove the excess liquid.
10. Add the streptavidin conjugated horseradish peroxidase (HRP) enzyme (DAKO) and incubate at room temperature for 30 min.
11. Wash in a TBS bath 2 times for 3 min each on shaker. Dry the slide area surrounding the tissue section to remove the excess liquid.
12. Add the second primary antibodies individually. If anti-RelB is used as the primary antibody, these will be anti–HLA-DR, anti-CD20, anti-CD23, and anti-CD68 and the appropriate control antibody to control tissue sections, and incubate at room temperature for 60 min.

Dendritic Cells 175

13. Wash in a TBS bath 2 times for 3 min each on shaker. Dry the slide area surrounding the tissue section to remove the excess liquid.
14. Add the appropriate biotinylated secondary antibody (anti-mouse Ig) and incubate at room temperature for 30 min. Also prepare at the same time Strep-ABC-alkaline phosphatase according to kit instructions.
15. Wash in a TBS bath 2 times for 3min each on shaker. Dry the slide area surrounding the tissue section to remove the excess liquid.
16. Add the streptavidin-conjugated alkaline phosphatase enzyme and incubate at room temperature for 30 min.
17. Wash in a TBS bath 2 times for 3 min each on shaker. Dry the slide area surrounding the tissue section to remove the excess liquid.
18. Add the diaminobenzidine chromogen and develop brown color for 2 to 5 min. Check color development under the light microscope each minute.
19. Wash in a TBS bath 2 times for 3 min each on shaker. Dry the slide area surrounding the tissue section to remove the excess liquid.
20. Add the Fast Red chromogen and develop red colour for 5 min. Again check colour development by microscopy.
21. Wash in a deionised water bath for 2 min. Dry the slide area surrounding the tissue section to remove the excess liquid.
22. Counter-stain for 45 to 60 s using Mayer's Haematoxylin.
23. Wash under running tap water for 2 to 3 min. Dry the slide area surrounding the tissue section to remove the excess liquid.
24. Mount and cover-slip using Faramount aqueous mounting media. For tips *see* **Note 7 (Figs. 3** and **4**).

3.4.3. Variations

3.4.3.1. STAINING FROZEN SECTIONS

1. The same basic protocol should be followed with the following variations: cut fresh.
2. Allow slides to dry at room temperature overnight.
3. Store slides in slide boxes (5 slides/box) at -20 or $-70°C$.
4. When planning to stain, remove box from freezer and bring to room temperature before opening to remove individual slides.
5. Let slides dry for 5 min on bench.
6. Fix in acetone for 4 min.
7. Dry for 5 min.
8. Proceed with **Subheading 3.4.2., step 2**. For tips, *see* **Note 8**.

3.4.3.2. STAINING CELL CYTOSPINS

1. The same basic protocol should be followed with the following variations: Re-suspend cells to give 10^5 cells/slide depending on size of cells in PBS (i.e., 10^6 cell/mL of PBS, larger cells 10^4–10^5).
2. Cytospin 100 µL of cells [~40 g for 5 min] in suspension onto vectorbond-coated slides using a cytospin centrifuge. Air dry slides for a minimum of 2 h, preferably

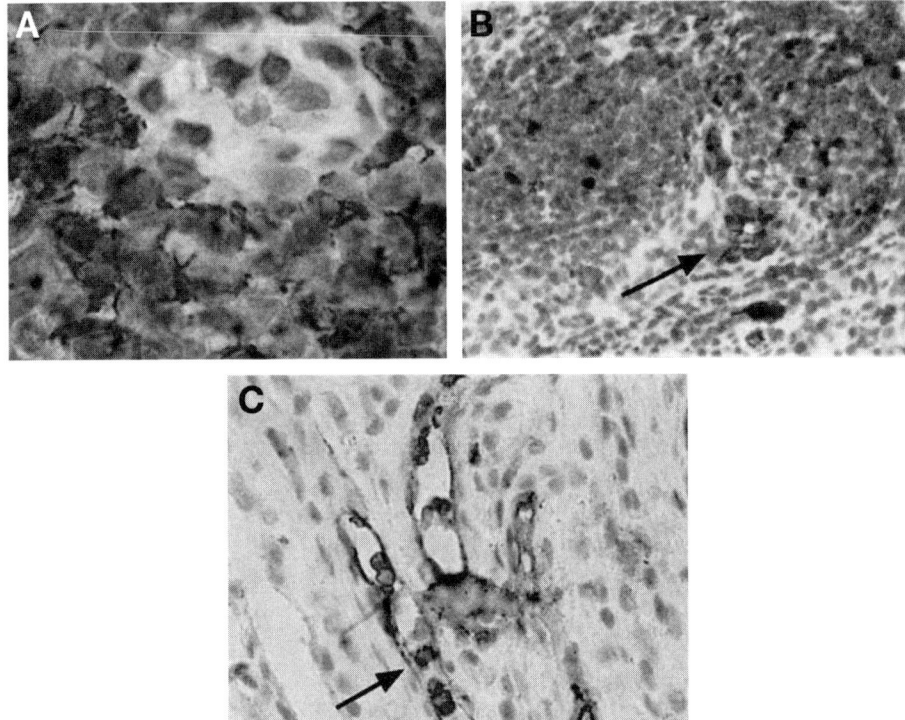

Fig. 3. Identification of synovial tissue DC by 2-color immunohistochemistry for RelB and HLA-DR. (**A**) After retrieval, paraffin embedded samples of RA ST from a patient with active untreated RA were stained with RelB (brown) and HLA-DR (red). (**B,C**) RA ST samples obtained serially from a patient undergoing treatment (**B**) before, and (**C**) after) for RA was stained with RelB (brown) and Ulex (red) to identify blood vessels. (Color illustration appears in insert following p. 138.)

overnight, before fixation with 4% paraformaldehyde for 30 min at 37°C OR ice with acetone for 4 min.
3. Wash cytospin for 5 min in TBS and proceed with immunohistochemistry as described above. For tips *see* **Note 9**.

4. Notes

1. From RA PB, the yield of PBMCs will be approx 10^6 cells/mL of blood (i.e., lower than the expected yield of $1.5–2 \times 10^6$/mL from normal PB). From RA SF and ST, the yield will vary greatly between donors. If the patient's disease is active, approx 2×10^6 cells/mL SF would be expected. Both RA PB and SF MNC contain a greater proportion of contaminating granulocytes than normal PBMC. By flow cytometry these cells have a characteristically high side and forward scatter and are excluded from the live gate (*see* **Fig. 1**).

Table 1
mAb Dilutions for Immunohistochemical Staining

Primary Abs	Fr/Pf/Cyto	Dilution	DAB brown/ fast red	Secondary	Dilution
MoXCD123	✔/X/✔	1:50–1:100	+++/+	RbXMo^{bio}	1:200
MoXCD123 Retrieve paraffin sections with citrate buffer	X/✔/X	1:50-1:100	+++	RbXMo^{bio}	1:200
MoXCD11c	✔/X/✔	1:100	+++/+++	RbXMo^{bio}	1:200
CD4 (NCL-CD4) for paraffin	X/✔/X	1:25–1:50	++/++	i. 2° Rb (whole Ab) X Mo	1:200
				ii. 3° Sw X Rb (F(Rb)$_2$^{bio})	1:200
CD4 for frozen sections and cytospins	✔/X/✔	1:25–1:50	+++/+/+++	Rb X Mo^{bio}	1:200

2. There are various sizes and types of MACS columns for cell separation depending on the cell number, and whether cells are being negatively or positively selected and there may be some variations in the size of the flow resistor and the volumes required to run through the column. Refer to the manufacturer's specifications for these variations. Use of the correct type and size of column and flow resistors will optimize the yield and purity of the cell preparation.
3. Starting from 20 to 25×10^6 myeloid-enriched non-T-cells, approx 0.5 to 1×10^6 DC and 2 to 3×10^6 monocytes can be sorted. However, yield and purity will depend on the sort parameters, and the efficiency of the instrument. From 85 to 99% purity can be achieved by cell sorting. The greater the purity the lower the yield. From RA SF non-T-cells, the yields are similar to 3-fold greater, as many SF samples contain very few B-cells and NK cells and are enriched in DC *(11)*. For this reason, and to reduce handling of the cells, myeloid-enrichment of SF non-T-cells is not routinely carried out.
4. It is essential to filter cell clumps from the suspension before flow cytometric analysis or sorting, to prevent blockages of the machine. Once the two color staining has been adequately compensated, the cytometer settings and sort protocol should be saved for routine use. There will be minimal if any variation in the compensation between donors, however the relative proportion of DCs and mono-

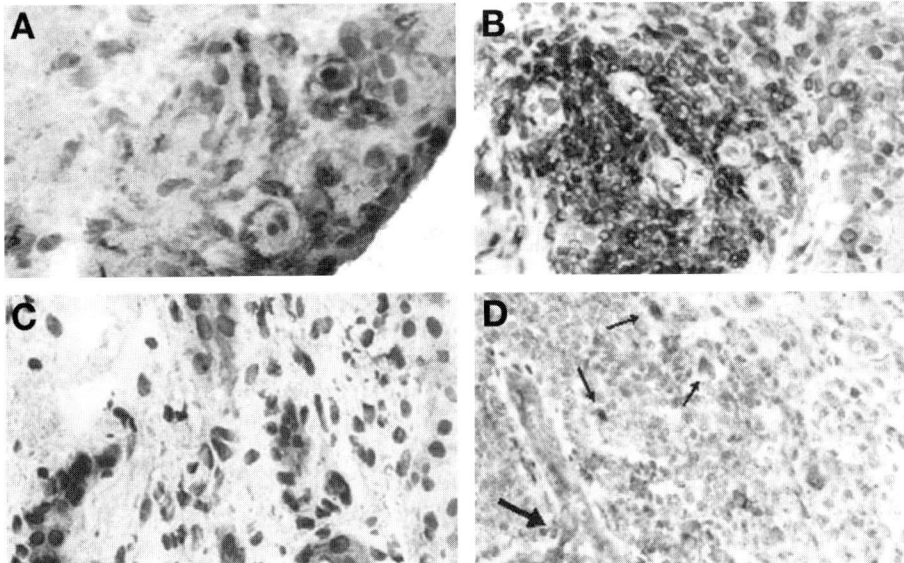

Fig. 4. CD11c and CD123 expression by DC in RA ST. Frozen sections of either normal (**A, C**) or RA ST (**B,D**) were stained with either CD11c (**A,B**) or CD123 (**C,D**) (brown). CD123 stains endothelial cells as well as DC. Perivascular CD123+ DC are noted only in the RA sample (small arrows). (Color illustration appears in insert following p. 138.)

cytes will vary. Keep the cells cool and well mixed while sorting to prevent adherence to the sort tube. Process the cells as soon as possible after sorting to maintain viability. For sterile sorting, the machine should be pre-rinsed with ethanol. For RNA work, cells can be sorted directly into Trizol reagent (Gibco). Equal numbers of particular populations can be sorted for semi-quantitative comparison, by setting the machine as required. When using unconjugated mAb, background staining can be reduced by subsequent incubation with a biotinylated antibody of the appropriate species specificity, followed by streptavidin-conjugated fluorochrome, rather than using fluorochrome-conjugated anti-mouse antibody.

5. Monocytes: the number of monocytes obtained depends on the amount of starting material (PBMCs), and the amount of CD14 magnetic beads used. The rule of thumb is that PBMC contain approx 10% monocytes, and each microliter of magnetic beads can yield 1×10^6 monocytes; a minimum of 50 µL of beads should be used for each separation (using buffy coat, less for fresh blood), and increased depending on number of monocytes required. The beads: buffer ratio should be at 1:4. If the columns are used properly (vigorously washed to remove negative fraction), the purity should be more than 95% and can be up to 99%. DC purity is dependent on purity of starting population (monocytes). Yield after 7 d is approx 50% of starting population.

6. When cutting and mounting sections, do not allow the oven temperature in **Subheading 3.4.1., step 2** to go below 58°C or above 72°C, as this may result in excess background staining on tissue sections. It is important that the sections are not permitted to dry out from **Subheading 3.4.1., step 8** onward. Slides with extra grip (e.g., Superfrost/Plus or Vectabond), should be used to prevent floating of sections during retrieval and staining. Microwave is an alternative to the autoclave *(12)*.
7. Antibody and streptavidin-conjugated enzyme dilutions are all done in TBS as follows:

Antibody	Dilution
Anti-RelB	1:1000
Rabbit Ig	1:10000
Anti-rabbit Ig biotinylated	1:200
Streptavidin-HRP	1:300
Anti-HLA-DR,-CD20,-CD68	1:50
Anti-CD23	1:40
Mouse Ig	1:200
Anti-mouse Ig	1:200

See manufacturer's specifications for the preparation and storage of the streptavidin, ABC-AO, and the chromogens. Background staining levels can vary between different tissues. If a tissue has high background staining this can be combated by:

- Increasing the length of the serum block in **Subheading 3.4.2., step 2**.
- Adding a serum block before the incubation with the second primary antibody in **Subheading 3.4.2., step 12**.
- Increasing the duration and concentration of the peroxidase block solution up to 3%. This may be necessary if the biopsy contains many cells with myeloperoxidase, such as granulocytes.
- Increasing the concentration of the serum components of the serum block to 20%
- Dilute the primary and secondary antibodies with serum from the same animal as the one that the second antibody raised in (e.g., swine sera for use with Sw X Rb or Rabbit sera for Rb X Mo]

It is extremely important that the tissue sections are not permitted to dry out through the entire protocol or the staining will not be successful. All incubations should be carried out at room temperature in a moist environment. For example, sit the slides on moist paper towelling in a covered box. Possible causes of high background staining include:

a. Slides dried out in autoclave or buffer level lower than tissue after autoclave retrieval (let autoclave cool slowly/ don't vent).
b. pH of EDTA buffer too high(check periodically).

c. Air drying of sections.
   d. Too long in DAB.
   e. Tissues with high peroxidase (e.g., skin, liver, kidney, thyroid, melanoma, glandular tissue, blood specimen with many PMNLs, eoxinophils). All of these with biotin can give alkaline phosphatase/red background (can use livamisol).
   f. Secondary Ab concentration too high.

   It is necessary to be extremely gentle with the issue sections as they are relatively weakly attached to the slides, therefore avoid any vigorous washing techniques and use gentle pipetting methods. Biopsies containing large areas of fat float off slides more readily. If the color development is weak, the duration of development can be extended. In some cases, staining is more successful if the development of the DAB chromogen is carried out before the addition of the second primary antibody (i.e., **Subheading 3.4.2., steps 19** and **21** followed by a TBS wash step are performed between **steps 11** and **12**.

8. Some of the mAb will need to be retitrated on frozen tissue. Note that the morphology of formalin-fixed tissue is always superior to that of frozen tissue. On the other hand, the range of mAb available is much greater for frozen sections. CMRF-44, CD123, CD11c, BDCA mAb, DC-LAMP, CD1a, CCL19, CCL20, CCL21, and CD83 are useful additional markers for detection of DC in frozen sections *(13)*. It is good practice to check the quality of frozen sections with a 'quick' H&E stain before proceeding with immunostaining.
9. Some of the staining parameters such as protein block duration and the mAb dilutions may need to be altered for the optimization of staining toxic paraformaldehyde fumes may be generated during incubation. It is good practice to stain the cytospins with diff quick to check whether the cytospins look good before proceeding with immunostaining.

References

1. Grouard, G., Rissoan, M. C., Filgueira, L., Durand, I., Banchereau, J., Liu, Y. J. (1997) The enigmatic plasmacytoid T cells develop into dendritic cells with interleukin (IL)-3 and CD40-ligand. *J. Exp. Med.*, **185,** 1101–1111.
2. Dzionek, A., Fuchs, A., Schmidt, P., et al. (2000) BDCA-2, BDCA-3, and BDCA-4: three markers for distinct subsets of dendritic cells in human peripheral blood. *J. Immunol.* **165**, 6037–6046.
3. Vuckovic, S., Gardiner, D., Field, K., et al. (2004) Monitoring dendritic cells in clinical practice using a new whole blood single-platform TruCOUNT assay. *J. Immunol. Methods* **284**, 73–87.
4. Thomas, R. and Quinn, C. (1996) Functional differentiation of dendritic cells in rheumatoid arthritis: Role of CD86 in the synovium. *J. Immunol.* **156**, 3074–3086.
5. Summers, K. L., Daniel, P. B., O'Donnell, J. L., and Hart, D. N. J. (1995) Dendritic cells in synovial fluid of chronic inflammatory arthritis lack CD80 surface expression. *Clin. Exp. Immunol.* **100**, 81–89.

6. Lopez, J. A., Bioley, G., Turtle, C. J.,. et al. (2003) Single step enrichment of blood dendritic cells by positive immunoselection. *J. Immunol. Methods* **274**, 47–61.
7. Zhou, L. J., Schwarting, R., Smith, H. M., and Tedder, T.F. (1992) A novel cell-surface molecule expressed by human interdigitating reticulum cells, Langerhans cells, and activated lymphocytes is a new member of the Ig superfamily. *J. Immunol.* **149**, 735–742.
8. Zhou, L. -J. and Tedder, T. F. (1995) Human blood dendritic cells selectively express CD83, a member of the immunoglobulin superfamily. *J. Immunol.* **154**, 3821–3835.
9. Zhou, L. -J. and Tedder, T. F. (1996) $CD14^+$ blood monocytes can differentiate into functionally mature $CD83^+$ dendritic cells. *Proc. Natl. Acad. Sci. USA*, **93**, 2588–2592.
10. O'Sullivan, B. J. and Thomas, R. (2002) CD40 Ligation conditions dendritic cell antigen-presenting function through sustained activation of NF-κB. *J. Immunol.* **168**, 5491–5498.
11. Thomas, R., Davis, L. S., and Lipsky, P. E. (1994) Rheumatoid synovium is enriched in mature antigen-presenting dendritic cells. *J. Immunol.* **152**, 2613–2623.
12. Boon, M. E. and Kok, L. P. (1994) Microwaves for immunohistochemistry. *Micron* **25**, 151–170.
13. Page, G., Lebecque, S., and Miossec, P. (2002) Anatomic localization of immature and mature dendritic cells in an ectopic lymphoid organ: correlation with selective chemokine expression in rheumatoid synovium. *J. Immunol.* **168**, 5333–5341.

II

ANIMAL MODELS OF ARTHRITIS

13

The Use of Animal Models for Rheumatoid Arthritis

Rikard Holmdahl

Abstract

Rheumatoid arthritis (RA) is a description of classical symptoms that can be grouped together to be recognized as a common entity. However, the disease is more heterogeneous and is likely composed of many different, distinct diseases *(1)*. This heterogeneity is of two types. First, the disease can be classified into several subtypes, and second, the underlying cause of each disease group may be attributable to different genetic and environmental factors, each contributing to a distinct disease phenotype. This is an important starting point when discussing animal models for RA. Thus, there is not one model for RA. There are, and must be, several different models with different symptoms and different pathogenic mechanisms, controlled by different genes.

Key Words: Animal model; RA.

Rheumatoid arthritis (RA) is a description of classical symptoms that can be grouped together to be recognized as a common entity. However, the disease is more heterogeneous and is likely composed of many different, distinct diseases *(1)*. This heterogeneity is of two types. First, the disease can be classified into several subtypes, and second, the underlying cause of each disease group may be attributable to different genetic and environmental factors, each contributing to a distinct disease phenotype. This is an important starting point when discussing animal models for RA. Thus, there is not one model for RA. There are, and must be, several different models with different symptoms and different pathogenic mechanisms, controlled by different genes.

As in RA, most animal models are dependent on both a permissive genetic background and various environmental factors. In contrast to RA, we know some of the factors that induce arthritis. Such factors are live bacteria *(2,3)* or bacterial components *(4)*, various adjuvants *(5–7)*, cartilage specific proteins *(8–12)*, and even certain more or less ubiquitous antigens *(13,14)*. There are also models in which the genetic influence is strong enough to induce "spontaneous" arthritis; these usually arise as a consequence of transgenic or genetic mutations with a high penetrance for disease *(15–18)*.

It is not straightforward to give general advice on which animal model should be used for a specific project, all the more so given the large number of models available for study. It is certainly dependent on the application, whether it will be used for understanding basic pathophysiology, specific disease mechanisms, or for validating diagnostic or therapeutic targets or whether it will be used as a drug-screening model. However, the following guidelines may be of value. First, for most purposes the model needs to mimic RA, or at least some aspects of RA. It is not productive to analyze a model created in the laboratory which does not reflect or has not been regarded as a disease or physiologic trait produced by nature. Thus, the criteria we use for diagnosing RA should also be applied to the animal model as well.

Second, for comparative studies it is important to study the same phase of the disease. For example, we usually diagnose RA several years after onset, whereas in most animal model experiments the experiment is completed long before that time, a mere 4 to 7 wk after disease induction. This is most likely to be the most important difference between animal models and human disease because we know that different genes and different disease pathways operate before and during onset as compared with the later chronic phase of disease. Thus, in many cases we need better models for this chronic relapsing phase both for understanding basic mechanisms and for developing new treatments.

Third, we need to control the genetics of the animal model. Just as for humans, genes have a major impact on the disease which an individual develops as well as which treatment he or she should be offered. We are entering an era of individualized therapies and in the future treatment will be unthinkable without understanding the genetic basis for disease in each individual patient. With respect to animal models, we have inbred strains. Each strain is analogous to one individual and so it is wise to address each problem in a defined background and, optimally, in two different genetic backgrounds. In addition, it is extremely important to use a genetically controlled strain as there is often confusion regarding which animal strains have actually been used. The introduction of genetically manipulated animals has not improved the situation and has given rise to an uncontrolled mix of embryonal stem cell derived genes with other backcrossed genes *(19,20)*. In addition, inbred stains from different breeders are often genetically contaminated *(21,22)*.

Fourth, the environmental conditions need to be controlled as they will influence disease penetrance. For autoimmune models environmental factors such as stress, caging, and dark–light cycles are of significant importance and often overlooked. Infectious agents may also be of importance but what we need is not sterile animals but animals with stable microbial exposure. We like to use them as models for humans who do not live in sterile environments.

Fifth, we need to compare our data. Therefore it is advisable to use the most common models and to accurately reproduce conditions used by other laboratories. This will facilitate assessment of the data from each model.

Finally, there are important ethical considerations when applying animal models for human RA. Because animal models should mimic the human disease, they unfortunately need to be chronic, which is likely to lead to more suffering in the animals. In this regard it is also advisable to avoid old models that are not very good and which inflict unnecessary suffering on animals, such as the classical adjuvant arthritis model (renamed mycobacteria induced arthritis [MIA]) which is a better model for studying mycobacterial induced inflammatory complications rather than for RA. If we avoid such models, we will make a significant contribution to avoiding the suffering of animals.

We would recommend applying some of the more commonly used models to address specific questions: as a first choice one could try the collagen induced arthritis (CIA) model. It is the most commonly used, and best described model and one which was used for validating tumor necrosis factor as a therapeutic target in RA *(23)*. Unfortunately, its acute phase is most commonly studied, even though there are variants of CIA characterized by a chronic relapsing disease (PIA). If rats can be used, another model of interest is the pristane induced arthritis model that is easy to induce, and which mimics RA surprisingly well. In addition there are a number of models that mimic different phases of the disease, including several spontaneous arthritis models, each of them likely to have clear advantages for studies of different aspects of human RA.

References

1. Holmdahl, R. (2000) Rheumatoid arthritis viewed using a headache paradigm. *Arthritis Res.* **2**, 175–178.
2. Abdelnour, A., Bremell, T., Holmdahl, R., and Tarkowski, A. (1994) Clonal expansion of T lymphocytes causes arthritis and mortality in mice infected with toxic shock syndrome toxin-1-producing staphylococci. *Eur. J. Immunol.* **24(5)**, 1161–1166.
3. Schaible, U. E., Kramer, M. D., Wallich, R., Tran, T., and Simon, M. M. (1991) Experimental Borrelia burgdorferi infection in inbred mouse strains: antibody response and association of H-2 genes with resistance and susceptibility to development of arthritis. *Eur. J. Immunol.* **21(10)**, 2397–2405.

4. Pearson, C. M. (1956) Development of arthritis, periarthritis and periostitis in rats given adjuvants. *Proc. Soc. Exp. Biol. Med.* **91**, 95–101.
5. Wooley, P. H., Seibold, J. R., Whalen, J. D., and Chapdelaine, J. M. (1989) Pristane-induced arthritis. The immunologic and genetic features of an experimental murine model of autoimmune disease. *Arthritis Rheum.* **32**, 1022–1030.
6. Vingsbo, C., Jonsson, R., and Holmdahl, R. (1995) Avridine-induced arthritis in rats; a T cell-dependent chronic disease influenced both by MHC genes and by non-MHC genes. *Clin. Exp. Immunol.* **99(3),** 359–363.
7. Vingsbo, C., Sahlstrand, P., Brun, J. G., Jonsson, R., Saxne, T., and Holmdahl, R. (1996) Pristane-induced arthritis in rats: a new model for rheumatoid arthritis with a chronic disease course influenced by both major histocompatibility complex and non-major histocompatibility complex genes. *Am. J. Pathol.* **149**, 1675–1683.
8. Trentham, D. E., Townes, A. S., and Kang, A. H. (1977) Autoimmunity to type II collagen: an experimental model of arthritis. *J. Exp. Med.* **146**, 857–868.
9. Glant, T. T., Mikecz, K., Arzoumanian, A., and Poole, A. R. (1987) Proteoglycan-induced arthritis in Balb/c mice. *Arthritis Rheum.* **30**, 201–212.
10. Carlsén, S., Hansson, A. S., Olsson, H., Heinegård, D., and Holmdahl, R. (1998) Cartilage oligomeric matrix protein (COMP)-induced arthritis in rats. *Clin. Exp. Immunol.* **114(3),** 477–484.
11. Cremer, M. A., Ye, X. J., Terato, K., Owens, S. W., Seyer, J. M., and Kang, A. H. (1994) Type XI collagen induced arthritis in the Lewis rat. Characterization of cellular and humoral immune responses to native types XI, V, and II collagen and constituent alpha-chains. *J. Immunol.* **153(2),** 824–832.
12. Lu, S., Carlsen, S., Hansson, A-S, and Holmdahl, R. (2002) Immunization of Rats with Homologous Type XI Collagen Leads to Chronic and Relapsing Arthritis with Different Genetics and Joint Pathology Than Arthritis Induced with Homologous Type II Collagen. *J. Autoimmun.* **18(3),** 199–211.
13. Verheijden, G. F., Rijnders, A. W., Bos, E., et al. (1997) Human cartilage glycoprotein-39 as a candidate autoantigen in rheumatoid arthritis. *Arthritis Rheum.* **40(6),**1115–1125.
14. Schubert, D., Maier, B., Morawietz, L., Krenn, V., and Kamradt, T. (2004) Immunization with glucose-6-phosphate isomerase induces T cell-dependent peripheral polyarthritis in genetically unaltered mice. *J. Immunol.* **172(7),** 4503–4509.
15. Iwakura, Y., Tosu, M., Yoshida, E., et al. (1991) Induction of inflammatory arthropathy resembling rheumatoid arthritis in mice transgenic for HTLV-I. *Science* **253**, 1026–1028.
16. Keffer, J., Probert, L., Cazlaris, H., et al. (1991) Transgenic mice expressing human tumour necrosis factor: a predicitive genetic model of arthritis. *EMBO J.* **10**, 4025–4031.
17. Kouskoff, V., Korganow, A. S., Duchatelle, V., Degott, C., Benoist, C., and Mathis, D. (1996) Organ-specific disease provoked by systemic autoimmunity. *Cell* **87(5)**, 811–822.

18. Sakaguchi, N., Takahashi, T., Hata, H., et al. (2003) Altered thymic T-cell selection due to a mutation of the ZAP-70 gene causes autoimmune arthritis in mice. *Nature* **426(6965),** 454–460.
19. Chabas, D., Baranzini, S. E., Mitchell, D., et al. (2001) The influence of the proinflammatory cytokine, osteopontin, on autoimmune demyelinating disease. *Science* **294(5547),** 1731–1735.
20. Blom, T., Franzen, A., Heinegard, D., and Holmdahl, R. (2003) Comment on "The influence of the proinflammatory cytokine, osteopontin, on autoimmune demyelinating disease. *Science* **299(5614),** 1845.
21. Dong, P., Hood, L., and McIndoe, R. A. (1996) Detection of a large RIII-derived chromosomal segment on chromosome 10 in the H-2 congenic strain B10.RIII(71NS)/Sn. *Genomics* **31,** 266–269.
22. Olofsson, P., Johansson, Å., Wedekind, D., Klöting, I., Klinga-Levan, K., and Holmdahl, R. (2004) Inconsistent susceptibility to autoimmunity in inbred LEW rats is due to genetic crossbreeding involving segregation of the arthritis-regulating gene Ncf1. *Genome* **83,** 765–771.
23. Williams, R. O., Feldmann, M., and Maini, R. N. (1992) Anti-tumor necrosis factor ameliorates joint disease in murine collagen-induced arthritis. *Proc.Natl.Acad.Sci.,USA* **89(20),** 9784–9788.

14

Collagen-Induced Arthritis in Mice

Richard O. Williams

Abstract

Collagen-induced arthritis in mice has been widely used to address questions of disease pathogenesis and to validate therapeutic targets for human rheumatoid arthritis. Arthritis is normally observed about 3 wk after immunization with autologous or heterologous type II collagen in complete Freund's adjuvant and susceptibility to the disease is strongly associated with major histocompatibility complex class II genes. The development of collagen-induced arthritis is associated with strong T- and B-cell responses to type II collagen and the chief pathological features of the disease include a proliferative synovitis with infiltration of polymorphonuclear and mononuclear cells, pannus formation, cartilage degradation, erosion of bone and fibrosis. Proinflammatory cytokines, such as tumor necrosis factor-α and interleukin—1β, are abundantly expressed in the arthritic joints of mice with collagen-induced arthritis and, as in human rheumatoid arthritis, blockade of these molecules is effective in reducing the severity of disease.

Key Words: Rheumatoid arthritis; autoimmunity; experimental animal models; adjuvant arthritis; collagen-induced arthritis; type II collagen; mice.

Introduction

There has been considerable progress in recent years in identifying the mediators that contribute to the pathogenesis of rheumatoid arthritis (RA), and a number of studies have pointed to a key role for tumor necrosis factor (TNF)-α in the disease process. Indeed, the success of infliximab, a chimeric anti-TNF-α monoclonal antibody (mAb) *(1–3)*, and etanercept, a soluble p75 TNF receptor-Fc fusion protein *(4,5)*, in the clinic confirms the importance of TNF-α in RA. However, the underlying cause of RA is still unknown and there remains an urgent need for more effective and durable remedies with low-toxicity profiles. It is for this reason that animal models of arthritis are being studied.

Animal models have been used in a wide variety of studies, including the evaluation of novel therapies, the identification of pathological mediators, the identification of genes associated with disease susceptibility, and in the search for markers of disease progression *(6)*. Collagen-induced arthritis (CIA) in mice has come to be the most widely used model for studies of therapeutic intervention and in general, these studies may be described as prophylactic (i.e., treatment before onset of arthritis) or therapeutic (i.e., treatment after onset of arthritis). These two different experimental approaches do not necessarily provide the same results. For example, a number of targeted therapies (e.g., anti-CD4, anti-IL-12, and CTLA4-Ig) have been shown to be effective when given around the time of immunization but are usually found to be much less effective when given in the face of an ongoing inflammatory response *(7–10)*.

Immunization of mice with heterologous (foreign) type II collagen usually leads to a relatively acute and self-remitting form of arthritis. In contrast, immunization with autologous (self) collagen results in a more protracted and fluctuating disease course that is probably more reminiscent of human RA than the heterologous CIA model *(11–13)*. However, autologous type II collagen is less arthritogenic than heterologous collagen and this has been attributed to the low affinity of a specific epitope of murine collagen (CII256-270) for I-A^q, which results in a low level of CII-specific T-cell activation *(14)*.

Both the autologous and heterologous CIA models bear many pathological similarities to RA *(15)*. For example, RA and CIA both show similar patterns of synovitis, pannus formation, erosion of cartilage and bone, fibrosis, and loss of joint mobility *(16)*. In addition, susceptibility to both RA and CIA is strongly associated with genes encoding major histocompatibility complex (MHC) class II molecules, implying that CD4$^+$ T-cells are involved in the pathogenesis of both forms of arthritis. Susceptibility to CIA is restricted to mouse strains bearing MHC types I-A^q and I-A^r, the mouse homologs of HLA-DQ, whereas in human RA certain subtypes of HLA-DR4 and DR1 are associated with disease susceptibility. Furthermore, humoral responses are thought to play a significant role in the pathogenesis of both CIA and RA *(15)* although there is a lack of convincing data pointing to a role for type II collagen specific autoantibodies in the majority of RA patients. An extremely important feature of CIA that makes it a valid model for RA is the expression of pro-inflammatory cytokines, including TNF-α and IL-1β, in the joints of mice with arthritis *(17)* and the fact that blockade of these molecules results in reductions in both the clinical and histological severity of disease *(18–25)*.

2. Materials
2.1. Purification of Type II Collagen

1. Powdered cartilage (*see* **Note 1**).
2. 4*M* guanidine-HCl and 0.05*M* Tris-HCl (pH 7.5).

3. 0.5*M* and 0.1*M* acetic acid.
 4. Sodium chloride (powder).
 5. 70% *(v/v)* formic acid.
 6. Pepsin from porcine gastric mucosa (3X crystalized; Sigma-Aldrich).
 7. 0.02*M* Na$_2$HPO$_4$ (pH 9.4).

2.2. Immunization of Mice

 1. Male DBA/1 mice, 8 to 12 wk of age (Harlan-Olac).
 2. Type II collagen.
 3. 0.1*M* acetic acid.
 4. *Mycobacterium tuberculosis* H37 RA (Difco).
 5. Incomplete Freund's adjuvant (IFA; Difco).
 6. Fentanyl 0.2 mg/mL and fluanisol 10 mg/mL (Hypnorm®).

2.3. Measurement of Anti-Collagen IgG

 1. Type II collagen.
 2. 0.05*M* Tris-HCl and 0.2*M* NaCl (pH 7.4).
 3. Nunc-Immuno microtitre plates (Nalge-Nunc).
 4. 0.05% *(v/v)* Tween-20 in phosphate buffered saline (PBS).
 5. 2% *(w/v)* bovine serum albumin (BSA) in PBS.
 6. Test sera.
 7. Standard serum sample.
 8. HRP-conjugated anti-IgG, IgG1 and IgG2a (BD Biosciences).
 9. TMB microwell peroxidase substrate system (Kirkegaard and Perry).
 10. 4.5*N* H$_2$SO$_4$.

2.4. Analysis of T-Cell Responses

 1. Type II collagen immunised DBA/1 mice.
 2. Cell strainers (70 µm; Falcon).
 3. Hank's balanced salt solution (HBSS).
 4. Complete medium: RPM1 1640 containing 10% *(v/v)* heat-inactivated fetal calf serum (FCS) or 1% *(v/v)* mouse serum, 100 U/mL penicillin, 100 µg/mL streptomycin, $2 \times 10^{-5}M$ 2-mercaptoethanol, and 20 m*M* L-glutamine.
 5. 0.05% *(v/v)* Tween-20 in PBS.
 6. 2% *(w/v)* BSA in PBS.
 7. Recombinant IL-5, IL-10, and IFN-γ (Peprotech).
 8. The following capture/biotinylated detect antibody pairs. IL-5, TRFK5/TRFK4; IL-10, JES5-16E3/JES5-2A5; IFNγ, R4-6A2/XMG1.2 (Immunokontact).
 9. Streptavidin-HRP (BD Biosciences).
 10. TMB microwell peroxidase substrate system (Kirkegaard and Perry).
 11. 4.5*N* H$_2$SO$_4$.
 12. [^{3}H]thymidine (Amersham).

3. Methods
3.1. Purification of Type II Collagen

The method of purification of type II collagen from cartilage is based on the studies of Miller *(26)* and Herbage et al *(27)*.

1. Powder cartilage in a liquid nitrogen freezer mill (Spex, Metuchen, NJ). If unavailable, the cartilage may be ground to a fine powder using a pestle and mortar placed in a bath of dry ice and liquid nitrogen.
2. To remove proteoglycans, suspend powdered cartilage in 5 volumes of $4M$ guanidine-HCl, $0.05M$ Tris-HCl (pH 7.5) for 24 h at $4°C$. Centrifuge at $14,000g$ for 1 h at $4°C$.
3. Discard supernatant and wash cartilage pellet with $0.5M$ acetic acid to remove guanidine-HCl. Centrifuge at $14,000g$ for 1 h at $4°C$.
4. To solubilize collagens, resuspend cartilage pellet in 20 volumes of $0.5M$ acetic acid. Adjust pH of the suspension to 2.8 using 70% formic acid. Add 1 g of pepsin for every 20 g of cartilage (wet weight). Leave stirring for 48 h at $4°C$.
5. Centrifuge at $14,000g$ for 1 h at $4°C$ and discard pellet. To precipitate type II collagen from the supernatant, add NaCl (powder) gradually with stirring to give a final concentration of $0.89M$. Leave to equilibriate overnight at $4°C$ then centrifuge at $14,000g$ for 1h at $4°C$.
6. Dissolve pellet in $0.1M$ acetic acid. Then inactivate residual pepsin by dialysing against $0.02M$ Na_2HPO4 (pH 9.4). The collagen will form a precipitate.
7. Centrifuge at $14,000g$ for 1h at $4°C$, then redissolve pellet in $0.1M$ acetic acid.
8. Dialyze against $0.1M$ acetic acid and freeze-dry. Store at $4°C$ in a dessicator.
9. Purity of the collagen can be assessed on a 5% sodium dodecyl sulfate (SDS)-polyacrilamide gel. In addition, the presence of contaminating proteoglycans (which may not be detected by gel electrophoresis) can be assessed according to the method of Ratcliffe *(28)*. In brief, 40 µL of sample is added to 250 µL of 1,9-dimethylmethylene blue in formate buffer (pH 3.5) in a 96-well microtitre plate. The absorbance is then read immediately at 600 nm using an enzyme linked immunosorbent assay (ELISA) plate reader. A standard curve is constructed by titrating a known concentration of chondroitin sulfate.

3.2. Induction and Assessment of Arthritis

1. Dissolve type II collagen at 4 mg/mL in $0.1M$ acetic acid overnight at $4°C$, with vigorous stirring. Collagen dissolved in this way may be stored at $-20°C$ (*see* **Note 2**).
2. To produce complete Freund's adjuvant (CFA), grind *M. tuberculosis* with a pestle and mortar to produce a fine powder, then suspend in IFA (approx 3 mg *M. tuberculosis*/ml of IFA). This should be carried out in a fume hood and with a face mask to prevent inhalation of *M. tuberculosis* powder.
3. Emulsify dissolved type II collagen with an equal volume of CFA on ice, using a syringe (preferably glass) or an Ultra-Turrax (IKA) in short bursts to prevent

heating. The emulsion should be thick enough not to drip out of the vessel when inverted.
4. Sedate mice by intraperitoneal injection of 100 µL of 10% *(v/v)* Hypnorm®, diluted in distilled water.
5. Shave rumps of mice using electric clippers to facilitate the injection.
6. Inject emulsion intradermally (as far as possible) at 2 or more sites at the base of the tail using a glass syringe or a latex-free syringe and a 27-gage needle. The needle becomes blunt easily and should be changed frequently. Each mouse should receive 0.1 mL of emulsion in total. The emulsion should be shallow enough to be visible under the skin (*see* **Note 3**).
7. Some workers boost the mice with a second intraperitoneal injection of 100 µg type II collagen in 100 µL of 0.1*M* acetic acid, 21 d after primary immunization. However, we have not found this to be necessary.
8. Monitor mice for arthritis every day from day 14 after immunization. The peak time of arthritis onset is around day 30.
9. To compare the clinical severity of arthritis a scoring system may be used where 0 = normal, 1 = slight swelling and/or erythema, 2 = pronounced swelling, 3 = ankylosis. Each limb is graded in this way, giving a maximum score of 12/mouse. In addition, paw-swelling can be monitored using calipers (Poco 2T, Kroeplin).
10. To compare histological severity, paws are removed at post mortem, fixed in buffered formalin (10% *[v/v]*), then decalcified in EDTA in buffered formalin (5.5% *[w/v]*). The tissues are then embedded in paraffin, sectioned and stained with haematoxylin and eosin. The severity of arthritis may be graded as mild, moderate or severe based on the following criteria: mild = minimal synovitis, cartilage loss and bone erosions limited to discrete foci; moderate = synovitis and erosions present but normal joint architecture intact; severe = synovitis, extensive erosions, joint architecture disrupted. Alternatively, the proportion of joints with erosions (defined as demarcated defects in cartilage or bone filled with inflammatory tissue) can be determined.

3.3. Measurement of Anti-Collagen IgG

Serum levels of anti-collagen IgG provide a marker of the magnitude of the humoral anti-collagen response whereas levels of IgG1 and IgG2a serve as extremely valuable in vivo markers of Th2 and Th1 responses, respectively.

1. Make up stock solution of type II collagen in 0.05*M* Tris-HCl, 0.2*M* NaCl (pH7.4) at 1 mg/mL. Aliquot and store at –20°C (*see* **Note 2**).
2. Coat ELISA plate with type II collagen at 2 to 5 µg/mL in 0.05*M* Tris-HCl, 0.2*M* NaCl (pH7.4) overnight at 4°C.
3. Block for 1 h at room temperature with 2% BSA.
4. Incubate test sera (diluted in PBS/Tween-20) for 2 h at room temperature. Levels of anti-collagen IgG may vary enormously between mice and it is important to serially dilute samples to ensure that comparisons are made based on the linear portion of the titration curve. A suggested starting dilution is 1/100, with 7 three-

to five-fold dilution steps. Include a standard serum sample on each plate. Pooled serum from collagen-immunized mice or affinity purified anti-collagen IgG can be used as a standard.
5. Wash 6X with PBS/Tween-20, then detect bound IgG with HRP-conjugated anti-mouse IgG, IgG1 or IgG2a.
6. Develop with TMB substrate. Stop reaction with 4.5N H_2SO_4 and read at 450 nm.

3.4. Analysis of T-Cell Responses

1. Proliferative responses may be measured by incorporation of [^{3}H]thymidine in response to stimulation of lymph node cells with type II collagen. Alternatively, cytokines in culture supernatants can be measured by ELISA.
2. Make up stock solution of type II collagen in 0.05M Tris-HCl, 0.2M NaCl (pH7.4) at 1 mg/mL. Collagen in solution can be kept at 4°C for up to 3 mo (for stimulation of T-cells only; *see* **Note 2**).
3. Remove draining (inguinal) lymph nodes from collagen-immunized mice.
4. Push through cell strainer using syringe plunger, then wash 3X in HBSS.
5. Resuspend at 5×10^6 cells/mL in complete medium and culture for 72 h (37°C; 5% CO_2) in the presence or absence of type II collagen (50 µg/mL).
6. Remove supernatant for measurement of IL-5, IL-10, and IFN-γ (*see* **Note 5**). Coat ELISA plates with relevant capture mAb (overnight at 4°C), block with 2% BSA (1 h at room temperature), then add supernatants (overnight at 4°C). Generate standard curve using appropriate recombinant cytokine at a range of 10,000 pg/mL to 14 pg/mL for IL-5 and IL-10 and 100,000 pg/mL to 137 pg/mL for IFN-γ. Wash 6X with PBS/Tween-20, then incubate with biotinylated detection mAb. Wash 6X with PBS/Tween-20, then add horseradish peroxidase (HRP)-conjugated streptavidin (1 h at room temperature) and develop with TMB substrate. Stop the reaction with 4.5N H_2SO_4 and read at 450 nm.
7. To determine the rate of T-cell proliferation, pulse cells with [^{3}H]thymidine and culture for a further 16 h. Harvest cells and assess for incorporation of radioactivity.

4. Notes

1. For immunization of mice, type II collagen from bovine, porcine, or chick cartilage is normally used. Alternatively, mouse collagen (derived from mouse sternums) may be used, which results in a more chronic relapsing form of arthritis, which is reported to be more similar to human RA than conventional CIA induced with heterologous collagen *(13)*. The yield and solubility of the type II collagen is greater when derived from young animals because of the reduced level of crosslinking. The sternum, nasal septum, or articular cartilage may be used. A good source is femoral head cartilage from a young calf. Peel off the cartilage from the surface of the bone using a scalpel wearing chain mail gloves and eye protection for safety. The cartilage is white. Avoid the underlying bone, which is pink. In the case of sternum or nasal septum the cartilage should be chopped into

small pieces. Type II collagen for immunization can also be purchased from Chondrex.
2. Once in solution, it is important for type II collagen to be maintained at low temperature to prevent denaturation. This is important for successful immunisation and for measurement of anti-collagen antibody levels whereas T-cell activity is less sensitive to the conformational state of the collagen molecule.
3. Two factors of importance in determining incidence of arthritis are the concentrations of *M. tuberculosis* in the adjuvant and collagen in the acetic acid. To establish the system, high concentrations can be used (e.g., 3 mg mycobacterium/mL of IFA and collagen at 4 mg/mL of acetic acid). This invariably induces high incidence but the arthritis is severe and acute in nature, therefore the amounts can subsequently be reduced once the system is known to be working.

   The most likely reasons for a failure to induce a high incidence of arthritis include the following: (i) The collagen preparation is of low purity or in a denatured form; (ii) there is concurrent infection in the mouse colony, the mice are immature (less than 8 wk of age), or females rather than males are used; (iii) the concentration of type II collagen or *M. tuberculosis* in the emulsion used for immunisation is insufficient, or the emulsion is not thick enough or is injected too deeply.
4. Following immunisation, mice will develop arthritis of varying degrees of severity and novel treatment regimes may produce unexpected adverse effects. Hence, mice should be monitored on a daily basis for signs of ill-health or distress. Clearly defined humane endpoints should be strictly enforced. For example, any mouse showing severe and sustained paw-swelling should be humanely killed. Any mouse that has lost 20% or more of its body weight should be humanely killed. Any mouse with severe lameness should be humanely killed. Any mouse with dyspnoea, ruffled fur, weakness, dehydration, or a hunched appearance should be humanely killed. Any mouse showing blistering at the injection site should be humanely killed. In addition, the duration of experiments involving arthritic animals should be minimised, compatible with the aims of the study.
5. In general, IL-5, IL-10, and IFN-γ can be measured in culture supernatants of collagen-stimulated T cells taken from mice with active disease, whereas IL-2 and IL-4 are difficult to detect, probably because of consumption of these cytokines by proliferating T-cells. An alternative is therefore to study cytokine expression by intracellular staining and FACS analysis. Another important factor is that Th2 cells are usually found in low abundance in DBA/1 mice immunized with CFA, although immunomodulatory treatments may influence the Th1:Th2 ratio.

References

1. Elliott, M. J., Maini, R. N., Feldmann, M., (1993) Treatment of rheumatoid arthritis with chimeric monoclonal antibodies to tumour necrosis factor α. *Arth. Rhuem.* **36**, 1681–1690.

2. Elliott, M. J., Maini, R. N., Feldmann, M., et al. (1994) Treatment with a chimaeric monoclonal antibody to tumour necrosis factor α suppresses disease activity in rheumatoid arthritis: results of a multi-centre, randomised, double blind trial. *Lancet* **344**, 1105–1110.
3. Elliott, M. J., Maini, R. N., Feldmann, M., et al. (1994) Repeated therapy with a monoclonal antibody to tumour necrosis factor α in patients with rheumatoid arthritis. *Lancet* **344**, 1125–1127.
4. Moreland, L. W., Baumgartner, S. W., Schiff, M. H., et al. (1997) Treatment of rheumatoid arthritis with a recombinant human tumor necrosis factor receptor (p75)-Fc fusion protein. *ew Engl. J. Med.* **337**, 141–147.
5. Weinblatt, M. E., Kremer, J. M., Bankhurst, A. D., Bulpitt, K. J., Fleischmann, R. M., Fox, R. I., et al. (1999) A trial of etanercept, a recombinant tumor necrosis factor receptor:Fc fusion protein, in patients with rheumatoid arthritis receiving methotrexate. *New Engl. J. med* **340**, 253–259.
6. Williams, R. O. (1998) Rodent models of arthritis: relevance for human disease. *Clin. Exp. Immunol.* **114**, 330–332.
7. Ranges, G. E., Sriram, S., and Cooper, S. M. (1985) Prevention of type II collagen-induced arthritis by in vivo treatment with anti-L3T4. *J. Exp. Med.* **162**, 1105–1110.
8. Hom, J. T., Butler, L. D., Riedl, P. E., and Bendele, A. M. (1988) The progression of the inflammation in established collagen- induced arthritis can be altered by treatments with immunological or pharmacological agents which inhibit T cell activities. *Eur. J. Immunol.* **18**, 881–888.
9. Webb, L. M., Walmsley, M. J., and Feldmann, M. (1996) Prevention and amelioratrion of collagen-induced arthritis by blockade of the CD28 co-stimulatory pathway: requirement for both B7-1 and B7-2. *Eur. J. Immunol.* **26**, 2320–2328.
10. Malfait, A.-M., Butler, D. M., Presky, D. H., Maini, R. N., Brennan, F. M., and Feldmann, M. (1998) Blockade of IL-12 during the induction of collagen-induced arthritis (CIA) markedly attenuates the severity of the arthritis. *Clin. Exp. Immunol.* **111**, 377–383.
11. Holmdahl, R., Jansson, L., Larsson, E., Rubin, K., and Klareskog, L. (1986) Homologous type II collagen induces chronic and progressive arthritis in mice. *Arth. Rheum.* **29**, 106–113.
12. Boissier, M. C., Feng, X. Z., Carlioz, A., Roudier, R., and Fournier, C. (1987) Experimental autoimmune arthritis in mice. I. Homologous type II collagen is responsible for self-perpetuating chronic polyarthritis. *Ann. Rheum. Dis.* **46**, 691–700.
13. Malfait, A. M., Williams, R. O., Malik, A. S., Maini, R. N., and Feldmann, M. (2001) Chronic relapsing homologous collagen-induced arthritis in DBA/1 mice as a model for testing disease-modifying and remission-inducing therapies. *Arth. Rheum.* **44**, 1215–1224.
14. Huang, J. C., Vestberg, M., Minguela, A., Holmdahl, R., and Ward, E. S. (2004) Analysis of autoreactive T cells associated with murine collagen-induced arthritis using peptide-MHC multimers. *Int. Immunol.* **16**, 283–293.

15. Holmdahl, R., Andersson, M. E., Goldschmidt, T. J., et al. (1989) Collagen induced arthritis as an experimental model for rheumatoid arthritis. Immunogenetics, pathogenesis and autoimmunity. *APMIS* **97**, 575–584.
16. Trentham, D. E. (1982) Collagen arthritis as a relevant model for rheumatoid arthritis: evidence pro and con. *Arth. Rheum.* **25**, 911–916.
17. Marinova-Mutafchieva, L., Williams, R. O., Mason, L. J., Mauri, C., Feldmann, M., and Maini, R. N. (1997) Dynamics of proinflammatory cytokine expression in the joints of mice with collagen-induced arthritis (CIA). *Clin. Exp. Immunol.* **107**, 507–512.
18. Thorbecke, G. J., Shah, R., Leu, C. H., Kuruvilla, A. P., Hardison, A. M., and Palladino, M. A. (1992) Involvement of endogenous tumor necrosis factor α and transforming growth factor β during induction of collagen type II arthritis in mice. *Proc. Natl. Acad. Sci. USA* **89**, 7375–7379.
19. Williams, R. O., Feldmann, M., and Maini, R. N. (1992) Anti-tumor necrosis factor ameliorates joint disease in murine collagen-induced arthritis. *Proc. Natl. Acad. Sci. USA* **89**, 9784–9788.
20. Piguet, P. F., Grau, G. E., Vesin, C., Loetscher, H., Gentz, R., and Lesslauer, W. (1992) Evolution of collagen arthritis in mice is arrested by treatment with anti-tumour necrosis factor (TNF) antibody or a recombinant soluble TNF receptor. *Immunology* **77**, 510–514.
21. Geiger, T., Towbin, H., Cosenti-Vargas, A., Zingel, O., Arnold, J., Rordorf, C., and Vosbeck, K. (1993) Neutralization of interleukin-1β activity in vivo with a monoclonal antibody alleviates collagen-induced arthritis in DBA/1 mice and prevents the associated acute-phase response. *Clin .Exp. Rheumatol.* **11**, 515–522.
22. Wooley, P. H., Dutcher, J., Widmer, M. B., and Gillis, S. (1993) Influence of a recombinant human soluble tumour necrosis factor receptor Fc fusion protein on type II collagen-induced arthritis in mice. *J. Immunol.* **151**, 6602–6607.
23. Van den Berg, W. B., Joosten, L. A., Helsen, M., and van de Loo, F. A. (1994) Amelioration of established murine collagen-induced arthritis with anti-IL-1 treatment. *Clin .Exp. Immunol.* **95**, 237–243.
24. Williams, R. O., Ghrayeb, J., Feldmann, M., and Maini, R. N. (1995) Successful therapy of collagen-induced arthritis with TNF receptor-IgG fusion protein and combination with anti-CD4. *Immunology* **84**, 433–439.
25. Joosten, L. A. B., Helen, M. M. A., van de Loo, F. A. J., and Van den Berg, W. B. (1996) Anticytokine treatment of established type II collagen-induced arthritis in DBA/1 mice: a comparative study using anti-TNFα, anti-IL-1α/β, and IL-1Ra. *Arth. Rheum.* **39**, 797–809.
26. Miller, E. J. (1972) Structural studies on cartilage collagen employing limited cleavage and solubilization with pepsin. *Biochemistry* **11**, 4903–4909.
27. Herbage, D., Bouillet, J., and Bernengo, J. C. (1977) Biochemical and physiochemical characterization of pepsin-solubilized type-II collagen from bovine articular cartilage. *Biochem. J.* **161**, 303–312.
28. Ratcliffe, A., Doherty, M., Maini, R. N., and Hardingham, T. E. (1988) Increased concentrations of proteoglycan components in the synovial fluids of patients with acute but not chronic joint disease. *Ann. Rheum. Dis.* **47**, 826–832.

15

Collagen-Induced Arthristis in Rats

Marie M. Griffiths, Grant W. Cannon, Tim Corsi, Van Reese, and Kandie Kunzler

Abstract

Collagen-Induced Arthritis (CIA) is a complex model of autoimmune-mediated arthritis that is regulated by multiple genetic and environmental factors. CIA is induced in rats by immunization with native type II collagen and develops joint pathology similar to that of rheumatoid arthritis. This chapter details methods for the extraction and purification of native type II collagen from sternal and articular cartilage, an arthritis induction protocol that has resulted in reproducible CIA expression in several rat strains from year to year and criteria for measuring clinical, radiographic and immunological outcome parameters characteristic of CIA.

Key Words: Rats; collagen; arthritis; inflammation; autoimmunity; animal model; autoantibody.

1. Introduction

Collagen-Induced Arthritis (CIA) is a Th1-mediated model of both chronic, inflammatory joint disease and autoimmunity targeting a major component of articular cartilage—type II collagen (CII). CIA is a self-limiting model with an explosive onset at 12 to18 d, peaking within a wk and gradually resolving by 35–42 d leaving obvious ankylosis in joints that were severely affected. CIA is restricted to the peripheral joints, affecting primarily fore- and rear-paws. Rarely are the knees involved. Onset of CIA requires induction of T-cell and B-cell immune responses to CII that are crossreactive with self CII in articular cartilage. The severity of CIA is substantially influenced by the level of inflammatory mediators and proteolytic enzymes released within the joint space from the rapidly infiltrating neutrophils and monocytes and from locally activated chondrocytes and synovial cells. Thus, CIA is regulated by both the innate and adaptive immune systems and is suitable for preclinical trials of both anti-inflammatory and immunomodulatory drug candidates. Aspects critical to the success of the model include the strain of rat selected, the species source and quality of type II collagen used for immunization and good animal husbandry techniques *(1–4)*.

2. Materials

2.1. Collagen

The species source and quality of CII used for immunization is critical for the induction of arthritis and consistent results. Native type II collagen (CII) is required. CII is available commercially but is expensive and of variable quality. Bovine and porcine CII are the most practical for use because highly arthritogenic, native CII can be prepared from the articular and sternal cartilages of these large farm animals in gram quantities (*see* **Subheading 3.1.** and **Note 1**). Listed below are the materials needed for the preparation of native CII (*see* **Note 2**). All working solutions are prepared fresh daily and held at 4°C. Stock solutions are stored at 4°C for no more than 1 wk or frozen.

1. $5.0N$ acetic acid.
2. $0.1M$ benzamidine-HCl (Sigma; cat. no. B6506) (inhibits endogenous endoproteinases).
3. DEAE-cellulose (Sigma; cat no. D6418 or D3764) for removal of residual pepsin.
4. $0.2M$ ethylene diamine tetraacetic (EDTA) acid-disodium salt dihydrate (pH 7.0–7.2). Inhibits metalloproteinases (Sigma; cat. no. E4884). Dissolve 37.22 g $Na_2EDTA \cdot 2H_2O$ in about 400 mL water. Bring into solution by drop-wise addition, with stirring, of $1.0M$ NaOH. Take to final volume of 500 mL with water. Keeps for several weeks at 4°C.
5. Guanidine-HCl for extraction of proteoglycans from cartilage (Sigma; cat. no. G7153)
6. $1.0N$ hydrochloric acid.
7. Pepsin (BioChemika 2X crystallized) (Sigma; cat. no. 77152) (solubilizes CII from cartilage).
8. Pepstatin A (Sigma; cat. no. P5318) (pepsin inhibitor).
9. Phenylmethanesulfonyl fluoride (PMSF) (Sigma; cat. no. P7626)—(pepsin inhibitor).
10. N-Ethylmaleimide (NEM) (Sigma; cat. no. E3876) (inhibits endogenous proteinases).
11. 1.0% Sodium azide (Sigma; cat. no. S2002) (inhibits bacterial growth). *Caution:* TOXIC!
12. Solid sodium chloride for fractional precipitation of CII.
13. $0.2M$ Na_2HPO_4 Stock solution. Precipitates form at 4°C. Warm to 37°C to resolubilize. Dilute 1:10 with cold, distilled water to provide $0.02M$ phosphate working solution.

14. 0.5*M* Tris-HCl buffer (pH 7.4) when diluted to 0.05*M* and at 25°C. Prepare stock solution using 9.7 g Tris(hydroxymethyl)amino methane (Trizma®base) (Sigma; cat. no. T4661) and 66.1 g Trizmahydrochloride (Sigma; cat. no. T3253). Store at –20°C as stock solution.
15. 4- to 6-L suction flask equipped with a 6-inch diameter Buchner funnel.
16. 1.5-inch dialysis tubing.
17. 2-, 4-, or 6-L Erlenmeyer flasks.
18. Glass fiber filters (Whatman GF/B).
19. 6-L plastic beakers for dialysis.
20. Scalpels with extra blades.
21. Tweezers and flat spatulas.
22. Large dish pans with ice for transport and cleaning of fresh cartilage and joints.
23. Cartilage soak buffer (Tris-BzAzE) : 0.05*M* Tris-HCl buffer (pH 7.4) 1 m*M* benzamidine-HCl, 0.01% sodium azide, and 0.02*M* EDTA.
24. Proteoglycan extraction buffer : 0.05*M* Tris-HCl buffer (pH 7.4), 4.0*M* guanidine-HCl, 10 m*M* EDTA, 1 m*M* benzamidine HCl, and 10 m*M* N-ethylmaleimide.
25. Collagen solubilization buffer : 0.5*N* HAc containing 0.2*M* NaCl.
26. 0.2*M* NaCl in 0.05*M* Tris-HCl buffer (pH 7.4).
27. 1.0*M* NaCl in 0.05*M* Tris-HCl buffer (pH 7.4).
28. 2.4*M* NaCl in 0.05*M* Tris-HCl buffer (pH 7.4).

2.2. Animals and Equipment

1. Animals: Because of the strong genetic regulation of this model only a few inbred rat strains consistently provide a high incidence (90–100%) of early onset, severe arthritis and uniform levels of CII auto-reactivity. The most commonly used inbred strains are DA, LOU, BB, and LEW *(5–7)*. Rats are ordered in at 6 to 7 wk of age and allowed to recover from transport and accommodate to the new surroundings and diet for 1 to 2 wk before being entered into an experiment (*see* **Note 3**).
2. Transponders: A reliable method of animal identification is the use of radio frequency transponders that are inserted subcutaneously between the shoulder blades of each animal and programmed with a unique identifier number. The transponders can be read at any time, without handling the animal, using a sensor wand connected to a computer tablet (both of which are provided by the vendor). For some studies it is advantageous to use transponders that provide both the rat identifier and body temperature (*see* **Note 4**).
3. Small animal anesthesia machine (*see* **Subheading 3.2.2.**).
4. Adjuvant: Incomplete Freund's adjuvant (IFA) is used for immunization. IFA is obtained from Difco Laboratories cat. no. 3113-60-5) in 10-mL sterile, glass ampules which are stored at 4°C. The vials are opened only once and unused portions discarded. The IFA must not be outdated or exposed to air for any extended time prior to use.
5. Syringes: Disposable 1-mL BD™ Tuberculin syringes equipped with removable, 25-gage, 5/8-inch needles (Becton Dickson; cat. no. 309626) for CII injections.

6. Mechanical homogenizer with rheostat speed control, variable volume (1–50 mL) and adaptable for ice cooling of the mixing tube (*see* **Note 5**).

2.3. Enayme Linked Immunosorbent Assay of Anticollagen Serum Antibody Titers

1. NUNC-Immuno-plate. F96 Maxisorp multi-well plates, 96-well, flat bottom, nonsterile, polystyrene. (cat. no. 442404). Clear lids for the coating step.
2. 8- and 12-well Multichannel pipets.
3. SIGMA*FAST*™OPD (Tablet) (Sigma; cat. no. P9187). *o*-Phenylenediamine. Enzyme substrate chromophore. Make fresh daily and store in the dark at 4°C. Light sensitive.
4. Phosphate-citrate buffer with urea hydrogen peroxide (Tablet) (Sigma; cat. no. P4560). OPD substrate buffer with hydrogen peroxide substrate. Use glass distilled water.
5. Collagen coating buffer : $0.15M$ potassium phosphate buffer (pH 7.6). Make fresh from stock solutions. Buffer A: $0.15M$ KH_2PO_4. Buffer B: $0.15M$ K_2HPO_4. Mix at ratio of 13 volumes of A to 87 volumes of B. Store stock and working buffers at 4°C.
6. Tween-20 (Sigma; cat. no. P1379).
7. Sterile 10X phosphate buffered saline (PBS). Obtain from Sigma (cat. no. D1408) or prepare using anhydrous reagents (2 g KCl, 2 g KH_2PO_4, 80 g NaCl, 11.5 g Na2HPO4/L of solution) and sterilize by autoclaving. Dilute 1:10 with sterile water for use.
8. PNT : PBS containing 0.05% Tween-20 and NaCl increased to $0.2M$.
9. Bovine serum albumin (BSA) (PENTEX).
10. Second Antibody. Peroxidase-conjugated goat IgG anti-rat IgG, whole molecule, affinity purified. Must be azide free as azide will inhibit the peroxidase (Sigma; cat. no. A9037). Avoid repetitive freeze–thaw cycles as this denatures the peroxidase enzyme.
11. Normal goat serum for diluent (Sigma; cat. no. G9023).
12. $2.5N$ H_2SO_4.
13. Microplate reader with 492-nm filter.
14. Native CII 100 µg/plate.

3. Methods

The methods described in this section delineate (1) the purification of native type II collagen (CII) from cartilage, (2) animal husbandry, (3) immunization and test protocols, and (4) clinical and laboratory outcome measures of arthritis severity.

3.1. Purification of CII From Cartilage

1. Young calves or pigs, 2- to 3-mo old, are euthanized and exsanguinated by a veterinarian. Source CII cartilages are sterna, xyphoids, small ribs, and the

articulating surfaces of ankles, knees and hips. Collect these on ice and wash free of blood with cold Tris-BzAzE. Cover the crude cartilages with Tris-BzAzE and freeze at –20°C for 4 to 18 h. Then, thaw for cartilage recovery but keep cold by working in an ice bucket (*see* **Note 6**).
2. To obtain joint cartilage, disarticulate the joints and use a scalpel to carefully dissect off the thin layer of cartilage covering the interior of the joint. Do not cut too deeply. The underlying bone is a source of contaminating type I collagen. It is detected by a change in texture. Collect slices in cold Tris-BzAzE.
3. To prepare rib, sterna, and xyphoid cartilage, thaw and soak overnight in Tris-BzAzE. Thoroughly clean away adherent tissue and perichondrium. Cut cleaned cartilage into 0.5- to 1.0-mm slices and collect in cold Tris-BzAzE.
4. Rinse cartilage slices with Tris-BzAzE. Collect with a sieve to obtain wet weight. Store frozen in 100 g aliquots covered with Tris-BzAzE. All volume ratios in the protocol refer to this original cartilage wet weight unless noted.
5. Stir cartilage slices in proteoglycan extraction buffer for 48 h at 4°C. Use 15 volumes of buffer/gram of cartilage (wet weight). It is convenient to work in 100 g lots. It is necessary to process cartilage for the full 48 h or there will be incomplete removal of proteoglycans and thus difficulty in solubilizing and purifying type II collagen (*see* **Note 7**).
6. Recover cartilage by filtration using Buchner funnel and glass fiber filters.
7. Wash cartilage slices on the funnel with large volumes of cold water using a minimum of 4 L/100 g of cartilage.
8. Stir cartilage slices overnight at 4°C in 0.5M HAc containing 0.01 EDTA and 10 mM NEM at a ratio of 1 L/100 g cartilage.
9. Recover slices by filtration.
10. Suspend cartilage slices in 0.5 HAc-0.2 NaCl (2 L/100 g cartilage)
11. Add pepsin at a ratio of 2 g/100 g cartilage (approx 1 mg/mL).
12. Stir slowly in cold room for 24 h.
13. Recover supernatant by centrifugation, 1 h, 10,000 rpm. (Discard cartilage pellet or re-extract with pepsin at one-half volume if solubilization appears to be incomplete).
14. To the supernatant, add pepstatin A (5µg/mL) and PMSF (1 mM).
15. Take supernatant to 0.86M NaCl final concentration with solid NaCl by slow addition with stirring in the cold room. Allow to equilibrate overnight. CII precipitates.
16. Collect precipitate by centrifugation, 1 h, 10,000 rpm.
17. Dissolve CII precipitate in Tris-0.2 NaCl. Bring pH to between 7.8 and 8.0 by slow, drop-wise addition of 1.0M NaOH on ice with stirring. This step inactivates residual pepsin and enhances solubility of the CII. CII from 100 g of cartilage will require 500 to 600 mL of Tris-0.2 NaCl and between 15 and 30 mL of base. The solution will be quite viscous.
18. Dialyze CII solution against Tris-0.2 NaCl. 10 volumes for 3 changes with 6 to 8 h between changes or overnight.
19. Clarify the CII solution by centrifugation and filtration.

20. Determine CII concentration (*see* **Note 8**).
21. Dilute CII solution to between 0.5 and 1.0 mg/mL with Tris-0.2 NaCl to decrease viscosity.
22. Pass the diluted CII solution through a DEAE column equilibrated with Tris 0.2 NaCl. A short, fat column, bed volume 100 mL, intitial flow rate 100 mL/h can be used for up to 1 g of CII in 2 L of Tris-0.2 NaCl. Collect effluent and combine with column wash (200 mL Tris 0.2 NaCl). CII passes directly through. Pepsin and proteoglycans are retained on column.
23. Dialyze the post-DEAE CII solution against 10 volumes of $0.02M$ Na_2HPO_4 for 3 buffer changes with 6 to 18 h between changes. The CII will precipitate (*see* **Note 9**). Pepsin and proteoglycans are soluble and any residual will remain in solution.
24. Collect CII by centrifugation and dissolve in $0.5M$ HAc-0.2 NaCl at 1 mg/mL.
25. Repeat **steps 6** and **7** (Second NaCl precipitation).
26. Dissolve CII in Tris-HCl $1M$ NaCl. Adjust pH to between 7.4 and 7.6 with $1.0M$ NaOH and adjust protein concentration to 1 mg/mL (*see* **Note 8**).
27. Dialyze CII solution against Tris-HCl $2.4M$ NaCl, 10 volumes for 3 changes with a minimum of 24 h total dialysis time. Types I and III collagen will precipitate. Type II collagen will remain in solution. Clarify by centrifugation and filtration. Keep the supernatant which contains CII. Discard pellet.
28. Repeat **steps 13** and **14** (second phosphate dialysis step).
29. Repeat **steps 6** and **7** (third NaCl precipitation).
30. Dissolve CII pellet in 500 to 800 mL $0.1N$ acetic acid.
31. Clarify by centrifugation ($75,000g$ for 1 h).
32. Dialyze first against 10 volumes of $0.1N$ acetic acid then against 10 volumes of H_2O for 3 changes or until CII precipitates out of solution.
33. Lyophilize and store desiccated in the dark at 4°C for short term or –70°C for long term.

3.2. Animal Husbandry

1. Housing: Because of unpredictable hormonal effects on CIA expression, a constant, stress-free, quiet housing situation is required. Rats are housed 2 to 3 per cage in barrier or conventional conditions and provided rodent chow breeder diet (Harlan Teklad; cat. no. 4058; *www.harlan.com*) and water ad lib. Room conditions are light cycles of 12 h (7:00 AM–7:00 PM), 30 to 50% humidity, and temperature 68 to 72°C. Comfortable bedding is preferred (e.g., Tek-Fresh, Harlan; cat. no.7099) or Shredded Aspen Shavings (Harlan; cat. no. 7093) to diminish pain and stress for the arthritic rats. The experimental rats are kept apart from any breeding colonies. Room access is restricted to necessary personnel and all traffic and noise in and near the room is minimized. Only two animal handlers immunize, manipulate, and score the animals. Only one staff person changes cages and maintains room sanitation during the experiment period. All personnel wear protective clothing including disposable masks, booties, gowns, and hats.

2. Anesthesia: All procedures are carried out under anesthesia. Rats are anesthetized by inhalation using a small animal anesthesia machine available from VetEquip Inc (Pleasanton; www.vetequip.com). The oxygen flowmeter is set at 2 L/min and isoflorane is vaporized at 1%. The rat is first placed in the induction chamber until anesthetized and then held in that state using a nose cone upon removal to a procedure table. This concentration of gases stabilizes the rat at stage 3 plane 2 which is the ideal surgical plane. Waste gas is collected through evacuation waste gas filters. The rat remains fully anesthetized during ear clipping, transponder placement, immunization, and blood collection. After completion of the procedure, the rat is returned to its cage and monitored until fully alert.

3.3. Test Protocol—Time Course

1. Day 0–14: Order rats. Plan for 8 to 10 wk of age at the time of immunization.
2. Day 0–7: Arrange rats in pans, insert transponders, and take presera.
3. Day 0 – 1: Shave back of rat to facilitate injections. Weigh rats, record baseline weights.
4. Day 1: Calculate total collagen needed for study [(total rat weight in kg) × (2 mg/kg)] plus 25% excess for spillage. Dissolve CII in cold $0.1N$ acetic acid at 2 mg/mL by overnight mixing on a rotator in the cold room. A very viscous, clear solution should result (*see* **Note 10**).
5. Day 0: Prepare a 1:1 emulsion of the dissolved CII (2 mg/mL) with cold IFA. Use a 50-mL disposable centrifuge tube, chilled in an ice bucket and a mechanical homogenizer (*see* **Note 10**). Take care not to denature or fragment the CII by excessive mechanical manipulation. The emulsion should peak within 60 s. Final concentration of CII in the emulsion is 1 mg/mL.
6. Day 0: For each rat, calculate the immunization dose (volume of emulsion) by multiplying the rats weight in kilograms by 2. For example, a rat weighing 175 g would receive 350 µg CII in 0.35 mL. of emulsion.
7. Day 0: With the rat anesthetized, inject the CII emulsion intradermally on the shaved back (up to 0.1/injection site) evenly over 4 sites: upper left, upper right, lower left, lower right. Inject anything over 0.4 mL along the center of the back. The injection site should show a small bleb in the skin, similar to a tuberculin skin test site. Do *not* allow the emulsion to be injected subcutaneously. Keep the emulsion on ice until injected (*see* **Note 11**).
8. Day 1–7: Observe animals daily. They should be healthy and mobile. Injection sites should be visible but not show evidence of infection. Note any early arthritis onset.
9. Day 7: Booster injection (100 µg CII/rat in 0.1 mL IFA-HAc emulsion). Injection site is the tail base (*see* **Note 12**).
10. Day 7–34: Observe daily for arthritis onset. Weigh and score arthritis severity 3 times/wk.
11. Day 35: Final weights and arthritis scores. Score each paw for ankylosis. Take joint tissues for histology and radiology. Take final blood samples for serum and anti-collagen antibody titers.

3.4. Clinical Evaluation of Arthritis Severity

Arthritis severity is evaluated by a visual scoring system that is weighted for cartilage surface area and joint size *(10)*. The severity of arthritis in the wrist, mid-forepaw, mid-hind paw, and ankle joints are scored as 0 = no arthritis, 1 = extreme redness and/or minimal swelling, 2 = medium swelling, 3 = severe swelling, 4 = severe swelling and nonweight bearing. The presence or absence of arthritis in the interphalangeal, metacarpophalangeal, and metatarsophalangeal joints of the 4 prime digits of each paw are scored as 0 = no arthritis and 1 = arthritis present. The fifth digit is scored with the mid-foot. The joint score is the sum of the scores of all involved joints. The maximum arthritis severity score is 20/paw and 80/rat.

Ankylosis can be used as a measure of residual joint destruction and are determined by the angle of joint mobility. Full mobility = 0; 45° angle = 1; immobile = 2. Weight loss (or failure to gain weight) is transient, usually about 10 to 15% from base line weight at day 7 and paralleling arthritis onset and resolution. For humane reasons, animals with an arthritis score of greater than 70 for more than 3 d or weight loss of > 20% should be sacrificed when this level of arthritis is identified.

3.5. Laboratory Evaluation of Arthritis Severity

1. Joint histology is used to characterize infiltrating cell types and detect the presence or absence of cartilage and bone loss. Hind paws are removed, placed in 10% formalin, decalcified, sectioned and stained with hematoxylin and eosin. Slides are read, blinded, for neutrophilic and lymphocytic infiltrates, edema and fibrosis as: trace, 1+, 2+, or 3+. The presence or absence of periosteal changes, erosions, or new bone formation is noted.
2. Joint destruction has been evaluated by radiology after sacrifice with a mammography unit (Mammomat-B) using 28 kvp and an exposure of 15 mAs. The limbs are amputated from the animal. The samples may be evaluated immediately or placed in formalin until studied. The limbs are positioned on the X-ray film for a single lateral radiograph. A composite score is taken for the entire limb. Structural changes of erosions and joint space narrowing are scored: 0 = no erosions or joint space narrowing, 1 = mild, reactive changes with general maintenance of joint space and bone architecture and no erosions, 2 = moderate, loss of joint space and definite erosions, 3 = severe, marked erosive and destructive change with loss of bone architecture. Periostitis is scored on the scale: 0 = normal, 1 = mild, 2 = moderate, 3 = severe. In general, we have limited our structural evaluation to the hind paws. The total radiograph score for each animal is the sum of the grades for both rear paws. Micro computed tomography (CT) scanning promises to replace radiology as the method of choice in the future for evaluation of joint destruction in animal models of arthritis; however, this scoring protocol will be valid for the newer visualization method.

3. Circulating autoantibodies are measured by enzyme linked immunosorbent assay (ELISA) in blood samples taken routinely at 35 d at sacrifice by cardiac puncture. Samples can also be taken earlier under anesthesia from tail veins or retro-orbital sampling. Blood samples are transferred to prelabeled B-D vacutainer, blood collection tubes (Beckton Dickson; cat. no. 36-6514). Remove needle before expelling blood into the collection tubes as rat red blood cells lyse very easily. We use outdated sterile SST™ Tubes with polymer barrier material and clot activator on walls obtained from the hospital laboratories. Tubes are allowed to clot at room temperature for 30 min and straw colored serum obtained by centrifugation ($14,000g$ for 5 min). Serum is immediately aliquoted, snap-frozen and stored at $-70°C$ for subsequent measurement of inflammatory cytokines using commercial kits and IgG collagen antibodies by ELISA.

3.6. Enzyme Linked Immunosorbent Assay for Measurement of Collagen Autoantibodies

1. Dissolve CII overnight in $0.1N$ acetic acid, on rotator in cold room at 2 mg/mL. To coat plate, dilute CII with fresh, cold phosphate coating buffer to 10 µg/mL.
2. Chill plates on ice. Add cold native CII solution, 1 µg/well in 100 µL volume, to coat the wells. Cover with clean lids and incubate at 4·C overnight.
3. Wash wells thoroughly three times with cold PNT.
4. Block wells with 0.25 mL cold 1% BSA overnight at $4°C$.
5. Wash wells thoroughly three times with cold PNT.
6. Dilute test sera in PNT with 0.5% BSA: 1:100, 1:1000, 1:10,000, 1:100,000.
7. Add 100 µL/well of each serum dilution in duplicate.
8. Control wells (triplicates) contain PNT, 0.5% BSA, and naïve rat serum at the same dilutions as the test sera.
9. Prepare a standard curve using a sera pool from high responder, CII-immunized rats, diluted as in **step 6** and add 100 µL/well in duplicate as for the test sera in **step 7**.
10. Cover and leave at $4°C$ overnight.
11. Wash wells thoroughly three time with cold PNT.
12. Add second antibody, diluted in PNT + 0.5% normal goat serum, 100 µL/well.
13. Incubate at room temperature for 1 h.
14. Wash wells thoroughly three times with room temperature PNT.
15. Add OPD substrate, 100 µL/well. Incubate at room temperature in the dark.
16. Stop color reaction by addition of 0.05 mL, $2.5N$ sulfuric acid/well.
17. Read Optical Density (OD) units for raw data.
18. Convert OD to Antibody Units (AU) / mL: AU = [(dilution × OD/1000)] for wells with OD readings > 0.1 and < 1.2.
19. Mean AU for each serum is normalized against the standard curve developed with the positive serum pool on each plate and assigned an arbitrary AU/mL concentration. Plate to plate assay variation is ± 10%. The theoretical assay range is 0.01 to100 AU (*see* **Note 15**).

4. Notes

1. CII Characteristics: The structure of the pepsin-solubilized, native CII used for arthritis induction is a triple helix composed of 3 α-chains with identical amino acid sequence (Gly-Pro-X) and post-translational modifications of short carbohydrate side chains. CII is freed from the cartilage by pepsin action on short terminal, nonhelical peptide chains which anchor it to the cartilage matrix. The helical portion is resistant to pepsin but easily digested in the denatured state. It is necessary to protect native CII from light and moisture because of its tendency to form chemical cross-links which decrease its solubility and arthritogenicity.
2. Reagents: Use Ultra-pure or ACS reagent grade. Most of the reagents are available from Sigma-Aldrich Chemical Company (*www.sigmaaldrich.com*). It is important to use fresh, ice-cold reagents and nano-pure (Super-Q™ System, Millipore Corp) or glass distilled water to maintain a semisterile milieu for the solubilized CII during the purification procedure. For example, Tris buffer at neutral pH supports rapid bacterial growth, a source of endotoxin. Glassware can be rinsed in 0.5*N* acetic acid to remove adherent, denatured collagen.
3. Animals: LEW rats develop a less inflammatory clinical presentation compared to DA and usually do not reach 100% incidence. DA rats show 100% incidence of rapid onset, severe disease that is characterized by a strong inflammatory response but moderate autoantibody titers. LOU and BB rats, also 100% incidence, are characterized by severe disease and very high autoantibody titers but a slightly later onset. BB rats are expensive and must be produced in house. DA, LOU, and LEW are readily available commercially as SPF animals. If less than 100% incidence is observed with these highly susceptible strains, experimental procedures should be reviewed to assure that the CII preparation and injection techniques are being properly performed. Resistant strains include F344 and BN. Reliable commercial sources include Harlan (*www.harlan.com*), Charles River Laboratories (*www.criver.com*), and Bantin & Kingman . If genetic equivalence from experiment to experiment is important a production colony can be established in house or an understanding that animals will come from the same facility can be maintained with the vendor.

    Rats are 8- to 10-wk of age at the time of immunization. Female or male can be used. DA males tend to an earlier onset whereas females develop a significantly higher arthritis severity compared to males. SPF rats are required because certain infections, particularly mycoplasma pulmonis and internal-external parasites, negatively affect the model whereas Killum Rat Virus aggravates the arthritis (unpublished observations). For a truly autoimmune trial, DA rats are also responsive to rat type II collagen.

    Several inbred congenic strains are now available that isolate small regions of the genome, from a resistant strain (F344, BN) onto the background of a highly susceptible strain (DA). These newly developed rat strains introgress 20- to 40-cM segments (Quantitative Trait Loci, QTL) or entire chromosomes (Chromosome substitution strains, Css) known to modulate arthritis severity and/or collagen autoimmunity. Some are available from the NIH Rat Repository where they were embryo-rederived and are now preserved as frozen embryos (*www.ors.od.nih.gov/dirs/vrp/ratcenter*). Currently, viable mating pairs are available only from university laboratories. Finally, new biotechnology companies will soon offer transgenic and knock-out rats with targeted gene disruptions that will be useful for certain research applications if provided upon the appropriate (CIA-susceptible) -genetic background (*www.genosys.com*).

4. Rat Identification Systems: Individual rats can be identified using an ear clip system which is usually adequate for most studies. We prefer to use IMI™-1000, sterile, implantable transponders obtained from BMDS (Biomedic Data Systems, Inc.; *www.BMDS.com*). We, and the company, have shown that the transponders do not affect the model. The transponders can be frozen with tissue samples to insure correct identification (10% breakage rate at –70°C).
5. Suitable choices are the Virtis HandiShear table top homogenizer or a small scale, handheld Brinkmann Polytron® Homogenizer. Both are available from several distributors
6. Collagen extraction and purification: This purification scheme consistently produces native type II collagen of high purity free of: extraneous proteins (by 230/260/280 n*M* absorption ratios and protein determinations under nondenaturing conditions), proteoglycans as measured by the carbazole method for analysis of uronic acid *(8)* and residual pepsin (by ELISA using mouse-anti pepsin immune serum). The final product is 99.9% native type II collagen by gel electrophoresis under denaturing and nondenaturing conditions and amino acid analysis *(9)*. Aspects critical to success in the CIA model are: (a) protection of the CII from denaturation by performing all procedures at 4°C; (b) protection of solubilized CII from endogenous proteolytic enzymes by incorporation of inhibitors; (c) control of pH; (d) patience during the dialysis to allow full equilibration of CII out of solution; and (e) very slowly raising the NaCl concentration during the fractional salt precipitation steps to avoid coprecipitation of other collagen types with CII.

A major goal of thorough cartilage pretreatment is to inhibit collagen digestion by endogenous proteinases. Some are not stable to freeze thaw. EDTA/benzamidine/azide/NEM inhibitors are used. A second goal is to remove as much contaminating protein as possible. The perichondrium is a source of type I collagen plus other proteins and lipids. Bone is a source of type I collagen and calcium which binds to EDTA. Bone can be avoided by being aware of its greater hardness and proximity to the marrow. Bacterial growth is a major concern as a source of contaminating proteins, proteolytic enzymes, lipopolysaccharides and endotoxin.

To prepare rat CII, remove the inner and outer ears, strip off the outer skin layer with scissors, and collect in cold BzAzE water. Freeze. Thaw and soak in BzAzE water overnight and strip off inner ear skin layer with a scalpel. Soak ears overnight in chloroform:methanol, 2:1 *(v/v)*, room temperature and in a fume hood. Then soak for 3 to 6 h with methanol for 2 changes followed by several BzAzE water washes to remove any residual methanol. Store frozen or transfer to proteoglycan extraction buffer.

7. Proteoglycan Extraction: Use several filters as they tear easily. With suction off, fill the funnel and gently stir with a glass test tube and turn on suction. Repeat several times until all of the calculated wash water is used. The cartilage slices will turn from translucent to opaque white when free of the proteoglycan extraction buffer. Small cartilage slices are difficult to see and can be recovered from glass fibers with tweezers or scraping with a flat spatula.
8. Protein Concentration Method: CII does not absorb ultraviolet light at 280 (only 230 nm). Because of helical nature, native CII does not react with usual protein detection reagents. To measure CII protein concentrations, denature the CII by heating to 60°C for 10 min in water bath and quick chill on ice for micro-BCA Assay or use 0.015% sodium dodecyl sulfate (SDS) (final) with the Bradford Reagent (*www.piercenet.com*).
9. Do not confuse the initial sodium phosphate precipitate with CII. Na_3PO_4 is granular in appearance, making the solution uniformly cloudy. It will go back into solution. CII precipitates later during the second or third dialysate buffer change with a more compact appearance leaving areas of clear solution.
10. Visually, it is difficult to discern when the collagen is in complete solution because it initially forms clear gel droplets which adhere to the sides of the tube and are difficult to see, giving the appearance of a clear solution. When the CII is fully dissolved, the solution is very highly viscous at this concentration.
11. Keep the emulsion on ice until injected. Always change needles and syringes between rats. Remove needle and slowly fill the syringe. This avoids air bubbles which are difficult to see in the white emulsion.
12. The booster injection at 7 d is an optional step. For genetic studies it brings incidences to 100%. However, for investigative drug studies, it may be preferable to omit this step and test against a single primary challenge comparing both incidence and severity.
13. Rats are visually scored for arthritis severity and weighed 3 times/wk for 35 d, starting on the day 7 after primary immunization. Although the study can be carried out by one technician, we prefer to use two, one of which (the scorer) is "blinded" to the experimental or "test" history of each rat for comparison studies such as genetic comparisons or drug testing. Animals are identified on cage cards by experiment number and rat number only. Any investigational drug dosing or experimental manipulation is performed by a one technician who maintains the procedure logs. The second technician uses a voice activated dictaphone to record his clinical evaluation scores of each rat. The tape is transcribed by the first technician such that the scorer does not know the clinical or treatment history of any given rat at the time he is evaluating its arthritis. It is possible to reproduce such scoring within 1% variation upon repeat 1-d comparisons.
14. The edema associated with arthritis can be evaluated by measuring hind paw swelling using a mercury displacement plethysmograph. This method does not distinguish between areas of inflammation (toe, ankle, and midfoot). It not as sensitive as the fully numerical scoring system described. It is useful for a rapid,

nonsubjective assessment of arthritis, particularly when only one technician is available. Constant tension calipers have also been used to measure swelling through the ankles and mid-paw regions. The method is tedious and repeat measurements are highly variable. Neither method detects ankylosis or paw deformity, a measure of joint destruction and cartilage loss.

15. ELISA assays measure circulating collagen antibody titers to insure adequate immunization. Any animal showing significantly lower anti-collagen antibody titers compared to the group (2 standard deviations below the mean) is removed from the study as a technical failure at the immunization step.

   ELISA assay conditions can be varied to suit a given laboratory's usual protocols. CII specific conditions are: the coating buffers, overnight solubilization of the CII, maintaining the plates at 4°C at all times and using cold reagents until the final second antibody step and color development. This is to maintain CII as a nondenatured antigen for the test *(3)*. The overnight incubation with test serum is convenient and provides a longer interaction time for complete binding of collagen antibody to nCII at 4°C. The water quality is critical as minor contamination with chloride will cause a color change which is uniform over the plate. If color is spotty over the plate, it is an indication of inadequate washing of wells or that the test serum and/or conjugate have aggregated and need to be clarified by centrifugation.

References

1. Griffiths, M. M. and Remmers, E.F. (2001) Genetic analysis of collagen-induced arthritis in rats: a polygenic model for rheumatoid arthritis predicts a common framework of cross-species inflammatory/autoimmune disease loci. *Immunol. Rev.* **184**, 172–183.
2. Myers, L. K., Myllyharju, J., Nokelainen, M., et al. (2004) Relevance of post-translational modifications for the arthritogenicity of type II collagen. *J. Immunol.* **1**, 2970–2975.
3. Griffiths, M. M., Cremer, M. A., Harper, D. S., McCall, S. and Cannon, G. W. (1992) Immunogenetics of collagen-induced arthritis in rats. Both MHC and non-MHC gene products determine the epitope specificity of immune response to bovine and chick type II collagens. *J. Immunol.* **149**, 309–316.
4. Wilder, R. L., Remmers, E. F., Kawahito, Y., et al. (1999) Genetic factors regulating experimental arthritis in mice and rats, in *Curr. Dir. Autoimmun.* (Theophilopolous, A. ed.) Karger, Basel, pp. 121–165.
5. Cremer, M. A., Griffiths, M. M., Terato, K. and Kang, A. H. (1995) Type XI and II collagen-induced arthritis in rats: characterization of inbred strains of rats for arthritis-susceptibility and immune responsiveness to type XI and II collagen. *Autoimmunity* **20**, 53–161.
6. Brahn, E., Banquerigo, M. L., Firestein, G. S., Boyle, D.L., Salzman, A. L,. and Szabo, C. (1998) Collagen induced arthritis: reversal by mercaptoethylguanidine, a novel anti-inflammatory agent with a combined mechanism of action. *J. Rheumatol.* **9**, 1785–1793.

7. Ye, X. Y., Tang, B., Ma, Z., Kang, A. H., Myers, L. K., and Cremer, M. A. (2004) The roles of interleukin-18 in collagen-induced arthritis in the BB rat. *Clin. Exp. Immunol.* **1**, 440–447.
8. Bitter, T. and Muir, H. M. (1962) A modified uronic acid carbazole reaction. *Anal. Biochem.* **4**, 330–334.
9. Stegemann, H. and Stalder, K. (1967) Determination of hydroxyproline. *Clin. Chem. Acta.* **18**, 267–273.
10. Kawahito, Y., Cannon, G. W., Gulko, P. S., et al. (1998) Localization of quantitative trait loci regulating adjuvant-induced arthritis in rats: evidence for genetic factors common to multiple autoimmune diseases. *J. Immunol.* **161**, 4411–4419.

16

Collagen Antibody Induced Arthritis

Kutty Selva Nandakumar and Rikard Holmdahl

Abstract

Rheumatoid arthritis (RA) is a polygenic and multifactorial disease. Many complex immunological and genetic interactions are involved in the final out come of the clinical disease. To understand the various disease pathways operating during the disease course, we need many different animal models. Collagen induced arthritis (CIA) is one of the widely used animal models sharing many pathological and histological similarities with RA and antibodies play an important role in the inflammatory phase of CIA. This chapter describes, in detail, an animal model for arthritis using CII specific monoclonal antibodies, the so-called collagen antibody induced arthritis (CAIA), which shares many characteristics of CIA. CAIA model provides an opportunity to study the inflammatory phase of arthritis without involving the priming phase of the immune response. CAIA can be used for not only studying inflammatory processes in arthritis and screening drug candidates controlling joint inflammatory phase but also as a model for studying common mechanisms involved in many antibody mediated diseases.

Key Words: Collagen induced arthritis; collagen antibody induced arthritis; collagen type II; monoclonal antibodies; rheumatoid arthritis.

1. Introduction

Serum from CII immunized rats and mice was shown to induce an acute arthritis and the active component was proposed to be IgG antibodies specific for CII *(1–3)* although other factors might also have been of importance *(4)*. Single monoclonal antibodies (mAbs) to CII was later shown to induce arthritis *(5,6)* and to bind to the cartilage surface *(7,8)*. A cocktail of monoclonal antibodies could be shown to increase arthritis severity *(9–18)*. These were specific for conformational (triple helical) epitopes on CII that are conserved through mice, rats, and humans *(13,19–22)*. In fact in both CII immunized rats and mice, and in RA, antibodies to the same dominant epitopes are increased and antibodies to these epitopes are arthritogenic and can now be used for

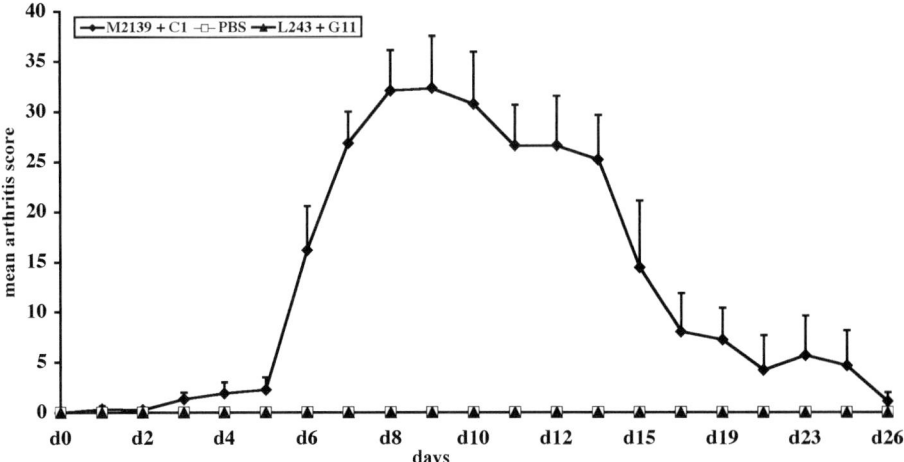

Fig. 1. A typical disease curve in BALB/c mice during antibody induced arthritis. A number of mice ($n = 12$) were injected with 2 mAbs cocktail iv on day 0. As an internal control equal number of mice were injected with PBS. For isotype antibody control a pair of irrelevant antibodies were injected into a group of mice ($n = 10$) of same age. All the mice received 50 μg of LPS injection on day 5 ip. Arthritis was scored as described in materials and methods for 30 d. None of the control mice developed arthritis.

induction of the CAIA model. This model is described in detail in the following section.

1.1. Collagen Antibody Induced Arthritis

CAIA is inducible with well-characterized mAbs to CII, which bind specified triple helical epitopes. It is a T-cell and B-cell independent arthritis caused by infiltration of polymorphonuclear neutrophils and macrophages activated through the aggregated Fc on the cartilage surface, interacting with Fc Receptors and activating both classical and alternative complement pathways.

A panel of monoclonal antibodies were generated by using the rat collagen type II (CII) immunized DBA/1 mice spleenocytes or lymph node cells by classical hybridoma technique *(5,23)*. These CII specific mAbs injected in vivo binds to the normal cartilage both in neonatal *(8,24)* and adult mice. Intravenous (iv) injection of between 3 and 9 mg of a cocktail of 2 mAbs M2139 (IgG2b;kd = 3×10^{-6}) and C1(IgG2a;kd = 1.9×10^{-3}) binding to J1 (MP*GERGAAGIAGPK-P* indicates hydroxy proline) and C1^1 (GARGLT) epitopes are arthritogenic *(14)*. Intraperitoneal (ip. injection of lipopolysaccharide (LPS) on day 5 enhances the incidence and severity of arthritis induced

Table 1
Genetic Variation in CAIA Incidence and Severity

Mice	% of Incidence		Mean max score ± SEM	
	A	L	A	L
BALB/c	89.3	100	5.0 ± 0.9	32.9 ± 3.0
B10.RIII	72.2	94.3	8.9 ± 1.3	13.0 ± 1.3
(BALB/cxB10.Q)F1	55.7	95	3.8 ± 0.7	20.3 ± 2.0
(DBA/1xB10.Q)F1	26.3	85.3	3.6 ± 1.1	14.7 ± 1.6
B10.Q	10	40	1.1 ± 0.6	4.5 ± 1.1
C3H.Q	6.6	26.7	0.3 ± 0.3	2.0 ± 1.0
DBA/1	0	100	0	15.1 ± 2.7
C57BL/6	0	21.4	0	0.5 ± 0.3
NFR/N	0	33.3	0	0.3 ± 0.2
RIIIS/J	0	14.2	0	1.4 ± 1.4
NOD.Q	0	0	0	0

Note: A group of normal mice from several different mice strains were injected with two mAbs on day 0 and all the mice received LPS on day 5. BALB/c ($n = 36$), B10.RIII ($n = 36$), (BALB/c × B10.Q)F1 ($n = 61$), (DBA/1 × B10.Q)F1 ($n = 34$), B10.Q ($n = 40$), C3H.Q ($n = 15$), DBA/1 ($n = 19$), C57BL/6 ($n = 14$), NFR/N ($n = 6$), RIIIS/J ($n = 7$), and NOD.Q ($n = 11$) mice were used in this experiment. N indicates the number of mice in each group, A indicates arthritis susceptibility at day 5 (before LPS injection), and L indicates maximal arthritis after LPS injection.

with the antibodies. A typical arthritis disease curve in BALB/c mice is depicted in **Fig. 1**. Another mAb CIIF4 binding to F4 epitope (ERGLKGHRGFT) was found to inhibit CAIA *(13)*, which might be because of the blocking of the binding of arthritogenic antibodies to CII via steric hindrance or the stromelysin cleavage site that colocalizes with this epitope. Intravenous transfer of CII specific mAbs induced more severe arthritis with a higher incidence than intraperitoneal transfer. Mice 4 mo and older were more susceptible to CAIA with the 2 mAbs compared with 2-mo old mice. However, when a cocktail of 4 mAbs binding to 4 different CII epitopes viz., J1, C1[1], D3 and U1 were used, CAIA could be inducible in young mice (8-wk-old) even with a total concentration of 2 mg of antibodies/mouse *(18)*. Male mice were more prone to arthritis with the antibody cocktail than females, however after LPS injection such a sex difference was not observed. Among the several different mouse strains tested (**Table 1**), BALB/c and B10.RIII mice were more susceptible to CAIA

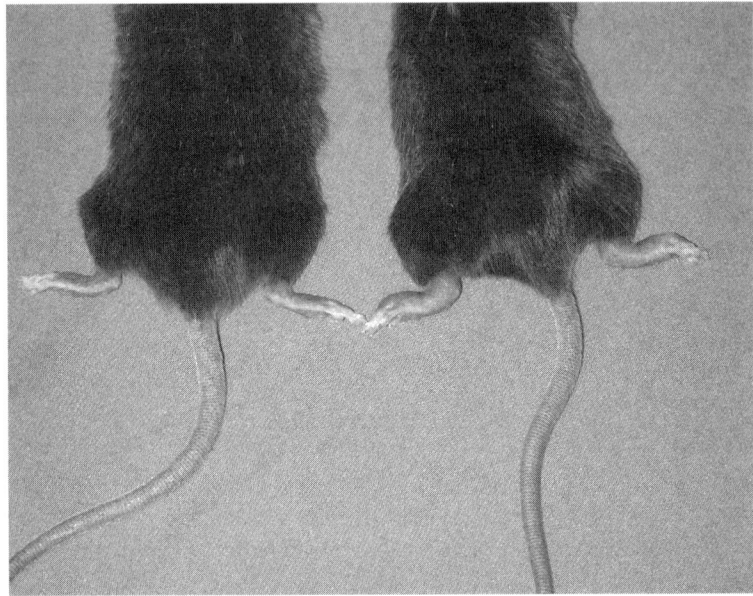

Fig. 2. Clinical disease developed with antibody transfer in B10.RIII mice on day 10. *Left:* normal mice; *right:* arthritic mice.

and NOD.Q was resistant. As shown in **Fig. 2**, B10.RIII mice showed high clinically severe arthritis with antibodies alone. Histology of the normal and arthritic joint sections (*see* **Fig. 3A,B**) revealed massive infiltration of granulocytes, bone erosion, pannus formation, and fibrin deposition in the arthritic mice. Depletion of neutrophils significantly reduced the incidence and severity of CAIA *(14)*. In addition, FcR γ-chain deficient mice were resistant to CAIA demonstrating the importance of FcγR I, III, and IV bearing cells in mediating the antibody mediated inflammation in the joints, whereas FcγRIIb (negative regulator of Ig synthesis) deficient mice were highly susceptible to CAIA. Single mAb can also induce arthritis in these mice *(6)*. Furthermore, both the classical and alternate pathway of complement activation was found to be essential for the antibody induced arthritis *(15)*. Plasminogen was found to be essential for induction of CAIA *(16)*.

2. Materials
1. CII M2139 and CII C1 B-cell hybridomas.
2. Medium: DMEM glutamax-I containing 10% Ig-free fetal bovine serum (FBS) medium (Gibco BRL) and kanamycin monosulfate, 100 mg/L (Sigma).

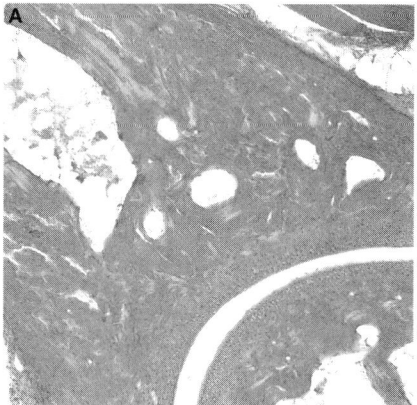 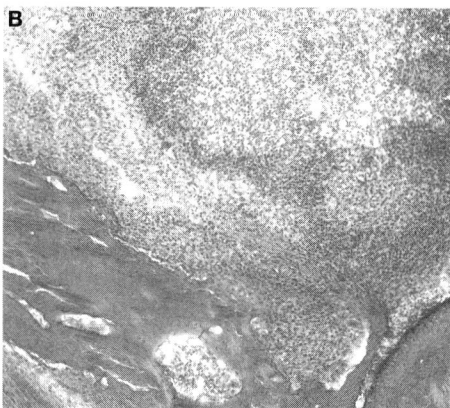

Fig. 3. Histology of normal (**A**) and arthritis (**B**) joint sections of B10.RIII mice. Magnification ×10. Haematoxylin staining of normal and arthritis paw of male mice on day 17. Massive infiltration of granulocytes, bone erosion, pannus formation and fibrin deposition was observed (*see* **Note 5**).

3. Integra cell line 1000 flasks, CL-1000 (Integra Biosciences).
4. Mini bioreactor Cellmax (GTF).
5. γ-Bind plus affinity gel matrix (Pharmacia).
6. Äkta purification system (Amersham Pharmacia Biotech AB).
7. Phosphate buffered saline (PBS) (pH 7.0): $0.14M$ Nacl, 2.7 mM KCl, 8.26 mM Na_2HPO_4, and 1.47 mM KH_2PO_4.
8. Syringe filters, 0.2 nm (Dynagard; Spectrum Laboratories).
9. Limulus amebocyte lysate (Pyrochrome).
10. Lipopolysaccharide from *Escherichia coli* serotype 055:B5 (Sigma).

3. Methods

The methods described below outline (1) purification of collagen specific monoclonal antibodies, (2) passive transfer of antibodies into mice, (3) clinical evaluation of arthritis developed, and (4) the statistical analyses of the results.

3.1. Purification of Monoclonal Antibodies

Collagen specific monoclonal antibodies were generated in large scale as culture supernatants using integra cell line 1000 flasks and the mini-bioreactor Cellmax (*see* **Note 1**) Antibodies were purified using γ-bind plus affinity gel matrix and Äkta purification system. Briefly, culture supernatants were centrifuged at 12,500 rpm for 30 min, filtered, and degassed before applying to the gel matrix. Antibodies were eluted using acetic acid buffer at pH 3.0 (*see* **Note 2**) and neutralized with $1M$ Tris-HCl (pH 9.0). The peak fractions were pooled and dialyzed extensively against phosphate buffered saline (PBS) (pH 7.0) with or without azide. The IgG content was determined either spectrophotometrically or by freeze-drying. The antibody solutions were filter sterilized using 0.2 μm syringe filters, aliquoted and stored at –70°C until used. Amount of endotoxin content in the antibody solutions prepared was found to be in the range of 0.02 to 0.08 EU/mg of protein as analyzed by limulus amebocyte lysate (Pyrochrome) method. Also, Tlr4-deficient (LPS nonresponder) mice in the BALB/c background were highly susceptible to arthritis induced with the CII antibody cocktail, excluding a crucial role for endotoxin in the antibody preparation *(14)*.

3.2. Passive Transfer of Antibodies

The cocktail of M2139 and C1 mAbs was prepared by mixing equal concentrations of each of the sterile filtered antibody solutions to get a final amount of 9 mg. Mice (*see* **Note 3**) were injected intravenously with 0.25 to 0.4 mL volumes of antibody solutions twice with a minimum of 3 h interval. As internal controls, mice received equal volumes of PBS. On day 5, lipopolysaccharide (25 or 50 μg/mice) was injected ip to all the mice. A pair of irrelevant antibodies of same subclass (mouse anti-human HLA-DRα, IgG2a (L243) and mouse anti-human parathyroid epithelial cells, IgG2b (G11) did not induce arthritis in the most susceptible mice strain, BALB/c *(14)*.

3.3. Clinical Evaluation of Arthritis

Mice were examined daily for the arthritis development before and after LPS treatment for a minimum of 21 d or until the inflammation disappeared. Scoring of animals was done blindly using a scoring system based on the number of inflamed joints in each paw, inflammation being defined by swelling and redness (*see* **Note 4**).

3.4. Statistical Analyses

All the mice were included for calculation of arthritis susceptibility and severity. The severity of arthritis was analyzed by Mann Whitney U test or Kruskal Wallis and the incidence by Chi Square or Fisher's exact test using the statview v5.0.1. Significance was considered when $p \leq 0.05$.

4. Notes

1. Cell lines were regularly monitored for production of mAbs by standard collagen specific enzyme linked immunosorbent assay (ELISA) method *(5)*.

2. Some mAbs tend to precipitate while purifying them. To avoid this, Glycine-Hcl buffer (pH 3.0) is recommended instead of acetic acid buffer (pH 3.0).
3. BALB/c mice were obtained from Jackson and B10.RIII strain was originated from Professor Jan Klein (Tubingen, Germany) stock. Among the various mice strains tested BALB/c and B10.RIII were found to be highly susceptible and NOD.Q mice was resistant to CAIA.
4. Extended scoring system was used to monitor arthritis development. In this scoring system each inflamed toe or knuckle gives one point, whereas an inflamed wrist or ankle gives five points, resulting in a score of between 0 and 15 (5 toes + 5 knuckles + 1 wrist/ankle) for each paw and between 0 and 60 points for each mouse.
5. In some mouse strains, cartilage erosion was also observed. Granulocyte population contained both macrophages and polymorphonuclear leukocytes (PMNL).

References

1. Stuart, J. M., Cremer, M. A., Townes, A. S., and Kang, A. H. (1982) Type II collagen-induced arthritis in rats. Passive transfer with serum and evidence that IgG anticollagen antibodies can cause arthritis. *J. Exp. Med.* **155**, 1–16.
2. Stuart, J. M. and Dixon, F. J. (1983) Serum transfer of collagen-induced arthritis in mice. *J. Exp. Med.* **158**, 378–92.
3. Cremer, M. A., Hernandez, A. D., Townes, A. S., Stuart, J. M., and Kang, A. H. (1983) Collagen-induced arthritis in rats: antigen-specific suppression of arthritis and immunity by intravenously injected native type II collagen. *J. Immunol.* **131**, 2995–3000.
4. Helfgott, S. M., Dynesius-Trentham, R., Brahn, E., and Trentham, D. E. (1985) An arthritogenic lymphokine in the rat. *J. Exp. Med.* **162**, 1531–1545.
5. Holmdahl, R., Rubin, K., Klareskog, L., Larsson, E., and Wigzell, H. (1986) Characterization of the antibody response in mice with type II collagen-induced arthritis, using monoclonal anti-type II collagen antibodies. *Arthritis Rheum.* **29**, 400–410.
6. Nandakumar, K. S., Andren, M., Martinsson, P., et al. (2003) Induction of arthritis by single monoclonal IgG anti-collagen type II antibodies and enhancement of arthritis in mice lacking inhibitory FcgammaRIIB. *Eur. J. Immunol.* **33**, 2269–2277.
7. Jonsson, R., Karlsson, A. L., and Holmdahl, R. (1989) Demonstration of immunoreactive sites on cartilage after in vivo administration of biotinylated anti-type II collagen antibodies. *J. Histochem. Cytochem.* **37**, 265–268.
8. Holmdahl, R., Mo, J. A., Jonsson, R., Karlstrom, K., and Scheynius, A. (1991). Multiple epitopes on cartilage type II collagen are accessible for antibody binding in vivo. *Autoimmunity* **10**, 27–34.
9. Terato, K., Hasty, K. A., Reife, R. A., Cremer, M. A., Kang, A. H., and Stuart, J. M. (1992) Induction of arthritis with monoclonal antibodies to collagen. *J. Immunol.* **148**, 2103–2108.
10. Terato, K., Harper, D. S., Griffiths, M. M., et al. (1995) Collagen-induced arthritis in mice: synergistic effect of *E. coli* lipopolysaccharide bypasses epitope speci-

ficity in the induction of arthritis with monoclonal antibodies to type II collagen. *Autoimmunity* **22**, 137–147.
11. Johansson, A. C., Hansson, A. S., Nandakumar, K. S., Backlund, J., and Holmdahl, R. (2001) IL-10-deficient B10.Q mice develop more severe collagen-induced arthritis, but are protected from arthritis induced with anti-type II collagen antibodies. *J. Immunol.* **167**, 3505–3512.
12. Svensson, L., Nandakumar, K. S., Johansson, A., Jansson, L., and Holmdahl, R. (2002) IL-4-deficient mice develop less acute but more chronic relapsing collagen-induced arthritis. *Eur. J. Immunol.* **32**, 2944–2953.
13. Burkhardt, H., Koller, T., Engstrom, A., et al. (2002) Epitope-specific recognition of type II collagen by rheumatoid arthritis antibodies is shared with recognition by antibodies that are arthritogenic in collagen-induced arthritis in the mouse. *Arthritis Rheum.* **46**, 2339–2348.
14. Nandakumar, K. S., Svensson, L., and Holmdahl, R. (2003). Collagen type II-specific monoclonal antibody-induced arthritis in mice: description of the disease and the influence of age, sex, and genes. *Am. J. Pathol.* **163**, 1827–1837.
15. Hietala, M. A., Nandakumar, K. S., Persson, L., Fahlen, S., Holmdahl, R., and Pekna, M. (2004) Complement activation by both classical and alternative pathways is critical for the effector phase of arthritis. *Eur. J. Immunol.* **34**, 1208–1216.
16. Li, J., Ny, A., Leonardsson, G., Nandakumar, K. S., Holmdahl, R., and Ny, T. (2005) The plasminogen activator/plasmin system is essential for development of the joint inflammatory phase of collagen type II-induced arthritis. *Am. J. Pathol.* **166**, 783–792.
17. Nandakumar, K. S. and Holmdahl, R. (2005) A genetic contamination in MHC-congenic mouse strains reveals a locus on chromosome 10 that determines autoimmunity and arthritis susceptibility. *Eur. J. Immunol.* **35**, 1275–1282.
18. Nandakumar, K. S., and Holmdahl, R. (2005) Efficient promotion of collagen antibody induced arthritis (CAIA) using four monoclonal antibodies specific for the major epitopes recognized in both collagen induced arthritis and rheumatoid arthritis. *J. Immunol. Methods* **304**, 126–136.
19. Schulte, S., Unger, C., Mo, J. A., et al. (1998) Arthritis-related B cell epitopes in collagen II are conformation-dependent and sterically privileged in accessible sites of cartilage collagen fibrils. *J. Biol. Chem.* **273**, 1551–1561.
20. Kraetsch, H. G., Unger, C., Wernhoff, P., et al. (2001) Cartilage-specific autoimmunity in rheumatoid arthritis: characterization of a triple helical B cell epitope in the integrin-binding-domain of collagen type II. *Eur. J. Immunol.* **31**, 1666–1673.
21. Wernhoff, P., Unger, C., Bajtner, E., Burkhardt, H., and Holmdahl, R. (2001) Identification of conformation-dependent epitopes and V gene selection in the B cell response to type II collagen in the DA rat. *Int. Immunol.* **13**, 909–919.
22. Bajtner, E., Nandakumar, K. S., Engstrom, A., and Holmdahl, R. (2005) Chronic development of collagen-induced arthritis is associated with arthritogenic antibodies against specific epitopes on type II collagen. *Arthritis Res. Ther.* **7**, R1148–R1157.

23. Mo, J. A., Bona, C. A., and Holmdahl, R. (1993) Variable region gene selection of immunoglobulin G-expressing B cells with specificity for a defined epitope on type II collagen. *Eur. J. Immunol.* **23**, 2503–2510.
24. Mo, J. A., Scheynius, A., Nilsson, S., and Holmdahl, R. (1994) Germline-encoded IgG antibodies bind mouse cartilage in vivo: epitope- and idiotype-specific binding and inhibition. *Scand. J. Immunol.* **39**, 122–130.

17

Arthritis Induced with Minor Cartilage Proteins

Stefan Carlsen, Shemin Lu, and Rikard Holmdahl

Abstract

Type XI collagen (CXI) and cartilage oligomeric matrix protein (COMP) are minor components in cartilage, shown to be arthritogenic. CXI is a heterotrimeric triple helical fibrillar collagen and intermingled in the collagen fibers with type II (CII). COMP is the major noncollagenous protein of cartilage and is a homopentamer, interacting with the collagen fibers with each of its subunits. Similar to CII, homologous rat CXI also induces a chronic arthritis in rats but with a different major histocompatibility complex (MHC) genetic control and pathogenesis. CXI induced arthritis ($C^{XI}IA$) is characterized by a more pronounced chronic relapsing disease course. The MHC allele of importance is the RT1f haplotype and, surprisingly, some of the CII associated MHC alleles like RT1a are less permissive. Immunization with COMP induces a severe but self-limited arthritis in strains with a genetic background resistant to most other forms of arthritis or even autoimmune models, the E3 rat. The MHC association also differs between the different models (CIA, CXI, and COMPIA). An autoimmune response to COMP is triggered despite the circulation of COMP fragments in both physiologic and arthritic states. The induction of arthritis in rats with CXI or COMP provides an arthritis models with a distinct pathogenesis as compared with other induced arthritis models.

Key Words: RA; arthritis; COMPIA; CIA; $C^{XI}IA$; PIA; EAE; COMP; CII; CXI; CIX; ELISA; rat.

1. Introduction

Apart for the abundant protein CII, which constitutes up to 80% dry weight of cartilage, several other proteins from cartilage have been reported to induce arthritis, with autoimmune features *(1–8)*. CXI and COMP are both minor content cartilage proteins and have unique characteristics as antigens in the arthritis model.

Collagen types IX (CIX) and XI (CXI) constitute between 10 and 15% and 1 and 5% dry weight of cartilage, respectively depending on source and age *(9)*.

They are composed of three distinct α-chains in equimolar amounts. CXI has structural similarities with the homotrimeric CII triple helix, but also shares the same gene, α3 (CXI) is identical with α1 (CII) gene. However the α3-chain has a higher degree of glycosylation, indicating a different posttranslational processing *(11)*. CIX belongs to the group of FACIT collagens, fibrillar associated collagens with interrupted triple helices, with three collagenous domains and four noncollagenous (NC) domains. The third NC creates a bend in the molecule and is also the site for a glycosaminoglycan (GAG) chain *(9)*. COMP is a major noncollagenous matrix protein in articular cartilage and the fifth member of the thrombospondin family *(12)*. It is a very large protein with pentamers of identical subunits at 86650 Da *(13)*, joined together close to the N-terminus and forming a five-stranded helical coil *(14)*.

Except for cartilage tissues *(15,16)* CIX, CXI and COMP are also found in the vitreous body of the eye *(17)*. CIX and CXI have also been found in the developing cornea, in vertebral disc, but can also be found in noncartilaginous tissues during embryogenesis *(18,19)*. However CIX differs in cartilage tissues, with respect to the size of NC4 and GAG chains *(20,21)*. COMP has also been detected at lower levels in tendon *(22,23)* and in human synovial tissue *(24)*.

It is believed that CXI is intercalated into the core of the microfibril, crosslinked with CII by head-to-tail bonds. Its function is probably to control fibril assembly and to regulate fiber size *(10,25)*. CIX is located on the surface of the fibril, and in an antiparallel orientation *(26)* and covalently bonded to either CII or CXI as well as other CIX molecules through the free collagen-3 and NC4 domain *(27,28)*. Proposed functions of CIX are that it may form a "spacer" for binding with other fibrils, maintaining and organizing the cartilage meshwork through interactions with other molecules.

The C-terminal domain of COMP is globular and has been shown to bind with each of its subunits to CII through Zn^{2+} dependent interactions *(29)*. Hence, its function is likely to be a bridge and it is thereby of importance for the formation and stability of cartilage. Altogether these proteins are important in the development and integrity of cartilage. This is further suggested by the inherited gene defects in α1(XI) and α2(XI), which cause lethal chondrodysplasia *(30)* and Stickler syndrome, respectively *(31)*. Transgenic mice expressing a truncated *(32)*, or lacking *(33)* the α1(CIX) chain develop mild forms of chondrodysplasia associated with osteoarthritis-like degeneration in old mice, but with the fibrils appearing normal. Mutations in COMP cause pseudochondrodysplasias and multiple epiphyseal dysplasia in humans *(34)*.

CXI were found to induce arthritis in rats *(8,35–37)*, but there is as yet no consensus that CIX is arthritogenic in rodents *(5–8)*. Most reports describe arthritis induced with heterologous CXI, inducing an acute, mild and self-limiting disease. With homologous CXI the disease course is more chronic *(37)*. Heterologous CXI of rat origin appears not to be arthritogenic in mice *(6,38)*.

Immunization with rat CXI emulsified in mineral oil induces arthritis in rats *(37)*. The susceptible strains are those with an f- or u-haplotype at the major histocompatibility complex (MHC) whereas the susceptible genetic background is similar to CIA. The onset of $C^{XI}IA$ is relatively late (5–7 wk) and the first phase of acute arthritis is similar to CIA. However, the disease progresses and develops into a lifelong inflammatory active chronic relapsing disease course (*see* **Fig. 1**) and in the pannus-tissue, follicular-like structures can be found. It is accompanied with a strong autoantibody response to CXI. Interestingly, there is a T-cell response but not a B-cell response to CII, reflecting that one of the α-chains in CXI is shared and presenting CII peptides but not conformational B-cell epitopes. However, the a-haplotype that permits a CII response, is not allowing a strong response to CXI, whereas the f-haplotype allows a CXI but not a CII response.

COMP is found circulating as smaller fragments in blood. An interesting feature of COMP is its increased release from cartilage during the erosion of the tissue seen in different forms of arthritis *(39–41)*. COMP is also released in rats developing chronic arthritis after induction with pristane, a synthetic low-molecular-weight adjuvant, and after induction with CII. The levels in serum correlate strongly with the occurrence of erosive arthritis *(42,43)*. Immunization with COMP emulsified in mineral oil induces arthritis in susceptible rat strains, prominently with u-haplotype *(44)*. COMP induced arthritis (COMPIA) has several unique characteristics as compared with other arthritis models. The genetic background determines the outcome differently as the E3 rat is susceptible to COMPIA (*see* **Fig. 1**), whereas it is resistant to most other arthritis models (CIA, $C^{XI}IA$, PIA, OIA), as well as to models for multiple sclerosis such as experimental autoimmune encephalomyelitis. In contradiction, the highly susceptible DA rat is resistant to COMPIA, even if a congenic strain with u-haplotype is utilized *(37,45–50)*. Both native and denatured COMP from one preparation induced arthritis equally well in contrast to CIA, where the protein needs to be in a native triple helical conformation *(44)*. Given that fragments of COMP circulate in blood from both normal and diseased patients *(40)*, this indicates that other epitopes are exposed in purified denatured COMP or that immune tolerance is incomplete. Overall, these models show different characteristics to all other known arthritis models in the rat, including models induced with cartilage proteins like CIA or various adjuvant-induced arthritides, such as PIA.

COMP and collagens can be obtained from Swarm rat chondrosarcoma (RCS), joint, tracheal, or nasal cartilage. Swarm RCS was first discovered in

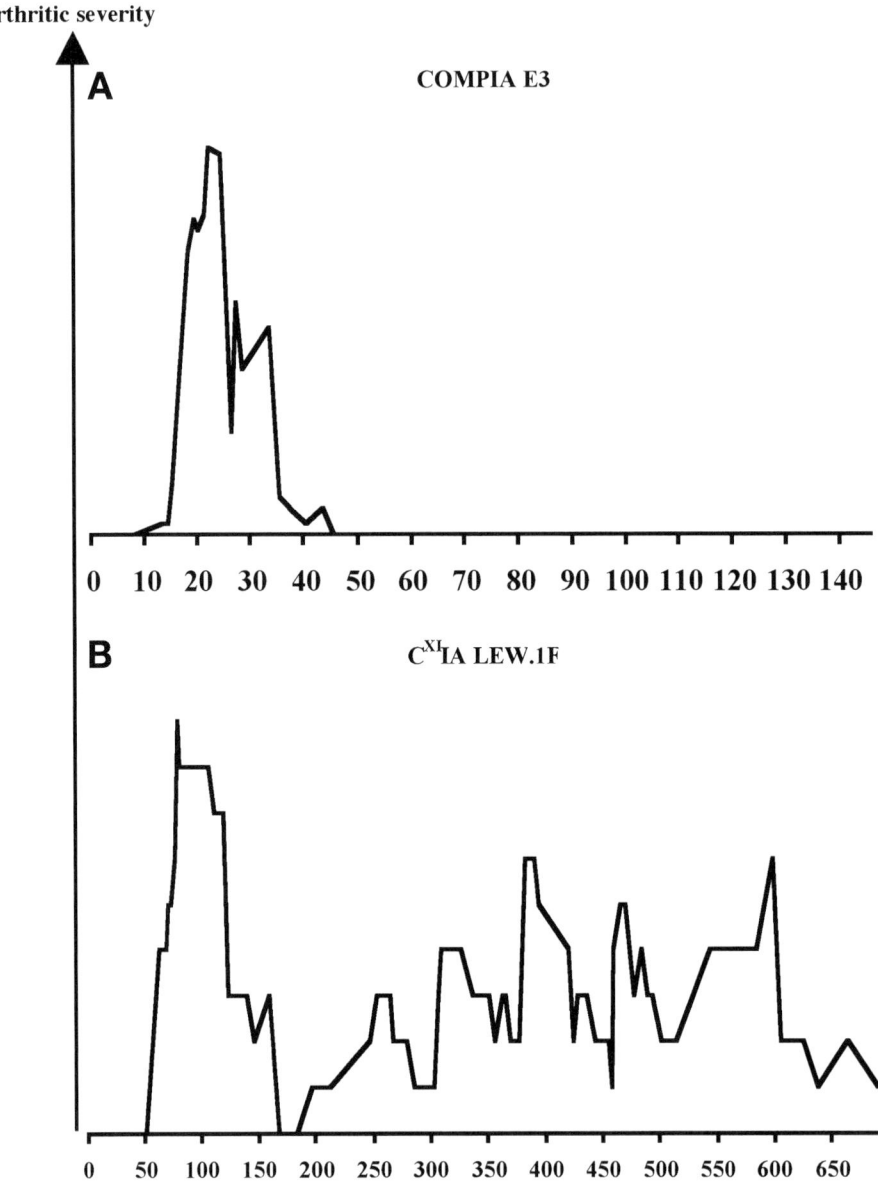

Fig. 1. Typical disease course and arthritis severity in rats from susceptible strains for different models. Severity was calculated as the mean of arthritic rats. (**A**) COMPIA gives an acute, but severe arthritis in 6 E3 female rats that later subsides around day 43. (**B**) CXI induced arthritis ($C^{XI}IA$) gives a chronic and much more severe arthritis in 3 male LEW.1F rats that continued with relapses until the experiment was terminated after more then 2 yr.

1972 as a spontaneous bone tumor arising in the Sprague–Dawley rat *(51)*. RCS consists of about 40% collagen and mainly CII, but also other minor cartilage proteins. COMP is extracted with an ethylene diamine tetraacetic acid (EDTA) solution and further purified by chromatography; we based our modified protocol on previously described methods *(15,29,44)*. Collagens are further purified with pepsin digestion and conventional desalting procedures, a modified protocol based on previously described methods *(38,52–56)*. Pepsin is used to cleave the intermolecular crosslinks between collagen molecules to increase the amount of soluble collagen molecules. The disadvantage from use of pepsin is derived from the loss of short NH2- and COOH-terminal sequences *(53)* and nonhelical sequences within the CIX molecule, giving a mixture of fragments referred to as low-molecular-weight and high-molecular-weight *(56)*. However, as an alternative method too keep the native conformation of collagens, lathyritic collagen can be prepared. Lathyrism is a dietary disease arising from regular consumption of seeds of *Lathyrus odoratus*, the sweet pea. The seeds contain β-aminopropionitrile (BAPN) and were isolated in 1957 by Seylene, which inactivates lysyl oxidase. The enzyme is essential for formation of cross-links between collagen molecules, rendering them soluble for extraction *(52,57)*.

2. Materials
2.1. Tissues

1. RCS, sternal xiphoid, joint, tracheal, or nasal cartilage.
2. Phosphate buffered saline (PBS).
3. Blunt-ended needle, 15 cm in length and 2 mm inner diameter.
4. Agraff's.

2.1.1. Lathyritic Tissues

1. 10X stock solution; 30 g/L β-aminopropionnitrile fumaratic salt in water (BAPN) (Sigma; cat. no. A3134) (*see* **Note 1** and **subheading 2.5.1**).

2.2. COMP Purification

1. Benzamidine Hydrochloride Hydrate (Sigma; cat. no. B6506).
2. 0.20M Phenylmethylsulfonyl fluoride (PMSF) (Sigma; cat. no. P 7626) in 100% anhydrous ethanol (*see* **Note 2**).
3. *N*-Ethylmaleimide (NEM) (Sigma; cat. no. E3876) (*see* **Note 2**).
4. Chromatography equipment with a variable ultraviolet (UV-)monitor (227 and 280 nm) (*see* **Note 3**).
5. DEAE Sepharose™ fast flow media (Amersham Biosciences) with a volume of approx 65 mL.
6. 5 mL HiTrap Heparin Sepharos™ high performance column (Amersham Biosciences).

7. 5 mL HiTrap Q Sepharos™ fast flow column (Amersham Biosciences).
8. Superose® 6 and a Superdex™ 200 (Amersham Biosciences) in series with a minimal bead volume of 25 mL each.
9. Amicon Ultra-15 centrifugal filter NMVL 30 kDa (Millipore).
10. Amicon Ultra-4 Centrifugal Filter NMVL 30 kDa (Millipore).
11. Preextraction buffer: 150 mM NaCl and 50 mM Tris-HCl (pH 7.4).
12. Extraction buffer: 150 mM NaCl, 50 mM Tris-HCl, and 10 mM EDTA (pH 7.4).
13. DEAE buffer A: 150 mM NaCl, 50 mM Tris-HCl, 2 mM EDTA, and 5 mM benzamidine (pH 7.4).
14. DEAE buffer B: 500 mM NaCl, 50 mM Tris-HCl, 2 mM EDTA, and 5 mM benzamidine (pH 7.4).
15. Heparin and Q buffer A: 150 mM NaCl, 10 mM Tris-HCl, and 2 mM EDTA (pH 7.4).
16. Heparin and Q buffer B: 500 mM NaCl, 10 mM Tris-HCl, and 2 mM EDTA (pH 7.4).
17. Gel filtration buffer: 7 mM NaCl, 20 mM Tris-HCl, and 10 mM EDTA (pH 7.4).

2.3. Protein Analysis

1. Sodium dodecyl sulfate polyacrylamide gel electrophoresis (SDS-PAGE) equipment.
2. SDS-PAGE minigels: 4 to 20% gradient.
3. Precipitation solution; 99.5% ethanol and 50 mM NaAc.

2.4. Collagen Purification

1. Dialyze tubing's MWCO 12,000 to 14,000, width 24 to 32 mm (Spectra/Por® 4) (Spectrum Laboratories, Inc).
2. 2.0 mg/mL pepsin (Pepsin A) (Sigma; cat. no P7012) in 0.5M acetic acid (HAc).
3. Guanidine buffer: 4.0M guanidine-HCl, 50 mM Tris-HCl (pH 7.5). Filtered through activated carbon.
4. 0.5M HAc.
5. CII precipitation buffer: 0.5M acetic acid and 0.9M NaCl.
6. CXI precipitation buffer: 0.5M acetic acid and 1.2M NaCl.
7. CIX precipitation buffer: 0.5M acetic acid and 2.0M NaCl.
8. 0.1M HAc.
9. Exclusion buffer (optional): 200 mM NaCl, 50 mM Tris-HCl (pH 7.4).
10. 5.0 M NaCl (optional).

2.5. Lathyritic Collagen Purification

1. Lathyritic extraction buffer: 1.0M NaCl and 50 mM Tris-HCl (pH 7.5).

2.6. Arthritic Procedures

1. Rat strains; DA, E3, DxEA, DxEB, DxEC, DxER, and congenic LEW (*see* **Note 4**).
2. Freund's incomplete adjuvant (FIA) (Difco) (*see* **Note 5**).
3. Microplates (immunolon II, Dynatech laboratories).

3. Methods

The methodology is divided into sequential parts describing (1) tissue collection and preparation, (2) protein purification and differentiation, (3) arthritis induction, (4) arthritis evaluation, and (5) antibody quantification.

3.1. Tissues

The following protocols can be used for cartilage matrix protein purification from different tissue sources and species. The purpose of these protocols is to extract different proteins with the highest yield as possible, from one set of tissue. Different amounts and preparations are needed for different protocols and tissues (*see* **Note 6**). Collection and preparation of tissues are described in **Subheadings 3.1.1.–3.1.4.**

3.1.1. Transplantation and Collection of RCS

RCS must be retained by transplantation into rats; strain and sex are of no importance. Suspend a part of the RCS in PBS and incise caudally and laterally. Inoculate 1 mL subcutaneously from the forward shoulder to the incision site, using a blunt-ended needle. Close with an agraff. Collect RCS after approx 3 wk or if necrosis develops in the incision site, by gentle cutting and separation from the skin. Keep the connective tissue surrounding the RCS prior to preparation, for easier handling.

3.1.2. Other Tissue Sources

Sternal xiphoid cartilage is the simplest source of cartilage from mice and rats. Cut and collect the outermost half; at least 200 pieces (4–5 g) are needed from young animals. For other tissues, cut out pieces from knee or hip cartilage. Take whole sections of trachea or nasal cartilage (*see* **Note 6**).

3.1.3. Lathyritic Tissues

Lathyrism is induced by administration of BAPN in transplanted RCS rats or young mice and rats, ad libitum in water at a concentration of 3 g/L during growth (*see* **Note 1**). Lathyritic RCS is collected as for ordinary RCS. Higher doses or longer exposure than 7 to 10 d in growing animals offers little advantage and increases animal mortality and morbidity.

3.1.4. Preparation of Tissues

Be careful to remove connective tissue and fat during preparation, since it contains type I collagen (CI). Additional removal from tissues other than RCS can be done by stirring in water with a homogenizer at low speed. For COMP purification from tissues other than RCS, freeze cleansed pieces in liquid nitrogen and grind to a fine powder. To purify lathyritic collagen from tissues other than RCS, do likewise but continue as described in **Subheading 3.2.2.1.**, in protein preparation procedures. If collagen is desired only from nonlathyritic tissues, treat all tissues directly as described in **subheading 3.2.2**. Then grind pieces to a fine powder. The powder is then ready for further treatment.

3.2. Protein Preparation Procedures

The next steps involve the extraction and purification of different proteins from the tissue selected. Do not discard any materials, until the preparation is complete. All steps should be done at 4°C, with the solutions and working material cold and/or kept on ice. This minimizes bacterial growth, enhances stability and decreases degradation and aggregation. For an overview, see methodological flow-scheme in **Fig. 2**.

3.2.1. Purification of COMP From RCS

This protocol is not suited for lathyritic tissues (*see* **Note 6**). The last three extraction steps must be completed in a row, without disruptions. Otherwise freeze pellet at –20°C. Time is a critical factor, because of protease activity. Use plastic ware as COMP has affinity for glass.

1. Start with 50 g RCS, dissected from surrounding connective tissue.
2. Use pre-extraction buffer to a volume of 5× the tissue weight *(v/w)*. Add 10 mM NEM, 5 mM benzamidine as salt, and 1 mM dissolved PMSF directly to the buffer during homogenization. Start with a smaller volume of buffer for better homogenization.
3. Extract in cold under stirring for 1 h. Centrifuge for 20 min at 38,000g. Extract the pellet again as before with protease inhibitors, but for 15 min. Repeat twice and discard supernatants.
4. Extract the pellet in extraction buffer containing 10 mM EDTA as in **step 3** but for 30 min and with 5 mM NEM, as in **step 3** but for 30 min and repeat once. Pool both supernatants and freeze pellet at –20°C.
5. COMP binds to DEAE Sepharose™. Elute with 6.7 column volumes (CV) linear gradient of buffer B. Collect 5 mL fractions. Usually COMP is eluted as the first peak.
6. Analyze with reduced and nonreduced SDS-PAGE. Use 50 µL and 150 µL concentrated by 10 volumes of precipitation solution. Precipitate at –80°C for 1 h and centrifuge at 10,000g for 20 min. Discard supernatant, dry the pellet completely and dissolve in SDS sample buffer. Reduce samples with 1/5 volume of 2-mercaptoethanol and incubate at 100°C for 1 min. *See* example of COMP migration in **Fig. 3**.
7. Pool fractions, concentrate and exchange buffer to Heparin and Q buffer A with Amicon Ultra-15.
8. Contaminants of thrombospondins bind to Heparin Sepharos™. Add the HiTrap Q column last in series with the Heparin column. Disconnect the Heparin column and elute COMP with a 100% step gradient of Heparin and Q buffer B.
9. Analyze as in **step 6**, but use only 10 µL.

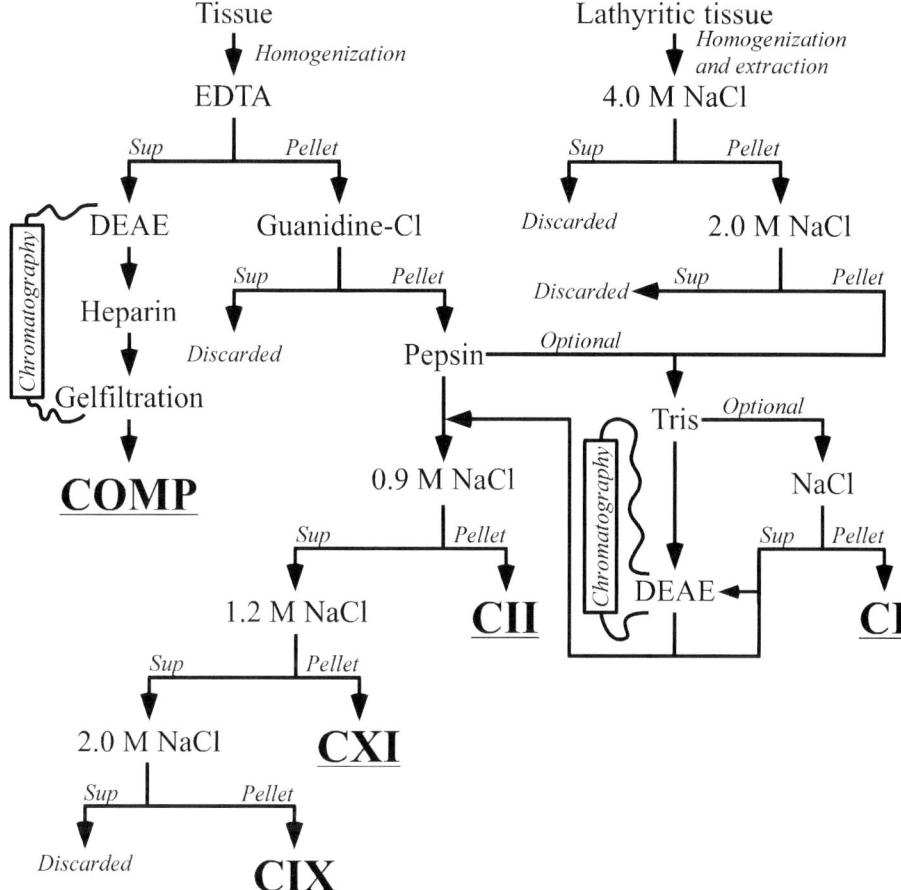

Fig. 2. Flow-scheme of protein purification of cartilage proteins. Only the important and critical steps are shown. Note that supernatant from guanidine extraction and CIX precipitation contains additional cartilage proteins that are not described here and are therefore marked as discarded.

3.2.1.1. GELFILTRATION CHROMATOGRAPHY

If purification of COMP is not satisfactory (i.e., contaminants are detected on SDS-PAGE or native COMP is required), the following steps are necessary.

1. The Superose® 6 will separate high and low weight contaminants. Superdex™ 200 will separate contaminants and denatured COMP fragments from native COMP. Be careful not to overload the columns and concentrate with Amicon Ultra-4.
2. Run isocratically 1 CV and collect 0.5-mL fractions. Save fractions in cold and analyze peaks with SDS-PAGE, as **Subheading 3.2.1., step 6**.

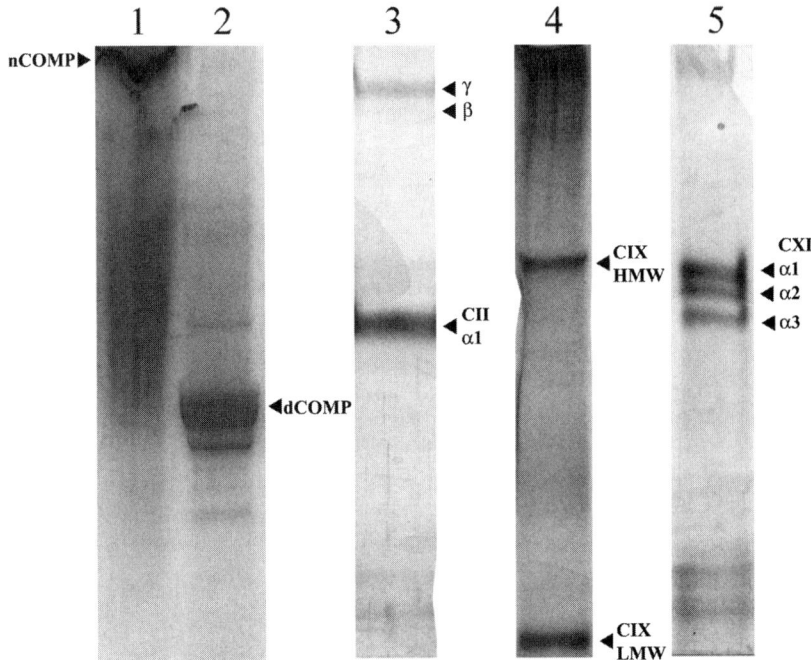

Fig. 3. Examples of RCS cartilage protein migration patterns in different SDS-PAGE, for comparison. *Lane 1*: native COMP (435 kDa) and; *lane 2:* denatured COMP (120 kDa) acquired by gel filtration purification. 10 µg of unreduced samples were loaded and later stained with Bio-Safe™ coomassie (Bio-Rad Laboratories, Inc). *Lane 3:* 2 µg reduced Collagen type II (140 kDa) with β- and γ-complexes of di- and trimeric α-chains respectively; *Lane 4:* 5 µg unreduced collagen type IX with high molecular weight (160 kDa) and lo- molecular-weight (50 kDa) fragments; *lane 5:* 2 µg reduced collagen type XI (160, 150, 140 kDa) was resolved and then gel stained with silver. *Lanes 1, 2,* and *4,* were run in a 4–20% gradient ready gel (Bio-Rad Laboratories, Inc). *Lanes 3* and *5* were run in a 5% 16-cm gel. Molecular weights, except native COMP, are estimated according to Precision Plus Protein™ standards, dual color (Bio-Rad Laboratories, Inc).

3. Pool fractions and concentrate with Amicon Ultra-15. COMP should be stored in buffer used for gel filtration. Determine the concentration with the Lowry method using BSA as standard. Estimation can be made at 280 nm with an absorbance coefficient of 1.20. Aliquot and store at –80°C for longer intervals, otherwise at 4°C. COMP degrades over time at 4°C and by repetitive thawing/freezing.

3.2.2. Collagen Extraction With Pepsin Digestion

Start with 200 g RCS or 4 batches of COMP pellet from **Subheading 3.2.1.**, **step 4**. Indicated times are the minimum time required, except when proteins

Cartilage Protein Induced Arthritis 235

are exposed to neutral buffers, as described below. In tissues other than RCS and joint cartilage, there is a considerable amount of CI. CI contamination can be reduced, but with great loss of other collagens. Additional removal of impurities and pepsin can be accomplished as described below. Use glassware as collagen has affinity to plastic.

1. Homogenize or stir in 3 *(v/w)* of guanidine buffer. Extract proteoglycans and other impurities by stirring the homogenate for 12 h. Centrifuge at 23,000g for 30 min. Keep the pellet and repeat this step twice.
2. Resuspend the pellet again with guanidine buffer and stir for 48 h. Centrifuge as before and resuspend pellet in 200 mL of ddH_2O. Wash the pellet three more times with water and then with 200 mL of 0.5M HAc.
3. Add 0.5 g of pepsin to 100 g tissue and digest collagen fibers with stirring for 24 h. Centrifuge as before and keep the supernatant.
4. Inactivate and precipitate pepsin by raising pH to 8.1 with NaOH. Centrifuge at 34,000g for 60 min. and keep the supernatant.
5a. Preparation for exclusion of impurities (optional): Dialyze the supernatant for 5 times 2 h each in tubing against exclusion buffer. Centrifuge as above and keep supernatant. If not clear, filtrate supernatant through several layers of gauze weaves.
5b. CI exclusion (optional): Measure the volume of the supernatant and add by dripping an equal volume of 5.0M NaCl solution slowly to supernatant under constant stirring. Stop adding the NaCl solution when the supernatant goes opaque or if the volume is used up. Stir for 24 h, centrifuge as above, and keep the supernatant.
5c. Pepsin and proteoglycan exclusion (optional): equilibrate the DEAE Sepharose™ column and supernatant if CI was excluded as in **step 5**, with exclusion buffer. Monitor collagen at wavelength 227 nm, run isocratically and collect the whole peak.
5d. Add crystalline NaCl to the supernatant or collected peak from **step 5b** and/or **5c** under gentle stirring to a total concentration of 4.4M. Allow to precipitate without stirring for 24 h. Centrifuge as before. Dissolve the pellet with 0.5M HAc and stir over night.

3.2.2.1. COLLAGEN DIFFERENTIATION

Different types of collagen can be separated by salt precipitation at different NaCl concentrations (*see* **Fig. 2**). Given that type IX (CIX) and type XI (CXI) collagen precipitates at concentrations close to each other, there will be some cross contamination. However, these can be separated by repetitive salt precipitation at selected molarities of the final dissolved pellet.

1. If the solution is not clear, centrifuge and filtrate through several layers of gauze. Dialyze in tubing against 5 l CII precipitation buffer with stirring for a total of 96 hr and at least 3 additional exchanges of buffer. Centrifuge at 34,000g for 60 min. and dissolve pellet with 0.5M HAc overnight.

2. Repeat step 1 with the supernatant, but with CXI precipitation buffer.
3. Repeat **step 1** with the supernatant, but with CIX precipitation buffer.
4. The dissolved collagens should be clear with a fluorescent blue color and viscous. If not, centrifuge and filtrate as before. Dilute the solution to lower viscosity in order to minimize losses. Dialyze the dissolved collagen 4 times for 2 h each against $0.1M$ HAc in 5 L.
5. Lyophilize the dialysate in a freeze-dryer. To obtain a fluffy-white, cotton like product, rotation-freeze the dialysate at $-70°C$ before storing in the freeze-dryer container.
6. Measure the weight and dissolve to a concentration of 5 mg/mL in $0.1M$ HAc. Store at $4°C$; time will not affect the collagen under these conditions.
7. Analyze as in **Subheading 3.2.1., step 6**, but use 5 times volume of SDS sample buffer for neutralization and load 10 µg. Examples of different collagen migration patterns are shown in **Fig. 3**.

3.2.2.2. COLLAGEN EXTRACTION OF LATHYRITIC TISSUE

Start with 300 g lathyritic RCS (*see* **Note 6**). Note that proteins are exposed in neutral buffer in the beginning. Hence, it is important to retain time and temperature.

1. Homogenize using 5 *(v/w)* of lathyritic extraction buffer. Extract collagen by stirring for 24 h and centrifuge at 23,000g for 30 min. Extract again but with 3 *(v/w)* and centrifuge as above and once more with 2 *(v/w)*. Pool these three supernatants.
2. Add crystalline NaCl to the poolate to a total concentration of $4M$ under stirring. When the salt has dissolved, let precipitate for 24 h without stirring.
3. Centrifuge at 34,000g for 60 min and dissolve the collagens in the pellet with $0.5M$ HAc until slightly viscous, stir overnight. Centrifuge as before and keep the supernatant.
4. Repeat **steps 2** and **3** with the supernatant, but with $2M$ NaCl.
5. Because of the high content of impurities **steps 5a** and **5b**, **Subheading 3.2.2.** is necessary thereafter as forpepsin-digested tissues.

3.3. Induction of Arthritis

Arthritis is induced in rats by a single dose of 150 µg proteins in 150 µL FIA (*see* **Notes 4** and **5**). All experiments should be performed on 8- to 16-wk-old rats. Antigens are diluted in solvent and emulsified in an equal volume of FIA.

3.4. Evaluation of Arthritis

Scoring is determined through assessment of the number of joints involved as has earlier been described in detail *(58)*. Inspection should be performed early before onset (day 10) and at least 3 times/wk (*see* **Note 7**). Two type of scoring systems are employed. In the simple system maximal score is 3 for rats per paw, which is equal to score 15 in the extended system. The paw can be divided into three parts; ankle, mid-paw with knuckles, and toe joints. Each part is given 1 in the simple or 5 points in the extended system. For toe and knuckle-joints a maximum of 2 points in the simple or 5 in extended system can only subsist per part. In the mid-paw points can also be given if other parts are affected, but then only with maximum as 1 in the simple or 5 points in the extended system (*see* **Note 8**). In combinations of different parts 2 points is maximum in the simple system until all 3 parts are affected. In the extended system each toe and knuckle is given 1 point until a maximum of 10 is reached. A swollen ankle joint should always be given a point of 5 in the extended system, despite the degree of swelling. After inflammation has settled deformity appears which can make scoring difficult; as a rule always score active inflammation and note stiff joints. The relation between extended and simple system are; 1 = 1, 2–10 = 2, 11–15 = 3.

3.5. Quantification of Antibodies in Serum

Sera are obtained from retro-orbital plexus or tail, and should be stored at –70°C until assayed. For the quantification of anti-COMP, CII, CIX, CXI, and pepsin reactive antibody in serum, micro-plates immunolon II should be used (*see* **Note 9**). Additional blocking with 1% BSA in PBS after coating of COMP is recommended. Details of further handling are described elsewhere *(59)*.

4. Notes

1. The stock solution is stable and easily diluted in ordinary drinking water. Take precautions as it is believed to be teratogenic.
2. PMSF and NEM are very unstable in the presence of water and should be added prior to use. PMSF stored in stock solution is stable for several months. Ethanol can be substituted with isopropanol, but with reduced solubility of PMSF and may effect protein extraction negatively. Take precautions, as PMSF is highly toxic.
3. Collagens have absorbance maximum at 227 nm, thus making it difficult to monitor at 280 nm. Because flow through should be collected during collagen purification, it is also possible to collect without monitoring and by volume instead.
4. Rats should be used and kept in a climate-controlled environment with 12 h light–dark cycles, housed in polystyrene cages containing wood shavings, standard rodent chow and water *ad libitum*. The rats strains DA, E3, DA-E3 recombinants; DxEA, DxEB, DxEC, and DxER and congenic LEW rats originate from Zentralinstitut für Versuchstierzuchin Hanover, Germany.).
5. The oil used in the emulsion will affect highly susceptible strains as DA and DxEA *(44)*. However, the arthritis is usually easy to discriminate, because it is affecting peripheral joints to higher degree, less severe and more acute.
6. All proteins have an optimal preparative window, as well as an upper and lower limit for maximum yield. All starting weights are optimized for RCS, you have to

extrapolate if you are using other tissues. For example, joint cartilage consists of 80% collagen, so it is wise to purify with half of the amount compared to RCS. In COMP purification, 50 g RCS is used because of time and effort savings, as a result of protease activity. Collagens are relatively insensitive to degradation, therefore more tissue can be prepared. However, a maximum of 200 g RCS is more practical. It is possible to leave out the COMP purification if needed, but for collagen the guanidine steps are needed in order to remove impurities. Lathyritic tissues contain soluble collagen molecules and a separate extraction method must be applied (*see* **Subheading 3.2.2.2.**). Hence, extraction of impurities is not possible, purified collagen will be more contaminated, making the chromatography step necessary. It will also make COMP purification difficult from lathyritic tissues. The yield of lathyritic collagens will decrease and for this reason 300 g lathyritic RCS is recommended.
7. The animals should be scored in a light area level with the eyes. It is wise to indicate swollen areas using diagram first and then later assign the correct score. Make notes if inflammation is unusually severe, other parts are involved and when inflammation subsides.
8. Apart from the knuckles the mid-paw does not contain joints, but edema may have migrated from joints in proximity and other tissues can be inflamed as well. This is only possible to define by histology.
9. Collagens have poor binding properties and in our experience immunolon II is the most suitable for direct coating without blocking.

Acknowledgments

The authors thank Dick Heinegård for help and valuable advice concerning the COMP purification protocol.

References

1. Verheijden, G. F., Rijnders, A. W., Bos, E., et al. (1997) Human cartilage glycoprotein-39 as a candidate autoantigen in rheumatoid arthritis. *Arthritis Rheum.* **40,** 1115–1125.
2. Glant, T. T., Mikecz, K., Arzoumanian, A., and Poole, A. R. (1987) Proteoglycan-induced arthritis in BALB/c mice. Clinical features and histopathology. *Arthritis Rheum.* **30,** 201–212.
3. Buzas, E. I., Mikecz, K., Brennan, F. R., and Glant, T. T. (1994) Mediators and autopathogenic effector cells in proteoglycan-induced arthritic and clinically asymptomatic BALB/c mice. *Cell. Immunol.* **158,** 292–304.
4. Yao, Z., Nakamura, H., Masuko-Hongo, K., Suzuki-Kurokawa, M., Nishioka, K., and Kato, T. (2004) Characterisation of cartilage intermediate layer protein (CILP)-induced arthropathy in mice. *Ann. Rheum. Dis.* **63,** 252–258.
5. Myers, L. K., Pihlajamaa, T., Brand, D. D., et al. (2002) Immunogenicity of recombinant type IX collagen in murine collagen-induced arthritis. *Arthritis Rheum.* **46,** 1086–1093.

6. Boissier, M. C., Chiocchia, G., Ronziere, M. C., Herbage, D., and Fournier, C. (1990) Arthritogenicity of minor cartilage collagens (types IX and XI) in mice. *Arthritis Rheum.* **33,** 1–8.
7. Cremer, M. A., Ye, X. J., Terato, K., Griffiths, M. M., Watson, W. C., and Kang, A. H. (1998) Immunity to type IX collagen in rodents: a study of type IX collagen for autoimmune and arthritogenic activities. *Clin. Exp. Immunol.* **112,** 375–382.
8. Novotna, J., Hulejova, H., Svoboda, T., Deyl, Z., and Adam, M. (1991) The role of cartilage minor collagens in inducing arthritis. *Z. Rheumatol.* **50,** 93–98.
9. van der Rest M, M. R. (1987) *Structure and function of collagen types.* Mayne R, B. R., ed., Academic Press, Orlando.
10. Eyre, D. R. and Wu, J. J. (1995) Collagen structure and cartilage matrix integrity. *J. Rheumatol. Suppl.* **43,** 82–85.
11. Thom, J. R. and Morris, N. P. (1991) Biosynthesis and proteolytic processing of type XI collagen in embryonic chick sterna. *J. Biol. Chem.* **266,** 7262–7269.
12. Oldberg, A., Antonsson, P., Lindblom, K., and Heinegard, D. (1992) COMP (cartilage oligomeric matrix protein) is structurally related to the thrombospondins. *J. Biol. Chem.* **267,** 22,346–22,350.
13. Zaia, J., Boynton, R. E., McIntosh, A., et al. (1997) Post-translational modifications in cartilage oligomeric matrix protein. Characterization of the N-linked oligosaccharides by matrix-assisted laser desorption ionization time-of-flight mass spectrometry. *J. Biol. Chem.* **272,** 14,120–14,126.
14. Efimov, V. P., Lustig, A., and Engel, J. (1994) The thrombospondin-like chains of cartilage oligomeric matrix protein are assembled by a five-stranded alpha-helical bundle between residues 20 and 83. *FEBS Lett.* **341,** 54–58.
15. Hedbom, E., Antonsson, P., Hjerpe, A., et al. (1992) Cartilage matrix proteins. An acidic oligomeric protein (COMP) detected only in cartilage. *J. Biol. Chem.* **267,** 6132–6136.
16. Fife, R. S. (1988) Identification of cartilage matrix glycoprotein in synovial fluid in human osteoarthritis. *Arthritis Rheum.* **31,** 553–556.
17. Nguyen, B. Q. and Fife, R. S. (1986) Vitreous contains a cartilage-related protein. *Exp. Eye Res.* **43,** 375–382.
18. Nah, H. D., Barembaum, M., and Upholt, W. B. (1992) The chicken alpha 1 (XI) collagen gene is widely expressed in embryonic tissues. *J. Biol. Chem.* **267,** 22,581–22,586.
19. Sugimoto, M., Kimura, T., Tsumaki, N., et al. (1998) Differential in situ expression of alpha2(XI) collagen mRNA isoforms in the developing mouse. *Cell Tissue Res.* **292,** 325–332.
20. Brewton, R. G., Wright, D. W., and Mayne, R. (1991) Structural and functional comparison of type IX collagen-proteoglycan from chicken cartilage and vitreous humor. *J. Biol. Chem.* **266,** 4752–4757.
21. Yada, T., Suzuki, S., Kobayashi, K., et al. (1990) Occurrence in chick embryo vitreous humor of a type IX collagen proteoglycan with an extraordinarily large chondroitin sulfate chain and short alpha 1 polypeptide. *J. Biol. Chem.* **265,** 6992–6999.

22. DiCesare, P., Hauser, N., Lehman, D., Pasumarti, S., and Paulsson, M. (1994) Cartilage oligomeric matrix protein (COMP) is an abundant component of tendon. *FEBS Lett.* **354,** 237–240.
23. Smith, R. K., Zunino, L., Webbon, P. M., and Heinegard, D. (1997) The distribution of cartilage oligomeric matrix protein (COMP) in tendon and its variation with tendon site, age and load. *Matrix Biol.* **16,** 255–271.
24. Di Cesare, P. E., Carlson, C. S., Stollerman, E. S., Chen, F. S., Leslie, M., and Perris, R. (1997) Expression of cartilage oligomeric matrix protein by human synovium. *FEBS Lett.* **412,** 249–252.
25. Eyre, D. R. (1991) The collagens of articular cartilage. *Semin. Arthritis Rheum.* **21,** 2–11.
26. Wu, J. J., Woods, P. E., and Eyre, D. R. (1992) Identification of cross-linking sites in bovine cartilage type IX collagen reveals an antiparallel type II-type IX molecular relationship and type IX to type IX bonding. *J. Biol. Chem.* **267,** 23,007–23,014.
27. Diab, M., Wu, J. J., and Eyre, D. R. (1996) Collagen type IX from human cartilage: a structural profile of intermolecular cross-linking sites. *Biochem. J.* **314(Pt 1),** 327–332.
28. van der Rest, M. and Mayne, R. (1988) Type IX collagen proteoglycan from cartilage is covalently cross-linked to type II collagen. *J. Biol. Chem.* **263,** 1615–1618.
29. Rosenberg, K., Olsson, H., Morgelin, M., and Heinegard, D. (1998) Cartilage oligomeric matrix protein shows high affinity zinc-dependent interaction with triple helical collagen. *J. Biol. Chem.* **273,** 20,397–20,403.
30. Li, Y., Lacerda, D. A., Warman, M. L., et al. (1995) A fibrillar collagen gene, Col11a1, is essential for skeletal morphogenesis. *Cell* **80,** 423–430.
31. Vikkula, M., Mariman, E. C., Lui, V. C., et al. (1995) Autosomal dominant and recessive osteochondrodysplasias associated with the COL11A2 locus. *Cell* **80,** 431–437.
32. Nakata, K., Ono, K., Miyazaki, J., et al. (1993) Osteoarthritis associated with mild chondrodysplasia in transgenic mice expressing alpha 1(IX) collagen chains with a central deletion. *Proc. Natl. Acad. Sci. U. S. A.* **90,** 2870–2874.
33. Fassler, R., Schnegelsberg, P. N., Dausman, J., et al. (1994) Mice lacking alpha 1 (IX) collagen develop noninflammatory degenerative joint disease. *Proc. Natl. Acad. Sci. U. S. A.* **91,** 5070–5074.
34. Briggs, M. D., Hoffman, S. M., King, L. M., et al. (1995) Pseudoachondroplasia and multiple epiphyseal dysplasia due to mutations in the cartilage oligomeric matrix protein gene. *Nat. Genet.* **10,** 330–336.
35. Cremer, M. A., Ye, X. J., Terato, K., Owens, S. W., Seyer, J. M., and Kang, A. H. (1994) Type XI collagen-induced arthritis in the Lewis rat. Characterization of cellular and humoral immune responses to native types XI, V, and II collagen and constituent alpha-chains. *J. Immunol.* **153,** 824–832.
36. Cremer, M. A., Griffiths, M. M., Terato, K., and Kang, A. H. (1995) Type XI and II collagen-induced arthritis in rats: characterization of inbred strains of rats for

arthritis-susceptibility and immune-responsiveness to type XI and II collagen. *Autoimmunity* **20,** 153–161.
37. Lu, S., Carlsen, S., Hansson, A. S., and Holmdahl, R. (2002) Immunization of rats with homologous type XI collagen leads to chronic and relapsing arthritis with different genetics and joint pathology than arthritis induced with homologous type II collagen. *J. Autoimmun.* **18,** 199–211.
38. Cremer, M. A., Terato, K., Seyer, J. M., Watson, W. C., O'Hagan, G. O., Townes, A. S., and Kang, A. H. (1991) Immunity to type XI collagen in mice. Evidence that the alpha 3(XI) chain of type XI collagen and the alpha 1(II) chain of type II collagen share arthritogenic determinants and induce arthritis in DBA/1 mice. *J. Immunol.* **146,** 4130–4137.
39. Forslind, K., Eberhardt, K., Jonsson, A., and Saxne, T. (1992) Increased serum concentrations of cartilage oligomeric matrix protein. A prognostic marker in early rheumatoid arthritis. *Br. J. Rheumatol.* **31,** 593–598.
40. Saxne, T. and Heinegard, D. (1992) Cartilage oligomeric matrix protein: a novel marker of cartilage turnover detectable in synovial fluid and blood. *Br. J. Rheumatol.* **31,** 583–591.
41. Saxne, T., Glennas, A., Kvien, T. K., Melby, K., and Heinegard, D. (1993) Release of cartilage macromolecules into the synovial fluid in patients with acute and prolonged phases of reactive arthritis. *Arthritis Rheum.* **36,** 20–25.
42. Vingsbo-Lundberg, C., Saxne, T., Olsson, H., and Holmdahl, R. (1998) Increased serum levels of cartilage oligomeric matrix protein in chronic erosive arthritis in rats. *Arthritis Rheum.* **41,** 544–550.
43. Larsson, E., Mussener, A., Heinegard, D., Klareskog, L., and Saxne, T. (1997) Increased serum levels of cartilage oligomeric matrix protein and bone sialoprotein in rats with collagen arthritis. *Br. J. Rheumatol.* **36,** 1258–1261.
44. Carlsen, S., Hansson, A. S., Olsson, H., Heinegard, D., and Holmdahl, R. (1998) Cartilage oligomeric matrix protein (COMP)-induced arthritis in rats. *Clin. Exp. Immunol.* **114,** 477–484.
45. Holmdahl, R., Vingsbo, C., Hedrich, H., et al. (1992) Homologous collagen-induced arthritis in rats and mice are associated with structurally different major histocompatibility complex DQ-like molecules. *Eur. J. Immunol.* **22,** 419–424.
46. Vingsbo, C., Jonsson, R., and Holmdahl, R. (1995) Avridine-induced arthritis in rats; a T cell-dependent chronic disease influenced both by MHC genes and by non-MHC genes. *Clin. Exp. Immunol.* **99,** 359–363.
47. Vingsbo, C., Sahlstrand, P., Brun, J. G., Jonsson, R., Saxne, T., and Holmdahl, R. (1996) Pristane-induced arthritis in rats: a new model for rheumatoid arthritis with a chronic disease course influenced by both major histocompatibility complex and non-major histocompatibility complex genes. *Am. J. Pathol.* **149,** 1675–1683.
48. Holmdahl, R. Kvick, C. (1992) Vaccination and genetic experiments demonstrate that adjuvant-oil-induced arthritis and homologous type II collagen-induced arthritis in the same rat strain are different diseases. *Clin. Exp. Immunol.* **88,** 96–100.
49. Lorentzen, J. C. and Klareskog, L. (1996) Susceptibility of DA rats to arthritis induced with adjuvant oil or rat collagen is determined by genes both within

and outside the major histocompatibility complex. *Scand. J. Immunol.* **44,** 592–598.
50. Kjellen, P., Issazadeh, S., Olsson, T., and Holmdahl, R. (1998) Genetic influence on disease course and cytokine response in relapsing experimental allergic encephalomyelitis. *Int. Immunol.* **10,** 333–340.
51. Smith, B. D., Martin, G. R., Miller, E. J., Dorfman, A., and Swarm, R. (1975) Nature of the collagen synthesized by a transplanted chondrosarcoma. *Arch. Biochem. Biophys.* **166,** 181–186.
52. Trelstad, R. L., Kang, A. H., Toole, B. P., and Gross, J. (1972) Collagen heterogeneity. High resolution separation of native (1(I) 2 2 and (1(II) 3 and their component chains. *J. Biol. Chem.* **247,** 6469–6473.
53. Miller, E. J. (1972) Structural studies on cartilage collagen employing limited cleavage and solubilization with pepsin. *Biochemistry (Mosc)* **11,** 4903–4909.
54. Bernard, M., Yoshioka, H., Rodriguez, E., et al. (1988) Cloning and sequencing of pro-alpha 1 (XI) collagen cDNA demonstrates that type XI belongs to the fibrillar class of collagens and reveals that the expression of the gene is not restricted to cartilagenous tissue. *J. Biol. Chem.* **263,** 17,159,17,166.
55. Reese, C. A. and Mayne, R. (1981) Minor collagens of chicken hyaline cartilage. *Biochemistry (Mosc)* **20,** 5443–5448.
56. Arai, M., Yada, T., Suzuki, S., and Kimata, K. (1992) Isolation and characterization of type IX collagen-proteoglycan from the Swarm rat chondrosarcoma. *Biochim. Biophys. Acta* **1117,** 60–70.
57. Miller, E. J. and Rhodes, R. K. (1982) Preparation and characterization of the different types of collagen. *Methods Enzymol.* **82 Pt A,** 33–64.
58. Holmdahl, R., Carlsen, S., Mikulowska, A., et al. (1997) Genetic analysis of mouse models for rheumatoid arthritis. In: *Human Genome Methods*.Adolph, K. W., ed., New York:CRC press LLC, pp. 215–238.
59. Holmdahl, R., Klareskog, L., Andersson, M., and Hansen, C. (1986) High antibody response to autologous type II collagen is restricted to H-2q. *Immunogenetics* **24,** 84–89.

18

Murine Antigen-Induced Arthritis

Wim B. van den Berg, Leo A. B. Joosten, and Peter L. E. M. van Lent

Abstract

Antigen induced arthritis is a unilateral T-cell driven model caused by direct injection of an antigen into the knee joint of a FCA preimmunized animal. The chronicity is determined by antigen retention in avascular structures of the joint through charge mediated binding or antibody mediated trapping. Cationicity of the antigen is a prerequisite in this model in the mouse and commercial mBSA is a suitable antigen. Cartilage erosive character is strongly enhanced in the presence of marked antibody titer. Concomitant boosting of the immune response with *Bordetella pertussis* adds to this. T-cell mediated flares can be induced by local or systemic rechallenge with low dose antigen, and display a strong erosive phenotype.

Key Words: T-cell arthritis; cartilage destruction; antibodies; flares.

1. Introduction

The model of antigen-induced arthritis has been developed in rabbits by Dumonde and Glynn some 40 yr ago (*1*). The model is based on local antigen injection in a joint of a hyperimmunized animal, and can be induced in any species, provided that proper immunity to a particular antigen can be mounted. Classic antigens used have included ovalbumin, bovine serum albumin (BSA), and fibrin. This model has since been developed in mice, rats, and guinea pigs. In contrast to autoimmune polyarthritis models, this type of arthritis remains confined to the injected joint, enabling comparison of arthritic biochemical and structural changes with a normal contra-lateral joint. Severity can be controlled for the dose of locally injected antigen and the arthritis has a well defined time of onset. This makes the model suited to kinetic studies into mechanisms of cartilage and bone destruction, also facilitated by greater accessibility and standardization of histology of the knee joint as compared with ankles. Controlled flares can be induced by antigenic rechallenge, which may mimic episodes of exacerbations and remissions seen in rheumatoid arthritis (RA) patients.

There has been a growing interest in murine versions of the model, resulting from the availability of inbred strains and numerous transgenic and knockout mice. The first description of the mouse mBSA (methylated bovine serum albumin) model dates back to 1977 *(2)*, and characterization of underlying principles has been done by our group in the early 1980s. Important principles emerging from these studies were the following:

1. Chronicity is dependent on sufficient antigen retention in joint tissues, in combination with strong T-cell mediated delayed type hypersensitivity.
2. Joints contain numerous avascular collagenous tissues such as cartilage, ligaments, and tendons, which allow for prolonged antigen retention by antibody-mediated trapping and charge mediated binding *(3,4)*.
3. Antigen retention by antibody mediated trapping is prominent in rabbits, resulting from high antibody levels to foreign antigens in that species, but appears to be insufficient in mice.
4. Sustained antigen retention, and therefore chronicity of arthritis, can be achieved in mice using cationic antigens *(5,6)*.

A search for putative natural cationic antigens revealed components of bacteria as proper arthritogens *(7)*. Intriguingly, the recent identification of cationic GPI (glucose phosphate isomerase) as the underlying autoantigen in KRN arthritis *(8)* fits nicely with this concept. For clarity reasons details of the mBSA arthritis only will be covered in this chapter. The model shows immune complex deposition in the cartilage, progressive cartilage and bone erosion, and immune infiltrates in the synovial tissue, all elements of great importance in human RA. The model can be used to study any of these aspects of the human disease.

2. Materials

1. Cationic antigen (e.g., methylated bovine serum albumin) (Sigma; cat. no. A1009).
2. Freund's complete adjuvant (Difco; cay. no. 0638).
3. 10X concentrated Saline (NaCl).
4. *B. pertussis* organisms (National Institute of Public Health, Bilthoven, The Netherlands).
5. 50-µL Hamilton syringe.
6. 30 1/2-gage disposable needle (BD Microlance 3).
7. C57Bl/6 mice, 8- to 10-wk of age.

3. Methods

3.1. Control of Cationicity of mBSA

mBSA suitable for induction of chronic arthritis needs to have a pI (isoelectric point) of at least 8.5. The positive charge is needed for proper retention of the protein in the injected knee joint, allowing for firm adherence to negatively charged ligaments and cartilage surfaces (*see* **Notes 1** and **2**). Naïve BSA has a pI of 4.5 and can be made cationic by methylation (mBSA) or amidation (aBSA). Detailed description goes beyond the scope of this chapter but technical procedures have been published *(5,6)*. The cationic nature can be confirmed with an isoelectric focusing gel (*see* **Fig. 1**). Of note, properly methylated BSA does not dissolve in saline.

3.2. Preparing the mBSA/FCA Emulsion

1. 20 mg antigen (mBSA) is dissolved in 9 mL sterile aquadest.
2. 1 mL of 10X saline is added to yield 10 mL of 2 mg/mL mBSA solution.
3. This mBSA solution is taken up in a 10-mL syringe, a 27 1/2-gage needle is placed on top, and the solution is injected with force in 10 mL of the oil preparation of Freund's complete adjuvant (FCA) in a small plastic vial.
4. This procedure creates the start of a white oil-water emulsion, which can then be easily emulsified to homogeneity by repeated intake and flushing with a 2-mL syringe (without needle).
5. All the solutions should be kept at 4°C before preparing the emulsion. The emulsion is at optimal quality when droplets placed on water float on the surface and remain intact.

3.3. Immunization of Mice

1. C57Bl/6 mice, 8- to 10-wk old, are immunized with 100 µg of methylated bovine serum albumin in 0.1 mL Freund's complete adjuvant divided over the front paws and both flanks (*see* **Notes 3–5**).
2. Mice also receive an intraperitoneal injection (ip) of 2×10^9 heat-killed *B. pertussis* bacteria in 1 mL of saline for optimal boosting of T-cell reactivity against the antigen.
3. Seven days later, mice are boosted with 100 µg of mBSA in 0.1 mL Freund's complete adjuvant, divided over two sites in the neck region.
4. Additional ip injection of *B. pertussis* is recommended (*see* **Note 6**).

3.4. Arthritis Induction by Intra-Articular Injection

1. Three weeks after the start of the immunization 60 µg of mBSA in saline (6 µL total volume) is injected into the cavity of the knee joint. Only the right knee joint is injected to induce a unilateral arthritis.
2. The mBSA solution is sterilized through a bacterial filter before injection.
3. As control, 6 µL of saline is injected into the left knee joint.
4. Mice are sedated shortly with isoflurane.
5. To enable proper injection, a small incision is made in the skin along the patellar ligament, to visualize the ligament and the patella (*see* **Fig. 1**).
6. A small needle attached to a Hamilton syringe is used to enter the knee joint space by sliding the tip halfway and nearly parallel to the ligament, moving it until it reaches the area underneath the patella. Make sure to have the opening of

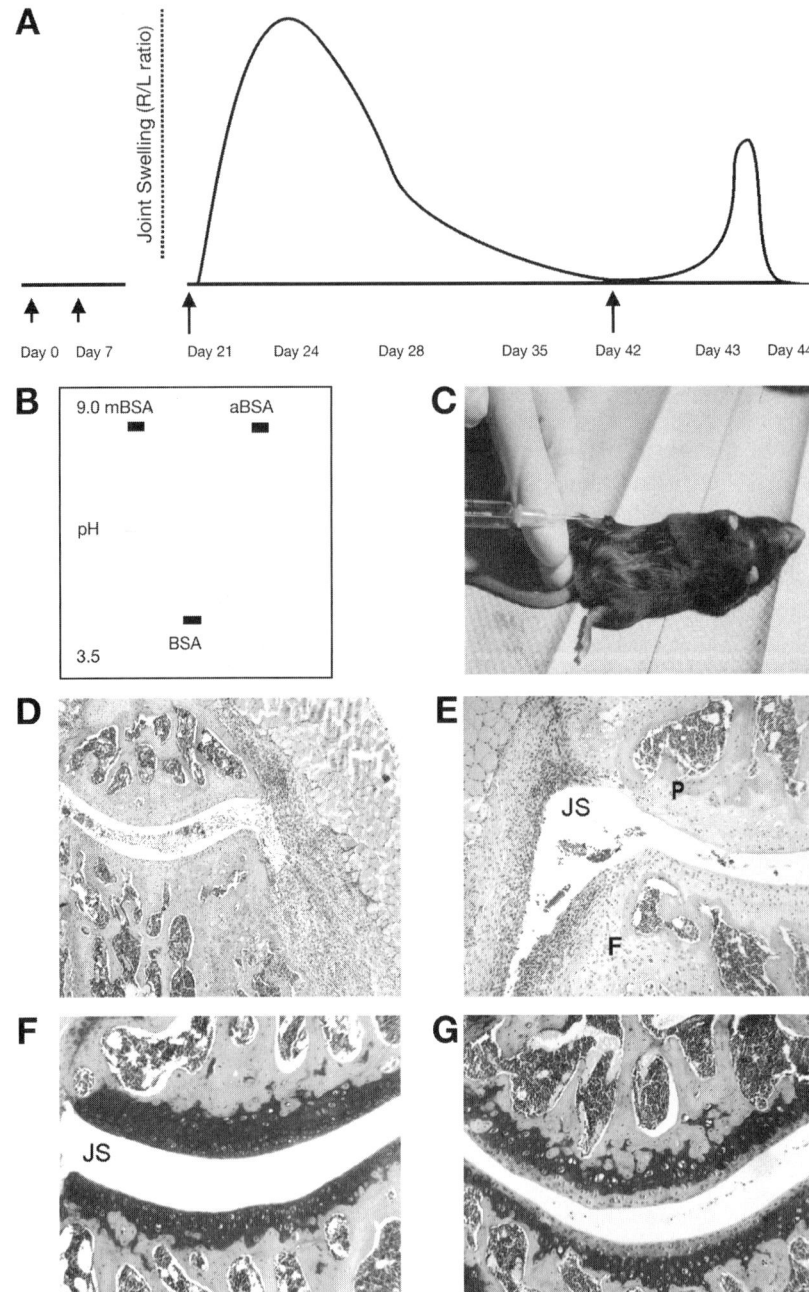

Fig. 1. (w) time course of antigen-induced arthritis (mBSA arthritis). Shortly after intra-articular injection of 60 μg of mBSA, joint swelling can be measured using ^{99m}Technetium uptake method *(9)*. The maximum swelling (R/L ratio around 1.8)

the needle at the upper side, to avoid needle artefacts on the patellar cartilage surface, if the latter is a readout site.
7. Upon injection of the 6 µL volume a symmetric butterfly-like swelling should be noted, with most of the volume going to the suprapatellar pouch.
8. Keep the needle inside the joint for a few seconds to let the fluid equilibrate and slowly withdraw the needle, avoiding leakage of injected fluid.
9. The small incision of the skin heals spontaneously in a day; do not clamp the wound.

3.5. Measurement of Joint Swelling

Joint inflammation can be quantified with the 99mTc-uptake method *(9)*. This methodology measures by external γ counting the accumulation of a small short-lived radioisotope at the site of knee joint inflammation. The radioisotope accumulates at higher concentration at the inflamed right knee joint as compared to the control left knee joint. Enhanced uptake reflects mainly tissue swelling, and the severity can be expressed as a right/left ratio. The noninvasive technology combined with the short half life of the radioisotope allows for consecutive measurements during the course of the arthritis. The number of allowable measurements is mainly dictated by the impact of the sedation, because repetitive episodes reduce the severity of arthritis.

3.6. Grading of Arthritis by (Immuno)Histology

1. At selected time points, mice are killed, whole knee joints (without skin) are removed by cutting the femur and tibia close to the knee.
2. Specimens are fixed for 4 days in 4% formaldehyde.
3. After decalcification in 5% formic acid the specimens are processed for paraffin embedding.

---

occurs around days 3 to 5 after induction of the unilateral arthritis. Seven days after antigen injection (day 28), joint swelling declines. Three weeks after the initial ia injection the arthritis can be reactivated by local injection of a small amount of antigen (1 µg mBSA). The flare shows a short swelling but can be highly destructive. (**B**) The isoelectric point of the cationic antigen should be at least 8.5. This to obtain a good local retention of antigen in the joint during the course of experimental arthritis. (**C**) Intra-articular injection of the antigen. A small incision is made in the skin and the antigen is injected with the smallest size needle available. The size is absolutely critical. (**D**) Histopathology at day 24 (day 3 of arthritis). Note the enhanced influx of inflammatory cells in the joint cavity. (**E**) Joint pathology at day 42. Note the chronic joint inflammation at this stage. T-cell foci and plasma cells are common features. (**F**) Normal knee joint, stained with safranin O to visualize cartilage proteoglycans. (**G**) Severe loss of matrix proteoglycans at day 28, 7 d after induction of arthritis.

4. Standardized frontal semi-serial sectioning (7 µm) spaced 10 sections apart is done, to obtain sections including patella, femur and tibia, as well as the menisci in one view.
5. Sections are stained with haematoxylin and eosin to score cellular infiltration, or with safronin O to evaluate proteoglycan loss in the articular cartilage. The patella-femural area is most suited, since spontaneous OA like pathology is generally absent there, but can be disturbing in tibial plateaus.
6. As a measure of erosive cartilage destruction, immunostaining can be performed with antibodies recognizing the proteoglycan breakdown neoepitope VDIPEN (*see* **Fig. 2**). This neoepitope reflects matrix metalloproteinase (MMP) activity in the articular cartilage (*see* **Note 7**) and is indicative of concomitant collagen breakdown and upcoming cartilage erosion *(10–12)*.

3.7. Measurement of Cartilage Metabolism

At early stages of the arthritis (up to 1 wk) (*see* **Note 8**) the impact of the arthritic process on cartilage metabolism can be measured with the patella assay.

1. Patellae with minimal surrounding tissue are isolated from arthritic and contralateral control knee joints, and incubated in standard culture medium in vitro for 2 h with the radioisotope ^{35}S-sulfate.
2. Tissue specimens are washed, fixed for 1 h in 4% formaldehyde and decalcified overnight in 5% formic acid.
3. The next day, the patella is punched from the tissue and dissolved in liquid scintillation fluid for counting of incorporated label.
4. Incorporated ^{35}S reflects chondrocyte proteoglycan synthesis and values can be expressed per patella, representing a defined anatomical entity *(13–15)*.

3.8. Induction of Flares of Arthritis

The chronicity of the arthritis depends on the degree of antigen retention in the joint and the accumulation, retention and local proliferation of mBSA-specific arthritogenic T-cells (*see* **Note 9**). mBSA arthritis is severe in the first 2 wk and subsides to a chronic smouldering arthritis thereafter.

1. Exacerbation can be achieved by local or systemic rechallenge with mBSA.
2. Local injection of 1 µg mBSA is sufficient to cause florid exacerbation, accompanied by severe and fast cartilage destruction (*see* **Note 10**).
3. Exacerbations can also be elicited by systemic, iv injection of 300 µg mBSA or oral dosing with 20 mg mBSA. Intriguingly, such treatments do not have an impact on the contralateral control joint, but only induce flares in a smouldering but hyperreactive area of the arthritic joint.
4. The nature of the flare is a strong T-cell dependent mBSA specific process *(16–18)*.

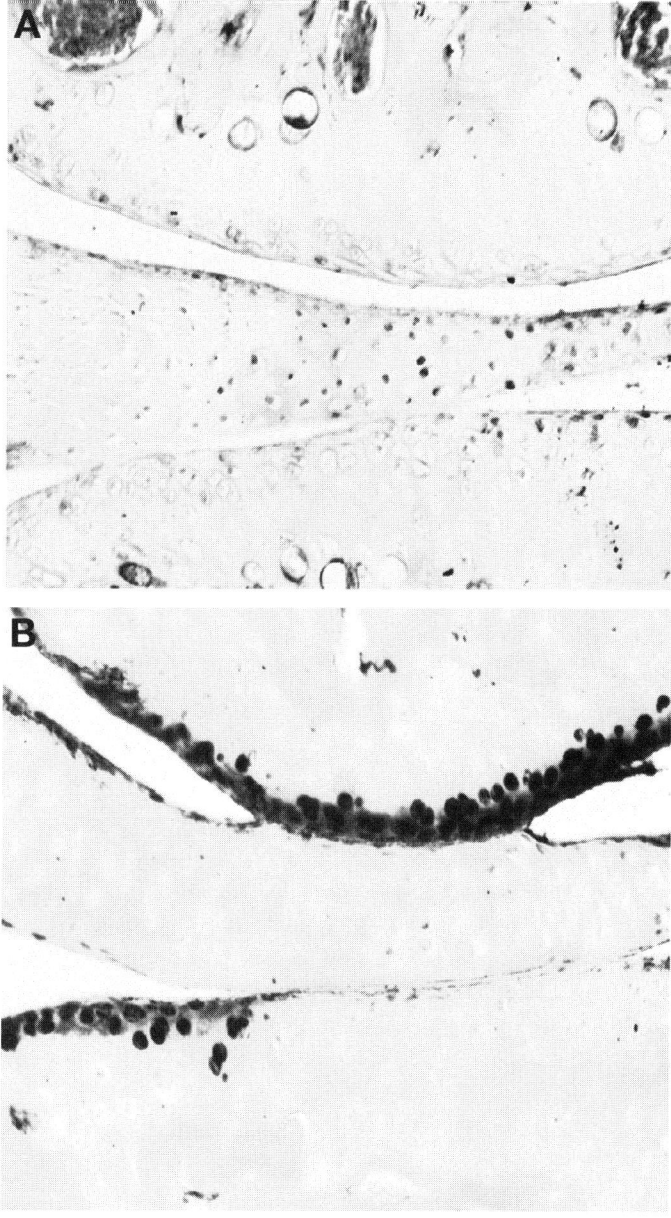

Fig. 2. Immunostaining for VDIPEN cartilage proteoglycan neoepitopes. Control (**A**) and arthritic joint (**B**), taken at day 7 after arthritis onset. Note the staining in the cartilage of the femur and tibia; in between part of the meniscus is visible showing minimal staining.

4. Notes

1. The model is described for mice, using the cationic antigen mBSA. Similar disease models can be developed using amidated forms of BSA (aBSA) or ovalbumin (aOA) as the antigen *(5,6)*. mBSA tends to aggregate upon injection into the joint space, with major sticking to the cartilage surface, whereas aBSA penetrates deeply into the cartilage *(6)*. With both antigens, immune complex formation occurs at the surface only.
2. As mentioned in **Subheading 3.1.** cationicity should be approx 8.5 to 9.0 ($\beta\pm$) for efficient retention in the murine knee joint. However, caution is warranted with the degree of amidation or methylation. If the pI is too high, cationic proteins become irritants, causing joint inflammation by nonspecific cell stimulation rather than immune activation.
3. The model can be induced in many species. Be aware of different balances between T- and B-cell components in various species. Rabbits are high antibody producers, whereas most mouse strains are modest responders. The relative contribution of immune complex-mediated pathology vs T-cell driven arthritis is high in rabbits, also showing more extensive plasma cell foci in the arthritic synovia.
4. The model of mBSA arthritis shows good severity and chronicity in C57Bl/6 mice and Balb/c mice, but poor activity in CBA mice *(2)*. The infiltrate is more granulocyte rich in Balb/c. Different sensitivities, including variation in the degree of cartilage damage has also been noted in various wild type mice and Fcγ receptor I, II, and III knockouts in different genetic backgrounds *(19,20)*. Booster injections are given at day 7 in the neck region. It is recommended to give these injections intradermally rather than subcutaneously. A safe procedure in the mouse is to go through the skin first and then to lift the bevel of the needle, reaching the skin from the inner side. Be careful not to immunize mice at an early age. Between weeks 6 and 8 the immune system matures rapidly, causing marked variations in response to immunization. It is recommended to use 8- to 10-wk-old mice. Less variation in anti-mBSA responses is noted in male mice. However, fighting is a problem and may be a reason for variation in disease phenotype. Optimal choice depends on local housing conditions.
6. To make a strong version of the model, with sufficient chronicity as well as progressive, erosive cartilage damage, it is recommended to include the *B. pertussis* organisms or a like bacteria as an additional ip adjuvant, both at days 0 and 7. This treatment causes systemic generation of high levels of interleukin (IL)-23, boosting the Th1 anti-mBSA response.
7. Cartilage proteoglycan loss is a common feature of all types of joint inflammation, and is a reversible process. When the mice are sufficiently boosted, the allergic reaction can cause major damage to the articular cartilage in the mBSA arthritis model, including surface erosion and occurrence of chondrocyte death. IL-1 is a major cytokine driving production of latent MMPs *(10,12 ,15)*. Studies

in FcγR knockouts reveal that immune complex-mediated activation is a crucial component of erosion, probably driving activation of IL-1-induced MMPs.
8. The patella can be used as a defined anatomical entity, to measure the impact of the arthritic process on cartilage metabolism. This is only feasible in the first week; thereafter major deformities are seen, including excessive cartilage apposition at the margins. When later time points are studied it is necessary to punch and isolate selectively the central part of the patella. Be aware of greater sensitivity of central chondrocytes to IL-1 effects.
9. As a variation of the classic model, mBSA can be directly injected in a knee joint of normal mice, followed by footpad administration of IL-1 to boost a local immune response *(21)*. This creates a local T-cell mediated arthritis and the model has been used in a range of studies in cytokine deficient mice. Be aware that the immune complex component is lacking.
10. When the destructive character of flares is the focus of study, it is recommended to make a mild version of the primary antigen-induced arthritis. This can be achieved by lowering the dose of intra-articular antigen to 20 µg, or limiting the boosting of the immune response, still creating a hypersensitive joint at day 42 with smouldering arthritis, but with less residual damage, allowing for a better window of flare mediated erosion. IL-1 and T-cell derived IL-17 appear dominant cytokines *(22,23)*.

References

1. Dumonde, D. C. and Glynn, L. E. (1962) The production of arthritis in rabbits by an immunological reaction to fibrin. *Br. J. Exp. Pathol.*, **43**, 373–383.
2. Brackertz, D., Mitchell, G. F., and Mackay, I. R. (1977) Antigen-induced arthritis in mice. I. Induction of arthritis in various strains of mice. *Arthritis Rheum.* **20**, 841–850.
3. Cooke, T. D. V., Hird, E. R., Ziff, M., and Jasin, H. E. (1972) The pathogenesis of chronic inflammation in experimental antigen induced arthritis. *J. Exp. Med.* **135**, 322–338.
4. Van den Berg, W. B., van Beusekom, H. J., van de Putte, L. B. A., Zwarts, W. A., and van der Sluis, M. (1982) Antigen handling in antigen-induced arthritis in mice: an autoradiographic and immunofluorescence study using whole joint sections. *Am. J. Pathol.* **108**, 9–16.
5. Van den Berg, W. B., van de Putte, L. B. A., Zwarts, W. A., and Joosten, L. A. B. (1984) Electrical charge of the antigen determines intraarticular antigen handling and chronicity of arthritis in mice. *J. Clin. Invest.* **74**, 1850–1859.
6. Van den Berg, W. B. and van de Putte, L. B. A. (1985) Electrical charge of the antigen determines its localization in the mouse knee joint: deep penetration of cationic BSA in hyaline articular cartilage. *Am. J. Pathol.* **121**, 224–234.
7. Mertz, A. K. H., Batsford, S. R., Curschella, E., Kist, M. J., and Gondolf, K. B. (1991) Cationic Yersinia antigen-induced chronic allergic arthritis in rats. A model for reactive arthritis in human. *J. Clin. Invest.* **87**, 632–642.

8. Matsumoto, I., Maccioni, M., Lee, D. M., et al. (2002) How antibodies to a ubiquitous cytoplasmic enzyme may provoke joint-specific autoimmune disease. *Nat. Immunol.* **3**, 360–365.
9. Kruijsen, M. W. M., van den Berg, W. B., van de Putte, L. B. A., and van den Broek, W. J. M. (1981) Detection and quantification of experimental joint inflammation in mice by measurement of 99 mTc-pertechnetate uptake. *Agents Actions* **11**, 640–642.
10. Van Meurs, J. B. J., van Lent, P. L. E. M., Singer, I. I., Bayne, E. K., van de Loo, F. A. J., and van den Berg, W. B. (1998) IL-1ra prevents expression of the metalloproteinase-generated neoepitope VDIPEN in antigen-induced arthritis. *Arthritis Rheum.* **41**, 647–656.
11. Van Meurs, J. B. J., van Lent, P. L. E. .M., Holthuysen, A. E. M., Singer, I. I., Bayne, E. K., and van den Berg, W. B. (1999) Kinetics of aggrecanase and metalloproteinase induced neoepitopes in various stages of cartilage destruction in murine arthritis. *Arthritis Rheum.* **42**, 1128–1139.
12. Van Meurs, J. B. J., van Lent, P. L. E. M., Stoop, R., et al. (1999) Cleavage of aggrecan at Asn341-Phe342 site coincides with the initiation of collagen damage in murine antigen-induced arthritis: A pivotal role for stromelysin-1 in MMP activity. *Arthritis Rheum.* **10**, 2074–2084.
13. Van den Berg, W. B., Kruijsen, M. W. M., and van de Putte, L. B. A. (1982) The mouse patella assay: an easy method of quantitating articular cartilage chondrocyte function in vivo and in vitro. *Rheumatol. Int.* **1**,165–169.
14. Van de Loo, A. A. J., Arntz, O. J., Otterness, I. G., and van den Berg, W. B. (1992) Protection against cartilage proteoglycan synthesis inhibition by anti-interleukin 1 antibodies in experimental arthritis. *J. Rheumatol.* **19**, 348–356.
15. Van de Loo, A. A. J., Joosten, L. A. B., van Lent, P. L. E. M., Arntz, O. J., and van den Berg, W. B. (1995) Role of Interleukin-1, Tumor Necrosis Factor I and Interleukin-6 in cartilage proteoglycan metabolism and destruction. Effect of in situ cytokine blocking in murine antigen- and zymosan-induced arthritis. *Arthritis Rheum.* **38**, 164–172.
16. Lens, J. W., van den Berg, W. B., and van de Putte, L. B. A. (1984) Flare-up of antigen-induced arthritis in mice after challenge with intravenous antigen. Studies on the characteristics of and mechanisms involved in the reaction. *Clin. Exp. Immunol.* **55**, 287–294.
17. Lens, J. W., van den Berg, W. B., van de Putte, L. B. A., Berden, J. H. M., and Lems, S. P. M. (1984) Flare-up of antigen-induced arthritis in mice after challenge with intravenous antigen: effects of pretreatment with cobra venom factor and antilymphocyte serum. *Clin. Exp. Immunol.* **57**, 520–528.
18. Lens, J. W., van den Berg, W. B., van de Putte, L. B. A., and Zwarts, W. A. (1986) Flare-up of antigen induced arthritis in mice after challenge with intravenous antigen: kinetic of antigen in the circulation and localization of antigen in the arthritic and noninflamed joint. *Arthritis Rheum.* **29**, 665–674.
19. Van Lent, P. L. E. M., Nabbe, K., Blom, A. B., et al. (2001) Role of activatory Fcγ RI and Fcγ RIII and inhibitory Fcγ RII in inflammation and cartilage

destruction during experimental antigen-induced arthritis. *Am. J. Pathol.* **159**, 2309–2320.
20. Nabbe, K. C. A. M., Blom, A. B., Holthuysen, A. E. M., et al. (2003) Coordinate expression of activating FcγRI and III and inhibiting FcγRII in the determination of joint inflammation and cartilage destruction during immune complex-mediated arthritis. *Arthritis Rheum.* **48**, 255–265.
21. Staite, N. D., Richard, K. A., Aspar, D. G., Franz, K. A., Galinet, L. A., and Dunn, C. J. (1990) Induction of an acute erosive monoarticular arthritis in mice by interleukin-1 and methylated bovine serum albumin. *Arthritis Rheum.* 33, 253–260.
22. Van de Loo, A. A. J., Arntz, O.J., Bakker, A. C., van Lent, P. L. E. M., Jacobs, M.J.M., and van den Berg, W. B. (1995) Role of Interleukin-1 in antigen-induced exacerbations of murine arthritis. *Am. J. Pathol.* **146**, 239–249.
23. Koenders, M. I., Lubberts, E., Oppers-Walgreen, B., et al. (2005) Blocking of IL-17 during reactivation of experimental arthritis prevents joint inflammation and bone erosion by decreasing RANKL and IL-1. *Am. J. Pathol.* **167**, 141–149.

19

Pristane-Induced Arthritis in the Rat

Peter Olofsson and Rikard Holmdahl

Abstract

A chronic relapsing arthritis develops after a single subcutaneous injection of small amounts of pristane; the pristane induced arthritis (PIA) model in the rat. PIA is characterized by a sudden onset of disease 2 wk after induction. The main pathological features of PIA include edema accompanied by an acute phase response, infiltration into the joint of mononuclear and polymorphonuclear cells, pannus formation, and erosion of cartilage and bone. PIA is a disease that is largely T-cell dependent that can be adoptively transferred by activated $CD4^+$ T-cells. PIA in rats is followed clinically by macroscopic scoring and is characterized by an early acute phase of severe inflammation after onset that eventually gradually disappears to be followed by less severe relapsing phases of new inflamed joints and increasing cartilage erosion and joint deformity. PIA can be diagnosed by biochemical analyzes in plasma that reflects systemic inflammation (α_1-acid glycoprotein (AGP) and IL-6) and cartilage erosion (COMP).

Key Words: RA; arthritis; PIA; rat; AGP; IL6; COMP; adoptive transfer.

1. Introduction

It was discovered many years ago that mycobacteria tuberculosum-containing adjuvant, the so-called complete Freunds adjuvant (CFA), by itself could induce severe arthritis in most rat strains. Subsequently, several different arthritogenic compounds have been identified from CFA. In the classical adjuvant arthritis model, commonly called "adjuvant arthritis," which gives a very severe systemic inflammatory response including severe and destructive arthritis, the mycobacteria are a requirement. To avoid confusion, this model has now been renamed mycobacteria-induced arthritis (MIA) *(1,2)*. The mineral oil component in CFA has also been found to be arthritogenic in itself, producing a disease with acute peripheral arthritis but lacking a systemic inflammatory response. This model is now called oil-induced arthritis (OIA) *(3–5)*. One of the most arthritogenic lipids in the mineral oil is pristane (2,6,10,14-tetramethylpentadecane). Pristane is a natural component in plants, being part of chlorophyll, and is therefore ingested and also absorbed by animals *(6)*. Thus, it is systemically widespread and therefore a natural "self" component of animals including rats and humans, as it passes through the intestine. Nevertheless, a single subcutaneous injection in susceptible rat strains produce chronic inflammatory disease. These rats develop pristane induced arthritis (PIA), a chronic relapsing arthritis sharing many features with rheumatoid arthritis (RA) which in fact fulfils most of the ARA criteria for RA *(7)* (**Table 1**) and is the animal model which most closely mimics these criteria *(8)*. In the mouse, several repeated intraperitoneal injections of pristane are needed and gradually producing a systemic chronic inflammatory disease, one of the symptoms being arthritis. The rat on the other hand does not develop clinically apparent disease after intraperitoneal (ip) injections.

The observation that arthritis can be induced by lipid adjuvants that are unable to bind to major histocompatibility complex (MHC) II molecules, renders adjuvant-induced arthritis in rats a particular interest. As in RA, an important question in these models is to understand how a nonspecific trigger of the immune system can precipitate in a joint inflammatory disease.

The collection of available adjuvant-induced arthritides, including pristane-induced arthritis (PIA) *(8)*, share many features but differ with regard to severity and chronicity of the disease. All these models are induced through intradermal or subcutaneous injections where it at least in the case of PIA in rats is not possible to induce disease through ip injection or oral administration.

Inflamed joints in PIA contain several inflammatory cell types, including macrophages, neutrophils, and T-cells. Besides the joint specific arthritis also systemic responses in PIA are observed. Thus, increased blood levels of acute phase reactants, α_1-acid glycoprotein (AGP) and interleukin (IL)-6, can be followed as a result of the arthritis induction *(9–11)*. Likewise the plasma levels of cartilage oligomeric matrix protein (COMP) can be monitored as a reflection of the ongoing cartilage erosion *(11–13)*.

There are several reports showing that both lymph node and spleen cells in adjuvant-induced arthritis can adoptively transfer disease after mitogen activation. However, as no arthritogenic oil is observed in the spleen *(14,15)* there is circumstantial evidence for a chain of events, in adjuvant induced arthritis, starting in the draining lymph nodes with activation of joint specific leukocytes that later assemble and possibly also proliferate and differentiate in the spleen *(16)*.

There is substantial evidence for a role of T-cells expressing $\alpha\beta$TCR in the induction and development of adjuvant arthritis *(5,17)*, whereas there is no evidence for an arthritogenic role of B-cells. Depletion of neutrophils as well as CD4 T-cells, but not CD8 T-cells or the complement system argues for an activation of autoreactive CD4 T-cells causing accumulation of neutrophils and macrophages as effector cells in peripheral joints of adjuvant-induced arthritis *(18–21)*. Attempts to transfer arthritis from rats with adjuvant arthritis with serum or IgG have so far failed *(4)*, in contrast with CIA where anti-CII antibodies can induce arthritis *(4,22,23)*. It is also possible that the IgG fraction in adjuvant-injected rats has a suppressive effect on disease development after transfer *(24)*. On the other hand, lymphocytes from adjuvant primed rats, which have been activated with T-cell mitogens in vitro, readily transfer severe arthritis to naive irradiated recipients *(4,23,25,26)*. Importantly, T-cells play a crucial role at different stages of the disease as suggested by experiments showing that administration of an antibody against the $\alpha\beta$TCR can both prevent induction, delay onset, and cure disease at later stages *(5,8,27)*.

Hence, adjuvant-induced arthritis in rats is a T-cell driven disease, where the primary induction of autoreactive T-cells seems to occur in draining lymph nodes. These cells later translocate to spleen and joints, causing inflammation and erosive arthritis.

2. Materials

2.1. Animals and Arthritis Induction

Rats should be kept in environmentally controlled facilities with stable conditions for treatment, caging, nutrition, infections, noise, and light–dark cycles. Although no distinct environmental factors have been identified, and the model seems to be very reproducible under different environmental conditions, it is always wise to keep the animals under as stable conditions as possible. Arthritis is induced by an intradermal or subcutaneous injection of pure pristane (2,6,10,14-tetramethylpentadecane alias Norphytane) (Sigma-Aldrich). For susceptibility to arthritis in various inbred rats strains refer to *(28)*.

2.2. Analysis of Blood Phenotypes, Radioimmuno Assay AGP

1. Heparin (5000 IE/mL) (Lövens Läkemedel).
2. 1 mg/mL Chloramine T (Sigma; cat. no. C9887) in 0.05M phosphate buffer (pH 7.4).
3. 2 mg/mL α1-acid glycoprotein (Zivic-Miller Laboratories) in phosphate buffered saline (PBS).
4. Polyclonal rabbit antibody against rat α1-acid glycoprotein (Agrisera). The dilution of the antibody is determined by titration experiments where the amount of antibody that precipitates 70% of the protein is chosen for the assays.
5. Bovine serum albumin (BSA).
6. 15% polyethylene glycol (PEG) 6000 (Sigma-Aldrich Fluka; cat. no. 81255) in RIA-buffer.

7. PBS: 5 mM NaH$_2$PO$_4$ and 0.15M NaCl (pH 7.4).
8. 0.5M phosphate buffer (pH 7.4).
9. 1 mg/mL NaHSO$_3$ in 0.05M phospate buffer (pH 7.4).
10. 2% BSA/0.02% NaN$_3$ in PBS.
11. RIA buffer: 0.1M phospate buffer, 0.1% BSA, and 0.02% NaN$_3$ (pH 7.4).

2.3. Analysis of Blood Phenotypes, Cartilage Oligomeric Matrix Protein

1. Enzyme linked immunosorbent assay (ELISA) plates, coating plate (96-well Immunoplate Nunc-Maxisorp) and incubation plate (common polysterene V-shaped bottom 96-well plate).
2. Rat COMP (400 μg/mL) prepared from rat chondrosarcoma and rabbit antiserum raised against rat COMP was purified according to a protocol described in Carlsen et al. *(29)*.
3. Substrate 4-Nitrophenyl phosphate disodium salt hexahydrate (Sigma; cat. no. N9389)
4. PBS: 5 mM NaH$_2$PO$_4$ and 0.15M NaCl (pH 7.4).
5. Coatbuffert: 0.05M Na$_2$CO$_3$ and 4M GuHCl (pure) (pH 10.0).
6. 4% Triton in 10 mM NaH$_2$PO$_4$ (pH 7.4).
7. 5X INK I: 0.7M NaCl, 0.04M Na$_2$HPO$_4$, 14 mM KCl, 7 mM KH$_2$PO$_4$, and 15 mM NaN$_3$ (pH 7.4).
8. INK II dilute INK I 5X with H$_2$O and add 2 mg/mL BSA and 0.05% Tween-20.
9. 0.8% sodim dodecyl sulfate (SDS) in INK I without Tween (pH 7.4).
10. 2 mg/mL BSA in PBS.
11. Pig anti-rabbit-alkaline phosphatase conjugated antobody (Dakocytomation; cat. no. D0306).
12. Substrate buffer, 194 mL diethanolamine (Merck; cat. no. 8.03116.1000), 202 mg MgCl$_2$ · 6H$_2$O in 2 L H$_2$O (pH 9.8).

2.4 Analysis of Systemic Immunological Reaction IL-6

1. IL-6 dependent cell line B9 (ATCC) *(30)*.
2. Sterile PBS: 0.14M, NaCl, 2.7 mM, KCl, 1.5 mM, KH$_2$PO$_4$, and 8.1 mM, Na$_2$HPO$_4$ (pH 7.4).
3. [^{3}H] TdR (Amersham Pharmacia).
4. Recombinant rat IL-6 (Research Diagnostics; cat. no. RDI-4016).
5. Complete culture media (Dulbeccos glutamax medium supplemented with streptomycin, penicillin, and 10% fetal calf serum [FCS]).

2.5. Adoptive T-Cell Transfer

1. Arthritis is induced by an intradermal injection of 500 μL pristane (2,6,10,14-tetramethylpentadecane alias Norphytane).
2. Dulbeccos Glutamax medium supplemented with streptomycin, penicillin, HEPES, 50 μM β-mercaptoethanol and 5% FCS.

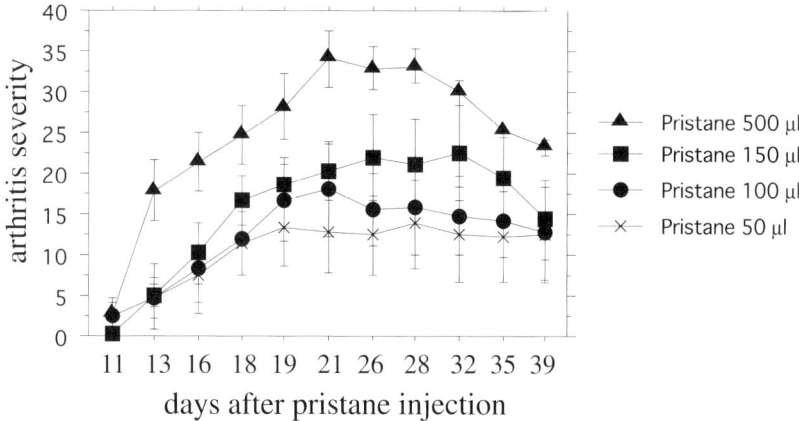

Fig 1. Titration of pristane dose. Different doses of pristane ranging from 50 to 400 µL diluted in olive oil to get a total injection volume of 400 µL were injected id at the base of the tail. Eight rats of both sexes were used in each group. A dose response effect is observed where we have chosen 150 µL of pristane to be the most suitable dose for inducing PIA in DA rats.

3. Erythrocyte lysis buffer 0.84% NH_4Cl (pH 7.4),
4. Sterile PBS: $0.14 M$, NaCl, 2.7 mM, KCl, 1.5 mM, KH_2PO_4, and 8.1 mM, Na_2HPO_4 (pH 7.4).
5. Concanavalin A (Sigma; cat. no. C 9604).

3. Methods
3.1. Induction and Evaluation of Arthritis

Arthritis is induced by a single intradermal or subcutaneous injection of pristane (2,6,10,14-tetramethylpentadecane), which means that no booster induction is necessary. Susceptible strains like DA and LEWIS rats have an onset of disease 2 wk after induction with development of severe inflammation and oedema that lasts for 2 to 3 wk when the ongoing inflammation slowly declines. Around day 60 new inflammation occurs causing a relapsing chronic phase of the disease. During the arthritis development massive cell infiltration of mononuclear and polymorphonuclear cells cause pannus formation and erosion of cartilage and bone. For inducing arthritis a volume of 50 to 500 µL is sufficient (*see* **Fig. 1**). We have, as a routine, used 150 µL for induction, resulting in a highly penetrant (95–100%) arthritis with high reproducibility in DA rats (*see* **Note 1**).

To evaluate the severity of PIA it is desirable to use a quantitative measurement of the arthritis development. Therefore the inflammation is monitored by a macroscopic scoring system taking into account all factors like oedema, redness, cartilage erosion, and bone destruction that follow during the course of diseases. The numerical scoring for the four limbs range from 0 to 15 (1 point for each swollen or red toe, 1 point for mid-foot digit and knuckle, 5 points for a swollen ankle [see **Note 2**]). The scores of the four paws are summarized for each animal yielding a maximum total score of 60. The rats should be observed 1 to 4 times/wk after pristane injection *(31)*. As a general measurement of the health stratus it is also advisable to follow the weight of the rats as this is often observed to decrease following active arthritis. Besides the severity of the arthritis scoring, other important features of the disease is the day of onset, judged as the first day of observable clinical signs of arthritis in the rats. Also the duration of ongoing arthritis is a valuable characteristic of the disease severity. A way to make a valuation of all these measurements of disease is to make a graph of the disease severity during the experimental period and to calculate the area under the graph to represent the sum of arthritis in each individual rat (*see* **Fig. 2**).

3.2. Analysis of Acute Phase Response AGP

α_1-acid glycoprotein (AGP) is an acute phase protein in rats that is produced and released mainly from the liver in response to inflammatory stress. Hence, AGP is a good plasma marker that can be used to monitor the systemic response to inflammation or trauma in the animals. Plasma levels of AGP will range from 0 to 10 mg/mL plasma with levels higher that 2 mg/mL of AGP indicating ongoing inflammation (*see* **Notes 3** and **4**).

3.2.1. Plasma Sample Preparations

Blood (500–1000 µL) is obtained by cutting the tip of the tail, and collecting it in tubes containing 10 µL of heparin in order to prevent blood coagulation. The plasma is separated from blood cells by centrifugation and stored at –70°C until assayed (*see* **Notes 3–6**). Prior to analysis, blood plasma, or blood serum should be diluted 10,000 times in RIA buffer.

3.2.2. ^{125}I Labeling of AGP

The labeling of purified rat AGP is performed by the Chloramine T method *(32)*.

1. Attach a Sephadex G-25 M column (Column PD-10) (Pharmacia LKB) to a tripod and equilibrate the column by applying 200 µL of 2% BSA in PBS. Allow the BSA solution to enter the column and wash the column with 20 mL PBS.
2. Mix 10 mg AGP protein dissolved in PBS (5 µL of 2 mg/mL AGP) with 2 to 4 µL ^{125}I (Amersham Pharmacia). Add 6 µL 0.5*M* phosphate buffer and 2 µL chloramine-T to start the radioiodination of AGP.

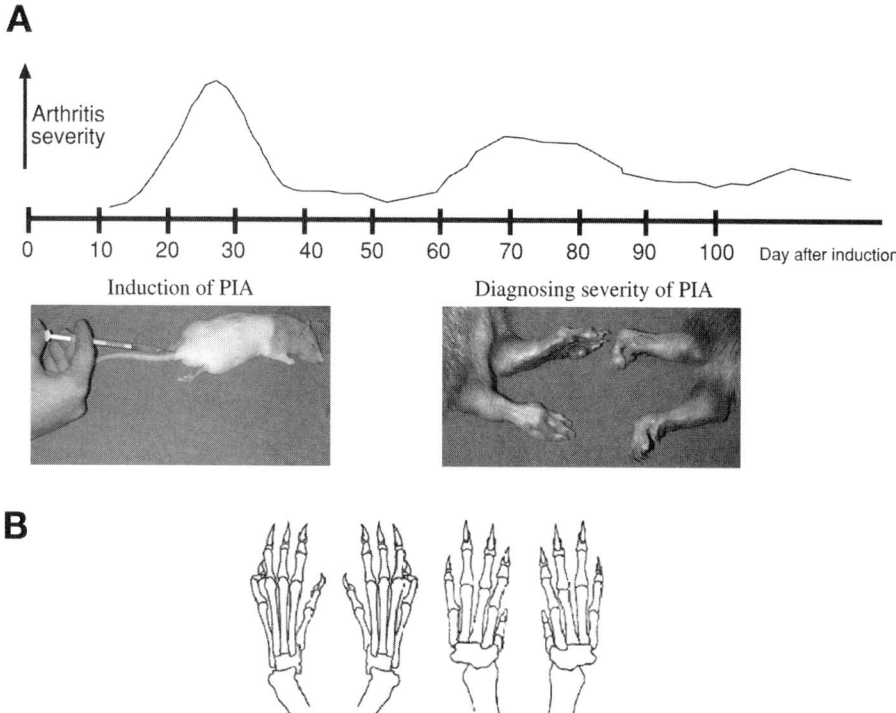

Fig 2. Induction and scoring of Pristane Induced Arthritis (PIA) in rats. (**A**) PIA is induced by intradermal injection with pristane. Susceptible rat strains like DA and LEWIS develop a severe arthritis with clinical onset 10 to 14 d after induction. The disease shows maximal clinical manifestation 16 to 36 d after induction, after which the disease subsides. PIA is characterized by being chronic relapsing and show relapses in the chronic phase (after >56 d) that are of less inflammatory severity that in the first acute phase of the disease, but includes more pronounced deformity of peripheral joints.(**B**) PIA in rats is best diagnosed clinically by making markings on schematic images of the arthritic joint in the rats. These drawings are valuable for evaluation of the arthritis severity, where the disease is numerically validated.

3. Incubate for 10 min on ice.
4. Stop the reaction by adding 50 µL NaHSO$_3$ (1 mg/mL in 50 mM phospate buffer) and 50 µL of 2% BSA/0.02% NaN$_3$ in PBS.
5. Purify the ^{125}I-labeled AGP from unincorporated iodine by separation on the prepared PD-10 column. Apply the iodination reaction to the column and eluate with PBS. Collect 5 to 10 drops/5-mL collection tubes (about 30 tubes in all). Analyse the fractions by measuring the amount of incorporated ^{125}I in AGP in 10 µL volume in a γ-counter (Beckman Instruments). For the radioimmunoassay,

pool the 3 collection tubes containing the peak of eluted ^{125}I-labeled AGP. Analyze the amount of incorporated radiolabel (usually 200–800 × 10^6 cpm/ml).

3.2.3. Radio Immuno Assay (RIA) (see **Notes 3** and **4**).

1. All samples should be analysed in duplicates.
2. In addition to the diluted plasma samples a range of APG standards (2000, 1000, 500, 250, 125, 63, 31, 16, 8, 4, and 2 ng/mL in RIA buffer) should be analysed as a reference.
3. In each reaction tube (3- to 5-mL tubes suitable for γ-counter) add 200 µL diluted sample (or standard reference), 200 µL diluted anti-rat AGP antibody (in RIA buffer) and 100 µL ^{125}I-AGP (0.2–0.3 × 10^6 cpm/mL in RIA buffer).
4. Incubate at 20 to 25°C overnight.
5. Add 300 µL BSA and 1.6 mL 15% PEG to each sample.
6. Centrifugate at 1500*g* for 20 min.
7. Discard the supernatant from the pellet by vacuum suction.
8. Analyze in γ-counter and calculate the concentrations of AGP in the blood samples using the produced standard curve as reference.

3.3. Analysis of Blood Phenotypes, COMP

The blood level of COMP is a good marker of cartilage erosion, as increased levels of cartilage specific COMP is circulating as a result of ongoing cartilage destruction. Plasma levels of COMP will range from 200 to 2000 ng/mL plasma with levels higher that 400 ng/mL of circulating COMP indicating cartilage destruction and turnover (*see* **Note 5**).

1. Coat the coating plates with rat COMP 0.2 µg/mL in coating buffer (50 µL/well).
2. To the incubation plate add 50 µL of 1/10 diluted plasma samples in 0.8% SDS in INK I and 50 µL rabbit anti-rat COMP antibody (diluted 1/1500).
3. Incubate over night at 4°C.
4. Wash the coating plate 3 times with PBS/0.1% Tween-20. Add 100 µL 2 mg/mL BSA in PBS as blocking agent for the coating plate to reduce the background binding.
5. Incubate 1 h at room temperature.
6. Wash the coating plate 3 times with PBS/0.1% Tween-20.
7. Transfer 50 µL/well of the sample/antibody incubation mixture to the coating plate. Incubate 1 h at room temperature.
8. Wash the coating plate 3 times with PBS/0.1% Tween-20
9. Add 50 µL/well of pig anti rabbit-alkaline phosphatase conjugated antobody (D0306 Dakocytomation) diluted 1/1000 in INK II. Incubate 1 h at room temperature.
10. Wash the coating plate 3 times with PBS/0.1% Tween-20.
11. Add 50 µLwell of the phospatase substrate (Paranitrophenylphosphate) 1 tablet/ 5mL substrate buffer.

12. Incubate in the dark at room temperature. Measure absorbance at 405 nm. The obtained absorbance values are invertedly correlated with the plasma concentration of COMP. Therefore, comparison with an internal positive control samples used as standard references are valuable in obtaining relative concentrations of plasma COMP in the rats.

3.4. Analysis of Systemic Inflammation Reaction, IL6

The blood levels of IL-6, measured by the B9 cell bioassay, is a sensitive measurement of the systemic effect of ongoing inflammation. The blood level of IL-6 is raised from being undetectable to levels of approx 200- to 00 pg/mL (*see* **Note 6**).

1. Dilute plasma samples 25 times in complete media and heat inactivate the samples at 65°C for 30 min.
2. The IL-6 dependent cell line B9 is maintained in complete media supplemented with recombinant rat IL-6 (20 pg/mL) to ensure survival of the cells.
3. Before starting the assay the cells are washed two times in complete medium.
4. Add 100 μL cells at 5×10^4/mL to the wells of a flat bottom 96-well culturing plate.
5. Add 100 μL of diluted plasma samples to each well. As reference standard samples of recombinant rat IL-6 (32, 16, 8, 4, 2, 1, 0,5, 0 pg/ml) should be analysed in the same plate as the samples.
6. Incubate the samples at 37°C at 5% CO_2 for 72 h.
7. Pulse the cells by adding 1 μCurie ^{3}H-thymidine and culture overnight.
8. Analyse the proliferation of ^{3}H-Thymidine by harvesting the cells in a Filtermate cell harvester (Packard Instruments) and measure the incorporation of [^{3}H] TdR in a matrix 96 direct β counter (Packard Instruments).
9. A high concentration of IL-6 in the rat plasma samples is reflected by an increased B9 cell proliferation (*see* **Note 6**).

3.5. Adoptive T-Cell Transfer

1. Induce arthritis in the PIA susceptible rats by an intradermal injection with 500 μL pristane.
2. At day of onset of disease (day 10–14) the rats are sacrificed and spleen (or draining lymph nodes, inguinal and lumbar lymph nodes) are surgically recovered (both spleen and lymph node cells can transfer disease).
3. The spleens (or lymph nodes) are homogenised through sieves and washed twice in PBS (pH 7.4).
4. The erythrocytes in the spleen samples are lysed by treatment with erythrocyte lysis buffer.
5. Both lymph nodes and spleen cells are cultured at 4×10^6/mL for 48 h in Dulbeccos Glutamax medium supplemented with streptomycin, penicillin, HEPES, 50 μM β-mercaptoethanol, 5% FCS and 3 μg/mL Con A at 37°C in humidified 5% CO_2 atmosphere incubator.

6. The mitogen stimulated cells are harvested, washed twice in PBS and injected intravenously (or ip) into donor rats at a concentration of 5 to 50×10^6 cells/animal (*see* **Note 7**).

4. Notes

1. Besides intradermal (id) injection of pristane we have also attempted to induce arthritis through subcutaneous (sc), intraperitoneal (ip) or intravenous (iv) routes. It has been found that only id or sc routes of administration provoked arthritis. By performing autopsies after sacrificing the rats day 45 after pristane injection, it was seen that all rats with arthritis had swollen draining lymph nodes (inguinal and lumbar nodes). The same lymph nodes in ip injected rats remained normal. Interestingly the most optimal induction of PIA is obtained when injection is performed intradermally into the skin at the base of the tail where the fur of the back is starting to grow. If the inductions are performed at the same location at the base of the tail we have noticed that sc induction induces a less severe arthritis compared to id injection. Therefore, care must be taken to inject animals in the same way for a meaningful comparison, since the injection is one major factor of variance in the experimental setup.
2. The use of scoring systems for evaluating the disease severity is an attempt at quantitation. There are some problems with numerical assessments of disease as the scoring is based on discrete values where for example the difference in arthritis severity between 1 and 2 is not equal to the difference between 25 and 26, as should be the case for a true quantitative measurement. To avoid statistical error when estimating the differences between groups of animals we recommend using non-parametric tools like the Mann-Whitney test for calculating the statistical significance levels in an experiment. Furthermore, the scoring system is set to value a swollen ankle joint as scoring 5. The question is then whether a mild swollen ankle joint should be given the same scoring number as a highly severely swollen ankle joint. In this case it is up to the researcher in charge of the experiment to make the decision. In both these questions it is advisable to occasionally perform a double blind scoring to evaluate differences in scoring evaluation between different researchers in a laboratory, especially if the researchers are involved in the same project where comparisons between their respective experiments will be performed.
3. Blood levels of both AGP, COMP, and IL-6 are equal in plasma and serum. Hence neither of these proteins is precipitated with the coagulated blood that is removed from the serum sample (unpublished observation).
4. There is a commercial reagent for detection of AGP (Cardiotech Services). However we have no experience with these reagents.
5. The circulating level of COMP is dependent on both cartilage destruction and normal cartilage turnover. Hence, the age of the of the animals is a strong influence of the results *(12)*. Thus great care must be taken to perform statistical comparisons of closely age-matched groups of animals.

Table 1
Comparison of PIA in Rats With American Rheumatism Association Criteria for RA

	RA	PIA
Early morning stiffness	+	ND
Arthritis of at least 3 areas (PIP-MC, wrist, elbow, knee, ankle, PIP-MC) >6 wk	+	+
Arthritis of hand joints >6 wk	+	+
Rheumatoid nodules	+	−
Symmetric arthritis	+	+
Serum rheumatoid factors (RF)	+	+
Radiographic changes	+	+
Classical RA (>4 criteria)	+	+
Additional observations		
MHC association	+	+
Anti-CII antibodies	+	−
Enthesopathy	(+)	+

Abbr: ND, not determined; PIP-MC, proximal interphalangeal metacarpophalangeal joint; PIP-MT, proximal interphalangeal metatarsophalangeal joint.

6. The level of IL-6 in plasma was determined by correlation to the standard curve of recombinant rat IL-6 included in each plate. Detection limit in the assay is 2.5 pg/mL.
7. Both iv and ip transfer of cells are possible as administration route for the ConA activated cells. It is our experience that iv cause more severe arthritis in the donor rats, but this administration needs higher accuracy in administration. The onset of adoptively transferred disease is earlier compared to PIA, usually at day 5 to 8 after cell transfer. Usually naive donor rats can be used to transfer PIA, but we have noticed that donor rats that have been irradiated (600 RAD) prior to transfer develop a more severe and also chronic disease *(33)*.

Acknowledgments

The authors thank Bo Åkerström and Maria Allhorn (Lund University, Sweden) for help with the AGP RIA assay. Tore Saxne, Mette Lindell and Dick Heinegård (Lund University, Sweden) for help with the COMP assay. Jens Holmberg (Lund University, Sweden) for help with the adoptive transfer protocols.

References

1. Pearson, C. M. (1956) Development of arthritis, periarthritis and perioscitis in rats given adjuvants. *Proc. Soc. Exp. Biol. Med.* **91**, 91–101.
2. Holmdahl, R., Lorentzen, J. C., Lu, S., Olofsson, P., Wester, L., Holmberg, J., and Pettersson, U. (2001) Arthritis induced in rats with nonimmunogenic adjuvants as models for rheumatoid arthritis. *Immunol. Rev.* **184**, 184–202.
3. Kleinau, S., Erlandsson, H., Holmdahl, R., and Klareskog, L. (1991) Adjuvant oils induce arthritis in the DA rat. I. Characterization of the disease and evidence for an immunological involvement. *J. Autoimmun.* **4**, 871–880.
4. Kleinau, S., and Klareskog, L. (1993) Oil-induced arthritis in DA rats passive transfer by T cells but not with serum. *J Autoimmun* **6**, 449–458.
5. Holmdahl, R., Goldschmidt, T. J., Kleinau, S., Kvick, C., and Jonsson, R. (1992) Arthritis induced in rats with adjuvant oil is a genetically restricted, alpha beta T-cell dependent autoimmune disease. *Immunology* **76**, 197–202.
6. Garrett, L. R., Chung, J. G., Byers, P. E., and Cuchens, M. A. (1989) Dietary effects of pristane on rat lymphoid tissues. *Agents Actions* **28**, 272–278.
7. Arnett, F. C., Edworthy, S. M., Bloch, D. A., et al. (1988) The American Rheumatism Association 1987 revised criteria for the classification of rheumatoid arthritis. *Arthritis Rheum.* **31**, 315–324.
8. Vingsbo, C., Sahlstrand, P., Brun, J. G., Jonsson, R., Saxne, T., and Holmdahl, R. (1996) Pristane-induced arthritis in rats: a new model for rheumatoid arthritis with a chronic disease course influenced by both major histocompatibility complex and non-major histocompatibility complex genes. *Am. J. Pathol.* **149**, 1675–1683.
9. Olofsson, P., Nordquist, N., Vingsbo-Lundberg, C., et al. (2002) Genetic links between the acute-phase response and arthritis development in rats. *Arthritis Rheum.* **46**, 259–268.
10. Svelander, L., Holm, B. C., Buchtt, A., and Lorentzen, J. C. (2001) Responses of the rat immune system to arthritogenic adjuvant oil. *Scand. J. Immunol.* **54**, 599–605.
11. Olofsson, P., Holmberg, J., Pettersson, U., and Holmdahl, R. (2003) Identification and isolation of dominant susceptibility loci for pristane-induced arthritis. *J. Immunol.* **171**, 407–416.
12. Vingsbo-Lundberg, C., Saxne, T., Olsson, H., and Holmdahl, R. (1998) Increased serum levels of cartilage oligomeric matrix protein in chronic erosive arthritis in rats. *Arthritis Rheum.* **41**, 544–550.
13. Wester, L., Olofsson, P., Ibrahim, S. M., and Holmdahl, R. (2003) Chronicity of pristane-induced arthritis in rats is controlled by genes on chromosome 14. *J. Autoimmun.* **21**, 305–313.
14. Kleinau, S., Dencker, L., and Klareskog, L. (1995) Oil-induced arthritis in DA rats: tissue distribution of arthritogenic 14C-labelled hexadecane. *Int. J. Immunopharmacol.* **17**, 393–401
15. Holm, B. C., Svelander, L., Bucht, A., and Lorentzen, J. C. (2002) The arthritogenic adjuvant squalene does not accumulate in joints, but gives rise to

pathogenic cells in both draining and non-draining lymph nodes. *Clin. Exp. Immunol.* **127**, 430–435.
16. Rodriguez-Palmero, M., Pelegri, C., Ferri, M. J., Castell, M., Franch, A., and Castellote, C. (1999) Alterations of lymphocyte populations in lymph nodes but not in spleen during the latency period of adjuvant arthritis. *Inflammation* **23**, 153–165
17. Yoshino, S., Schlipkoter, E., Kinne, R., Hunig, T., and Emmrich, F. (1990) Suppression and prevention of adjuvant arthritis in rats by a monoclonal antibody to the alpha/beta T cell receptor. *Eur. J. Immunol.* **20**, 2805–2808
18. Larsson, P., Holmdahl, R., Dencker, L., and Klareskog, L. (1985) In vivo treatment with W3/13 (anti-pan T) but not with OX8 (anti-suppressor/cytotoxic T) monoclonal antibodies impedes the development of adjuvant arthritis in rats. *Immunology* **56**, 383–391
19. Pelegri, C., Paz Morante, M., Castellote, C., Castell, M., and Franch, A. (1995) Administration of a nondepleting anti-CD4 monoclonal antibody (W3/25) prevents adjuvant arthritis, even upon rechallenge: parallel administration of a depleting anti-CD8 monoclonal antibody (OX8) does not modify the effect of W3/25. *Cell Immunol.* **165**, 177–182
20. Pelegri, C., Morante, M. P., Castellote, C., Franch, A., and Castell, M. (1996) Treatment with an anti-CD4 monoclonal antibody strongly ameliorates established rat adjuvant arthritis. *Clin. Exp. Immunol.* **103**, 273–278
21. Santos, L. L., Morand, E. F., Hutchinson, P., Boyce, N. W., and Holdsworth, S. R. (1997) Anti-neutrophil monoclonal antibody therapy inhibits the development of adjuvant arthritis. *Clin. Exp. Immunol.* **107**, 248–253.
22. Stuart, J. M., Cremer, M. A., Townes, A. S., and Kang, A. H. (1982) Type II collagen-induced arthritis in rats. Passive transfer with serum and evidence that IgG anticollagen antibodies can cause arthritis. *J. Exp. Med.* **155**, 1–16
23. Taurog, J. D., Sandberg, G. P., and Mahowald, M. L. (1983) The cellular basis of adjuvant arthritis. I. Enhancement of cell- mediated passive transfer by concanavalin A and by immunosuppressive pretreatment of the recipient. *Cell Immunol.* **75**, 271–282.
24. Ulmansky, R., and Naparstek, Y. (1995) Immunoglobulins from rats that are resistant to adjuvant arthritis suppress the disease in arthritis-susceptible rats. *Eur. J. Immunol.* **25**, 952–957.
25. Taurog, J. D., Sandberg, G. P., and Mahowald, M. L. (1983) The cellular basis of adjuvant arthritis. II. Characterization of the cells mediating passive transfer. *Cell Immunol.* **80**, 198–204.
26. Svelander, L., Mussener, A., Erlandsson-Harris, H., and Kleinau, S. (1997) Polyclonal Th1 cells transfer oil-induced arthritis. *Immunology* **91**, 260-265.
27. Carlson, B. C., Jansson, A. M., Larsson, A., Bucht, A., and Lorentzen, J. C. (2000) The endogenous adjuvant squalene can induce a chronic T-cell-mediated arthritis in rats. *Am. J. Pathol.* **156**, 2057–2065.
28. Heidrich, H. J. (1990) *Genetic monitoring of inbred strains of rats*, Gustav Fischer Verlag, Stuttgart, New York.

29. Carlsen, S., Hansson, A. S., Olsson, H., Heinegard, D., and Holmdahl, R. (1998) Cartilage oligomeric matrix protein (COMP)-induced arthritis in rats. *Clin. Exp. Immunol.* **114(3),** 477–484.
30. Aarden, L. A., De Groot, E. R., Schaap, O. L., and Lansdorp, P. M. (1987) Production of hybridoma growth factor by human monocytes. *Eur. J. Immunol.* **17**, 1411–1416.
31. Holmdahl, R., Carlsén, S., Mikulowska, A., et al. (1998) In: *Human Genome Methods*, Adolph, K. W., ed., , New York:, CRC Press LLC, New York, USA. pp. 215–238.
32. McConahey, P. J., and Dixon, F. J. (1980) Radioiodination of proteins by the use of the chloramine-T method. *Methods Enzymol.* **70**, 210–213.
33. Holmberg, J., Tuncel, J., Yamada, H., Lu, S,. Olofsson, P., and Holmdahl, R. (2006) Pristane, a non-antigenic adjuvant, induces MHC class II-restricted, arthritogenic T cells in the rat. *J. Immunol.* **176(2),** 1172–1179.

20

The K/BxN Mouse Model of Inflammatory Arthritis

Theory and Practice

Paul Monach, Kimie Hattori, Haochu Huang, Elzbieta Hyatt, Jody Morse, Linh Nguyen, Adriana Ortiz-Lopez, Hsin-Jung Wu, Diane Mathis, and Christophe Benoist

Abstract

Mice expressing the KRN T cell receptor transgene and the MHC class II molecule A^{g7} (K/BxN mice) develop severe inflammatory arthritis, and serum from these mice causes similar arthritis in a wide range of mouse strains, owing to pathogenic autoantibodies to glucose-6-phosphate isomerase (GPI). This model has been useful for the investigation of the development of autoimmunity (K/BxN transgenic mice) and particularly of the mechanisms by which anti-GPI autoantibodies induce joint-specific imflammation (serum transfer model). In this chaper, after a summary of findings from this model system, we describe detailed methods for the maintenance of a K/BxN colony, crossing of the relevant TCR and MHC genes to other strain backgrounds, evaluation of KRN transgenic T cells, measurement of anti-GPI antibodies, induction of arthritis by serum transfer, and clinical and histological evaluation of arthritis.

Key Words: Rheumatoid arthritis; glucose-6-phosphate isomerase; GPI; anti-GPI; serum transfer; autoantibodies; mouse model; KRN; methods.

1. Introduction

The K/BxN T cell receptor (TCR) transgenic mouse model of spontaneous inflammatory arthritis was described in 1996 *(1)*. Arthritis appeared serendipitously when the KRN transgene was crossed into mice carrying the H-2^{g7} haplotype of the major histocompatibiity locus. Subsequent work has shown that autoimmunity is initiated by TCR recognition of a peptide derived from the glycolytic enzyme glucose-6-phosphate isomerase (GPI) in the context of the major histocompatibility complex (MHC) class II molecule H2-A^{g7}, resulting in production of high titers of autoantibodies to GPI *(2)*. These antibodies

From: *Methods in Molecular Medicine, vol. 136: Arthritis Research, Volume 2*
Edited by: A. P. Cope © Humana Press Inc., Totowa, NJ

are essential for the production of arthritis, and serum (or anti-GPI antibodies) purified from serum, or monoclonal anti-GPI antibodies) of K/BxN mice induces arthritis in a wide range of mouse strains *(3,4)*. This final stage requires no input from the adaptive immune system, but involves several players of the innate immune system: lymphocytes and the cytokines they produce are necessary to generate the high titers of anti-GPI but become dispensable for the effector phase, which relies instead on responses and mediators from neutrophils, mast cells, or the vascular endothelium. Many genes required for or influencing the severity of arthritis in this model have been identified. Breeding of the KRN transgene and MHC restriction element to genetically-deficient backgrounds has allowed assessment of genetic effects on either the breakdown of tolerance to GPI or on subsequent effector mechanisms. More easily tractable has been genetic analysis isolated to the effector phase by use of the serum transfer model. By transfer of K/BxN serum into gene-deficient strains, arthritis induced by anti-GPI antibodies has been shown to require mast cells *(5)*, neutrophils *(6)*, the alternative pathway of complement activation *(7)*, the C5a receptor *(7)*, the low-affinity IgG receptor FcγRIII *(7,8)*, the cytokines interleukin (IL)-1 and tumor necrosis factor (TNF) *(9)*, the integrin LFA-1 *(10)*, and the MHC-like Fc receptor FcRn *(11)*. Mechanisms by which these factors interact to produce joint-specific inflammation are being actively investigated in several laboratories. The system has also allowed the relatively facile analysis of natural allelic variants of key genes, identifying several quantitative trait (QT) regions, and pinpointing variation in the C5 and IL-1β genes as influencing susceptibility to arthritis among inbred mouse strains *(12,13)*. Overall, the K/BxN serum transfer system very likely involves the same effector mechanisms as the model in which arthritis is induced by transfer of a cocktail of monoclonal antibodies (mAbs) directed against type-II collagen (cII) *(14–16)*. The K/BxN system does have the important advantage, however, of not requiring the complementary injection of LPS needed for activity of the anti-cII mAbs. Thus, effects of genetic mutations or pharmacological intervention on the effector phase can be evaluated directly, without interference from effects on the TLR signaling pathway.

A current working model of the pathogenesis of arthritis in this system is shown in **Fig. 1**.

In KRN transgenic mice also carrying the H-2^{g7} haplotype, arthritis appears very stereotypically around 30 ± 5 d of age. In some backgrounds or genetic variants where the disease is particularly aggressive, the first signs can be observed between 22 and 25 d of age; conversely, in genetic backgrounds that are less susceptible, the first signs may only be visible between 35 and 45 d of age. Arthritis in the K/BxN model is predominantly distal: ankles, wrists, associated tendon sheaths, and digits are always affected, the knees less so, whereas

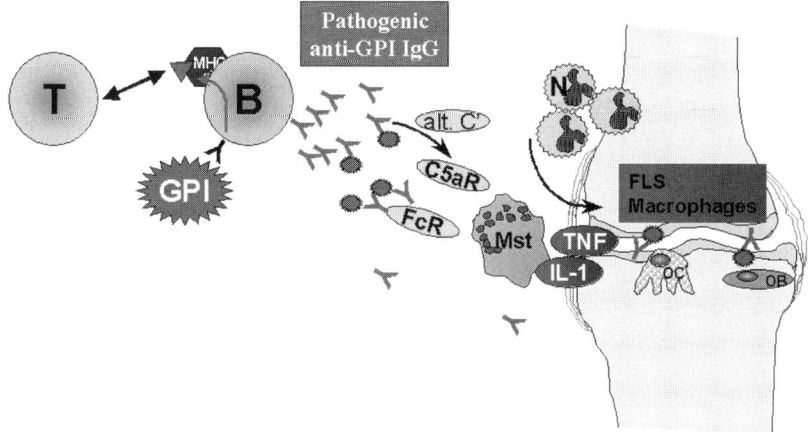

Fig. 1. Model of arthritis induction by anti-GPI antibodies. T-cell tolerance is broken, and T-cells reactive against a peptide derived from GPI, presented by the A^{g7} MHC molecule, provide potent help for those B-cells whose GPI-binding receptor make them particularly good antigen-presenting cells. These B-cells convert to plasma cells and produce high titers of high-affinity anti-GPI antibodies of the IgG1 isotype. These antibodies form immune complexes in the circulation and, more importantly, in the joint tissues, activating mast cells and complement (via the alternative pathway), leading to the influx of neutrophils. A chronic inflammatory state then arises, similar to that seen in human rheumatoid arthritis, that features marked expansion of macrophages and fibroblast-like synoviocytes (FLS) in the synovial tissue, production of TNF and IL-1, ongoing neutrophil recruitment into synovial fluid, erosion of cartilage, and erosion of bone via activation of osteoclasts (OC). Arthritis in the K/BxN model also features prominent growth of new bone around inflammatory lesions, mediated in part by osteoblasts (OB).

hips and shoulders normally show no involvement. Histological lesions of the spine are sporadic and limited to one or two levels. In addition, there is a degree of stochastic variability in the topography of the joints affected in any individual mouse, particularly when disease is induced by limited amounts of serum.

2. Breeding of KRN T Cell Receptor Transgenic Mice and K/BxN Arthritic Mice

The overall operation of a colony of K/BxN model involves two sets of breedings:

1. A core line in which the KRN transgene is maintained on an inbred background where the MHC haplotype is not $H2^{g7}$, and the mice are thus free of arthritis. Although fairly healthy, these mice are somewhat immunocompromised, because

the TCR transgene does limit their T-cell repertoire. Slightly delayed growth is not unusual. On such backgrounds (we have used either C57Bl/6 or B10.BR), animals are reasonably fertile. It is possible to generate animals carrying two copies of the transgene, indicating that the KRN transgene does not induce a homozygous lethal mutation, as do a number of transgenes. On the other hand, we found these animals to have significantly reduced breeding performance (Hergueux, J., unpublished work); thus, and although it would appear theoretically advantageous to use such homozygous animals in breedings that would produce 100% transgenic offspring, we have found that the colonies perform better in the long term when involving only heterozygous KRN mice, even at the cost of having to genotype breeders and of obtaining only 50% arthritic offspring.

2. To generate cohorts of arthritic animals, KRN TCR transgenic mice are crossed to mice carrying the stimulating MHC haplotype. We typically use NOD/Lt mice for this purpose. NOD females are good breeders with large litters (often >10 offspring), and the hybrid vigor of the BxNOD F1 combination is favorable to the offspring's health. In practice, a proven KRN male is rotated weekly between 2 or 3 cages, with 2 NOD females in each. These KRN males can be bred productively from 7 wk of age onwards, and usually remain fertile until 10 mo of age. The size of such a colony obviously depends on the experimental requirements for serum or arthritic experimental groups, but should probably involve 6 or more such rotated breeders (1 male replaced every month) for regular generation of mice and arthritogenic serum. NOD females do succumb to diabetes between 12 and 25 wk of age, and must be replaced when signs of diabetes are present (i.e., soiled fur, wet litter, and early weight loss). We have recently begun to use NOD. $E\alpha 16$ mice *(17)* in such breedings: the $E\alpha 16$ transgene encodes an MHC-II molecule that protects NOD mice from diabetes, but does not seem to affect the parameters of arthritis or anti-GPI titers in the progeny (Tran, V., unpublished work). It is also possible to breed KRN$^+$ females to NOD males, but litter sizes are smaller in this combination.

Over the course of 1 yr, it was noticed that the K/BxN serum produced in different animal facilities can have somewhat different titers. In particular, serum produced from one K/BxN colony housed in conditions of high health safety had a mean titer 5-fold lower than that produced in another colony at the same time (Morse, J., and Hattori, K., unpublished work). This difference was not the result of genetic drift, nor could it be attributed to any clear bacteriological agent. This has been a very rare occurrence, however, and most serum production colonies operated by various investigators generate K/BxN serum of high titer and arthritogenic activity.

3. Genotyping

The KRN transgene is integrated within 5 cM of the *Tyr* locus on Chr 7, from the original B6xSJL background on which the KRN transgene was generated *(1)*, most likely on the chromosome from the SJL parent which carries the *agouti* allele at the *Tyr* locus. We have strived to keep the loci together over long periods: it allows a preselection of the animals to genotype, as most agouti-gray animals type positive (but not all, as some recombinants do appear, and coat color should *not* be used alone to identify KRN$^+$ mice). Genotyping can be done by flow cytometric analysis or by polymerase chain reaction (PCR).

For flow cytometry, KRN$^+$ mice are detected by staining blood cells for CD4 and Vβ6; in KRN$^+$ mice, most CD4$^+$ cells are Vβ6$^+$, compared with only about 10% of CD4$^+$ cells in KRN$^-$ mice. In practice, one recovers one or two drops of blood, from the lateral tail vein of an immobilized mouse, into a 1.5-mL tube containing 45 µL FACS wash buffer (PBS + 30 mM HEPES [pH 7.4], 3% heat-inactivated horse serum, and 0.1% sodium azide) and 5 µL heparin (from 5000 U/mL stock). Mix, then store on ice. Add 1 mL red blood cell (RBC) lysis buffer (10 µM HEPES [pH 7.3], 0.15M NaCl, and 0.1 mM ethylene diamine tetraacetic acid [EDTA]), mix well. Leave at room temperature for 4 to 10 min. Centrifuge and discard supernatant. Repeat if the pellet appears red. Resuspend cells in FACS wash buffer containing diluted antibodies: 1:50 fluorescein isothiocyanate (FITC)-conjugated anti-mouse Vβ6 TCR (Pharmingen; cat. no. 01364C) and 1:25 PE-conjugated anti-mouse CD4 (Caltag; cat. no, RM2504-3). Leave on ice for 30 min and then wash once with FACS wash buffer. Resuspend in 100 to 200 µL of 1% paraformaldehyde in phosphate buffered saline (PBS) for flow cytometry.

For PCR typing, 1 µL of DNA (derived from a 100 µL proteinase K digestion of a 2-mm piece of tail: 50 µg/mL proteinase K added fresh from 400X stock to buffer of 10 mM Tris [pH 8.3], 50 mM KCl, 0.01 mg/mL gelatin, 0.045% NP-40, and 0.045% Tween-20, incubated at 50°C overnight) is amplified in a 25 µL reaction, using 0.5 U Taq polymerase, 3 mM Mg, 1M betaine, and otherwise standard conditions for Taq amplification. One of two sets of oligos can be used: TCR α primer sequences giving a 146-bp product (forward 5'-AGG TCC ACA GCT CCT TCT GA-3'; reverse 5'-GTA TTG GAA GGG GCC AGA G-3'), or TCR β primers giving a 227-bp product (forward 5'-GGG CAA AAA CTG ACC TTG AA-3'; reverse 5'-GAG CCT GGT TGT TTG TGG AT-3'). We typically use only the β primers. PCR conditions are the same with either primer set: 95°C for 5 min; then 30 cycles of 95°C for 1 min, 60°C for 1 min, 72°C for 1 min; and finally 72°C for 5 min. Products are detected using 1% agarose gels impregnated with ethidium bromide.

4. Clinical Evaluation of Arthritis

Ankle thickness is measured with a caliper (this laboratory uses the Kafer dial thickness gauge with flat anvils) (Long Island Indicator; cat. no. 21-790-1) placed across the ankle joint at the widest point (the malleoli) (*see* **Fig. 2A**). Measuring should be done quickly. Once both sides of the caliper touch the

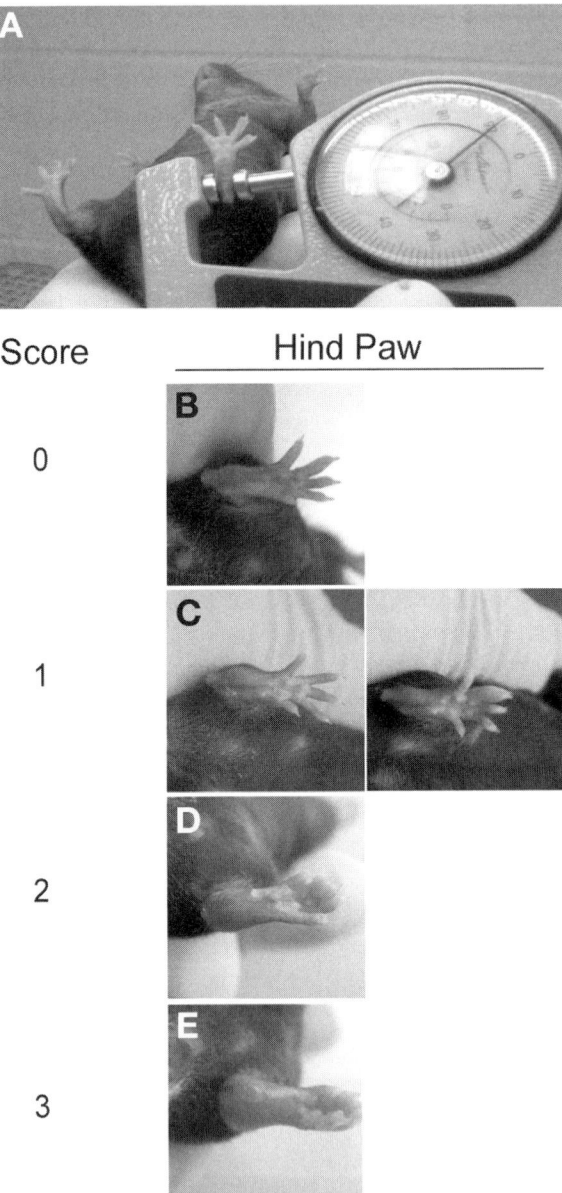

Fig. 2. Evaluation of arthritis severity. (**A**) Demonstration of technique for measuring ankle thickness using a precision caliper. (**B–E**) Examples of hind paws assessed as having clinical severity (**B**) 0 (nonarthritic), (**C**)1 , (**D**)2 , or (**E**) 3 (maximum). Note that clinical severity of 1 can represent either mild swelling of the ankle and surrounding tendon sheaths, or focal swelling of one or more digits.

ankle, the meter should be read immediately. The swelling of ankles is caused by inflammatory fluid and soft tissue. Failure to read the meter immediately results in squeezing of this pliable tissue, therefore under-estimating the ankle thickness. When uncertain about the first measurement, let the mice rest for 5 min before taking the measurement again so that the swelling returns to the original size. With these precautions, the measurement of ankle swelling is a surprisingly sensitive indicator: in the hands of an experienced investigator, an increase of 0.1 mm from one day to the next is an unmistakable sign of incipient arthritis. For optimal precision, it is preferable to measure both hind limbs. For the serum transfer model, the ankle thickness is measured before serum injection, daily or at least every other day during the first week, and every other day afterward until the disease severity plateaus or starts to dwindle (around 12–15 d). Because of variation of ankle width even among unmanipulated animals of similar age, it is suggested to use ankle thickening (the difference between ankle thickness at day 0 and at later time points) instead of ankle thickness to analyze data. For the spontaneous arthritis that develops in the transgenic mice, measurements are first taken at 22 d of age (i.e., before the onset in essentially all cases), then every 2 to 3 d thereafter, typically until 40 d of age. To obtain a "clinical index," each limb is also scored on a scale of 0–3, where "0 = indicates no observable swelling; 1 = indicates one involved digit or mild swelling of the foot and ankle, but where the foot maintains its original V shape; 2 = indicates that the long edges of foot are parallel with each other, with disappearance of the normal V shape; and 3 = indicates inversion of the V shape by expansion of the ankle and hindfoot to greater than the width of the forefoot (*see* **Fig. 2**). The four limb scores are added together to give the clinical index (maximum 12 points).

5. Induction of Arthritis by Serum Transfer

The induction of arthritis by transfer of arthritogenic immunoglobulins provides a robust tool to analyze the impact of genetic variations or therapeutic interventions on the disease process. With serum transfer, the onset of arthritis is very robust and follows a reproducible course in genetically identical animals, although there is some inter-individual variability. We have reported *(4)* that arthritis can be induced by transfer of a cocktail of monoclonal antibodies (mAbs), which in theory would provide a good replacement for serum from arthritic K/BxN mice. Unfortunately, the disease induced by mAbs did not prove, in the long run, to be as robust as that induced by serum transfer; in addition, we have prepared some large batches of anti-GPI mAbs which were very poor at inducing arthritis, even though they had respectable anti-GPI titers. The cause of these problems (e.g., contamination from tissue culture, and possible drift in the post-translational modifications of the hybridomas over time) has not been sorted out, but we have not adopted monoclonal transfer in routine use, and do not recommend it.

Arthritis is induced by intraperitoneal injection of serum from 8-wk-old K/BxN mice on days 0 and 2, 150 µL/injection (intravenous [iv] injection can also be used, particularly when wishing to analyze very early aspects of the response, but intraperitoneal [ip] injection is perfectly adequate in general practice). It is advised to use at least 5-wk-old males or 6-wk-old females as the recipients. Complement is critical for this model, and mice younger than these ages usually are relatively deficient in serum complement factors, which results in decreased incidence and severity of arthritis. For B6 mice, the first signs of arthritis and ankle thickening are detectable after 2 to 3 d, and maximum severity of arthritis usually occurs between days 8 and 10 after serum transfer. Male mice tend to reach the plateau of severity faster, but both males and females reach the same level of ankle thickening. There are different sensitivities to serum transferred arthritis among various inbred strains of mice, which translate as different times of onset and/or extent of joint swelling. The sensitivities of some common strains are listed in **Table 1**. At the end of a serum transfer experiment, it is advised to collect serum from each individual mouse and examine its serum anti-GPI titer, particularly in the event that some mice do not develop severe arthritis. This allows one to confirm that all mice received accurate ip injection of serum (misinjection into the bladder can occur), and also can be helpful in ensuring that the mechanism for arthritis-resistance in a given strain is not caused by a shortened half-life of the anti-GPI antibodies.

6. Crossing KRN and H-2^{g7} to Other Genetic Nackgrounds

The spontaneous K/BxN transgenic model can be used to study the effects of other genes on the immunological or effector phases of disease. This is usually done by crossing the transgene to mice deficient for a particular gene X of interest ($X^{-/-}$). Because the KRN TCR transgene and the MHC class II H-2^{g7} allele are both required for arthritis, and because breeding performance of arthritic animals is quite poor, the breeding strategy we have used is to breed a gene-deficient line independently with KRN TCR transgenic mice (on the H-2^b B6 background) and with B6 H-2^{g7} congenic mice. This first cross generates KRN$^+$X$^{+/-}$ and H-$2^{g7/b}$X$^{+/-}$ animals, respectively. Intercrossing these mice can be performed at this stage, but with only a very low yield of the desired mice (1/16), so we usually resort to an additional round of "gene enrichment" to produce more favorable breeders. Typically, and depending on the viability/fertility of $X^{-/-}$ mice, H-$2^{g7/b}$X$^{+/-}$ mice are further intercrossed to generate H-$2^{g7/g7}$X$^{+/-}$ or H-$2^{g7/g7}$X$^{-/-}$ mice, and/or KRN$^+$X$^{+/-}$ mice are intercrossed to generate KRN$^+$X$^{-/-}$ animals (this second round of "preliminary crossing" thus

Table 1
K/BxN Serum-Ttransferred Arthritis in Diverse Inbred Strains

Strain	Max. clinical index	Day of onset[a]	Max. ankle thickening (mm)	No. of mice examined	Histological evaluation	
					Inflammation	Cartilage destruction
Balb/c	4.0	1.2 (1–2)	1.5	11	(++)	(++)
C57Bl/10	3.9	1.3 (1–3)	1.4	4	(++)	(+)
PL	3.8	2.0 (1–4)	1.4	8	(++)	(++)
C57Bl/6	3.5	2.9 (1–6)	1.5	19	(++)	(+)
DBA/1	3.5	2.8 (2–5)	1.4	4	(+)	(+)
CBA	3.5	3.5 (2–5)	1.1	4	(++)	(+)
MRL/Mp	2.9	3.8 (2–4)	0.9	4	(++)	(+++)
NZW	2.9	4.8 (1–14)	0.7	4	(++)	(+++)
C3H/He	2.2	6.4 (3–12)	1	5	(+/–)	(+/–)
SJL	2.2	12.4 (2–28)	0.6	12	(+)	(+/–)
129/Sv	1.6	3.5 (3–4)	0.3	4	(+/–)	(+/–)
DBA/2	ND	ND	0	4	(–)	(–)
FvB/N	ND	ND	0	4	(–)	(–)
NZB	ND	ND	0	4	(–)	(–)
NOD	ND	ND	0	4	(–)	(–)

[a] Mean (range).

Note: Different strains of mice were transferred with 7.5 µL/g body weight of serum from K/BxN mice on days 0 and 2. Two mice were killed at day 7 for histology, and the others followed until days 12–28. The "day of onset" is the first day of overt disease on one limb. Reprinted with permission from **ref. 12**.

enriches the frequency of the $H\text{-}2^{g7}$ and X^- chromosomes in the final crosses). Finally, $KRN^+X^{+/-}$ mice are crossed to $H\text{-}2^{g7/g7}X^{+/-}$ (or $H\text{-}2^{g7/g7}X^{-/-}$) to generate experimental mice ($KRN^+H\text{-}2^{g7/b}X^{-/-}$) and control littermates ($KRN^+H\text{-}2^{g7/b}X^{+/+}$ or $KRN^+H\text{-}2^{g7/b}X^{+/-}$). We have also used a similar strategy for crossing to mouse strains deficient for two genes. Screening for the KRN transgene and MHC haplotype in these breeding schemes is most readily done by flow cytometry of peripheral blood. KRN^+ mice are detected as above. $H\text{-}2^{g7/b}$ and $H\text{-}2^{g7/g7}$ can be distinguished by staining with monoclonal antibodies that distinguish A^{g7} from A^b (typically mAbs 10.2.16 and Y-3P, respectively).

7. Analysis of KRN Transgenic T-Cells

In order to evaluate T-cell tolerance in this model, key markers can be examined in thymocytes. Perturbations in thymocyte negative selection or positive selection can be manifested as changes in the subpopulations delineated by CD4 and CD8 expression. Selection of T-cells bearing the KRN TCR can be evaluated by the expression of the transgene-encoded TCR-β chain (Vβ6) vs other variable regions from endogenous TCR loci (e.g., Vβ8), in the CD4 single positive T-cell subset. In addition, thymocyte development can be evaluated based on the expression of the maturation markers CD44 and CD5 in various populations: $CD4^-CD8^-$ double negative (DN), $CD4^+CD8^+$ double positive (DP), $CD4^+CD8^-$ single positive (SP) and $CD4^-CD8^+$ SP populations. Combinations of labeled antibodies that we have found to be effective for such analysis, using a Cytomation MoFlo instrument, include: (1) CD8α-FITC, Vβ6-PE, Vβ8-Bio/SA-PE-Cy7, CD3-APC, and CD4-PE-TR; and (2) CD8α-FITC, CD4-PE, and CD5(or CD44)-APC. Some of the populations in K/BxN mice are not as well defined as they are in non-TCR transgenic mice. Thus, we find it useful to stain cells from non-TCR transgenic mice in parallel with cells from experimental mice, and to set some of the gating (e.g., $Vβ6^+$, $Vβ8^+$, $CD44^{high}$) based on the populations in non-TCR transgenic mice. Thymi from K/BxN mice display an altered CD4/CD8 profile compared with non-TCR transgenic mice, the most prominent differences being a decrease in the percentage of CD4 SP thymocytes and an increase in the percentage of DN thymocytes. Thymocyte subsets in K/BxN mice are approx 20% DN, 75% DP, 3% CD4 SP, and 2% CD8 SP. In the CD4 SP subset, the majority (approx 60–70%) of cells express the transgenic Vβ6 TCR, and notably, several percent of cells express both transgenic and endogenous TCRs ($Vβ6^{int} Vβ8^+$). Levels of Vβ6 and CD3 expression are very tightly correlated. In nontransgenic mice, levels of CD5 increase as thymocyte differentiation progresses to the SP stage. In contrast, CD5 expression is already high on DN thymocytes in K/BxN mice, perhaps resulting from early expression of and signaling through the transgenic TCR. In order to evaluate T-cell activation in peripheral lymphoid organs, we examine spleen and lymph node suspensions for expansion of CD4 T-cells, expression of transgenic and endogenous TCR chains, and expression of CD44, CD62L, CD25, and CD69. We also evaluate the regulatory T-cell subset as defined by $CD4^+$ $CD69^-$ $CD25^+$ T-cells. Effective combinations of labeled antibodies include (1) CD4-PE-TR, Vβ6-PE, B220-Bio/SA-PE-Cy7, CD69-FITC, and CD25-APC; and (2) CD4-PE-TR, Vβ6-PE, B220-Bio/SA-PE-Cy7, CD62L-FITC, and CD44-APC. For the markers we have examined, similar results have been obtained from spleen and lymph node cells. In addition, we have not detected significant differences between cells from lymph nodes draining inflamed joints (e.g., axillary) vs nondraining lymph nodes (e.g., cervical). The distribution of Vβ6, Vβ8, and CD3 on $CD4^+$ T-cells in the periphery is similar to that in CD4 SP thymocytes. One difference is that the percentage of T-cells expressing both transgenic and endogenous TCRs is slightly higher in the periphery than in the thymus. $CD4^+$ and $CD8^+$ populations remain at fairly low proportions in the periphery (approx 3-7% CD4$^+$, 1–2% CD8$^+$). CD4$^+$ T-cells display clear evidence of activation, with the early activation marker CD69 upregulated on approx 20% of cells, and the activated/memory cell marker CD44 upregulated on approx 60% of cells. Generally, characteristics of the peripheral T-cell compartment are consistent between 2 and 10 wk of age, with some populations expanding over time (e.g., CD4$^+$ T-cells, dual TCR expressors, and CD44^{high} cells). In both thymus and periphery, the frequency of CD4$^+$ cells in adult K/BxN mice is lower than what is usually found with mice expressing MHC class II-restricted TCR transgenes. Before thymic tolerance of KRN T-cells is broken between 2 and 3 wk of age in K/BxN mice, the number of CD4$^+$ cells is very low. GPI-specific T-cells can also be followed by staining with an MHC tetramer (A^{g7} MHC molecule presenting a GPI-derived peptide) *(18)*.

8. Screening K/BxN Serum for Anti-GPI Antibodies

Levels of anti-GPI antibodies in individual or pooled K/BxN sera can be measured by enzyme linked immunosorbent assay (ELISA). Microtiter plates are coated with the recombinant fusion protein mouse GPI-GST (*see* below), 5 µg/mL in phosphate buffered saline (PBS), at 4°C overnight or longer. Plates are then blocked with 1% bovine serum albumin (BSA) in PBS (with 2 changes to effect washing as well) for 30 min at room temperature, then with serum dilutions in PBS for 1 h. The linear range of anti-GPI activity in K/BxN serum is typically between dilutions of 1:3,000 and 1:1,000,000. After incubation with serum, plates are washed 3 times with PBS, then incubated with alkaline phosphatase-conjugated F(ab')$_2$ fragments of goat-anti-mouse IgG (Jackson ImmunoResearch), typically diluted 1:2000 in PBS, again for 1 h at room temperature. After 3 more washes with PBS, phosphatase substrate is added. We use substrate pellets (Sigma), dissolved in 8% diethanolamine, 2.4 mM MgCl2 (pH 9.8); plates are read at 405 nm with the linear range typically apparent after 5 to 10 min. This ELISA assay can also be adapted to confirm transfer of anti-GPI antibodies, or measure their half-life, after transfer of K/BxN serum into other mice. The starting serum dilution should then be lower (i.e., 1:100 or 1:300). To produce recombinant GPI, mouse GPI cDNA was cloned into the pGEX-4T-3 vector (Amersham/Pharmacia), which creates an N-terminal GST fusion protein. The manufacturer's standard protocol for induction and one-step purification of the fusion protein (Amersham/Pharmacia) has produced good purity, excellent yields, and no significant problems with solubility. Briefly, a 10 mL overnight culture is diluted into 1 L of 2X YT medium and grown at 20 to 30°C until the A$_{600}$ is between 0.5 and 2. Then, IPTG is added to a final concentration of 0.1 mM, and incubation is continued for an additional 4 h. The culture is centrifuged, the supernatant removed, and the pellet of cells placed on ice, then resuspended in 50 mL cold PBS. This suspension is sonicated in short bursts until partial clearing is apparent, then Triton X-100 added to a final concentration of 1%, and the suspension incubated on ice for 30 min. Then, the suspension is centrifuged at 12,000g for 10 min and the supernatant applied to 2 mL of a 50% slurry of glutathione sepharose 4B (Amersham/ Pharmacia) preequilibrated with PBS. The lysate/bead mixture is incubated for 30 min at room temperature with rotation, then the beads recovered and washed 3 times with PBS (10 mL/wash) by low-speed centrifugation (500g, 5 min). After the third wash, 1 mL of elution buffer (0.154 g reduced glutathione/ 50 mL of 50 mM Tris-HCl [pH 8.0]) is added to the 1 mL of packed beads, incubated for 10 min at room temperature, and the supernatant recovered after low-speed centrifugation. Elution is repeated twice and the three elutions pooled, dialyzed against PBS, and stored at $-20°C$ after determination of concentration by ultraviolet (UV) absorption at 280 nm (assuming approximate A_{280} of 1.4 for a 1 mg/mL solution).

9. Histology of Ankles

Ankle joints can be prepared for either paraffin or frozen sectioning. In both cases, we typically remove the skin, then cut above the toes and 5 to 10 mm above the ankle joint in order to isolate the ankle and midfoot joints. For paraffin sectioning, after fixation in 4% paraformaldehyde/PBS overnight at $4°C$, the bones must be decalcified. This can be done rapidly with formic acid or nitric acid, over 48 h with Kristensen's solution (a mixture of 20% 8N formic acid and 80% 1N Sodium Formate), or over 2 wk with 0.375M EDTA (pH 7.5) at $4°C$ (4 mL/ankle, changed 2 times/wk). It is wise to use the latter, gentler method if there is any chance of using the tissue for immunohistochemistry to detect targets of unknown stability in acid. Decalcified ankles are then processed for paraffin sectioning and subsequent staining by standard protocols. Frozen sections may be required to detect certain targets by immunostaining (in our experience, C3 and GPI have not been detectable on paraformaldehyde-fixed paraffin sections, but IgG is readily detected). Sectioning of undecalcified bone requires special techniques to preserve tissue structure. Starting with ankles frozen in standard OCT medium, we use commercially-available (Instrumedics) equipment and reagents to obtain sections on pieces of tape, then transfer the sections to slides ("4x" from the Instrumedics line) precoated with a UV-activatable adhesive *(5,7)*. Immunofluorescence staining of frozen sections is then done by a standard protocol: fixation for 5 min in cold acetone, rehydration with PBS, blocking with PBS containing 4% BSA and 0.1% Tween-20, washing and dilution of primary and secondary antibodies in PBS containing 0.5% BSA and 0.1% Tween 20.

References

1. Kouskoff V., Korganow, A. -S., Duchatelle, V., Degott, C., Benoist, C., and Mathis, D. (1996) Organ-specific disease provoked by systemic autoimmunity. *Cell* **87,** 811–822.
2. Matsumoto, I., Staub, A., Benoist, C., and Mathis, D. (1999) Arthritis provoked by linked T and B cell recognition of a glycolytic enzyme. *Science* **286,**1732–1735.
3. Korganow, A. S., Ji, H., Mangialaio, S., et al. (1999) From systemic T cell self-reactivity to organ-specific autoimmune disease via immunoglobulins. *Immunity* **10,** 451–461.
4. Maccioni, M., Zeder-Lutz, G., Huang, H., et al. (2002) Arthritogenic monoclonal antibodies from K/BxN mice. *J. Exp. Med.* **195,** 1071–1077.
5. Lee, D. M., Friend, D. S., Gurish, M. F., Benoist, C., Mathis, D., and Brenner, M. B. (2002). Mast cells: a cellular link between autoantibodies and inflammatory arthritis. *Science* **297,** 1689–1692.
6. Wipke, B.T. and Allen, P.M. (2001) Essential role of neutrophils in the initiation and progression of a murine model of rheumatoid arthritis. *J. Immunol.* **167,** 1601–1608.
7. Ji, H., Ohmura, K., Mahmood, U., et al (2002) Arthritis critically dependent on innate immune system players. *Immunity* **16,** 157–168.
8. Corr, M. and Crain, B. (2002) The role of FcγR signaling in the K/B x N serum transfer model of arthritis. *J. Immunol.* **169,** 6604–6609.
9. Ji, H., Pettit, A., Ohmura, K., et al. (2002) Critical roles for interleukin 1 and tumor necrosis factor alpha in antibody-induced arthritis. *J. Exp. Med.* **196,** 77–85.
10. Watts, G. M., Beurskens, F. J., Martin-Padura, I., et al (2005). Manifestations of inflammatory arthritis are critically dependent on LFA-1. *J. Immunol.* **174,** 3668–3675.
11. Akilesh, S., Petkova, S., Sproule, T. J., Shaffer, D. J., Christianson, G. J., and Roopenian, D. (2004) The MHC class I-like Fc receptor promotes humorally mediated autoimmune disease. *J. Clin. Invest.* **113,** 1328–1333.
12. Ji, H., Gauguier, D., Ohmura, K., et al.. (2001) Genetic influences on the end-stage effector phase of arthritis. *J. Exp. Med.* **194,** 321–330.
13. Ohmura, K., Johnsen, A., Ortiz-Lopez, A., et al. (2005) Variation in IL-1β gene expression is a major determinant of genetic differences in arthritis aggressivity in mice. *Proc. Natl. Acad. Sci. USA* **102,** 12,489–12,494.
14. Watson, W. C., Brown, P. S., Pitcock, J. A., and Townes, A. S. (1987) Passive transfer studies with type II collagen antibody in B10.D2/old and new line and C57Bl/6 normal and beige (Chediak-Higashi) strains: evidence of important roles for C5 and multiple inflammatory cell types in the development of erosive arthritis. *Arthritis Rheum.* **30,** 460–465.
15. Kagari, T., Doi, H., and Shimozato, T. (2002) The importance of IL-1 beta and TNF-alpha, and the noninvolvement of IL-6, in the development of monoclonal antibody-induced arthritis. *J. Immunol.* **169,** 1459–1466.
16. Kagari,T., Tanaka, D., Doi, H., and Shimozato, T. (2003) Essential role of Fc gamma receptors in anti-type II collagen antibody-induced arthritis. *J. Immunol.* **170,** 4318–4324.

17. Le Meur, M., Gerlinger, P., Benoist, C., and Mathis,D. (1985) Correcting an immune-response deficiency by creating E alpha gene transgenic mice. *Nature* **316,** 38–42.
18. Stratmann, T., Martin-Orozco, N., Mallet-Designe, V., et al. (2003) Susceptible MHC alleles, not background genes, select an autoimmune T cell reactivity. *J. Clin. Invest.* **112,** 902–914.

21

Analysis of Arthritic Lesions in the Del1 Mouse
A Model for Osteoarthritis

Anna-Marja Säämänen, Mika Hyttinen, and Eero Vuorio

Abstract

Osteoarthritis (OA) is characterised by progressive erosion of articular cartilage with a number of associated degenerative processes within the joint. Animal models of OA provide the only feasible way to systematically study the development and progression of OA, in order to understand the molecular events, and to develop tools for prevention and therapy of OA. Gene manipulation techniques have provided opportunities to generate transgenic mouse models for OA. In heterozygous Del1 mice, incorporation of *Col2a1* transgenes with a short deletion mutation results in production of shortened proα1(II) collagen chains and a phenotype resembling human OA. This chapter describes techniques and practical aspects of preparation and processing of skeletal samples for radiological, histological, and molecular biologic analyses that have been used to monitor the development of knee OA in Del1 mice. A simple histological grading system to evaluate the progression of OA lesions, and examples of other degenerative alterations in the knee joint structures are presented. Semiquantitative microscopic techniques are described for the analysis of proteoglycan distribution based on safranin O staining of glycosaminoglycans, and for the analysis of collagen matrix based on birefringence of polarized light. Reference is also made to an experimental setup for correlating voluntary running activity of mice with OA score.

Key Words: Transgenic mouse; osteoarthritis; cartilage; bone; type II collagen; proteoglycan; mRNA; safranin O; polarized light microscopy; joint loading; knee joint degeneration; animal model.

1. Introduction

Systematic studies on human osteoarthritis (OA) are virtually impossible because of the slow and inconspicuous nature of the disease (e.g., disease progression, a poor healing capacity of articular cartilage, and the unpredictability of when and where the first changes in articular cartilage will take place).

However, an ideal animal model for OA should repeat as many of the features of the human disease as possible (including slow progression). Analysis of animal models for OA is met with similar challenges. In practice this means that frequent monitoring of the model animals is needed to determine the time points that will provide the most interesting and relevant information of the different stages of articular cartilage degeneration and responsiveness.

This chapter reviews the approaches we have used to monitor the development of OA changes in the knee joints of Del1 mice. Mice heterozygous for the Del1 locus harbor type II collagen transgenes which produce shortened $pro\alpha 1(II)$ collagen chains capable of interfering with production of normal cartilage collagen fibrils thereby making articular cartilage less resistant to normal wear *(1–6)*. Production and characterization of Del1 mice has been described in an earlier volume of this series of *Cartilage and Osteoarthritis (7)*. Although some of the basic techniques used for monitoring OA changes are repeated in the present chapter, the aim of this chapter is to extend the description of the analytical tools available.

Although X-ray techniques are available for in vivo monitoring of the development of radiologic changes in the mouse knee joint, their use is often restricted by practical issues. Transgenic mice are mostly maintained in pathogen free facilities, where X-ray facilities are rarely available. In most cases, anesthesized mice must be transported to an outside X-ray facility, after which they must be housed in a nonpathogen free facility, which may affect the outcome of long-term experiments. A second weakness of radiologic monitoring of mouse OA, as with human disease, is the difficulty of recognizing the earliest lesions of articular cartilage by X-ray. With these considerations in mind, we have chosen to use systematic histologic grading of mouse knee joints in order to identify the first stages of articular cartilage defects. This naturally requires a large colony of experimental animals. Therefore, the approach is feasible for mice but less so for larger animals. An added advantage of this approach is the accumulation of a large number of serial sections through the knee joints for subsequent analyses based on molecular biologic and other data. We have also used radiographic analyses, but only on isolated hind limbs, which can be easily aligned for radiographs in an outside X-ray facility. This makes it possible to correlate X-ray data with knee joint histology and subchondral bone structure, for which a semiquantitative microscopic technique based on birefringence of polarized light is also described. Furthermore, a semiquantitative microscopic technique using safranin O-stained sections is described for the analysis of proteoglycan content and distribution in articular cartilage.

Finally, a technique used to monitor the effect of voluntary running activity of mice on the development of OA is described. As mice voluntarily run 4 to 6 km every night, it is not surprising that this loading affects the integrity of articular cartilage. Again, the small size of the mouse makes it realistic to house even one hundred mice in individual cages for 1 yr and thereby obtain data on their individual running behaviour, and to correlate these to their genotype, OA score and X-ray findings as well as other analytical data.

2. Materials

2.1. Radiographic Analysis

1. Diagnostic film designed for mammography imaging (Min-RL) (Kodak,).
2. Faxitron X-ray cabinet system (Faxitron X-ray Corp.).

2.2. Microscopic Analysis

1. Phosphate buffered saline (PBS): To prepare 10X PBS stock solution weigh 80 g of NaCl, 21.6 g $Na_2HPO_4 \cdot 7 H_2O$, 2.0 g KH_2PO_4, and 2.0 g KCl. Fill to 1000 mL with dH_2O and autoclave.
2. Fixation solution: freshly prepared 4 % paraformaldehyde in 1X PBS. To make 100 mL heat 90 mL of distilled water to 60°C on a heating platform in a hood, add 4 g of paraformaldehyde and a drop of $2N$ NaOH (*see* **Note 1**). Close the container with aluminium foil. Turn off the heating, but keep the container still on the heating platform with constant stirring to dissolve the paraformaldehyde. Do not allow the temperature to rise above 65°C! Add 10 mL of 10X PBS. Cool the paraformaldehyde solution on ice and then filter through filter paper.
3. Biopsy cassettes: Uni-Cassette 4088 (Tissue-Tek, Miles).
4. Decalcification solution: $0.1M$ phosphate buffer (pH 7.0), 10 % Na_2 ethylene diamine tetraacetic acid (EDTA) (*see* **Note 2**).

Prepare stock solutions: $1M$ NaH_2PO_4, $0.5M$ Na_2HPO_4, and 20 % Na_2EDTA. To dissolve Na_2EDTA, adjust the pH to between 7.4 and 7.6 by adding solid NaOH with constant stirring. Autoclave all stock solutions for longer storage. Store at room temperature. To prepare decalcification solution, add 30.5 mL of $1M$ NaH_2PO_4, 218.5 mL of $0.5M$ Na_2HPO_4, 500 mL of 20 % Na_2EDTA, and fill to 1000 mL with dH_2O. Autoclave for longer storage.

5. Dehydration solutions: 40, 50, and 70% ethanol, (technical grade), and xylene.
6. Paraffin: Histowax, melting point 56 to 58°C (Histolab Products Ab).
7. Dehydration automat (Pathcenter).
8. Embedding unit and stainless steel molds soaked in 5% glycerol in 70% ethanol.
9. Rotating microtome (Leica RM 2155) (Leica Microsystems).
10. Disposable microtome blades (e.g., Tissue-Tek Accu-Edge Disposable Microtome Blades (Sakura Finetek).
11. Cardboard boxes, size $1.0 \times 20 \times 30$ cm.
12. SuperFrost Plus Microscope Slides (Menzel Gläser, Göttingen, Germany).
13. Delafield's hematoxylin: Mix 16 g of hematoxylin (Fluka Riedel de Haën; cat. no. 51260) and 100 mL of 94% ethanol. Dissolve hematoxylin in ethanol by heating slightly and slowly. Add 1600 mL of saturated ammonium sulphate into

solution. Leave in an open container for 3 to 5 d. Filter, and add 400 mL of glycerol (87%) (Merck; cat. no. 4094) and 400 mL of 94 % ethanol. Leave in daylight for 1 to 2 mo. Store in the dark after the solution is ready. Filter just before every use (*see* **Note 3**).
14. 1% eosin solution in ethanol: Mix 25 g of Eosin Y (BDH; cat. no. 34197) and 500 mL of dH_2O and add 2000 mL of 94 % ethanol. Just before use for staining, add 0.5 mL glacial acetic acid per 100 mL of dye solution and filter.
15. 1% HCl, 70% ethanol: Mix 1400 mL of 94% ethanol and 600 mL of dH_2O. Then pour 20 mL of strong HCl slowly into ethanol solution. Remember to do in this order to prevent excess heating of the solution.
16. 0.5 % Safranin O (Fisher Sci. Co.) in $0.1M$ NaAc (pH 4.6). Dissolve 2.5 g of safranin O in 500 mL of sodium acetate buffer, allow to dissolve overnight, filter before use. Store in dark.
17. Testicular hyaluronidase (Sigma; EC 3.2.1.35, Type IV-S from bovine testes), 1000 U/mL, prepare freshly in $0.1M$ Na-phosphate (pH 6.9).
18. DPX mounting medium (Difco).
19. Cover slips (Knittel-Gläser).
20. Monochromator 492 nm (Spindler & Hoyer).
21. Neutral density filter set (Schott Glaswerke).

2.3. Isolation of Total RNA From Skeletal Tissues

1. Liquid Nitrogen.
2. Tissue pulverization unit (*see* **Note 4**).
3. Diethyl pyrocarbonate (DEPC) (Sigma) (*see* **Note 2**).
4. GITC-solution: $4M$ Guanidinium isothiocyanate (GITC), 25 mM sodium citrate (pH 7.0), 0.5% sodium dodecyl sulfate (SDS), 0.7 % β-mercaptoethanol, 0.1% antifoam A. To prepare the solution, add 50 g of GIT (Fluka BioChemica), 0.5 g of SDS, 2.5 mL of $1M$ sodium citrate (pH 7.0), 0.7 mL of β-mercaptoethanol, 0.33 mL of 30 % antifoam A (Sigma; cat. no. A-5758), and fill with dH_2O to 100 mL. Stir overnight at room temperature, filter through glass fibre paper and a 0.45 μm filter (Nalgene). Adjust pH to 7.0 with $1N$ NaOH (*see* **Note 5**).
5. $5.7M$ CsCl, 25 mM sodium citrate (pH 5.0), DEPC-treated: to prepare the solution, add 95.8 g of CsCl (Technical grade, molecular biology certified, IBI Shelton Scientific Inc.), 2.5 mL of $1M$ Sodium citrate (pH 5.0) (25 mM), 0.2 mL of DEPC and fill with dH_2O to 100 mL, leave in a hood overnight and autoclave.
6. Ultracentrifuge tubes, Polyallomer, 13 × 51 mm (Beckman Instruments Inc.), ultracentrifuge (Beckman Optima LE-80K), swing-out rotor (Beckman SW 55TI).
7. DEPC-treated dH_2O, add 0.2 mL of DEPC in 100 mL of dH_2O, leave in a hood overnight and autoclave.
8. $3M$ NaAc (pH 5.5): prepare 100 mL of buffer, add 0.2 mL of DEPC, leave in a hood overnight and autoclave.

9. SET buffer: 10 mM Tris-HCl (pH 7.5), 4 mM Na$_2$EDTA, 1% SDS. DEPC-treated: add 5 mL of 20% SDS (1 %) prepared in DEPC-treated water, 0.8 mL of 0.5M DEPC-treated Na$_2$EDTA (pH 7.5), and 1 mL of 1M DEPC-treated Tris-HCl (pH 7.5) (10 mM). Fill with dH$_2$O to 100 mL and autoclave.
10. Proteinase K, RNase-free (Promega), 20 mg/mL, dissolve in DEPC-treated dH$_2$O and store at –20°C in small aliquots.
11. RNase free microcentrifuge tubes, pipet tips and glass pasteur pipets.
12. Disposable scalpels.

3. Methods
3.1. Radiography of Hind Limbs

1. Dissect the hindlimbs free of skin. Remove the fat and most of the muscle, but leave a little to protect the bone and joint structures. Wrap the hindlimbs in a piece of paper towel moistened in PBS, and place them on ice until radiography has been performed. Then process the limbs for histology.
2. Take X-rays both in the anterior-posterior and lateral projection. This is advisable, as possible varus deformity can be observed only in anterior–posterior projection, while possible ectopic ossification (ossicles) deep to tendon can be observed in lateral projection. Fix the limbs on the plastic sheet in a proper orientation using tape at the peripheral sites, and place the sheet on a light-tight envelope containing the X-ray film. Take the radiographs with a Faxitron X-ray cabinet or equivalent X-ray device suitable for small specimen radiography.
3. To analyze the radiographs either examine them using a magnifying glass or scan the films and analyze the enlarged images using image analysis software (e.g., Adobe® PhotoShop) (Adobe Systems Inc.).

3.2. Microscopic Analyses
3.2.1. Standard Histology

1. Bend the knee at a 90° angle, and place it into a biopsy cassette with the patella of the bent knee facing to the corner of the cassette to maintain the proper angle during fixation. If you do not wish to study the histology of the tarsal or hip joints, cut the tibial and femoral bones at the level of mid-diaphysis. This will help considerably in processing of the tissue sections in later steps.
2. Fix the samples for 24 h in freshly prepared 4% paraformaldehyde in PBS on a rocking platform at 4°C.
3. Rinse in tap water: first rinse briefly, then 3 times for 10 min each on a rocking platform (*see* **Note 6**).
4. Decalcify the samples in 10% Na$_2$EDTA in PBS (pH 7.4) at room temperature on a rocking platform until they are soft. Change the solution 1 to 2 times/wk. Decalcification time depends on the age of the animal. The status can be tested with a thin needle (e.g., 22 gage) at peripheral parts of the bone, or by X-ray. The following is a rough estimate for decalcification times:

Age (mouse)	Decalcification time
5 d	1 d
10 d	2–3 d
1 mo	1–2 wk
> 3 mo	2–4 wk

When the decalcification is complete, rinse the samples thoroughly in tap water on a rocking platform, 4 times for 10 min each.

5. Dehydrate the samples in increasing concentrations of ethanol on a rocking platform at room temperature before embedding in paraffin: 40% ethanol 1 h, 50% ethanol 1 h, 70% ethanol 1 h. Tissue samples can be stored for a short period of time (a few days or weeks) in 70% ethanol at 4°C if necessary before embedding in paraffin. From this step onward, it is advisable to carry out dehydration under vacuum in an automat (Pathcenter, Shandon): 70% ethanol 1 times for 1 h, 97% ethanol 5 times for 1 h each, xylene 2 times for 50 min each, and paraffin 4 times for 60 min each at 60°C.

6. Embed the tissues in paraffin either in sagittal plane, the lateral side facing down and taking care that the longitudinal plane of the patella is horizontal, or in anterior–posterior plane, with the longitudinal plane of the tibial bone adjusted horizontally and femur pointing upwards. The embedding plane depends on what you wish to analyze: soft tissue calcification can best be analyzed in sagittal sections, while simultaneous analysis of erosions on both medial and lateral condyles, and osteophytes can best be analyzed in anterior–posterior sections.

7. Using a rotating microtome, cut 5 µm serial sections through the entire block, or at least through the tibial condyles, if you have the tissue in anterior-posterior orientation. Collect the sections into cardboard boxes and store at 4°C.

8. Pick every 20th section on a microscope slide, and dry overnight at 37°C or for 1 h at 50°C to fix the sections on glass.

9. Stain the sections with H&E according to the following steps: Xylene 2 times for 5 min each; absolute ethanol 2 times for 2 min each; 96% ethanol 2 times for 2 min each; distilled water for 30 s; Delafield's hematoxylin for 9 min; and running tap water for 5 min. For removal of excess dye: 1% HCll 70% ethanol for 4 to 5; running tap water for 10 min; 1% eosin in ethanol 30 s; 96% ethanol 2 times for 2 min each; absolute ethanol 2 times for 2 min each; and xylene 3 times for 5 min each. Finally, mount the sections with DPX and allow to dry in a hood.

10. Examine the sections by light microscopy for characterization of knee joint degeneration.

3.2.1.2. GRADING OF DEGENERATIVE LESIONS

Grading of cartilage erosion in knee joints of Del1 mice is based on the classification method modified from Wilhelmi and Faust *(2,8,9)*. Every 20th section of each tissue block (*see* **Subheading 3.2.1.**, **step 9**) is stained with H&E, examined microscopically for the articular cartilage erosion, and graded according to the depth of erosion penetrating from the articular surface. Alter-

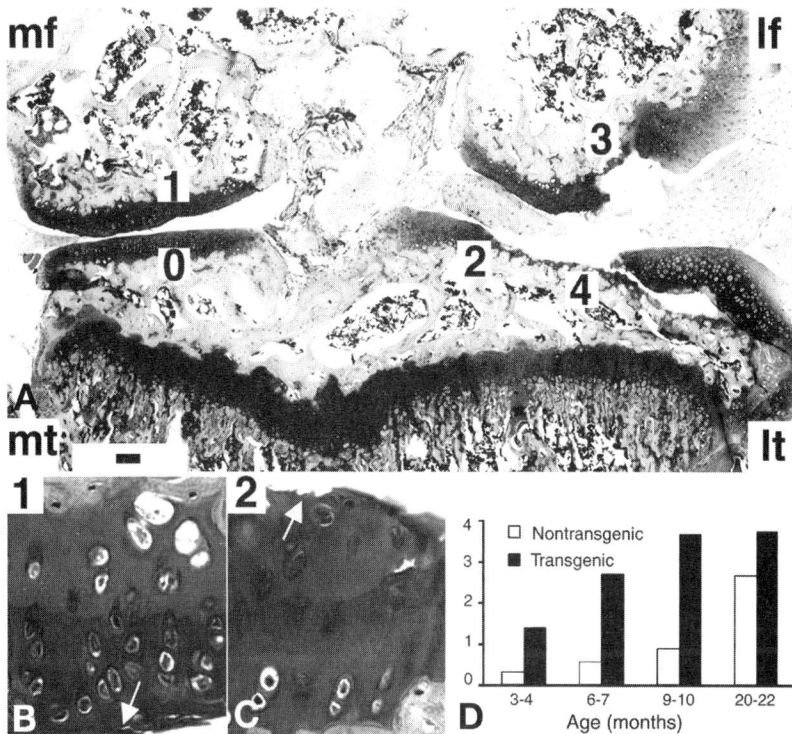

Fig. 1. Grading of osteoarthritic lesions in Del1 mice. Safranin O stained frontal section of contact area in tibiofemoral joint. (**A**) Lower row of images show higher magnifications of areas of articular cartilage with grade 1 (**B**) and grade 2 (**C**) lesions. mt, mf denote the medial tibia and femur; lf and lt are corresponding lateral compartments. Note detachment of cartilage on lt surface. OA grades: 0 = intact articular cartilage, 1 = small superficial lesions not affecting the intermediate zone, 2 = lesions extending to the intermediate zone, 3 and 4 = lesions reaching the deep cartilage and subchondral bone, respectively. Scalebar, 100 µm applies to (**A**). Cartilage erosion progresses faster in transgenic Del1 mice than in nontransgenic control mice (**D**). Columns represent the average grade values within each age group. Number of animals per group is between 3 (age 20–22 mo) and 36 (age 3–4 mo).

natively, the sections can be stained with safranin O (*see* **Subheading 3.2.2.1.**) (*see* **Fig. 1**). If more than one grade is observed within the sections, the greater score value is given to the joint sample (*see* **Note 7**).

Knee joint degeneration of Del1 mice is characterized by a number of other alterations that are observed in radiographical and histological analysis of the knee joints (*see* **Fig. 2**). Radiography provides a noninvasive method to follow

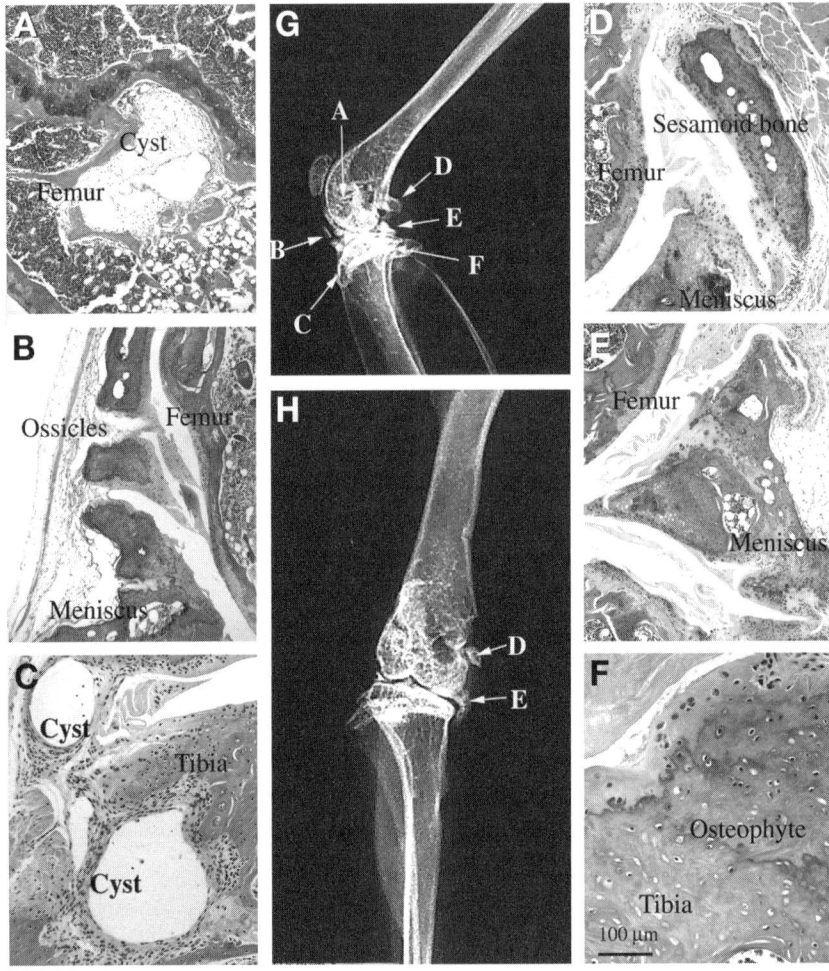

Fig. 2. Additional osteoarticular defects observed in 15- to 20-mo-old Del1 mice. Ossification of menisci is part of the normal aging process in mice, but it takes place earlier in transgenic Del1 mice. In Del1 mice the menisci are severely fragmented and disorganized. (**B**) Anterior meniscus, (**F**) posterior meniscus. Osteophytes (**F**) are commonly observed at the age of 15 mo. Cysts can be observed in subchondral bone of both femoral and tibial heads (**A,C**). Ossicles (ectopic ossifications) are observed deep to tendon, anterior to patello-femoral groove (**B**), and sesamoid bones posterior to femoral head (**D**). Hindlimb radiographs of a 15-mo-old Del1 mouse were taken both in sagittal (**G**) and in anterior–posterior projection (**H**). The arrows in panels G and H point to sites where the alterations described in panels A through F can be seen in the radiographs. Varus deformity is observed in anterior-posterior projection, while the ossicles (**B**) are observed only in the sagittal projection. **A–F** are H&E stained sagittal sections. Bar = 100 µm in (**F**) (also applies to **A–D**).

up the progression of knee joint degeneration, but is less sensitive, and observation of the early changes requires a radiologist's trained eye. The late degenerative changes observed in the histological sections and/or in radiographs involve cyst formation in subchondral bone, fragmentation, and calcification of menisci, ectopic ossifications in soft tissues, tendon, and sesamoid bones, osteophyte formation, and subchondral sclerosis. Ossification of menisci is part of the normal aging process in mice, but it takes place earlier in transgenic Del1 mice. In Del1 mice the menisci are severely fragmented and disorganized. Osteophytes are commonly observed at the age of 15 mo. Cysts can be observed in subchondral bone of both femoral and tibial heads. Ossicles (ectopic ossifications) are observed deep to tendon, anterior to patello-femoral groove. Sesamoid bones can be seen posterior to the femoral head. The presence of subchondral sclerosis, osteophytes, meniscal calcification, and soft tissue calcification, have been used for radiographical scoring of OA progression in other mouse models *(10,11)*.

3.2.2. Safranin O Staining for Semiquantitative Analysis of Glycosaminoglycans

For characterization of glycosaminoglycan distribution, stain the sections with safranin O. If you wish to do a semiquantitative analysis of proteoglycans in tissue sections, include 0.5% safranin O in the fixation solution and in the following decalcification solution to inhibit proteoglycan extraction from the tissue (*see* **Subheading 3.2.1., steps 2** and **4**) *(12)*. To cut down dye consumption (expensive!), reduce the sample size to a minimum. Embed tissue blocks in paraffin manually (*see* **Subheading 3.2.1., steps 6** and **7**), as the dye will partially dissolve in xylene and paraffin, and hence also stain other blocks. Cut blocks into 3-µm sections to keep absorbance values in a linear region. Safranin O stains sulphated glycosaminoglycans deep purple red (*see* **Note 8**).

3.2.2.1. SAFRANIN O STAINING

Xylene 2 times for 5 min each; absolute ethanol 2 times for 2 min each, 96% ethanol 2 times for 2 min each; distilled water for 30 s; safranin O staining solution 1for 0 min; running tap water for 10 min; 96% ethanol 2 times for 2 min each; absolute ethanol 2 times for 2 min each; and xylene 2 times for 5 min each. Finally, mount the sections with DPX and allow to dry in a hood.

For semiquantitative analysis of glycosaminoglycan content and distribution in different zones of articular cartilage, safranin O stained sections are analyzed with a scanning microspectrophotometer *(13)* or Peltier-cooled digital camera *(14)*. Recently, we have used a 12-bit Photometrics CH 250 camera (Photometrics) and IPLab software (BD Biosciences) because of the ease of use, high linearity and speed of analysis (*see* **Note 9**). The software has powerful scripting capabilities, which allow one to perform large series of image processing and camera control procedures with a single command.

3.3.2.2. Densitometry With a Peltier Cooled Digital Camera

1. Select appropriate magnification and align the microscope. Any research grade microscope will do. The microscope must have either a rotary specimen stage or possibility for the camera rotation to align the image adequately.
2. Insert monochromator (491 nm ± 1%) into the light path to optimize safranin O absorbance.
3. Keep fixed lamp voltage, condensor height and field diaphragm aperture settings and possible intermediate magnification lens positions. The aim of these steps is to standardize the intensity of the incident light. Stabilization of the lamp of the microscope takes about 20 min.
4. Take a blank background image from the optical field without a specimen on the microscope stage. If the intensity value reaches the maximum of the camera, either shorten the exposure time or reduce the illumination intensity.
5. Calibrate camera grey level response with a neutral density filter set. Typically, a 12-bit imaging system can be calibrated between 0 and 3.6 absorbance units. Insert each individual neutral density filter into light path, grab an image and correct it against the optical background image with the Flat Fielding algorithm of the IPLab software. An image without any filter is equivalent to zero absorbance. Do not change the exposure time after this step (*see* **Note 10**).
6. Measure the mean grey level intensity for each corrected image. Make a calibration scatterplot from the mean camera grey (abscissa) and filter absorbance value (ordinate) pairs with KaleidaGraph (Synergy Software) or any software capable of 10-based-logarithmic polynomial fitting (*see* **Fig. 3**). It is informative to use logarithmic scaling for abscissa, because of the logarithmic nature of ordinate, i.e. absorbance. This visualizes easily if saturation of the camera occurs below some absorbance. The fitting function for each pixel has the form:

$$\text{Absorbance}_{pix} = m_0 + m_1 \times {}^{10}\log(\text{Grey}_{pix}) .$$

Constants m_0 and m_1 can be obtained from the fit.

7. We are using IpLab software and a simple self-made measuring script. It grabs the image of interest, corrects it with the optical background with flat fielding, and finally converts it to absorbance (*see* **Fig. 4**). Each pixel in the image can be converted easily with constants m_0 and m_1 as variables in pixel arithmetic process (*see* **Note 11**).
8. Measure safranin O absorbance profile. Define a quadratic region of interest from articular cartilage surface to osteochondral junction. Plot the mean absorbance for every row of pixels. The profile can be fractioned to zones either in image processing system or in a Microsoft Excel spreadsheet.
9. Export profile in numerical form to Excel for further calculations of the mean distributions for each group in the experiment.

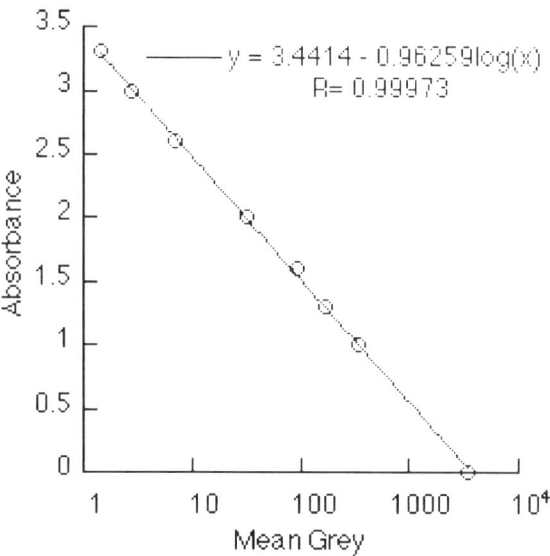

Fig. 3. Calibration of Photometrics CH 250 camera grey level response between 0 and 3.5 absorbance units.

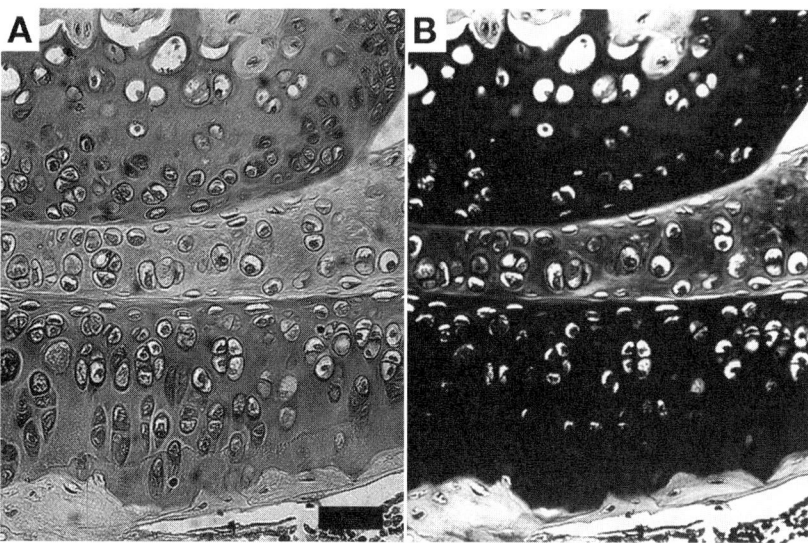

Fig. 4. Conventional, uncalibrated grey-level image of safranin-O stained articular cartilage imaged in white light (**A**) and the same image after conversion to absorbance (491 nm monochromatic light) (**B**). Bar = 50 µm.

3.2.3. Polarized Light Microscopy for Analysis of Matrix Collagen

Preparation of unstained tissue sections for polarized light microscopy (*2,15*). To analyze the orientation/content of collagen fibrils, prepare the paraffin sections from **Subheading 3.2.1., step 9** as follows: xylene 4 times for 10 min each; absolute ethanol 2 times for 2 min each; 96% ethanol 2 times for 2 min each; distilled water 2 times for 2 min each; digestion with testicular hyaluronidase, 1000 U/mL in 0.1M sodium phosphate (pH 6.9) for 18 h at 37°C in a moist chamber.

Perform the following treatments gently as the sections may detach from the glass: distilled water 3 times for 3 min each, performed in horizontal plane; 96% ethanol 2 times for 2 min each; absolute ethanol 2 times for 2 min each; xylene 3 times for 5 min each. Finally, mount the sections with DPX and allow to dry in a hood.

A sophisticated method to analyze collagen induced birefringence of unstained tissue sections has been developed by using a computer-assisted quantitative polarized light microscopy (*14–16*) Birefringence of cross polarized light by collagen networks can be estimated by measuring intensity differences between the incident light entering the polarizer, and transmitted light after passing through an unstained section and crossed analyzer (*16*) (*see* **Note 12**). The intensity difference can be converted to birefringence values according to Fresnel's equation (*16*). It must be remembered that the optimal detection sensitivity of the crossed polarizer system can be achieved for directions perpendicular and parallel to the surface oriented collagen fibril at the same time, if the cartilage surface is at 45° to the crossed polarizer and analyzer (*see* **Note 13**). Kiraly and coworkers have described in detail the critical factors in specimen processing and their effect on birefringence analysis (*12*). Careful removal of paraffin and glycosaminoglycans is extremely important.

Practical procedure: we have used the same microscope and camera as described in **Subheading 3.2.2.2.** The microscope must be equipped with a monochromator, two polarizers, and a rotatable lambda/4 phase shifter matched with the monochromator wavelength. The gamma axis of the phase shifter must be at 45° position between the crossed polarizers.

1. Align the microscope, decide on the suitable magnification and optimize illumination. Adjust polarizer and analyzer axis exactly perpendicular to each other. Remove dust as carefully as possible from lens surfaces. After this step, the optical settings cannot be changed without recalibration of the system.
2. Use rotatable lambda/4 phase shifter as a calibration specimen.
3. If you are studying low birefringence, calibrate and optimize your system performance between 0 to 5 α degrees. In case of high birefringence, calibrate between 0 and 10 or 0 and 20 -up to 45°.

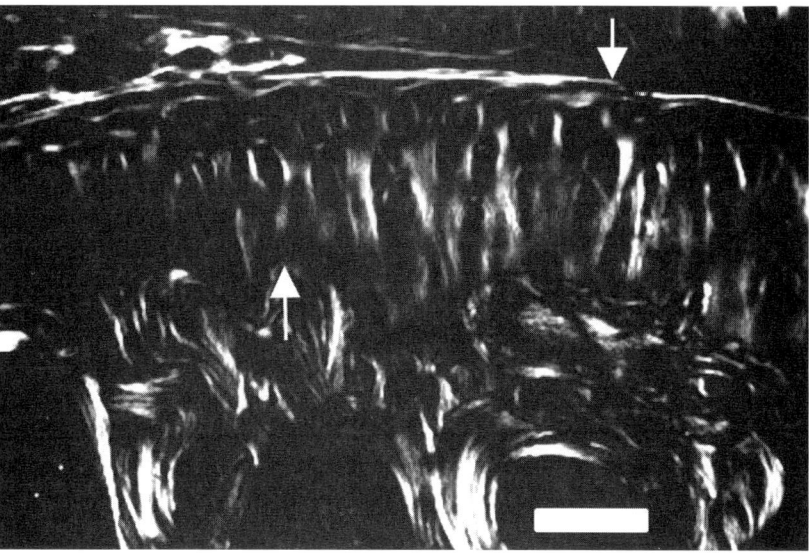

Fig. 5. Plane polarized light microscopic image of articular cartilage and subchondral bone plate of Del1 mouse. Upper and lower arrows point to cartilage surface and osteochondral junction, respectively. Bar = 50 µm.

4. Choose the optimal exposure time. It must be close to the saturation point of the camera for the largest rotation angle of the lambda/4 phase shifter.
5. During calibration, remove specimen from the microscope stage. Take a blank image for the crossed polarizers, and measure the mean grey intensity of the image. Then, rotate the $\lambda/4$ phase shifter for example a single one-degree step. Measure the corresponding mean gray value for the image. Continue calibration until the image intensity saturates.
6. Make a scatterplot of rotation angle (=α) (abscissa) and measured mean gray intensity value (ordinate) pairs. Make a $\sin^2$-function based fit for the data with KaleidaGraph (Synergy Software) or corresponding software:

$$I = a + bI_0 \sin^2(\alpha) \qquad (1)$$

where I is emergent light and I_0 is incident light intensity in grey level, accordingly. Numerical values for constants a and b can be obtained from the fit function.
7. In measuring process, individual pixel grey levels are first converted to alpha values using the calibration equation (1) in the following form:

$$(\alpha) = \arcsin((I - a)/(bI_0))^{0.5} \qquad (2)$$

8. Finally, birefringence (B) is calculated from α values for every pixel (*see* **Fig. 5**):

$$B = ((\alpha/\lambda)/180)/l \qquad (3)$$

In the formula (3), λ denotes the monochromator wavelength, and l is section thickness.

10. Define a quadratic region of interest from articular cartilage surface to osteochondral junction. Plot the mean birefringence for every row of pixels. The average distribution profile can be divided into zones either with IPLab or Microsoft Excel spreadsheet.

Image acquisition and transformation to birefringence (**steps 7** and **8**) can easily be made with IPLab software (BD Biosciences) utilizing its scripting and pixel arithmetic capabilities. All the constants needed in eqs. (2) and (3) can be given as preset variables.

3.2.3.1. Effect of Loading on the Development of OA

An additional advantage of mouse models for OA is the possibility of monitoring voluntary running activity of mice in long-term experiments. For this purpose transgenic and control mice are housed in individual cages where they have free access to a running wheel, the movements of which are continuously monitored by computer (*17*). Mice have been shown to run voluntarily 4 to 6 km each night! The running behaviour and progression of OA defects can then be compared between transgenic and control mice, and also with mice that have been housed in ordinary cages. Joint loading by voluntary running allows us to test in vivo the capacity of genetically modified articular cartilage to withstand increased long-term stress. Characterization of OA changes at the end of the experiment can be performed using methods described in this chapter. Surprising and contradictory findings have been made in different types of genetically modified models of OA in response to long-term voluntary running. For example, lifetime voluntary running has been shown to reduce the severity of knee OA in mice in which one allele of the *Col2a1* gene has been inactivated *(18)*. In contrast, a similar running program increased the severity of knee OA in the lateral compartment of Del1 transgenic mice *(19)*. The differences between these two types of mutations in response to running probably reflect distinct mechanisms by which OA develops in these mice.

3.3. Isolation of Total RNA From Skeletal Tissues

RNA isolation from tissues rich in fibrillar collagen and proteoglycans and low in cell density is a difficult task. Another problem in the isolation of RNA from mouse knee joints is the small sample size. The sampling method presented here was adopted as a compromise because dissecting articular cartilage from the knee epiphyses was too time consuming, and the RNA yield too low to allow Northern hybridizations of individual samples (*see* **Fig. 6**). The total RNA yield isolated from articular cartilage samples is generally less than 10 µg/joint in contrast to that of 100 to 150 µg isolated from epiphyseal blocks.

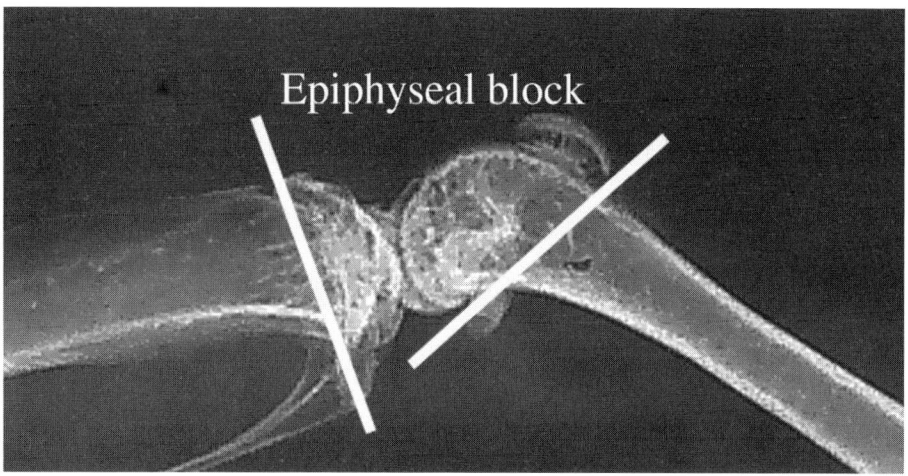

Fig. 6. Tissue sampling for RNA isolation. Knee epiphyses are cleared from soft tissue, and the knee joint is cut with a scalpel at standardized sites on tibial and femoral heads as indicated by the white lines in the sagittal radiograph of a hindlimb. The epiphyseal block thus contains patella, ligaments, menisci, subchondral bone, and growth plates in addition to articular cartilage.

The results obtained with these samples represent the average changes taking place in the articular cartilage, subchondral bone, growth plate, menisci, synovium, and ligaments (*see* **Fig. 6**). Hence the changes observed between nontransgenic and transgenic groups must be verified and localized by immunohistochemistry. An alternative for epiphyseal blocks is to collect tibial condyles dissected just above the growth plate, which better represents the articular cartilage compartment with a smaller influence from subchondral bone and other joint tissues (yield 20–30µg/joint). For extraction of RNA from knee epiphyses, quickly and carefully dissect the hindlimb free of muscle, and cut with a scalpel at the level of the tibial and femoral growth plates (*see* **Fig. 6**). Punch a small hole with an injection needle in the cap of microcentrifuge tube to keep the tube from exploding when moving it from the liquid nitrogen container to the freezer. Place the sample in a microcentrifuge tube and freeze the sample in liquid nitrogen and store it at −70°C, until processed for RNA isolation. Isolate total RNA using guanidium isothiocyanate (GIT) and CsCl density gradient centrifugation using a modification of a method presented by Chirgwin et al. 1989 (*20*) (*see* **Notes 14** and **15**). This method yields good quality RNA suitable for Northern analysis, RNase protection assays, and any of the reverse transcription polymerase chain reaction (RTPCR)-based methods available for the determination of specific mRNA levels.

1. Pulverize the sample in liquid nitrogen (see **Note 4**).
2. In a hood, soak the dispersing tool for 30 min in active DEPC-treated dH_2O (0.2 mL of DEPC in 100 mL of dH_2O) before homogenization. Transfer the powder with spatula (cooled in liquid nitrogen) into a 10-mL tube containing 2 mL of GITC solution on ice, and homogenize with Ultra Turrax. Lift the tool up from the sample and rinse the foam with 1 mL of GITC to the sample tube to fill the volume to 3 mL (see **Note 5**).
3. Remove the tissue debris by centrifugation (table top centrifuge at $3000g$ for 15 min).
4. During centrifugation, prepare the ultracentrifuge tubes: pipet 2 mL of $5.7M$ CsCl solution to the bottom of the tubes.
5. Layer the supernatant from **step 3** on top of the CsCl solution in the ultracentrifuge tube. Fill and balance the tubes with GIT (to within 0.01 g), leaving less than 2 mm empty space on top of the tube.
6. Centrifuge for 18 to 21 h at $175,000g$ at $20°C$ (Beckman Optima LE-80K Ultracentrifuge and SW 55 TI swing-out rotor).
7. Remove supernatant by suction half-way down with a glass capillary pipet.
8. Switch to a clean aspiration pipette and remove the remaining supernatant by suction. Be careful not to touch the inner surface of the tube to avoid any contamination with RNases that migrate to the top fraction of the gradient during centrifugation and may be bound to the inner surface.
9. After suction, invert the tubes to drain with a quick turn on a soft tissue or a paper towel.
10. Keep the tubes inverted, and cut off with a scalpel the bottom of tube containing the RNA pellet. Invert and transfer the tube bottoms on ice. Dissolve one sample at a time on ice with ice-cold solutions (**steps 11–12**):
11. Rinse the pellet once with 200 µL of 95 % DEPC-treated ethanol.
12. Add 200 µL of DEPC-treated dH_2O to dissolve the pellet. This takes a while. Pipet up and down. Pellet looks like cellophane. Remove the solution into a microcentrifuge tube on ice. Repeat dissolving step with 100 µL of DEPC-treated dH_2O. The RNA sample is now in 300 µL of DEPC-treated dH_2O on an ice bath.
13. Because proteoglycans also migrate to the same fraction as RNA, remove them by digestion with proteinase K: Add 500 µL of SET buffer + 20 µL of RNase-free proteinase K (20 mg/mL). Incubate 1 to 3 h at $37°C$.
14. Divide the sample into two Eppendorf tubes, 400 µL in each, add 400 µL of phenol/chloroform (1:1) into each tube and vortex for 30 s at room temperature. Work in a hood as chloroform is possibly a carcinogen.
15. Separate the phases by centrifugation for 2 to 3 min in a microcentrifuge at full speed and transfer the water phase into a new tube. Carefully avoid taking any of the lower phase.
16. Add 1:10 volume $3M$ NaAc (pH 5.5), and vortex.
17. Add 1 mL of 95% ethanol, vortex carefully, and keep overnight at $-20°C$ (or 1 h at $-70°C$).
18. Centrifuge at full speed for 15 min at $4°C$, decant the supernatant and dry the precipitate in a speed vac.

19. Dissolve in DEPC-treated dH2O (100 µL).
20. Take a 1 µL aliquot for determination of RNA amount (1 µL + 99 µL dH_2O) by ultraviolet (UV) absorbtion at 260/280 nm.
21. Store RNA at –70°C.

4. Notes

1. Paraformaldehyde is an irritating and allergenic compound; during weighing always wear a protective face mask to prevent inhalation, and work with the solution in a hood.
2. DEPC-treatment: When performing *in situ* hybridizations, add 2 mL of DEPC for 1000 mL of decalcification solution and distilled water to be used for rinsing. Mix the solutions, and let stand in a hood overnight. Autoclave the next day. Also prepare ethanol dilutions in DEPC-treated dH_2O, (2 mL/L). DEPC is irritant and possibly a carcinogen: avoid eye and skin contact and inhalation, wear protective clothes and gloves, and work in a hood when handling active DEPC. DEPC degrades to ethanol and CO_2 in water solution.
3. Ready-to-use instant hematoxylin (Shandon; cat. no. 9990107) can be used to replace Delafield's hematoxylin.
4. To pulverize the tissue we use a stainless steel unit composed of a cylinder, a bottom piece and a piston, that are cooled by soaking in liquid nitrogen. Soak also the sample in liquid nitrogen, and place between the bottom piece and piston, and pulverize by hitting the piston with a shot hammer 2 to 3 times.
5. Work in a hood and be aware that any heating of GITC will result in formation of highly toxic cyanide gas.
6. Fix the samples in paraformaldehyde at 4°C for 24 h on a rocking platform. Do not prolong the fixation time, as it will decrease the signal in immunohistochemistry and *in situ* hybridization.
7. Simultaneously with the screening for cartilage erosion of the tissue sections, it is advisable to make notes about the different tissue compartments that are observed within each section (e.g. in which sagittal section the tibial or femoral condyles are seen, whether the section is from lateral or medial condyle, or if patella, patellar tendon or cruciate ligaments are present in the sections). This helps in later selection of tissue sections for the immunohistochemical analysis to be made at certain tissue compartments. It is also possible to estimate from these notes whether the tissue had been embedded in the proper orientation. In samples embedded properly in horizontal plane, the patella, patellofemoral groove and cross ligaments appear in the same sagittal sections.
8. Do not counterstain with light green! The stain behaves like a green monochromator. It absorbs safranin O signal and causes bias.
9. The applicability of 8-bit videocameras (capable of reproducing 256 shades of gray) is very limited in proteoglycan densitometry because absorbance differences within a single image can easily exceed the dynamic range of the camera. This occurs already with 3-µm-thick paraffin sections. We recommend twelve to sixteen bit devices and very high sensitivity for accurate analysis of high stain absorbances.

10. Ensure that automatic exposure time mode, automatic gain control, histogram normalization and automatic contrast adjustments of the imaging system are disabled. They can cause significant bias.
11. **Steps 6–8** can practically be performed with any image processing software. For example, an excellent NIH-Image freeware supports density calibrations.
12. The approach cannot be used directly for picrosirius red stained sections because incident light absorption to the stain itself causes bias.
13. Similar theoretical and practical approaches can be utilized if the detection sensitivity of the system needs to be expanded to find other collagen fibril directions than preferential parallel or perpendicular directions *(14)*. If absolute orientation independence is needed, more sophisticated instrumentation and computational approaches for collagen orientation, anisotropy and birefringence analysis have been reported *(21)*.
14. For RNA isolation, all plastic and glass material needs to be RNase-free. Incubate the glassware at 180°C for at least 2 h, and gas sterilize plastic pipet tips and microcentrifuge tubes. Treat all the solutions with DEPC, add 2 mL of DEPC into 1 L of solution, allow to stand overnight at room temperature and autoclave. GITC solution does not need to be DEPC-treated, as the reagent itself inhibits RNase activity.
15. Advantages of GITC-CsCl density gradient centrifucation method: RNA isolated with this method is stable and tolerates several freeze-thaw cycles. We have succesfully used such RNA for RT-PCR, Northern hybridization and RNase protection analyses even after 5 to 10 yr of storage in DEPC-treated dH_2O solution at –70°C. The method is suitable particularly for isolation of total RNA from tissues rich in fibrillar collagen and/or low in cell density, such as cartilage, bone, cornea, and skin. mRNA isolated by this method has proven to give excellent results in different cDNA array analyses (e.g., Affymetrix chips even when RNA samples have been stored for several years at –70°C). Our attempts to purify RNA from cartilage tissues have failed with all silica-based microspin column kits we have tested. The problem is that these tissues are rich in collagen fibres and proteoglycans, and have relatively low cell density, which altogether result in low quality and quantity of RNA. Also the kits that do not include DNAse treatment, are ineffective in removing all DNA. Single-step isolation methods have given better results *(22)*, but the RNA often needs further purification if used as a template for quantitative RT-PCR or cDNA chip arrays. RNA purified by a single-step procedure is less resistant to repeated freeze-thaw cycles than RNA purified through CsCl density gradient centrifugation. Disadvantages: the method is time consuming, and requires expensive equipment such as an ultracentrifuge and a titanium rotor. Only four to six samples can be processed at a time, depending on the rotor. Furthermore, CsCl is rather expensive.

References

1. Metsäranta, M., Garofalo, S., Decker, G., Rintala, M., de Crombrugghe, B., and Vuorio, E. (1992) Chondrodysplasia in transgenic mice harboring a 15-amino acid

deletion in the triple helical domain of proα1(II) collagen chain. *J. Cell Biol.* **118**, 203–212.
2. Säämänen, A. -M. K., Salminen, H. J., de Crombrugghe, B., et al. (2000) Osteoarthritis-like lesions in transgenic mice harboring a small deletion mutation in type II collagen gene. *Osteoarthritis Cart.* **8**, 248–257.
3. Salminen, H., Perälä, M., Lorenzo, P., Saxne, T., Heinegård, D., Säämänen, A. -M., and Vuorio, E. (2000) Up-regulation of cartilage oligomeric matrix protein at the onset of articular cartilage degeneration in a transgenic mouse model for osteoarthritis. *Arthritis Rheum.* **43**, 1742–1748.
4. Salminen, H., Vuorio, E., and Säämänen, A. -M. (2001) Expression of Sox9 and type IIA collagen during attempted repair of articular cartilage damage in a transgenic mouse model for osteoarthritis. *Arthritis Rheum.* **44**, 947–955.
5. Salminen, H. J., Säämänen, A. -M. K., Vankemmelbeke M. N., Auho, P. K., Perälä, M. P., and Vuorio, E. I. (2002) Differential expression patterns of matrix metalloproteinases and their inhibitors during development of osteoarthrosis in a transgenic mouse model. *Ann. Rheum. Dis.* **61**, 591–597.
6. Helminen, H. J., Säämänen, A. -M., Salminen, H., and Hyttinen, M. M. (2002) Transgenic mouse models for studying the role of cartilage macromolecules in osteoarthritis. *Rheumatology* **41**, 848–856.
7. Säämänen A. -M. and Vuorio, E. (2004) Generation and use of transgenic mice as models of osteoarthritis. iIn: *Methods in Molecular Medicine, Vol. 101: Cartilage and Osteoarthritis, Volume 2: Structure and In Vivo Analysis*, De Ceuninck, F., Pastoureau, P., Sabatini, M., eds., Totows, NJ:Humana Press, pp. 1–23.
8. Wilhelmi, G. and Faust, R. (1976) Suitability of the C57 black mouse as an experimental animal for the study of skeletal changes due to ageing, with special reference to osteo-arthrosis and its response to tribenoside. *Pharmacology* **14**, 289–296.
9. Helminen, H. J., Kiraly, K., Pelttari, A., et al. (1993) An inbred line of transgenic mice expressing an internally deleted gene for type II procollagen *(COL2A1). J. Clin. Invest.* **92**, 82–595.
10. Ameye, L., Aria, D., Jepsen, K., Oldberg, Å., Xu, T., and Young, M. (2002) Abnormal collagen fibrils in tendon of biglycan/fibromodulin-deficient mice lead to gait impairment, ectopic ossification, and osteoarthritis. *FASEB J.* **16**, 673–680.
11. Evans, R. G., Collins, C., Miller, P., Ponsford, F. M., and Elson, C. J. (1994) Radiological scoring of osteoarthritis progression in STR/ORT mice. *Osteoarthritis Cart.* **2**, 103–109.
12. Kiraly, K., M. Lammi, Arokoski, J., et al. (1996) Safranin O reduces loss of glycosaminoglycans from bovine articular cartilage during histological specimen preparation." *Histochem. J.* **28**, 99–107.
13. Kiviranta, I., Jurvelin, J., Tammi, M., Säämänen, A. -M., and Helminen, H.J. (1985) Microspectrophotometric quantitation of glycosaminoglycans in articular cartilage sections stained with safranin O. *Histochemistry* **82**, 249–255.
14. Hyttinen, M. M., Töyräs, J., Lapveteläinen, T., et al. (2001). Inactivation of one allele of the type II collagen gene alters the collagen network in murine articular cartilage and makes cartilage softer. *Ann. Rheum. Dis.* **60**, 262–268.

15. Kiraly, K., Hyttinen, M. M., Lapveteläinen, T., et al. (1997) Specimen preparation and quantification of collagen birefringence in unstained sections of articular cartilage using image analysis and polarizing light microscopy. *Histochem. J.* **29**, 317–27.
16. Arokoski, J. P., Hyttinen, M. M., Lapveteläinen, T., et al. (1996) Decreased birefringence of the superficial zone collagen network in the canine knee (stifle) articular cartilage after long distance running training, detected by quantitative polarised light microscopy. *Ann. Rheum. Dis.* **55**, 253–264.
17. Lapveteläinen, T., Tiihonen, A., Koskela, P., et al. (1997) Training a large number of laboratory mice using running wheels and analyzing running behavior by use of a computer-assisted system. *Lab. Anim. Sci.* **47**, 172–179.
18. Lapveteläinen, T., Hyttinen, M., Lindblom, J., et al. (2001) More knee joint osteoarthritis (OA) in mice after inactivation of one allele of type II procollagen gene but less OA after lifelong voluntary wheel running exercise. *Osteoarthritis Cart.* **9**, 152–160.
19. Lapveteläinen, T., Hyttinen, M. M., Säämänen, A. -M., et al. (2002) Lifelong voluntary joint loading increases osteoarthritis in mice housing a deletion mutation in type II procollagen gene, and slightly also in non-trangenic mice. *Ann. Rheum. Dis.* **61**, 810–817.
20. Chirgwin, I. M., Przybyla, A. E., MacDonald, R. J., and Rutter, W. J. (1989) Isolation of biologically active ribonucleic acid from sources enriched in ribonuclease. *Biochemistry* **18**, 5294–5299.
21. Rieppo, J., Hallikainen, J., Jurvelin, J. S., Helminen, H. J., and Hyttinen, M. M. (2003) Novel quantitative polarization microscopic assessment of cartilage and bone collagen birefringence, orientation and anisotropy. *Trans. Orthop. Rec. Soc.* **28**, 570.
22. Chomczynski, P. and Sacchi, N. (1987) Single-step method of RNA isolation by acid guanidinium thiocyanate-phenol-chloroform extraction. *Anal. Biochem.* **162**, 156–159.

III

APPLICATION OF NEW TECHNOLOGIES TO DEFINE NOVEL THERAPEUTIC TARGETS

22

Gene Expression Profiling in Rheumatology

Tineke C. T. M. van der Pouw Kraan, Lisa G. M. van Baarsen,
François Rustenburg, Belinda Baltus, Mike Fero,
and Cornelis L. Verweij

Abstract

In the last decade, the analysis of gene expression in tissues and cells has evolved from the analysis of a selected set of genes to an efficient high throughput whole-genome screening approach of potentially all genes expressed. Development of sophisticated methodologies such as microarray technology allows an open-ended survey to identify comprehensively the fraction of genes that are differentially expressed between samples and that define the samples' unique biology. By a global analysis of the genes that are expressed in cells and tissues of an individual under different conditions and during disease, we can build up "gene expression profiles (signatures)" which characterize the dynamic functioning of the genome under pathophysiological conditions. This strategy also provides the means to subdivide patients that suffer from a complex heterogeneous disease into more homogeneous subgroups. Such discovery-based research identifies biological processes that may include new genes with unknown function or genes not previously known to be involved in this process. The latter category may hold surprises that sometimes urge us to redirect our thinking. We have used microarrays to disclose the heterogeneity of rheumatoid arthritis (RA) patients at the level of gene expression of the affected synovial tissues. Analysis of the expression profiles of synovial tissues from different patients with RA revealed considerable variability, resulting in the identification of at least two molecularly distinct forms of RA tissues. One is characterized by genes that indicate an active inflammatory infiltrate with high immunoglobulin production, whereas the other type shows little immune activation and instead shows a higher stromal cell activity. These results confirm the heterogeneous nature of RA and suggest the existence of distinct pathogenic mechanisms that contribute to RA. The differences in expression profiles provide opportunities to stratify patients for intervention therapies based on molecular criteria.

Key Words: Microarray; mRNA; DNA; labeling; hybridization; scanning; microarray; rheumatoid arthritis.

1. Introduction
1.1. Microarrays

Microarrays consist of a solid support on which DNA fragments derived from individual genes are placed in an ordered array. These arrays are hybridized with fluorescent cDNA probes prepared from total cellular mRNA. If the total amount of RNA is insufficient, linear amplification of mRNA can be applied, yielding amplified aRNA that can be reverse transcribed into DNA *(1,2)*. Labeling of either total RNA or aRNA is accomplished by incorporation of the reactive amine derivative of dUTP, 5-(3-Aminoallyl)-2'-deoxyuridine 5'-triphosphate, during first strand cDNA synthesis *(3,4)*. After removing free nucleotides the aminoallyl labeled cDNA is chemically coupled to a fluorescent dye (i.e., Cy3 or Cy5). Hybridization of the cDNA probes to microarrays results in specific base pairing with the corresponding gene sequence at known locations on the microarray. Following washing, the specific hybridization of the cDNA probes to each DNA spot is quantitated using a confocal scanning device. Two styles of microarrays are used most commonly. One consists of oligonucleotides that are produced by *in situ* oligonucleotide synthesis using photolithographic masking techniques, commercially available from Affymetrix^{TM}. These so-called DNA chips, are typically hybridized with a single sample probe at a time (one channel analysis). In this design, genes are represented by 11 to 16 oligonucleotides (25-mers), each including the perfect match and a mismatch that is identical except for a single base mismatch in its centre. In order to compare the mRNA expression profiles of two samples, the two sample probes are hybridized to separate arrays. Another type of microarray consists of longer sequences (approx 60–2000 bp) of cDNA (PCR products) or oligonucleotides with each element representing a distinct gene printed on glass microscope slides *(3,5)*. Two fluorescent probes, each labeled with a different fluorescent dye (typically Cy3 and Cy5) are simultaneously hybridized to these microarrays. Here we describe the procedure that is used for a two-color analysis protocol using microarrays fabricated at Stanford University by robotic spotting of polymerase chain reaction (PCR) fragments from cDNA clones (*see* **Figs. 1** and **2A**). Two RNA samples are labeled separately with different fluorescent dyes, Cy3 and Cy5. In this format the relative mRNA expression in two samples is directly compared for each gene on one microarray. To transfer data into information we have to go through several steps of data processing for further analysis. First, the scanned images have to be transformed into a gene expression matrix. Subsequently different algorithms (e.g., Cluster, Statistical analysis) can be applied to analyse the data.

1.2. Experimental Design and Choice of Reference

The first step in an experimental design is the choice of the control or the reference sample. If the aim of the experiment is to compare expression profiles of cells/tissues from a group of patients, the use of a common reference mRNA pool is ideal because it allows comparisons of all patients to any other patient. The main purpose of a common reference sample is to give a signal at all spots that are printed on the array to be able to calculate a red to green ratio for each mRNA species that is present in the sample of interest. This ratio may therefore be arbitrary, but it does allow comparison of all samples that were hybridized together with the same reference sample. For this purpose we use a common reference mRNA sample that consists of a mixture of mRNAs isolated from 11 different cell lines *(9)* supplemented with RNA from synovial tissue, fibroblasts, and activated peripheral blood mononuclear cells (PBMCs). When a common reference is used the data are centred to allow a more intuitive visual interpretation and pattern recognition of the data (*see* **Fig. 2B**). After centring, the ratio of gene X is expressed relative to the median expression ratio of gene X across all arrays (in log space).

In another experimental design the ratio may not be arbitrary, if for instance different conditions of a cell line (i.e., treated vs non-treated) are hybridized to the same array, or when time point zero is used as a reference, in a time course experiment. In this case no centring is required because the ratio reflects the actual ratio between treated versus untreated, or in case of a time-course experiment, the actual change compared to timepoint zero. However, a disadvantage of this design is the fact that not all genes may be expressed in the reference sample. Therefore, a mixture of treated and nontreated samples or a mixture of the samples of all time points may be a better option for the reference sample, because it will cover all relevant spots. An additional advantage of using mixtures of all samples that comprise the experiment is that future experiments may be performed with the same reference and can thus be compared with the previous experiments. It is therefore wise to prepare ample reference material. More detailed information on experimental design can be obtained from *(6,7)*.

1.3. Data Analysis

Microarrays measure the relative mRNA abundance indirectly by measuring the intensity of the fluorescence of each spot on the array for the two different fluorescent dyes (i.e., for each optical wavelength). Next, the scanning images have to be transformed into a red-to-green ratio for each gene on the array. The spots corresponding to genes on the microarray should be identified, their boundaries determined, the fluorescence intensity from each spot measured and compared to local background intensity of each spot and to the

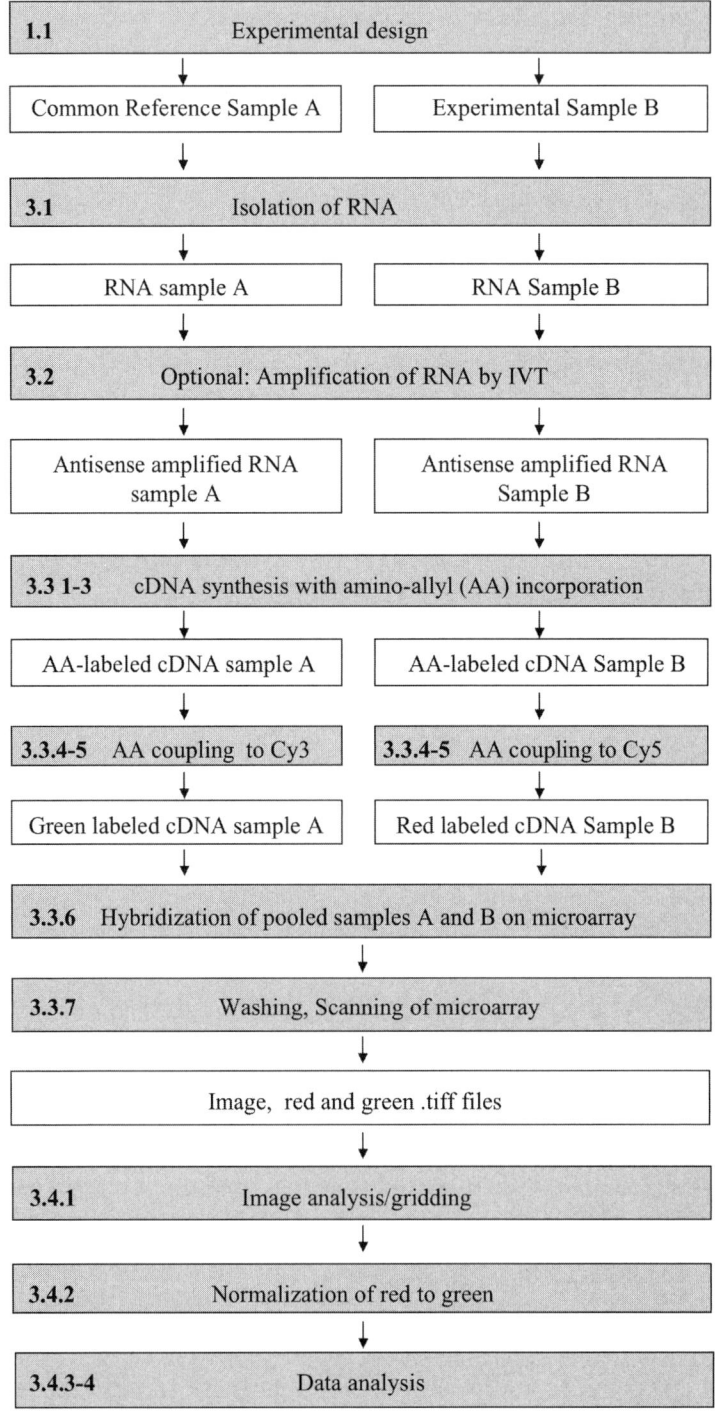

intensities from the other channel. The software for adjustment of the grid for spots and initial image processing is freely available (ScanAlyze at *http:// rana.lbl.gov/EisenSoftware*), whereas more automated programs are commercially available (i.e. Genepix from Axon instruments and Imagene from BioDiscovery).

1.3.1. Cluster Analysis

The goal of clustering is to group together objects (genes or samples [patients]) with similar properties. This can be viewed as the reduction of the dimensionality of the system. Clustering of expression profiles has been used for grouping genes as well as samples/arrays, so-called two-dimensional (2D), two-way, or unsupervised clustering. Without information about the identity of the samples, these are organized based solely on the overall similarity in their gene expression patterns (unsupervised, *see* **Fig. 2B** and **3**). In one-dimensional (1D) clustering, usually only the genes are clustered and this is referred to as supervised clustering. In this case the arrays are kept in a predetermined position, for instance categories of clinical parameters, or in the analysis of kinetic experiments *(9)*. The clustering of genes allows identification of coregulated and functionally related groups, which is particularly interesting. Eisen et al. *(10)* have developed a hierarchical clustering algorithm and visualization software package, which is currently one of the most frequently used tools for expression profile clustering and data visualization.

An interesting application of this approach is the clustering of tumors to find new possible tumour subclasses. Alizadeh et al. *(11)* performed unsupervised hierarchical clustering on expression data of lymph node biopsies from diffuse

---

Fig. 1. Flow diagram of experimental procedures for microarray technology. Numbers indicate the sections in which the procedures are described. From a reference sample and an experimental sample RNA is isolated. If the yield is less than 25 µg, RNA is amplified by in vitro transcription, resulting in anti-sense aRNA. Next, the samples are labeled with 5-(3-Aminoallyl)-2'-deoxyuridine 5'-triphosphate (aminoallyl-dUTP) during cDNA synthesis. Aminoallyl-labeled cDNA is then chemically coupled to Cy3 (green) or Cy5 (red). Next, both samples are pooled and hybridized to the same microarray. During washing excess dyes are removed and after scanning an image is generated. In a spot analysis program a grid is placed on the image that identifies the boundaries of the spots and generates the spot and local background intensities for the red and green signal. These raw data are stored in a database. To adjust for labeling and hybridization efficiencies the red signal is adjusted to the green signal (normalization), such that the average ratio of all spots on the array becomes equal to one. The normalized ratios can then be used for further analysis such as clustering and statistical analysis.

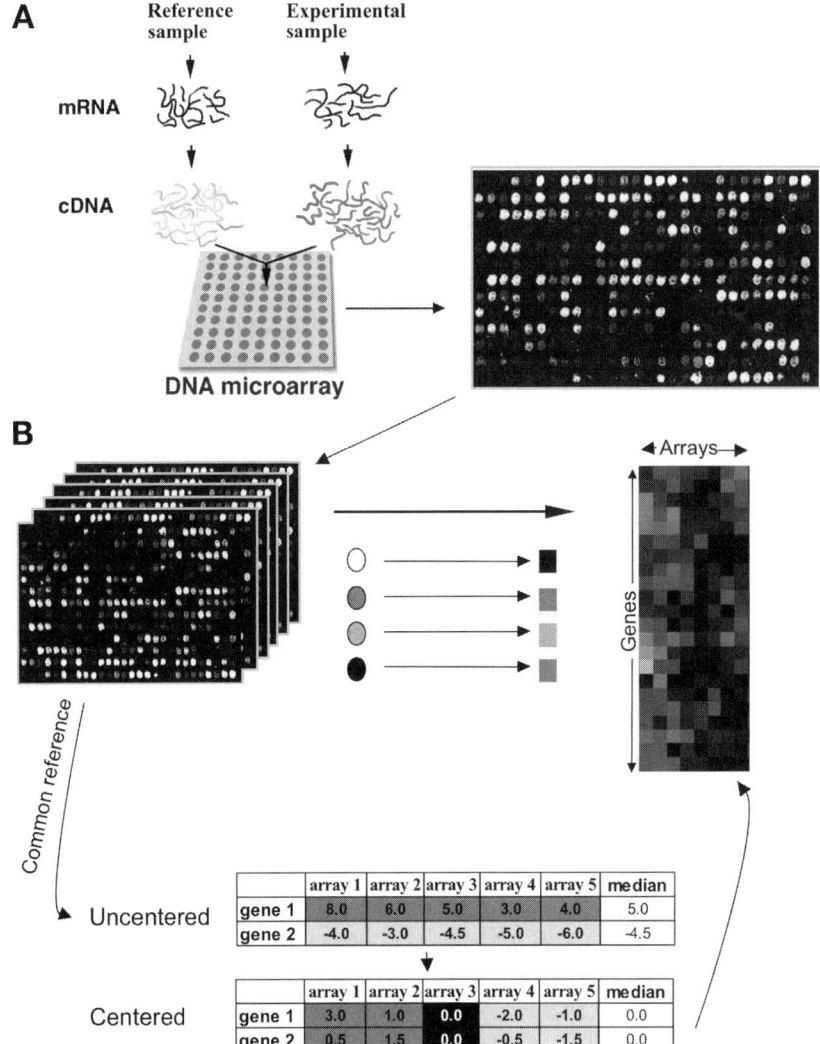

Fig. 2. Experimental set-up and analysis. (**A**) RNA is isolated from an experimental sample and a control or reference sample and labeled with different fluorescent dyes. Both samples are then mixed together and hybridised to the same microarray, resulting in a red to green ratio for each spotted element on the microarray. Scanning of the microarrays in generates an image of all spots. (**B**) For comparison of several samples a clustering algorithm can be used. Note the difference in color representation of spots on the array and the corresponding squares in the cluster diagram. If a common reference is used for all samples that does not allow direct comparison between the samples, then the data need to be centered first (i.e., ratio's are expressed relative to the median expression level across all arrays).

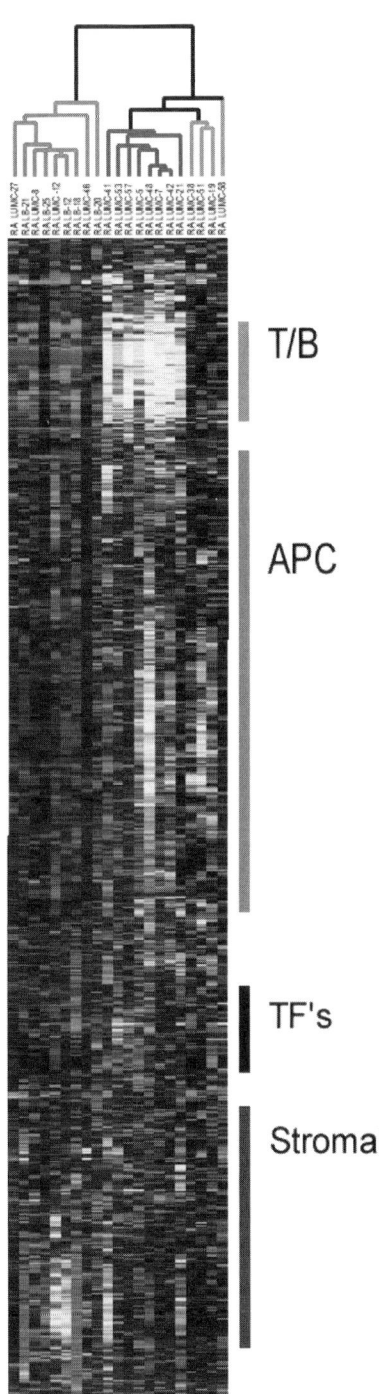

Fig. 3. RA tissues show a remarkable heterogeneity. Unsupervised clustering of RA tissues results in the identification of at least two major subclasses; one is defined by the red branch, the RA^{high} tissues, whereas the other major class can be subdivided into the RA^{low} (blue branch) and the RA$^{interm.}$ (green branch) tissues. Bars to the right of the panel identify the locations of a category of clustered genes with a related expression profile (T/B, T/B-cell cluster; APC, antigen-presenting cell cluster; TF, transcription/signaling factors cluster; Stromal, stromal cell-related gene cluster). Reproduced with permission from **ref. 12**.

large B-cell lymphoma (DLBCL) patients. This analysis revealed two groups of patients, one group showed characteristics of a gene expression profile resembling germinal centre B-cells, whereas the other group showed an expression profile characteristic of activated peripheral blood B-cells. This distinction was shown to be a good predictor for survival. Seventy percent of the germinal centre B-cell-like group reached a 5-yr survival, whereas only 20% of the activated B-like DLBCL patients survived for 5 yr.

In a study on breast cancer aimed to classify patients, hierarchical clustering also provided important information on the pathology of tumor development *(8)*. This study revealed at least four subgroups of patients with different genetic programs that were informative concerning the biology of the tumors.

We performed unsupervised hierarchical clustering of genes expressed in synovial tissues from RA patients and identified two main groups of patients (RA^{high} and RA^{low}) with clusters of correlated genes that made the distinction between the two groups. An interpretation of biological processes that take place revealed that the gene expression profile in RA^{high} tissues is indicative for an adaptive immune response with a high expression of genes representing the presence and activation of T- and B-cells and antigen presenting cell activity, whereas the second group showed little immune activation and instead showed a higher expression of genes suggestive for fibroblast dedifferentiation and extracellular matrix remodelling *(12)*. These results confirm the heterogeneous nature of RA and suggest the existence of distinct pathogenic mechanisms that contribute to RA (*see* **Fig. 3**).

2. Materials
2.1. Reagents

1. Amino-allyl dUTP (Biochemika; cat. no. A 0410).
2. Anchored Oligo-dT20-VN (Operon/Qiagen).
3. Cot-1 DNA (Invitrogen; cat. no.15279-011).
4. Cy3 mono-reactive dye pack (Amersham; cat. no. PA 23001).
5. Cy5 mono-reactive dye pack (Amersham; cat. no. PA 25001).
6. dATP, dCTP, dGTP, dTTP (100 m*M* each) 10 µmoL (Promega; cat. no. U1330).
7. Dimethyl sulfoxide (DMSO) (Aldrich; cat. no. D 8418)
8. 100% Ethanol absolute (Sigma; cat. no. E7023).
9. 96% Ethanol (Nedalco; cat. no. 104,000,0020E).
10. Humid chamber (Aldrich; cat. no. H6644).
11. Hybridization chamber (Ambion; cat. no. 10040).
12. Labtop cooler (Nalgene; cat. no. 5115-0012).
13. Message AMP aRNA kit (Ambion; cat. no. 1750).
14. Microcon YM-30 (Millipore; cat. no. 42410).
15. PAX gene blood RNA kit (PreAnalytix; cat. no. 762134)

16. PAX gene blood RNA tubes (PreAnalytix, QI 762125).
17. Poly N (6-mer) 1OD (Operon/Qiagen; cat. no. SP200).
18. Poly-A RNA (Sigma; cat. no. P 9403).
19. Polytron T8 probe 5 mm (IKA technologies).
20. Qiagen RNase-free DNase set (Qiagen; cat. no. 79254).
21. Qiaquick PCR purification kit (Qiagen; cat. no. 28104).
22. Recombinant RNAse inhibitor 40U/µL (Promega; cat. no. N2511).
23. Ribonuclease H, *Escherichia coli* (cloned) 10U/µL (Ambion; cat. no. 2292).
24. Rnase Zap (Biochemika; cat. no. 83930).
25. RNeasy mini kit 250 reactions (Qiagen; cat. no. 74106).
26. Slide box (Aldrich; cat. no. Z 374385).
27. Slide chamber (rack + dish + cover) (Shandon Lipshaw; cat. no. 121).
28. SSC Buffer Concentrate 20X (Biochemika; cat. no. 93017).
29. Superscript II RNA-se H RT (Invitrogen,; cat. no. 18064-071).
30. TE buffer: 10 mM Tris HCl, 1 mM EDTA
31. Trizol (Invitrogen; cat. no. 15596-026).
32. RNAse free water 1 L (Biochemika; cat. no. 95284).
33. Yeast tRNA (Invitrogen; cat. no. 15401-011).

2.2. Equipment

1. Heat block for 0.5 to 1.5-mL tubes (Techne Dri Block®; cat. no. BB-2D).
2. PTC-200 Peltier Thermal cycler for 0.5-mL tubes (MJ Research).
3. Thermomixer Comfort for 0.5 to 1.5-mL tubes (Eppendorf).
4. Centrifuge for 0.5- to 2-mL tubes Biofuge pico (Heraeus).
5. Centrifuge for slide racks Multifuge 3 S-R (Heraeus).
6. DNA Micro array scanner (Agilent).
7. Lab-on-a-chip 2100 BioAnalyzer (Agilent).
8. ND-100- Spectrophotometer (Nanodrop technologies).
9. Ultraviolet (UV) Crosslinker (Stratagene).

3. Methods

3.1. RNA Isolation (see Note 1)

3.1.1. Total RNA Isolation From Tissue Samples

Synovial tissue samples were obtained at arthroscopy or surgical resection. Synovial biopsy samples were obtained from multiple regions using a 2.5-mm grasping forceps (Storz). Arthroscopy was performed under local anaesthesia using a small bore, 2.7-mm arthroscope (Storz) and using an infrapatellar skin portal for macroscopic examination of the synovium and a second suprapatellar portal for the biopsy procedure *(12)*. Synovial tissue obtained by surgical resection (approx 1 g) was dissected and quickly frozen in liquid nitrogen and stored at –80°C.

Total RNA was isolated from the tissues by TRIzol reagent according to the manufacturers' instructions. After the last washing step the RNA was suspended in 40 μL RNase-free water to obtain sufficiently concentrated RNA.

3.1.2. Total RNA Isolation From Peripheral Blood Mononuclear Cells

After isolation, cells were stimulated and total RNA was isolated using RNeasy mini kit (Qiagen) according to manufacturers' instructions. A DNase step using Qiagen RNase-free DNase was included in the protocol to clear the fraction of genomic DNA present in the sample. RNA was eluted using 45 μL of 50°C RNase free water (*see* **Note 2**).

3.1.3. Total RNA Isolation From Whole Blood

2.5 milliliters of blood was drawn in PAXgene RNA isolation tubes. These tubes contain a buffer to ensure immediate lyses of the blood cells and stabilization of the RNA. After blood collection the tubes are incubated for 2 h at room temperature after which the tubes can be stored at –20 or –80°C until RNA isolation is performed. Samples are thawed at RT for 2 h prior to RNA isolation. RNA is then isolated using PAXgene RNA isolation kit according to the manufacturers' instructions with all centrifuge steps performed at RT (*see* **Note 3**).

1. Incubate the blood in PAXgene blood collection tubes at room temperature for 2 h while gently shaking at roller bank.
2. Store the tubes at –20 or –80°C until further processing (processing must occur within 7 mo!).
3. Thaw the tubes for 2 h at room temperature on a roller bank.
4. Centrifuge the PAXgene tubes for 10 min at 3000*g* using a swing-out rotor (*see* **Note 4**).
5. Add 5 mL of RNase-free water to the pellet, and close the tube using a fresh secondary Hemoguard closure.
6. Thoroughly resuspend the pellet by vortexing, and centrifuge for 10 min at 3000*g*.
7. Remove and discard the entire supernatant, decant the tube on a paper towel (*see* **Note 5**).
8. Add 360 μL Buffer BR1 to the pellet and replace Hemoguard closure. Thoroughly resuspend the pellet by vortexing. Move to the fume hood for **steps 9–28**.
9. Pipet the sample into a 1.5 ml microcentrifuge tube using a 200 μl filtertip.
10. Wash the tube with 300 μL buffer BR2 to collect the remaining sample and add this to the 1.5-mL microcentrifuge tube.
11. Add 40 μL proteinase K to the mixture. Mix by vortexing, and incubate for 10 min at 55°C, 1400 rpm, using the Thermomixer Comfort (*see* **Note 6**).
12. Centrifuge for 5 min at ≥10,000*g* in a microcentrifuge and transfer the supernatant to a fresh 1.5-mL microcentrifuge tube (*see* **Note 7**).

13. Add 350 µL 100% ethanol. Mix by vortexing, and centrifuge 1 to 2 s to remove drops from the inside of the tube (*see* **Note 8**).
14. Apply 700 µL sample to the PAXgene column sitting in a 2-mL processing tube, and centrifuge for 1 min at >8000g
15. Place the PAXgene column in a new 2-mL processing tube, and discard the old processing tube containing the flow-through.
16. Apply the remaining sample to the PAXgene column, and centrifuge for 1 min at >8000g.
17. Place the PAXgene column in a new 2-mL processing tube, and discard the old processing tube containing flow-through
18. Pipet 350 µL Buffer BR3 into the PAXgene column. Centrifuge for 1 min at >8000g. Discard the flow-through by suction.
19. Add 10 µL DNase I stock solution from RNase-free DNase Kit to 70 µL Buffer RDD. Mix by gently flicking the tube, and centrifuge briefly to collect residual liquid from the sides of the tube (*see* **Note 9**).
20. Pipet the DNase I incubation mix (80 µL) *directly onto* the PAXgene-column membrane, and incubate for 15 min at room temperature.
21. Add 350 µL buffer BR3 to the PAXgene spin column, and centrifuge for 1 min at >8000g.
22. Place the PAXgene column in a new 2-mL processing tube, and discard the old processing tube containing flow-through.
23. Apply 500 µL buffer BR4 (*see* **Note 10**) to the PAXgene column, and centrifuge for 1 min at >8000g.
24. Place the PAXgene column in a new 2-mL processing tube, and discard the old processing tube containing flow-through.
25. Add another 500 µL Buffer BR4 to the PAXgene column. Centrifuge for 3 min at maximum speed to dry the PAXgene column membrane.
26. Discard the tube containing the flow-through, place the PAXgene column on a 1.5-mL Eppendorf tube and centrifuge for 1 min at full speed.
27. To elute, discard the tube containing the flow-through, transfer the PAXgene column to a 1.5-mL elution tube, pipet 45 µL buffer BR5 directly onto the PAXgene column membrane and incubate 2 min at room temperature.
28. Centrifuge for 2 min at >8000g and repeat elution **steps 27** and **28**, using another 45 µL buffer BR5.
29. If necessary, pool eluates from columns.
30. Incubate the sample at 65°C for 5 min, followed by subsequent cooling on ice and measure RNA concentration and purity using spectrophotometry (mark the tube and store RNA at −80°C until use).

RNA samples were analysed on a Lab-on-a-Chip or the ND-1000 Spectrophotometer and quantified. The quality of the RNA was regarded as good when the intensity of the 28S peak exceeded the intensity of the 18S peak. A linear amplification of the RNA is performed if the yield is less then 25 µg.

3.2. Linear Amplification of Total RNA (see Fig. 4)

An amount of less then 25 µg total RNA is insufficient to successfully carry out a hybridization based on cDNA labeling. Therefore, if less than 25 µg RNA is isolated, an amplification step is required. We used the amplification Message Amp. aRNA kit (Ambion), which is based on the Eberwine method (*see* **Fig. 4**) *(2)*. In short, mRNA is transcribed to cDNA (first strand synthesis) using an oligo-dT primer containing a T7 promoter sequence. Next, the second DNA strand is synthesized. Finally, double-stranded cDNA is in vitro transcribed by T7 RNA polymerase to yield multiple copies of aRNA (the linear amplification step). We used 1 µg of RNA as input and followed the protocol according to the manufacturers' instructions allowing 5 h for the in vitro transcription. The Ambion kit contains the reagents for first strand synthesis, second strand synthesis and in vitro transcription of cDNA. Typically the yield from 1 µg RNA is approx 15 µg of amplified aRNA. *Please note that after amplification of RNA, antisense aRNA is obtained that is able to hybridize to the complimentary strand on cDNA microarrays (containing PCR products) but not to single stranded oligonucleotide microarrays.* In the latter case, the aRNA needs to be directly labeled with aminoally during the in vitro transcription step, instead of during cDNA synthesis of aRNA (not included in our protocols). Because of a gene-specific difference in amplification efficiency, amplified RNA samples can only be compared to other amplified RNA samples.

3.2.1. Protocol MessageAmp^{tm} aRNA kit (V0309A; Updated 27-04-2004)

All incubation steps are performed in a thermal cycler block using thin wall tubes. When the incubation temperature is ≥37°C a heated lid is used. Do *not* use a heated lid when incubation temperature is <37°C. It is preferable to take between 100 and 1000 ng of total RNA as input (max 5 µg) in a maximal volume of 11 µL. Put samples into the thermal cycler after it has reached its exact target temperature.

3.2.2. Day 1: First Strand Synthesis

1. Thaw your samples, T7 primer, dNTP's and 10X First strand buffer on ice.
2. Set the temperature of the thermal cycler on 70°C.
3. Prepare the following mix:RNA sample (max 11 µL) and 1 µL oligo T7 primer. Add RNase free water to a final volume of 12 µL.
4. Spin down and incubate samples 10 min at 70°C in thermal cycler.
5. Put on ice-water for 5 min and spin down.
6. Meanwhile, set the temperature of the thermal cycler on 42°C and prepare reverse transcriptase master mix at room temperature, use per reaction: 2 µL 10X first

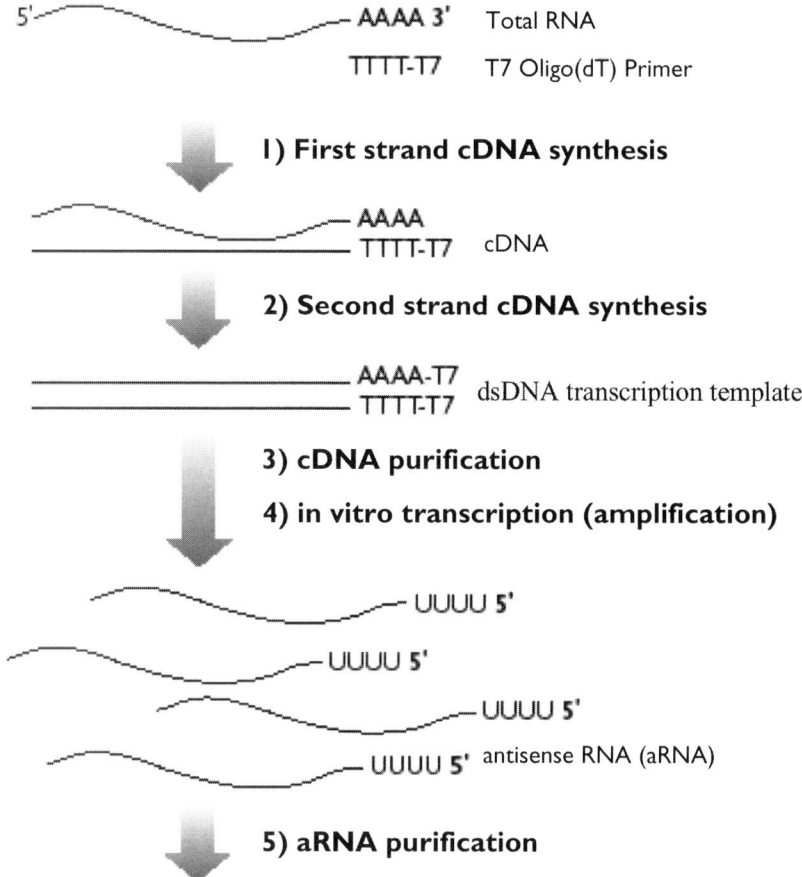

Fig 4. Overview of the protocol for amplification of mRNA. First strand cDNA is synthesized using an oligo-dT primer containing a T7 promoter sequence. After synthesis of the second DNA strand, double stranded cDNA is in vitro transcribed to aRNA by T7 RNA polymerase. This diagram was kindly provided by Ambion Inc.

strand buffer; 1 μL ribonuclease inhibitor; 4 μL dNTP mix; and 1 μL reverse transcriptase. Prepare mastermix for your total number of samples at room temperature. Mix, spin down, and put on ice.
7. Add 8 μL mastermix to each sample and mix by pipetting.
8. Put samples in thermal cycler at 42°C and incubate for 2 h.

3.2.2.1. SECOND STRAND SYNTHESIS

1. Thaw the 10X second strand buffer and dNTP mix on ice and prepare second strand (SS) mastermix for the amount of sample (+5%); use per reaction: 63 µL RNase free water; 10 µL 10X second strand buffer; 4 µL dNTP mix; 2 µL DNA polymerase; and 1 µL RNAse H *(don't use an RNA pipet!!)*. Mix by pipetting, spin down, and put on ice.
2. Then after completion of the first strand synthesis, add 80 µL SS mastermix to each sample *(don't use an RNA pipet!)*.
3. Mix by pipetting, spin down and incubate for 2 h at 16°C in the thermal cycler.
4. If desired; samples can be immediately stored at –20°C until the next day.

3.2.2.2. cDNA PURIFICATION

1. Preheat RNase free water for at least 10 min at 50°C.
2. Take 1 tube and 1 cDNA filter cartridge per cDNA sample and label the tubes.

The following steps are performed in the fume hood:

3. Pipet 50 µL cDNA binding buffer from the Message Amp kit on top of the filter cartridge and incubate for 5 min at room temperature.
4. Meanwhile, pipet 250 µL DNA binding buffer to the cDNA samples, vortex and spin down.
5. Pipet cDNA sample mix on the prewetted filter cartridge using a 200 µL filtertip, centrifuge 1 min at 10,000g and discard flow through.
6. Add 500 µL cDNA wash buffer to the filter cartridge and centrifuge 1 min at 10,000g, aspirate, and flow through.
7. Centrifuge 1 min at 10,000g to dry the filter cartridge and to remove traces of alcohol.
8. Put filter cartridge in a clean labeled 2-mL elution tube.
9. To elute your cDNA, pipet 10 µL preheated (50°C) RNase free water on top of the filter, incubate 2 min at room temperature and centrifuge for 2 min at 10,000 rpm.
10. Repeat **step 9**.
11. Discard filter in waste container; the tubes contain about 16 µL cDNA solution.
12. Measure in each sample the volume using a pipet; volume should be between 16 and 18 µL.
13. If necessary, adjust the volume of each cDNA sample to 16 µL with RNase free water and store cDNA at –20°C until the next day or at –80°C for a longer time period.

3.2.3. Day 2: In Vitro Transcription to Synthesize aRNA

1. Set the temperature of the thermal cycler on 37°C.
2. Thaw the reagents needed for the in vitro transcription on ice and put them at room temperature once they are thawed (if the buffer is too cold reagents will precipitate).
3. Prepare a master mix if you have more than 1 sample (make 5% extra), containing per sample: 4 µL each of 75 mM ATP, CTP, GTP nucleotide vial ; 4 µL 75 mM

UTP solution; 4 μL T7 10X reaction buffer; and 4 μL T7 enzyme mix. Prepare at room temperature.
4. Pipet 24 μL mastermix to the samples, close the tubes, flick the tubes and spin down. Total volume is now 40 μL.
5. Incubate for 5 h in thermal cycler at 37°C (heated lid), after 5 h temperature set temperature at 4°C.
6. Put the samples on ice and set the temperature of the thermal cycler again on 37°C.
7. To digest the contaminating DNA pipet 2 μL DNase I at the top of the tube wall, flick the tube to mix and spin down and incubate for 30 min at 37°C in thermal cycler (heated lid).

3.2.3.1. aRNA Purification

1. Preheat 100 μL RNase free water per sample for at least 10 min at 50°C.
2. Take 1 clean tube and 1 filter cartridge per aRNA sample and move all materials to fume hood.
3. Pipet 100 μL RNA binding buffer on the filter and incubate for 5 min at room temperature.
4. Pipet 58 μL RNA elution solution to each aRNA sample (total volume 100 μL), vortex gently and add 350 μL aRNA binding buffer to each sample and vortex gently.
5. Transfer the samples to a new 1.5-mL tube and add 250 μL 100% ethanol to each sample.
6. Mix by pipetting, apply the 650 μL mix on the prewetted filter, spin 1 min at 10,000g and discard flow through.
7. Add 650 μL aRNA wash buffer to the filter, spin 1 min at 10,000g and aspirate the flow though.
8. Spin 1 min at 10,000g to remove all traces of ethanol and move the dry filter to a fresh 2 mL labeled tube.
9. Apply 50 μL preheated RNase free water on top of the filter and incubate 2 min at room temperature. Spin for 2 min at 10,000g and apply another 50 μL preheated RNase-free water on top of the filter. Incubate for 2 min at room temperature.
   Spin for 2 min at 10,000 rpm and discard filter.
   Measure the concentration and quality of the aRNA on the Agilent BioAnalyzer or NanoDrop ND-1000 spectrophotometer and store the aRNA at –80°C.

3.3. Indirect Labeling and Hybridization (see Fig. 2A)

The following is a slight modification by Mitch Garber, Anatoly Urisman and subsequently by Lisa van Baarsen of a protocol developed by Joe DeRisi (UCSF) and Rosetta Inpharmatics (Kirkland, WA). Original document can be obtained at www.microarrays.org.

3.3.1. cDNA Synthesis

Per array hybridization set up two separate reaction vials for labeling: one for the experimental sample to be labeled with Cy3 (green) and one for the reference sample to be labeled with Cy5 (red). In first instance both samples are labeled with aminoallyl-dUTP during cDNA synthesis by reverse transcriptase, followed by chemical coupling of the aminoallyl group to Cy3 and Cy5 for the experimental and reference sample respectively.

1. For total RNA, mix 25 µg of the RNA with 5 µg of anchored oligo-dT [$(dT)_{20}$-VN] (Operon, HPLC purified) in a total volume of 18 µL in 1.5-mL tubes. For aRNA, use 5 µg of the amplified cRNA with 10 µg of Poly D(N)6 (Operon) primer in a total volume of 18 µL in 1.5-mL tubes.
2. Heat to 70°C for 10 min in thermomixer at 300 rpm.
3. Cool on ice water for 5 min.
4. While cooling, make a nucleotide master mix for all your samples together. Nucleotide Mix for one reaction is as follows:
   a. 6.0 µL 5X room temperature buffer.
   b. 0.6 µL 50X dNTP stock solution (50X dNTP stock solution using a 4:1 ratio aminoallyl dUTP to dTTP [*see* Note 12]) (*see* note 11).
   c. 3.0 µL 0.1*M* dithiothreitol (DTT).
   d. 1.5 200 U/µL Superscript II RT (Gibco).
   e. 0.5 40 U/µL RNasin (Gibco, optional).
   f. 10 µL each 100 m*M* dATP, dGTP, dCTP (Pharmacia).
   g. 8 µL 100 m*M* aminoallyl-dUTP. Dissolve 10 mg aminoallyl-dUTP in 170 µL water. Add approx. 6.8 µL 1N NaOH (Sigma; cat. no. #A0410). Final pH is roughly 7.0 using pH paper.
   h. 2 µL 100 m*M* dTTP.
5. Pipet 11.6 µL nucleotide mix in a labeled 0.5 mL thin wall tube for both the Cy3 and Cy5 samples.
6. Spin the RNA samples down and add the 18 µL of sample or reference to the right labeled 0.5-mL thin wall tubes.
7. Incubate reaction for 1 h at 42°C in a PCR apparatus with heated lid.
8. Add additional 1 µL reverse transcriptase and continue incubation at 42°C for an additional 1 h.

3.3.2. Hydrolysis of RNA

1. Degrade RNA by addition of 15 µL of 0.1*N* NaOH. Incubate at 70°C for 10 min if aRNA is used or 30 min if total RNA is used as template.
2. Neutralize by addition of 15 µL 0.1*N* HCl. To continue with the amino-allyl dye coupling procedure, all Tris must be removed from the reaction to prevent the monofunctional NHS-ester Cy-dyes from coupling to free amine groups in solution.
3. Carefully add 450 µL RNase-free water to each reaction.

3.3.3. Cleanup

1. Apply the 500 µL neutralized, diluted reaction mix on a labeled Microcon-30 filter (Amicon) using a p200 filtertip, spin at 10,000g for 7 min and discard flow through.
2. Pipet 450 µL water on the filter, spin at 10,000g for 7 min and discard flow through.
3. Repeat washing **step 2**.
4. Concentrate the probe to 10 µL by centrifugation in steps of 30 s. Check after each step the fluid level in the filter. Never let the membrane dry completely!!
5. If there is less then half a circle of fluid on top of the filter, invert the microcon in a clean labeled tube and spin at 10,000g for 1 min to recover the aminoallyl labeled cDNA probe; adjust to 10 µL with water.
6. Samples can be stored at –20 or –80°C.

3.3.4. Coupling of Aminoallyl-Labeled cDNA to Cy3 or Cy5 ()

1. Add 0.5 µL 1M sodium bicarbonate (pH 9.0) (50 mM final) (*see* **Note 13**) to the 10 µL aminoallyl-labeled cDNA and transfer the sample (final volume is 10.5 µL) to the right aliquot of dye (Cy3 for experimental sample and Cy5 for reference) (*see* **Note 14**).
2. Incubate 1 h at room temperature in the dark. Mix every 15 min. For this purpose, the Thermomixer Comfort can be used.

3.3.5. Quenching and Cleanup

Before combining Cy3 and Cy5 samples for hybridization, unreactive NHS-ester Cy dye must be quenched to prevent cross coupling.

1. Add 4.5 µL 4M hydroxylamine.
2. Let reaction incubate 15 min in the dark (Thermomixer). To remove unincorporated/quenched Cy dyes, proceed with Qia-Quick PCR purification kit (QIAGEN). Method described below is as specified by the manufacturer.
3. Combine Cy3 and Cy5 reactions and add 70 µL RNase free water.
4. Add 500 µL Buffer PB and apply total sample to a Qia-quick column and spin at 10,000g for 30 to 60 s.
5. Re-apply flow-though for optimal binding again on the column and spin at 10,000g for 30 to 60 s.
6. Save flow-through in a tube in case you want to measure it (= wash 1).
7. Add 700 µL buffer PE and spin for 30 to 60 s.
8. Save flow-through in a tube in case you want to measure it (= wash 2).
9. Wash again with 700 µL buffer PE and spin 30 to 60 s.
10. Transfer column to a new eppendorf tube without a lid and spin at high speed for 1 min to dry the column.
11. Transfer spin unit to a fresh-labeled eppendorf tube, add 30 µL buffer EB on the centre of the filter and incubate for 3 min at room temperature.

12. Spin at high speed for 1 min.
13. Repeat elution **Subheading 3.3.5., steps 11** and **12** with another 30 μL of buffer EB.

3.3.6. Hybridization

If poly-L-lysine-coated glass sides were used to manufacture the microarrays a post-processing step of the microarrays is required before they can be used for hybridization. For a protocol see (*http://cmgm.stanford.edu/pbrown/protocols/3_post_process.html*).

Before hybridization a washing step is performed combined with irrelevant DNA and RNA to prevent nonspecific hybridization.

1. Add to your sample: 420 μl TE; 20 μL (=20 μg) Cot DNA (Gibco; cat. no. 15279-011); 2 μL (=20 μg) PolyA RNA (10 μg/μL [Sigma; cat. no.P9403]); and 2 μL (=20 μg) tRNA (10 μg/μL [Gibco-BRL; cat. no.15401-011]).
2. Apply the 500 μL of sample on a fresh Microcon-30 filter.
3. Select the appropriate probe volume from **Table 1**. Spin at 10,000*g* for 7 min to a volume of 28 μL or less (half a circle of fluid on filter).
4. Invert microcon in a fresh tube and spin at 10,000*g* for 1 min.
5. Measure probe volume; if a cover slip of 22 × 60 mm is used, adjust the sample probe volume to 28 μL by adding TE or speedvac to a volume of 28 μL.
6. Add to the 28 μL of probe: 5.95 μL 20X SSC, and 1.05 μL 10%SDS. Wipe the tip with dry clean gloves to remove access of SDS.
7. Heat sample for 2 min at 100°C in thermo block.
8. Centrifuge for 15 min at 10.,000*g*.
9. Take 1 hybridization chamber per sample and fill the 2 reservoirs with 5 μL of 3X SSC *just before placing the array in the chamber*.
10. Place the array in the chamber, make sure that the array is not in contact with the wall of the chamber.
11. Place the entire probe volume on the array and cover it very carefully with the cover slip (*see* **Note 15**).
12. Close the hybridization chamber and place it in a 65°C water bath (keep the chamber horizontal).
13. Hybridize for 14 to 18 h.

3.3.7. Washing and Scanning Arrays

1. Prepare the wash solutions (*see* **Note 16**) in slide chambers as indicated in **Table 2**. Avoid adding excess SDS. The wash chamber 1A and 2 should each have a slide rack ready. Place chamber 1A in a 37°C water bath.
2. Blot dry chamber exterior with paper towels and aspirate any remaining liquid.
3. Unscrew chamber; aspirate the holes to remove last traces of water bath liquid.
4. Place arrays, singly, in rack, inside wash chamber 1A (maximum 4 arrays at a time). Allow cover slip to fall, or *carefully* use forceps to aid cover slip removal

Table 1
Coverslip Sizes With Corresponding Hybridization Volumes

Cover Slip Size (mm)	Total Hyb volume (µL)	Probe & TE (µL)	20X SSC (µL)	10% Sodium dodecyl sulfate (SDS) (µL)
22 × 22	15	12	2.55	0.45
22 × 40	25	20	4.25	0.75
22 × 60	35	28	5.95	1.05

Table 2
Wash Solutions

Wash	Description	Volume (mL)	Temp.	Rack
1A	2X SSC, 0.03% SDS	500	37°C	yes
1B	2X SSC	500	RT	no
2	1X SSC	500	RT	yes
3	0.2X SSC	500	RT	no

if it remains stuck to the array. *DO NOT AGITATE* until cover slip is safely removed. Then agitate for 2 min.
5. Remove array by forceps, rinse in a wash chamber 1B *without* a rack, and transfer to the wash chamber 2 in a new rack. This step minimizes transfer of SDS from wash 1 to wash 2.
6. Wash arrays by submersion and agitation for 2 min in wash chamber 2, then for 2 min in wash 3 (transfer the entire slide rack this time).
7. Spin dry by centrifugation in a slide rack in a Beckman GS-6 tabletop centrifuge at 1000 rpm for 3 min.
8. Optional: remove dust with air pressure.
9. Scan the arrays at the appropriate wavelengths for the Cy3 (532 nm) and Cy5 (632 nm) signal with a confocal scanner (Agilent). Images are saved as two separate .tiff files for Cy3 and Cy5.

3.4. Data Storage and Analysis

3.4.1. Analysis of Image Files

Gridding: Tiff files are analysed by the spot analysis program Genepix Pro 3.0. Spots with low intensity are flagged automatically and spots with irregular shapes are hand-flagged.

3.4.2. Processing of Data

The image files (Tiff files), genepix grid (.gps) and data files (.gpr) are loaded in the Stanford Microarray Database (SMD) *(14)* at *http://genome-www5.stanford.edu*. In SMD the gene names are connected to the spot-locations on the array. Normalization is performed: the red signal intensity and red background are adjusted to the green signal by global normalization such that the average log ratio becomes 1. The local background is subtracted from the intensity signal for both channels and the red to green ratio is calculated and log2 transformed.

Quality criteria: Spots were used for which the mean intensity was 1.6 times the median local background for both channels. Only genes for which 80% of the experiments showed datapoints that passed the flagging- and quality control criteria were used.

3.4.3. Clustering

Cluster (M. Eisen, *http://rana.lbl.gov/EisenSoftware.htm* *[6]*) is used to center the genes, that is, genes are expressed relative to their median expression level across all experiments. After centering a selection of genes can be made that show a certain degree of variation in expression across experiments, for instance those genes that show at least a twofold difference (absolute value 1 in log 2) from the median expression level in at least 4 experiments. Next the genes (and arrays) are hierarchically clustered (average linkage) by which genes and arrays with a correlated profile are placed in adjacent rows (genes) and columns (arrays). Results were visualized with treeview *(http://rana.lbl.gov/EisenSoftware.htm)*. Graphic representations of treeview files were made in Adobe Illustrator. Classification of samples were thus performed after clustering the arrays (unsupervised clustering). In case the groups are predetermined (for instance patients vs controls), supervised clustering can be performed, by which the arrays remain in their original file location.

3.4.4. Statistical Analysis

Determination of the genes that are expressed at significantly different levels between groups of patients/arrays was performed by Statistical Analysis of Microarray (SAM) data. (*http://www-stat.stanford.edu/~tibs/SAM*) *(15)*. SAM adjusts for multiple testing by performing data permutations that provides an estimate of the False Discovery Rate. A q-value is generated for each gene that indicates the lowest False Discovery Rate at which a gene is called significant.

It is similar to the familiar *p* value, adapted to the analysis of a large number of genes.

4. Notes

1. Gloves should be worn whenever handling biological samples. Dispose of all biological solid waste in biological hazardous waste container. Dispose of all liquid biological waste by adding bleach and disposing it down into the sink; followed by running water for a couple of minutes.
2. RNA more quickly dissolves in 50°C water.
3. RNA is isolated according to the PAXgene Blood RNA Kit Handbook, April 2001, 1017455 04/2001; supplemented with adjustments from the PAXgene development team and from http://cmgm.stanford.edu/pbrown/protocols/short_paxgene_protocol.html.
4. PreAnalytiX allows the tubes to be centrifuged in the range of 3000g to 5000g.
5. Incomplete removal of the supernatant will inhibit lyses and dilute the lysate, which will affect conditions for binding RNA to the PAXgene column.
6. Do not mix BR2 and Proteinase together before adding to sample!!
7. Avoid transfer of small debris remaining in supernatant after centrifugation; this will affect the yield.
8. Do not exceed more than a 2 second spin; this may result in pelleting of the nucleic acid.
9. Do not vortex DNase I, only gently flick the tube to mix, because DNase is sensitive to physical denaturation.
10. Ensure BR4 is diluted with ethanol.
11. 1X dNTP final concentration during labeling: 500 µM each dATP, dCTP, dGTP; 400 µM aminoallyl-dUTP; and 100 µM dTTP.
12. Altering the ratio of aminoallyl-dUTP to dTTP will affect the incorporation of Cy dye.
13. Check 1M stock solutions periodically for fluctuations in pH.
14. Monofunctional NHS-ester Cy3 and Cy5 dye is supplied as a dry pellet. Each tube is sufficient to label 10 reactions under normal conditions. Dissolve dry pellet in 20 µL DMSO. Aliquot 2 µL into 10 tubes that are then dried in vacuum (25 min) and stored desiccated at 4°C. NHS-ester conjugated Cy dye is rapidly hydrolyzed in water, therefore, do not store in DMSO or water. Decreasing the number of aliquots/dye tube may increase your signal.
15. Use 2 forceps, one to prevent the coverslip from sliding of the microarray and the other one to carefully lower the coverslip down on the microarray.
16. Prepare wash buffers in 1- to 5-L bottles by diluting the 20X SSC and adding 10% SDS.

Acknowledgments

This study was financially supported in part by the Howard Hughes Medical Institute, a grant from the National Cancer Institute, a research grant from the Netherlands Organization for Scientific Research, the Dutch Arthritis Foundation and the Centre for Medical Systems Biology (CMSB), a centre of excellence approved by the Netherlands Genomics Initiative/Netherlands Organization for Scientific Research (NWO). We are grateful to Drs Pat Brown and David Botstein in whose laboratories most of the work described in this report was performed.

References

1. Poirier, G. M. and Erlander, M. G. (1998) Postdifferential display: parallel processing of candidates using small amounts of RNA. *Methods* **16**, 444–452.
2. Van Gelder, R. N., von Zastrow, M. E., Yool, A., Dement, W. C., Barchas, J. D., and Eberwine, J. H. (1990) Amplified RNA synthesized from limited quantities of heterogeneous cDNA. *Proc. Natl. Acad. Sci. USA* **87**, 1663–1667.
3. Hughes, T. R., Mao, M., Jones, A. R., et al. (2001) Expression profiling using microarrays fabricated by an ink-jet oligonucleotide synthesizer. *Nat. Biotechnol.* **19**, 342–347.
4. Randolph, J. B. & Waggoner, A. S. (1997) Stability, specificity and fluorescence brightness of multiply-labelled fluorescent DNA probes. *Nucleic Acids Res.* **25**, 2923–2929.
5. Brown, P. O. and Botstein, D. (1999) Exploring the new world of the genome with DNA microarrays. *Nat. Genet.* **21**, 33–37.
6. Churchill, G. A. (2002) Fundamentals of experimental design for cDNA microarrays. *Nat. Genet.* **32 Suppl:490-5.**, 490–495.
7. Yang, Y. H. and Speed, T. (2002) Design issues for cDNA microarray experiments. *Nat. Rev. Genet.* **3**, 579–588.
8. Perou, C. M., Jeffrey, S. S., van de, R. M., et al. (1999) Distinctive gene expression patterns in human mammary epithelial cells and breast cancers. *Proc. Natl. Acad. Sci. USA* **96**, 9212–9217.
9. Boldrick, J. C., Alizadeh, A. A., Diehn, M., et al. (2002) Stereotyped and specific gene expression programs in human innate immune responses to bacteria. *Proc. Natl. Acad. Sci. USA* **99**, 972–977.
10. Eisen, M. B., Spellman, P. T., Brown, P. O., and Botstein, D. (1998) Cluster analysis and display of genome-wide expression patterns. *Proc. Natl. Acad. Sci. USA* **95**, 14,863–14,868.
11. Alizadeh, A. A., Eisen, M. B., Davis, R. E., et al. (2000) Distinct types of diffuse large B-cell lymphoma identified by gene expression profiling. *Nature* **403**, 503–511.
12. van der Pouw Kraan TC, Van Gaalen, F. A., Kasperkovitz, P. V., et al. (2003) Rheumatoid arthritis is a heterogeneous disease: Evidence for differences in the

activation of the STAT-1 pathway between rheumatoid tissues. *Arthritis Rheum.* **48**, 2132–2145.
13. Brumbaugh, J. A., Middendorf, L. R., Grone, D. L., and Ruth, J. L. (1988) Continuous, on-line DNA sequencing using oligodeoxynucleotide primers with multiple fluorophores. *Proc. Natl. Acad. Sci. USA* **85**, 5610–5614.
14. Gollub, J., Ball, C. A., Binkley, G., et al. (2003) The Stanford Microarray Database: data access and quality assessment tools. *Nucleic Acids Res.* **31**, 94–96.
15. Tusher, V. G., Tibshirani, R., and Chu, G. (2001) Significance analysis of microarrays applied to the ionizing radiation response. *Proc. Natl. Acad. Sci. USA* **98**, 5116–5121.

23

Differential Display Reverse Transcription-Polymerase Chain Reaction to Identify Novel Biomolecules in Arthritis Research

Manir Ali and John D. Isaacs

Abstract

Differential display is one of the simplest techniques for discovering novel transcripts when comparing gene expression in biological systems. The method can be carried out on small amounts of total RNA and permits the simultaneous comparison of multiple independent samples in a single experiment. The methodology is versatile in that it allows the researcher to adapt the existing protocol by varying the selection of oligonucleotide primers used in the PCR. The putative differentials, which are isolated from the polyacrylamide gels as cDNA fragments of approx 100 to 500 bases in size, can be instantly recognized after sequencing by searching the nucleotide databases. However, one of the drawbacks is the isolation of false positives and hence the need to confirm the results of the screen by another method. Once the true differentials have been identified, further downstream work is also required to recognize which splice variant of the transcript gives rise to the expression differences and whether the gene expression results are corroborated at the protein level.

Key Words: Differential display; arthritis; biomarkers; novel therapeutic targets; comparative gene expression; DNA-free RNA; reverse transcription; arbitrarily-primed polymerase chain reaction; polyacrylamide gel electrophoresis; transcriptional profiles; cDNA isolation; novel transcripts; reverse-northern analysis.

1. Introduction

There is currently a lack of biomolecules that allow discrimination between persistent and self-limiting synovitis as well as the different forms of arthritis. These markers could also act as novel therapeutic targets for inflammatory disease. So that when a patient first presents in the clinic the discriminatory markers would allow therapies to be tailored according to the patient's need and so result in a more effective regime for the treatment of the rheumatic condition. In an effort to identify such markers a number of methodologies are available that encompass the study of genomics and proteomics. These include two-dimensional (2D) protein gel electrophoresis combined with mass spectrometry, subtractive hybridization, serial analysis of gene expression (SAGE), differential display reverse transcription-polymerase chain reaction (DDRT-PCR) *(1,2)* and the analysis of cDNA microarrays. Although all of these approaches do not have any predetermined expectations of what will be discovered, the results from such investigations often form the platform for subsequent hypothesis-driven research. Of the techniques, the latter two have been the most popular in arthritis research to date. The analysis of cDNA microarrays provides a high-throughput approach that is best for identifying gene expression signatures and overall effects on pathways that may shed some light on disease pathogenesis *(3–5)*. The limitation of array technologies is that the findings of such research are entirely dependent on which cDNAs are on the grid in the first place. The differential display approach, however, has the ability to isolate novel biomolecules since it can also detect rare transcripts as well as splice variants that would otherwise be overlooked. A drawback is the isolation of false positive clones *(6,7)* and the requirement to confirm the results by another method. However the design of more focused strategies for the display protocol such as targeted differential display offers several exciting possibilities that would reduce the false positive output. The differential display technique for arthritis research has already been shown to be applicable to clonal populations of cells (anergic T-cells *[8]*), primary cell cultures (synoviocytes *[9,10]*) as well as the more complex, tissue specimens that have been obtained either at postmortem or during routine surgical procedures (cartilage *[11–13]* and joint tissue *[14,15]*). More recently the demonstration of laser mediated microdissection of synovial sections for differential display offers new possibilities for analysing unmanipulated cell subsets *(16)*. Biomolecules identified using the differential display technique include semaphorin E *(9)*, the kinesin-like protein CENP-E *(10)* and inhibitor of differentiation-1 (ID1) *(15)* for rheumatoid arthritis and the serine protease HtrA *(11)*, tumor necrosis factor (TNF)-α convertase enzyme TACE *(12)* and osteopontin *(13)* for osteoarthritis as novel therapeutic targets that may be tested in the future for the modulation of these arthritic diseases.

This chapter describes the basic differential display protocol although many variations to this versatile technique exist *(17)*. The differential display technique consists of three stages (*see* **Fig. 1**). The first part is to generate a transcriptional profile for each sample to be compared using RT-PCR and polyacrylamide gel electrophoresis (PAGE). The second is to isolate and iden-

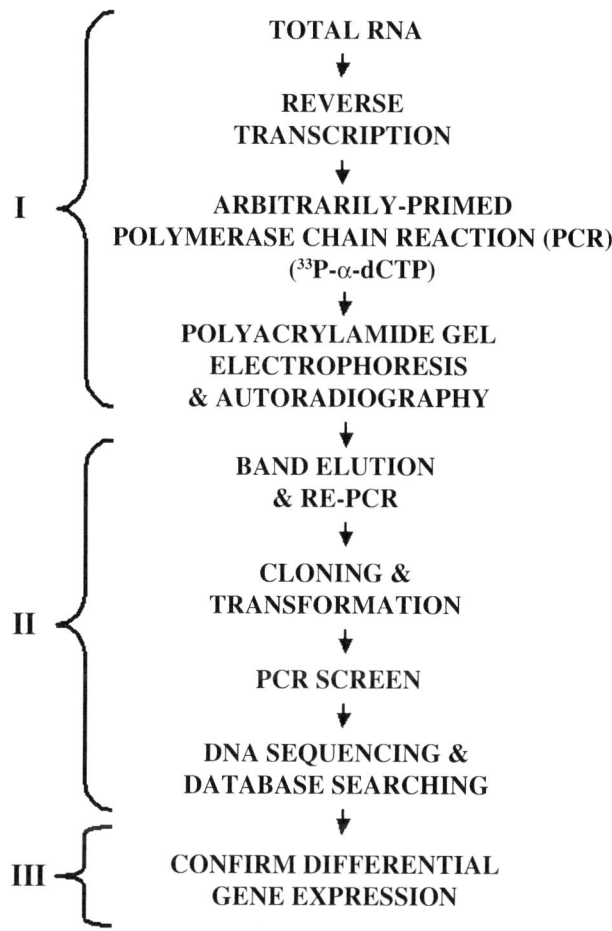

Fig. 1. Overview of the differential display protocol. The three stages of the method are depicted; (I) generating a transcriptional profile, (II) isolating the putative differentials and (III) confirming the true differentials.

tify the putative differentials and the third component is to confirm the true differentials. Briefly, once the RNA concentrations have been standardized, the samples participate in cDNA synthesis using reverse transcriptase and one of a set of four degenerate oligonucleotide primers $(dT)_{12}VG$, $(dT)_{12}VC$, $(dT)_{12}VA$, or $(dT)_{12}VT$ where V is G, C, and A *(18)*. In theory, each of the degenerate oligonucleotides transcribes from a quarter of all the mRNAs in the sample by binding to the poly (dA) tails located at the 3'-ends of the transcripts. Each mixture of single-stranded cDNAs is used as a template in a series of PCRs with radiolabeled nucleotide ^{33}P-α-dCTP, the degenerate oligonucleotide used in the first-strand synthesis and one of a set of random decamers that can anneal to any cDNA. Hence, the method allows the selective amplification of arbitrarily-primed cDNAs so that by varying the random primer in the PCR, different populations of cDNA can be amplified. Most of the amplified cDNAs arise from oligo dT-decamer priming and so represent the 3'-untranslated regions of transcripts. The PCR products from each sample to be compared are migrated simultaneously through a denaturing polyacrylamide gel by electrophoresis. The gel is dried and exposed to X-ray film, so that under autoradiography a profile is generated that represents the variety of amplified cDNAs in each PCR. Visual inspection of the autoradiogram identifies differences between the banding patterns as gene expression differences between the samples. The cDNA bands are eluted from the gel and amplified in the PCR as before but without radioactive incorporation. Given that the bands often contain more than one cDNA moiety *(19,20)*, the PCR products are cloned into a plasmid vector for transformation into bacteria. Multiple clones are selected and screened for the presence of insert DNA using the PCR with vector-sequence specific primers. Only the recombinants are sequenced and the nature of the DNA identified by searching the nucleotide databases at the National Center for Biotechnology Information *(http://www.ncbi.nlm.nih.gov)*. Because of the isolation of false positives, a high-throughput screening strategy is required for testing the putative differentials. Reverse-northern analysis is recommended as a confirmatory step *(8,21)*. Briefly, multiple identical cDNA grids are manufactured containing immobilized PCR products that have been derived from each putative differential. The samples to be compared are radiolabeled with ^{33}P-α-dCTP using the random prime labeling method before hybridizing onto identical gridded membranes. After several stringent washes, the membranes are exposed to phosphor-imaging screens. Once the images are developed, the analysis software overcomes the issue of sensitivity in detecting the true differentials.

2. Materials

2.1. General Reagents for the Polymerase Chain Reaction and Agarose Gel Electrophoresis

1. Taq DNA polymerase kit (Promega) contains 10X PCR buffer, 25 m*M* MgCl$_2$, and Taq DNA polymerase (5 U/µL). Stored at –20°C.
2. Deoxynucleoside triphosphates (dNTPs) (100 m*M* of each of dATP, dTTP, dCTP, and dGTP) (Promega). Stored at –20°C.
3. Sterile water.
4. Agarose.

5. 10X TBE: 0.89 M Tris, 0.89 M boric acid, and 0.02 M ethylene diamine tetraacetic acid (EDTA) (pH 8.3).
6. Ethidium bromide (EB) solution (10 mg/mL) (Promega). Store in foil wrapped bottle. Mutagen.
7. DNA loading dye: 40% glycerol, 60% sterile water, 0.00005% xylene cyanol, and 0.00005% orange G.
8. Standard 100-bp DNA ladder (New England Biolabs). Stored at –20°C.

2.2. Preparation of Total RNA

1. TRIzol™ reagent (Invitrogen). Store at 4°C. *Caution: Toxic.*
2. Chloroform/isoamylalcohol (24:1). *Caution: Toxic.*
3. Isopropyl alcohol.
4. 70% ethanol made with diethylpyrocarbonate (DEPC)-water.
5. DEPC-water. Add 0.1 mL DEPC to 100 mL distilled water and mix. Leave at room temperature overnight. Autoclave and allow to cool. DEPC is an irritant.
6. RNase-inhibitor (40 U/μL) (Roche). Store at –20°C.
7. DNA-free™ kit (Ambion). Store at –20°C.
8. Oligonucleotide $(dT)_{15}$ primer. Store at –20°C.
9. Superscript™ II reverse transcriptase kit (Invitrogen) contains 5X RT buffer, 0.1 M dithiothreitol (DTT), and Superscript™ II reverse transcriptase (200 U/μL). Store at –20°C.
10. Oligonucleotide primers p53F (dGTACTCCCCTGCCCTCAACA) and p53R (dCTGGAGTCTTCCAGTGTGAT). Store at –20°C.

2.3. Differential Display Reverse Transcriptase Polymerase Chain Reaction

1. DEPC-water. DEPC is an irritant.
2. Superscript™ II reverse transcriptase kit (Invitrogen). Store at –20°C.
3. Oligonucleotide primers $(dT)_{12}VG$, $(dT)_{12}VC$, $(dT)_{12}VA$, and $(dT)_{12}VT$ where V is A, C and G. Store at –20°C.
4. Random decamer. Store at –20°C.
5. ^{33}P-α-dCTP (10 mCi/mL) (Amersham Biosciences). Store at 4°C. *Caution: Radioactive.*
6. SequaGel™-6 (National Diagnostics). *Caution: Neurotoxin.*
7. 25% ammonium persulfate. Store at –20°C.
8. TEMED (N,N,N',N'-tetramethylethylenediamine). *Caution: Toxic.*
9. Formamide loading dye (96% deionized formamide, 4% 0.5 M EDTA [pH 8.0], 0.05% bromophenol blue, and 0.05% xylene cyanol). Store at –20°C. *Caution: Toxic.*

2.4. Identification of Putative Differentially Expressed Gene(s)

1. Glycogen (20 mg/mL) (Roche). Store at –20°C.
2. 3 M NaAc (pH 5.2) (sodium acetate).

3. 100%, 85%, and 70% ethanol.
4. Oligonucleotide primers $(dT)_{12}VG$, $(dT)_{12}VC$, $(dT)_{12}VA$, and $(dT)_{12}VT$. Store at –20°C.
5. Random decamer. Store at –20°C.
6. TA cloning™ kit (Invitrogen) contains TA cloning vector, pCR™2.1, 10X ligase buffer and T4 DNA ligase (4 U/μL). Store at –20°C.
7. INVαF 'one shot'™ competent cells (Invitrogen). Store at –80°C. Biological hazard.
8. LB-agar (Lennox L broth) (Sigma). Dissolve 20 g in 1 L distilled water. Add 15 g bacto-agar and autoclave. Add the antibiotic and pour the plates. Once set, store the plates at 4°C.
9. Kanamycin solution (5 mg/mL). Stored at –20°C. *Caution: Toxic.*
10. Oligonucleotide primers M13F (dACGTTGTAAAACGACGGCCAG) and M13R (dTCACACAGGAAACAGCTATGAC). Store at –20°C.
11. QIAquick™ PCR purification kit (Qiagen).
12. Reagents for DNA sequencing include BigDye™ terminator cycle sequencing kit, store at –20°C (Applied Biosystems), SequaGel™-4.25, neurotoxin (National Diagnostics), 10% ammonium persulfate. Store at –20°C and deionized formamide loading dye (8 mL deionized formamide added to 2 mL dextran blue solution (25 mM EDTA [pH 8.0], 50 mg/mL dextran blue). Store at –20°C. *Caution: Toxic.*

2.5. Confirming Differential Gene Expression by Reverse-Northern Analysis

1. Oligonucleotide primers M13F and M13R. Store at –20°C.
2. cDNA specific primer (approx 21mer). Store at –20°C.
3. Oligonucleotide primers GAPDHF (dACCACAGTCCATGCCATCAC) and GAPDHR (dTCCACCACCCTGTTGCTGTA). Store at –20°C.
4. QIAquick™ PCR purification kit (Qiagen).
5. Reagents for dot-blotting include 20X SSC (3 M NaCl, 0.3 M trisodium citrate dihydrate [pH 7.0] Adjust pH with 1 M HCl), 0.4 M NaOH (sodium hydroxide), 2 M NH$_4$Ac (ammonium acetate), and 2X SSC, 0.00005% bromophenol blue.
6. Random prime labeling kit (Rediprime™ II) (Amersham Biosciences). Store at –20°C.
7. ^{33}P-α-dCTP (10 mCi/mL) (Amersham Biosciences). Store at 4°C. *Caution: Radioactive.*
8. G-50 microspin columns (Pharmacia Biotech).
9. Human Cot-1 DNA (1 mg/mL) (Roche). Store at –20°C.
10. Poly (dA-dT) (1 mg/mL) (Pharmacia Biotech). Store at –20°C.
11. DIG Easy Hyb buffer (Roche).
12. 0.1% sodium dodecyl sulfate (SDS), 0.2X SSC.

3. Methods

The methods section describes the preparation of high-quality total RNA, the differential display RT-PCR, the identification of putative differentials, and a protocol for confirming the true differentials (*see* **Note 1**).

3.1. Preparation of Total RNA

3.1.1. Extraction of Total RNA From Cultured Cells and DNase Treatment (see **Notes 2** and **3**)

1. Add 1 mL TRIzol™ reagent to 10^7 cells in suspension or a 10 cm² dish for cells as a monolayer (*see* **Notes 4** and **5**). Lyse the cells by repetitive pipetting.
2. Transfer the homogenate into a tube and proceed with chloroform/isoamylalcohol extraction, isopropyl alcohol precipitation and the ethanol wash according to the instructions in the TRIzol™ protocol (Invitrogen). Air-dry the pellet and dissolve in DEPC-water, then add RNase-inhibitor.
3. Proceed with the instructions for the DNA-free™ kit (Ambion) to remove contaminating genomic DNA. If using the RNA on the same day, incubate on ice, otherwise store at –80°C (*see* **Note 6**).

3.1.2. Quality Control Check on the RNA

This 3-stage protocol involves first-strand cDNA synthesis using an oligonucleotide $(dT)_{15}$ primer and reverse transcriptase, followed by PCR using gene specific primers that span an intron in human genomic DNA and analysis of the reactions after migrating through an agarose gel by electrophoresis.

3.1.2.1. REVERSE TRANSCRIPTION

1. Add 1 µL total RNA to 1 µL oligonucleotide $(dT)_{15}$ primer (10 pmol/µL) and 10 µL DEPC-water, then incubate at 65°C for 5 min and rapidly chill on ice.
2. Make a master mix for $n + 1$ samples containing 4 µL 5X RT buffer, 2 µL DTT (0.1 *M*) and 1 µL dNTPs (10 m*M* each) for each sample. Dispense 7 µL of the master mix into each tube and incubate at 42°C for 5 min before adding 1 µL Superscript™ II reverse transcriptase (200 U/µL). Continue incubating at 42°C for a further 60 min, then heat to 65°C for 10 min.
3. Store the samples at 4°C until required. For long term storage keep in the –20°C freezer.

3.1.2.2. POLYMERASE CHAIN REACTION (*SEE* **NOTE 7**)

1. Prepare a master mix for $n + 1$ samples containing 1 µL 10X PCR buffer, 0.6 µL MgCl$_2$ (25 m*M*), 1 µL dNTPs (2 m*M* each), 1 µL of each oligonucleotide primer p53F (10 pmol/µL) and p53R (10 pmol/µL) as well as 0.25 µL Taq DNA polymerase (5U/µL) and 4.15 µL sterile water for each sample in a tube. Dispense 9 µL into a tube and add 1 µL cDNA. As a control add 1 µL sterile water instead of cDNA.
2. Thermocycle at 95°C for 5 min, followed by 94°C for 30 s, 55°C for 30 s, and 72°C for 30 s for 40 cycles with a final extension at 72°C for 5 min.

3.1.2.3. AGAROSE GEL ELECTROPHORESIS

1. Prepare a 1.5% gel with agarose, 0.5X TBE and 0.25 µg/mL ethidium bromide (*see* **Note 8**).

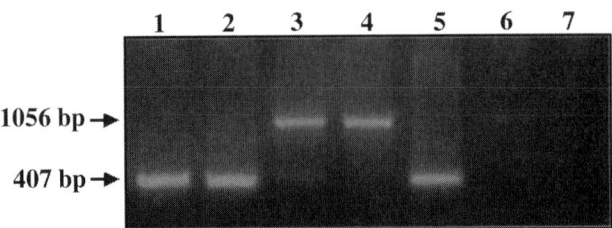

Fig. 2. Photograph depicting DNA fluorescence of ethidium bromide stained agarose gel under ultraviolet illumination. The lanes represent the migration of amplified PCR products after electrophoresis. The PCR has been carried out with p53 primers on cDNA that had been synthesized from T-cell RNA after DNase treatment (*lanes 1* and *2*), T-cell RNA without DNase treatment (*lane 3*) and PBMC RNA after DNase treatment as a control (*lane 5*). *Lane 4* represents the PCR product after using human genomic DNA as template. *Lane 6* corresponds to the PCR product after using a 1 in 500 dilution of cDNA synthesized from the DNase-untreated RNA and *lane 7* signifies the no DNA negative control.

2. Add DNA loading dye to the PCR products as well as the DNA standard ladder and load onto the gel. Electrophorese in 0.5X TBE running buffer. Visualize the DNA bands under ultraviolet illumination.
3. The PCR products migrate through the gel to give a band of 407 bp in the presence of cDNA (*see* **Fig. 2**). However the presence of a 1056-bp band reveals the presence of genomic DNA that must be removed by further DNase treatment of the RNA.

3.2. Differential Display Reverse Transcriptase Polymerase Chain Reaction

3.2.1. Standardizing the RNA Between Samples

There is no distinctive way of RNA standardization. However, when working with cultured cells it is recommended to start with the same cell numbers for each of the comparative conditions. To get an idea of the RNA concentration, an optical absorbance reading at 260 nm can be taken with only 1µl of the sample on the Nanodrop™ spectrophotometer (Nixor Biotech). Alternatively, the samples can be standardized against the abundance of a housekeeping transcript by quantitative real-time RT-PCR *(22)*.

3.2.2. First-strand cDNA Synthesis Using Reverse Transcriptase (see **Note 9**)

Reverse transcription of the RNA is carried out as described before in **Subheading 3.1.2.1.**, except that four oligonucleotide primer mixtures are used, $(dT)_{12}VG$, $(dT)_{12}VC$, $(dT)_{12}VA$, or $(dT)_{12}VT$, instead of the oligonucleotide

$(dT)_{15}$ primer. Theoretically, this helps to reduce the complexity of the banding patterns on the gels by synthesizing cDNA from a quarter of the sample during each reaction (see **Note 10**).

3.2.3. Radioactive Polymerase Chain Reaction Using Taq DNA Polymerase and ^{33}P-α-dCTP (see **Notes 11–13**)

1. Dispense 2 µL of the cDNA into the bottom of a PCR tube. Make up a master mix for $n + 1$ samples containing 0.5 µL 10X PCR buffer, 0.3 µL $MgCl_2$ (25 mM), 0.2 µL dNTPs (5 mM dATP, 5 mM dTTP, 5 mM dGTP, and 0.625 mM dCTP), 0.2 µL oligonucleotide primer mixture used in the reverse transcription reaction (10 pmol/µL), 0.2 µL random decamer (10 pmol/µL) (see **Notes 14** and **15**), 0.1 µL Taq DNA polymerase (5 U/µL) and 1.48 µL sterile water for each PCR. Aliquot 0.02 µL ^{33}P-α-dCTP (10 mCi/mL) for each reaction into the master mix and distribute 3 µL into each tube containing the cDNA.
2. Thermocycle at 95°C for 2 min, followed by 94°C for 30 s, 42°C for 2 min (41°C for $(dT)_{12}VA$ or $(dT)_{12}VT$ primers) and 72°C for 30 s for 40 cycles with a final extension at 72°C for 5 min (see **Note 16**).
3. The PCR products can be stored at 4°C overnight or in the –20°C freezer for up to 2 wk.

3.2.4. Polyacrylamide Gel Electrophoresis and Autoradiography (see **Note 17**)

1. Assemble the gel casting apparatus according to the manufacturer's instructions. Prepare a 6% polyacrylamide solution using SequaGel-6, add ammonium persulfate and TEMED, before pouring between the sandwiched plates. Insert the flat edge of the shark-toothed comb into the gel and allow to set for at least 2 h. Take out the comb and reinsert so that the teeth just touch the surface of the gel. Prerun the gel for 20 min in 1X TBE buffer.
2. Meanwhile, add 5 µL formamide loading dye to each sample and heat to 96°C for 4 min before chilling on ice. Once the prerun is complete, load 2 µL of the sample into each well so that reactions with the same primer set are adjacent to one another. The remaining sample can be stored in the freezer for 2 wk. Electrophorese until the bromophenol blue dye reaches the bottom of the gel.
3. Dismantle the gel apparatus and blot the gel with 3MM chromatography paper. Dry under vacuum suction, then place into a cassette. Mark the gel with radioactive ink then overlay with an X-ray film and close (see **Note 18**). Develop the autoradiogram after 48 h exposure to the gel (see **Fig. 3**) *(17)*.

3.3. Identification of Putative Differentially Expressed Gene(s)

3.3.1. Band Elution and Re-Polymerase Chain Reaction

1. By visual inspection of the autoradiogram, select the altered bands between different samples that were used with the same primer set. Orientate the developed autoradiogram onto the dried gel using the radioactive ink marks and with a fine-

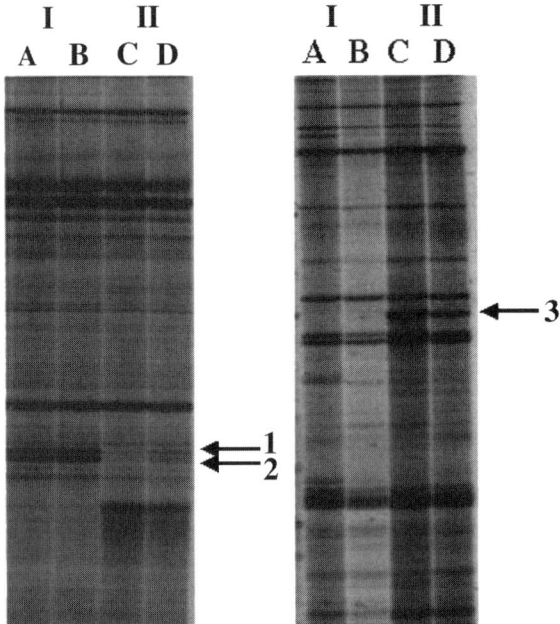

Fig. 3. Autoradiograms showing differential display RT-PCR gels. The lanes correspond to duplicate PCRs using cDNA synthesized from RNA that has been isolated from a resting T-cell clone (*lane 1*) and peptide-induced anergy of the clone (*lane 11*). The images depict that bands 1 and 2 represent cDNAs whose transcripts are down-regulated during the anergic process, whereas band 3 corresponds to cDNAs that are derived from up-regulated transcripts. Reprinted with permission from ref. **17**.

needle pierce the film and the dried gel at the corners where the band lies. Cut out the slice joining the needle marks from the dried gel with a scalpel. Place the dried gel into 100 µL sterile water and seal the tube lid with parafilm before boiling for 10 min.
2. Add 90 µL of the supernatant to 2.5 µL glycogen (20 mg/mL), 10 µL NaAc (pH 5.2) (3 M) and 450 µL 100% ethanol. Mix and incubate at –80°C for 30 min. Microfuge at 4°C for 15 min then wash the pellet with 85% ethanol, air dry and dissolve in 10 µL sterile water.
3. Prepare a master mix for $n + 1$ samples containing 2 µL 10X PCR buffer, 1.2 µL MgCl$_2$ (25 mM), 2 µL dNTPs (2 mM each), 0.8 µL oligo (dT) primer mixture used before (10 pmol/µL), 0.8 µL random decamer used before (10 pmol/µL), 0.4 µL Taq DNA polymerase (5 U/µL) and 8.8 µL sterile water for each reaction. Dispense 16 µL into a PCR tube and add 4 µL eluted cDNA. As a control, add 4 µL sterile water instead of cDNA.

Differential Display RT-PCR

4. Thermocycle using the same parameters that produced the gel fingerprint from which the cDNA band was eluted.

3.3.2. Cloning the cDNA Fragments Into a Plasmid Vector and Transformation Into a Bacterial Host

1. Following ethanol precipitation of the PCR products dissolve the pellet in 10 µL sterile water.
2. Ligate the purified PCR products into the TA cloning vector, pCR™2.1 according to the manufacturer's protocol (Invitrogen).
3. Transform the ligation mixture into INVαF 'one shot'™ competent cells according to the manufacturer's instructions (Invitrogen). Spread the culture onto an LB-agar plate containing kanamycin (50 µg/mL) and incubate at 37°C overnight.

3.3.3. Direct Polymerase Chain Reaction to Screen for Recombinants (see **Note 19**)

1. Set up a master mix for $n + 1$ samples containing 1.5 µL 10X PCR buffer, 0.9 µL MgCl$_2$ (25 mM), 1.5 µL dNTPs (2 mM each), 0.3 µL of each of the vector sequence specific primers M13F (10 pmol/µL) and M13R (10 pmol/µL), 0.12 µL Taq DNA polymerase (5 U/µL), and 10.38 µL sterile water for each reaction. Dispense 15 µL of the mixture into each well of a PCR plate. Using a sterile tip touch a colony of the transformed bacteria and place into a well containing the master mix. Bearing in mind that the cDNA fragments from which the clones are derived often contain more than one cDNA moiety, we recommend picking 4 colonies from each plate.
2. Thermocycle at 95°C for 2 min, followed by 94°C for 30 s, 54°C for 30 s, and 72°C for 30 s for 35 cycles, with a final extension at 72°C for 5 min.
3. Add DNA loading dye to 5 µL PCR product and migrate through a 1.5% agarose gel by electrophoresis. Visualize the DNA under ultraviolet light. In the absence of a recombinant DNA insert, the PCR product reveals a 210-bp band.
4. Purify the remaining 10µl PCR product from the clones containing a recombinant insert using the QIAquick™ PCR purification kit according to the manufacturer's instructions (Qiagen).

3.3.4. DNA Sequencing and Database Searching (see **Note 20**)

1. Sequence the purified PCR products with either M13F or M13R primers using the BigDye™ terminator cycle sequencing kit and the ABI PRISM™ 377 DNA sequencer according to the manufacturer's instructions (Applied Biosystems).
2. Analyze the gel run using the ABI PRISM™ DNA sequence analysis packages. The chromatograms indicate the quality of the sequence. The simple text files can be used to highlight the insert DNA sequence for pasting into the BLAST algorithm at the NCBI. This permits the identity of the cDNA to be identified against the nucleotide databases.

3.4. Confirming Differential Gene Expression by Reverse-Northern Analysis (see Note 21)

3.4.1. Manufacture of cDNA Grids on Nylon Membranes (see Notes 22 and 23)

1. Prepare a master mix for $n + 1$ samples containing 3 μL 10X PCR buffer, 1.8 μL $MgCl_2$ (25 mM), 3 μL dNTPs (2 mM each), 0.45 μL of one of the vector sequence specific primers M13F or M13R, 0.38 μL Taq DNA polymerase (5 U/μL), and 19.92 μL sterile water for each clone. Dispense 28.55 μL into a PCR tube, add 0.45 μL cDNA specific primer (approx 21mer) (10 pmol/μL) which has been designed so that the expected amplification avoids the poly (dA) tail (*see* **Note 24**) and 1 μL of the purified PCR product (*see* **Subheading 3.3.3.**) from which the cDNA specific primer is derived.
2. Thermocycle at 95°C for 2 min, followed by 40 cycles at 94°C for 30 s, 54°C for 30 s and 72°C for 30 s, with a final extension at 72°C for 5 min.
3. To check for a PCR product, electrophorese an aliquot through a 1.5% agarose gel.
4. Purify the remaining PCR product using the QIAquick™ PCR purification kit according to the manufacturer's instructions (Qiagen).
5. Dot-blot the purified PCR products onto a nylon membrane using the 96-well dot-blot apparatus (12 cm × 9cm) (Fermentas) according to the manufacturer's instructions. Repeat the dot-blotting procedure to create multiple, identical cDNA grids, one for each sample to be tested.

3.4.2. Sample Labeling, the Hybridization Screen and Image Analysis

1. Purify the previously prepared oligonucleotide (dT)$_{15}$-primed cDNA (*see* **Subheading 3.1.2.1.**) for each sample to be tested using the QIAquick™ PCR purification kit according to the manufacturer's protocol (Qiagen).
2. Radiolabel the cDNA with ^{33}P-α-dCTP (10 mCi/mL) according to the instructions supplied in the random prime labeling kit (Amersham Biosciences).
3. Purify the radiolabeled cDNA fragments from the unincorporated nucleotides using a G-50 microspin column according to the manufacturer's instructions (Pharmacia Biotech). A value for radioactive emissions is taken, so that comparative samples are only taken through to the hybridization screen if the counts are within 10% of each other. Add human Cot-1 DNA, poly (dA-dT) and DIG Easy Hyb buffer to the radiolabeled samples, denature for 10 min then allow to cool slowly to 45°C.
4. Prehybridize the gridded cDNA membranes at 45°C for 1 h in DIG Easy Hyb buffer, then hybridize the radiolabeled probe onto the membrane at 45°C overnight. Rinse the membranes once at room temperature and 4 times at 65°C, for 15 min each time, with 0.1% SDS, 0.2X SSC. Place the membranes inside the cassette and expose to a phosphor screen for 5 d. Develop the image by scanning the phosphor screen (*see* **Fig. 4**) (*8*).
5. Using image analysis software a numerical value for each dot on the grid is assigned which directly reflects the amount of hybridized cDNA in the sample

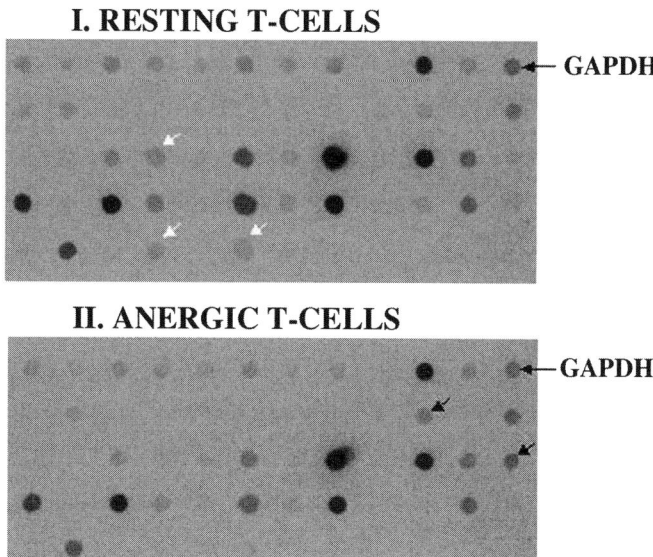

Fig. 4. Phosphor images depicting the confirmation of differential display results by reverse-northern blotting. The blot represents 55 unique cDNAs isolated by a differential display screen as well as the standard housekeeping control, GAPDH. cDNAs derived from transcripts that are down-regulated in T-cell anergy are illustrated by white arrows, whereas cDNAs synthesized from up-regulated transcripts are highlighted by black arrows. Reprinted with permission from **ref. 8**.

that was tested. These results are expressed in excel spreadsheet format. By relating all the numerical values from the grid to the corresponding value for the control cDNA that is derived from a housekeeping transcript, the values for each cDNA can be compared from one sample to the next. Hence for each of the molecules the confirmation of differential gene expression between samples can be assessed.

4. Notes

1. The differential display protocol from extraction of total RNA to confirming the true differentials takes 26 working d. The following is a breakdown of the timecourse. The preparation and quality control inspection of RNA takes 3 d. The differential display RT-PCR and polyacrylamide gel electrophoresis takes 2 d with a further 2 d for the exposure of the X-ray film. The isolation of the putative differentials from the gel bands, cloning, transformation and identification of the recombinants takes 3 d. DNA sequencing and analysis takes another 3 d. Before the construction of a cDNA grid containing the putative differentials without poly (dA) tails, a cDNA specific primer has to be designed and ordered. This may

take up to 4 d. Finally, the reverse-northern analysis takes 3 d with a further 6 d for the development and analysis of the images.
2. We advise using sterilized filter barrier tips during all the pipetting stages of the RNA extraction protocol as well as for dispensing RNA.
3. We always find that there are residual amounts of contaminating genomic DNA after the RNA extraction protocol and so recommend the DNase treatment of all samples using the DNA-free™ kit.
4. Avoid pre-washing the cells before adding the TRIzol™ reagent as this can cause RNA degradation.
5. The TRIzol™ reagent can also be used to extract total RNA from biopsied tissue. Briefly, 1 mL of reagent can be added to 50 mg tissue and transferred to a Ribolyser™ tube (Thermo Hybaid) containing ceramic beads. Load the tubes into the Ribolyser™ cell disrupter (Thermo Hybaid) and follow the instructions to homogenize the samples. Then proceed with **step 2** of the protocol (**Subheading 3.1.1.**).
6. Store the RNA as aliquots in several tubes so that the number of freeze-thaw cycles are minimized. Otherwise some deterioration of diluted RNA samples can occur.
7. We recommend the PCR amplification of a ubiquitous molecule such as p53.
8. An alternative to TBE buffer and ethidium bromide staining for agarose gel electrophoresis of DNA is TAE buffer (50X stock contains 2 M Tris-acetate, 50 mM EDTA [pH8.0]) and SYBR green dye (Invitrogen). This stain is less mutagenic and more sensitive than ethidium bromide.
9. An alternative to Superscript™ II reverse transcriptase (Invitrogen) that also synthesizes cDNAs is PowerScript™ reverse transcriptase (Clontech). Both enzymes are mutant derivatives of Moloney murine leukaemia virus reverse transcriptase that lack RNase H but retain polymerase activity to permit the synthesis of longer cDNAs.
10. Variations on oligonucleotide (dT)-primed cDNA include using random hexamers *(23–25)*, arbitrary primers (RNA arbitrarily-primed PCR, RAP-PCR) *(2,10,16,26,27)* or specific primers (targeted differential display *(28,29)*) in the first strand synthesis reaction.
11. For each sample to be analysed the PCR should be performed in duplicate. This will help to check for reproducibility of the banding patterns on the gel *(18)*.
12. An alternative to Taq DNA polymerase (Promega) that avoids the need for a hot start is Advantage™ DNA polymerase (Clontech). This polymerase contains an antibody that is denatured at 95°C before the PCR can proceed. This avoids mispriming in the PCR that would otherwise generate false positives.
13. The original DDRT-PCR protocol used ^{35}S-α-dATP as the radionucleotide in the PCR *(1)*. Since then it has been suggested that this radionucleotide is heat labile and hazardous aerosols are released during the PCR reaction *(30)*. Hence we now avoid the use of ^{35}S-α-dATP.
14. The original protocol used decamers that had been generated at random by the computer, as oligonucleotides in the PCR *(1)*. To obtain specificity in the reaction, these decamers need to be nonpalindromic and have a GC content >50%.

Theoretically, it has been suggested that to represent all the 15,000 transcripts that are supposedly expressed in a cell would require at least 25 decamers *(31)*.
15. An alternative strategy uses a multimer that has been designed against the corresponding nucleotide sequence of a protein motif or domain, instead of the random decamer. These oligomers can either be a sequence-specific primer or a degenerate oligonucleotide. There are multiple examples of this approach called targeted differential display to identify homologous molecules *(12,23–29,32,33)*.
16. Altering the thermocycle parameters in the PCR will affect its stringency and hence the banding patterns on the gel. Consequently by lowering the annealing temperature *(31)* or increasing the number of PCR cycles *(34)*, the number of bands on the gels will increase.
17. Although the original method recommends denaturing the PCR samples before loading on a denaturing polyacrylamide gel *(1)*, an alternative protocol suggests migrating non-denatured samples through non-denaturing gels *(31)*. This would probably reduce the complexity of the banding patterns on the gels as a result of minimizing the presence of denatured single-stranded cDNAs.
18. If radioactive ink is not available, use a fine-needle for puncturing the X-ray film and dried gel concurrently just before closing the cassette.
19. For direct PCR from a bacterial colony, the cells are optimal after an overnight incubation at 37°C. The ability of the cells to serve as templates in the PCR is dramatically diminished even after 1 d storage at 4°C. For the analysis of colonies after 1 d storage, prepare a fresh overnight culture for each colony to be tested, then with a sterile tip stab the culture and spike the PCR mixture. Proceed with the PCR as described.
20. For DNA sequencing of PCR products, the ABI PRISM™ 377 DNA sequencer (Applied Biosystems) or the MegaBACE™ capillary sequencer (Amersham Biosciences) provide a high throughput approach because they allow the simultaneous analysis of multiple fluorochromes in each sequencing reaction.
21. Other methods that have been used to confirm differential gene expression include northern blotting *(9,14,35)*, quantitative real-time PCR *(36,37)* and in situ hybridisation *(10,38,39)*. However, all of these approaches are not as high throughput as reverse-northern analysis and only permit the analysis of one differential at a time. Nevertheless, a recent adaptation of quantitative TaqMan™ real-time PCR into a microfluidic card system (Applied Biosystems) offers the simultaneous comparison of multiple cDNAs in a low-density 96-well or 384-well array format.
22. As a control for comparing gene expression between samples, a dot on the grid containing the cDNA for a housekeeping transcript such as GAPDH is recommended. For this, the PCR is carried out using oligonucleotides GAPDHF and GAPDHR at an annealing temperature of 55°C for 30 cycles. After agarose gel electrophoresis and EB staining, a 451-bp product is visualized under ultraviolet light. It has been suggested that more than one standard control may be required as the housekeeping genes are sometimes differentially regulated in biological systems *(22,40–43)*.

23. A manual method for dot-blotting PCR products onto membranes is described that uses a 96-well gridded apparatus and vacuum suction, however the introduction of solid pin blot replicators as well as robotic gridders offers a higher throughput approach for the manufacture of multiple identical grids.
24. The cDNA specific primer is designed using the Primer3 program, so that a PCR product of approximately 200 bp is amplified whilst avoiding the poly (dA) tract in the insert. This oligonucleotide primer should be approximately 21 bp.

Acknowledgments

The authors would like to thank the West Riding Medical Research Trust and the Yorkshire Arthritis Trust for financial support.

References

1. Liang, P. and Pardee, A. B. (1992) Differential display of eukaryotic messenger RNA by means of the polymerase chain reaction. *Science* **257**, 967–971.
2. Welsh, J., Chada, K., Dalal, S. S., Cheng, R., Ralph, D., and McClelland, M. (1992) Arbitrarily primed PCR fingerprinting of RNA. *Nucleic Acids Res.* **20**, 4965–4970.
3. Van Der Pouw Kraan, T. C. T. M., Van Gaalen, F. A., Kasperkoitz, P. V., et al. (2003) Rheumatoid arthritis is a heterogeneous disease. Evidence for differences in the activation of the STAT-1 pathway between rheumatoid tissues. *Arthritis Rheum.* **48**, 2132–2145.
4. Aidinis, V., Plows, D., Haralambous, S., et al. (2003) Functional analysis of an arthritogenic synovial fibroblast. *Arthritis Res. Ther.* **5**, R140–R157.
5. Olsen, N., Sokka, T., Seehorn, C. L., Kraft, B., Maas, K., Moore, J., and Aune, T.M. (2004) A gene expression signature for recent onset rheumatoid arthritis in peripheral blood mononuclear cells. *Ann. Rheum. Dis.* **63**, 1387–1392.
6. Callard, D., Lescure, B., and Mazzolini, L. (1994) A method for the elimination of false positives generated by the mRNA differential display technique. *BioTechniques* **16**, 1096–1103.
7. Debouck, C. (1995) Differential display or differential dismay? *Curr. Opin. Biotech.* **6**, 597–599.
8. Ali, M., Ponchel, F., Wilson, K. E., et al. (2001) Rheumatoid arthritis synovial T cells regulate transcription of several genes associated with antigen-induced anergy. *J. Clin. Invest.* **107** 519–528.
9. Mangasser-Stephan, K., Dooley, S., Welter, C., Mutschler, W., and Hanselmann, R.G. (1997) Identification of human semaphorin E gene expression in rheumatoid synovial cells by mRNA differential display. *Biochem. Biophys. Res. Commun.* **234**, 153–156.
10. Kullmann, F., Judex, M., Ballhorn, W., et al. (1999) Kinesin-like protein CENP-E is upregulated in rheumatoid synovial fibroblasts. *Arthritis Res.* **1**, 71–80.
11. Hu, S. -I., Carozza, M., Klein, M., Nantermet, P., Luk, D., and Crowl, R. M. (1998) Human HtrA, an evolutionary conserved serine protease identified as a

differentially expressed gene product in osteoarthritic cartilage. *J. Biol. Chem.* **273**, 34406–34412.
12. Patel, I. R., Attur, M. G., Patel, R. N., et al. (1998) TNF-α convertase enzyme from human arthritis-affected cartilage: isolation of cDNA by differential display, expression of the active enzyme, and regulation of TNF-α. *J. Immunol.* **160**, 4570–4579.
13. Attur, M. G., Dave, M. N., Stuchin, S., et al. (2001) Osteopontin. An intrinsic inhibitor of inflammation in cartilage. *Arthritis Rheum.* **44**, 578–584.
14. Yamada, E., Ishiguro, N., Miyaishi, O., et al. (1997) Differential display analysis of murine collagen-induced arthritis: cloning of the cDNA-encoding murine ATPase inhibitor. *Immunology* **92**, 571–576.
15. Sakurai, D., Yamaguchi, A., Tsuchiya, N., Yamamoto, K., and Tokunaga, K. (2001) Expression of ID family genes in the synovia from patients with rheumatoid arthritis. *Biochem. Biophys. Res. Commun.* **284**, 436–442.
16. Judex, M., Neumann, E., Lechner, S., et al. (2003) Laser-mediated microdissection facilitates analysis of area-specific gene expression in rheumatoid synovium. *Arthritis Rheum.* **48**, 97–102.
17. Ali, M., Markham, A.F., and Isaacs, J.D. (2001) Application of differential display to immunological research. *J. Immunol. Methods* **250**, 29–43.
18. Liang, P., Averboukh, L., and Pardee, A.B. (1993) Distribution and cloning of eukaryotic mRNAs by means of differential display: refinements and optimization. *Nucleic Acids Res.* **21**, 3269–3275.
19. Mathieu-Daude, F., Cheng, R., Welsh, J., and McClelland, M. (1996) Screening of differentially amplified cDNA products from RNA arbitrarily primed PCR fingerprints using single strand conformation polymorphism (SSCP) gels. *Nucleic Acids Res.* **24**, 1504–1507.
20. Smith, N. R., Li, A., Aldersley, M., et al. (1997) Rapid determination of the complexity of cDNA bands extracted from DDRT-PCR polyacrylamide gels. *Nucleic Acids Res.* **25**, 3553–3554.
21. Martin, K. J., Kwan, C. P., O'Hare, M. J., Pardee, A.B., and Sager, R. (1998) Identification and verification of differential display cDNAs using gene-specific primers and hybridization arrays. *BioTechniques* **24**, 1018–1026.
22. Schmittgen, T. D. and Zakrajsek, B. A. (2000) Effect of experimental treatment on housekeeping gene expression validated by real-time, quantitative RT-PCR. *J. Biochem. Biophys. Methods* **46**, 69–81.
23. Donohue, P. J., Alberts, G. F., Guo, Y., and Winkles, J.A. (1995) Identification by targeted differential display of an immediate early gene encoding a putative serine/threonine kinase. *J. Biol. Chem.* **270**, 10351–10357.
24. Chuaqui, R. F., Englert, C. R., Strup, S. E., et al. (1997) Identification of a novel transcript up-regulated in a clinically aggressive prostate carcinoma. *Urology* **50**, 302–307.
25. Ryoo, H. M., Hoffmann, H. M., Beumer, T., et al. (1997) Stage-specific expression of Dlx-5 during osteoblast differentiation: involvement in regulation of osteocalcin gene expression. *Mol. Endocrinol.* **11**, 1681–1694.

26. Stone, B. and Warton, W. (1994) Targeted RNA fingerprinting: the cloning of differentially-expressed cDNA fragments enriched for members of the zinc finger gene family. *Nucleic Acids Res.* **22**, 2612–2618.
27. Tohonen, V., Osterlund, C., and Nordqvist, K. (1998) Testatin: a cystatin-related gene expressed during early testis development. *Proc. Natl. Acad. Sci. USA* **95**, 14208–14213.
28. Yoshikawa, T., Xing, G. Q., and Detera-Wadleigh, S.D. (1995) Detection, simultaneous display and direct sequencing of multiple nuclear hormone receptor genes using bilaterally targeted RNA fingerprinting. *Biochim. Biophys. Acta* **1264**, 63–71.
29. Liu, G., Takano, T., Matsuzuka, F., Higashiyama, T., Kuma, K., and Amino, N. (1999) Screening of specific changes in mRNAs in thyroid tumors by sequence specific differential display: decreased expression of c-fos mRNA in papillary carcinoma. *Endocrine J.* **46**, 459–466.
30. Clinton, M. and Scougall, R.K. (1995) Detection and capture of ^{35}S-labeled gas released from reaction tubes during differential display PCR. *BioTechniques* **19**, 798–799.
31. Bauer, D., Muller, H., Reich, J., Ahrenkiel, V., Warthoe, P., and Strauss, M. (1993) Identification of differentially expressed mRNA species by an improved display technique (DDRT-PCR). *Nucleic Acids Res.* **21**, 4272–4280.
32. Dominguez, O., Ashhab, Y., Sabater, L., Belloso, E., Caro, P., and Pujol-Borrell, R. (1998) Cloning the ARE-containing genes by AU-motif-directed display. *Genomics* **54**, 278–286.
33. Utans-Schneitz, U., Lorez, H., Klinkert, W. E., Da Silva, J., and Lesslauer W. (1998) A novel rat CC chemokine, identified by targeted differential display, is upregulated in brain inflammation. *J. Neuroimmunol.* **92**, 179–190.
34. Guimaraes, J. M., Lee, F., Zlotnik, A., and McClanahan, T. (1995) Differential display by PCR: novel findings and applications. *Nucleic Acids Res.* **23**, 1832–1833.
35. Sun, Y., Hegamyer, G., and Colburn, N.H. (1994) Molecular cloning of five messenger RNAs differentially expressed in preneoplastic or neoplastic JB6 mouse epidermal cells: one is homologous to human tissue inhibitor of metalloproteinases-3. *Cancer Res.* **54**, 1139–1144.
36. Rajeevan, M.S., Ranamukhaarachchi, D. G., Vernon, S.D., and Unger, E. R. (2001) Use of real-time quantitative PCR to validate the results of cDNA array and differential display PCR technologies. *Methods* **25**, 443–451.
37. Lee, Y. H., Tokraks, S., Pratley, R. E., Bogardus, C., and Permana, P. A. (2003) Identification of differentially expressed genes in skeletal muscle of non-diabetic insulin-resistant and insulin-sensitive Pima Indians by differential display PCR. *Diabetologia* **46**, 1567–1575.
38. Toki, H., Namikawa, K., Su, Q., Kiryu-Seo, S., Sato, K., and Kiyama, H. (1998) Enhancement of extracellular glutamate scavenge system in injured motorneurons. *J. Neurochem.* **71**, 913–919.

39. Bryant, Z., Subrahmanyan, L., Tworoger, M., et al. (1999) Characterization of differentially expressed genes in purified Drosophila follicle cells: toward a general strategy for cell type-specific developmental analysis. *Proc. Natl. Acad. Sci. USA* **96**, 5559–5564.
40. Kim, J. W., Kim, S. J., Han, S. M., et al. (1998) Increased glyceraldehyde-3-phosphate dehydrogenase gene expression in human cervical cancers. *Gynecol. Oncol.* **71**, 266–269.
41. Goidin, A., Mamessier, A., Staquet, M. J., Schmitt, D., and Berthier-Vergnes, O. (2001) Ribosomal 18SRNA prevails over glyceraldehyde-3-phosphate dehydrogenase and beta-actin genes as internal standard for quantitative comparison of mRNA levels in invasive and non-invasive human melanoma cell subpopulations. *Anal. Biochem.* **295**, 17–21.
42. Hamalainen, H. K., Tubman, J. C., Vikman, S., et al. (2001) Identification and validation of endogenous reference genes for expression profiling of T helper cell differentiation by quantitative real-time PCR. *Anal. Biochem.* **229**, 63–70.
43. Glare, E. M., Divjak, M., Bailey, M. J., and Walters, E. H. (2002) β-Actin and GAPDH housekeeping gene expression in asthmatic airways is variable and not suitable for normalising mRNA levels. *Thorax* **57**, 765–770.

24

Two-Dimensional Electrophoresis of Proteins Secreted from Articular Cartilage

Monika Hermansson, Jeremy Saklatvala, and Robin Wait

Abstract

Two-dimensional electrophoresis (2DE) is a powerful method for separation of complex mixtures of proteins. The standard procedure is not, however, well suited to analysis of articular cartilage, which contains high concentrations of proteoglycans, the polyanionic glycosaminoglycan chains of which interfere with isoelectric focusing. We have developed a method for selective removal of proteoglycans by precipitation with cetylpyridinium chloride, after which the residual cartilage proteins are amenable to conventional 2DE analysis. Using this method, reproducible 2D-patterns can be obtained from proteins secreted by articular cartilage. The separated proteins may then be visualized by metabolic radiolabeling and silver staining, digested in gel with trypsin, and identified by tandem mass spectrometry.

Key Words: Two-dimensional electrophoresis; cartilage; aggrecan; proteomics; cetylpyridinium chloride; osteoarthritis.

1. Introduction

Both rheumatoid arthritis (RA) and osteoarthritis (OA) are characterized by extensive destruction of articular cartilage. Very little is known about the normal control mechanisms of cartilage matrix turnover, or why they are dysregulated in arthritic disease. In principle, a proteomic approach could illuminate these processes by enabling unbiased comparison of patterns of protein expression in normal and diseased tissue.

Historically, the main technology platform for proteomic studies has been two-dimensional electrophoresis (2DE), a powerful method for separation of complex mixtures of proteins, which achieves high resolution by combining an initial fractionation by intrinsic charge (isoelectric focusing), with an orthogonal separation by apparent mass using sodium dodecyl sulfate polyacrylamde gel electrophoresis (SDS-PAGE) *(1,5,7)*. However, conventional 2DE-based proteomic approaches are not particularly suitable for analysis of articular cartilage, which contains relatively few cells (chondrocytes), embedded in an excess of extracellular matrix. This matrix is composed of a network of type II collagen fibers, hyaluronan, and proteoglycans, of which the most abundant is aggrecan. The protein cores of the proteoglycans are substituted with abundant sulfated glycosaminoglycan chains, predominantly chondroitin sulfate and heparan sulfate. These anionic molecules interfere with the initial isoelectric focusing step of 2DE. A further difficulty is that the major extracellular matrix components are abundant and turn over slowly, so are likely to obscure alterations in low-abundance regulatory molecules.

We have developed a twofold strategy for proteomic analysis of secreted proteins from articular cartilage, in which metabolic radiolabeling is used to distinguish proteins newly synthesized by chondrocytes from the bulk extracellular matrix, and proteoglycans are selectively depleted from the culture medium prior to isoelectric focussing. This procedure is summarized in the flow chart (*see* **Fig. 1**).

Proteoglycans are removed by precipitation with the cationic detergent cetylpyridinium chloride (CPC) *(6)*. Verification of elimination of the bulk of the proteoglycan is achieved by quantitation of residual glycosaminoglycan, using the dimethyl methylene (DMB) blue assay *(3)*. This method is based on the metachromatic shift in absorption maximum that occurs when DMB is complexed with sulfated GAGs. After removal of proteoglycans reproducible patterns of expressed proteins are obtained by conventional 2DE using immobilized pH gradients. Protein spots may be visualized by autoradiography and by mass spectrometry-compatible silver staining *(8)*, excised, digested in gel, and identified by mass spectrometry.

We have used this method to compare expression of secreted proteins by cultured explants from normal and osteoarthritic articular cartilage in culture *(4)*. An example of a silver stained gel, and the corresponding autoradiographic pattern obtained from osteoarthritic cartilage, is shown in **Fig. 2**. The annotation (*see* **Fig. 2B**) indicates proteins identified by tandem mass spectrometry. The method is also applicable to investigation of cartilage protein expression in response to age, topography, cytokine treatment, and for monitoring the phenotype of chondrocytes during differentiation/dedifferentiation.

2. Materials
2.1. Preparation of Cartilage Explant Conditioned Medium

1. Cartilage: At least 1 g of human articular cartilage obtained from joint replacement surgery following OA or trauma. The method is equally applicable to non-

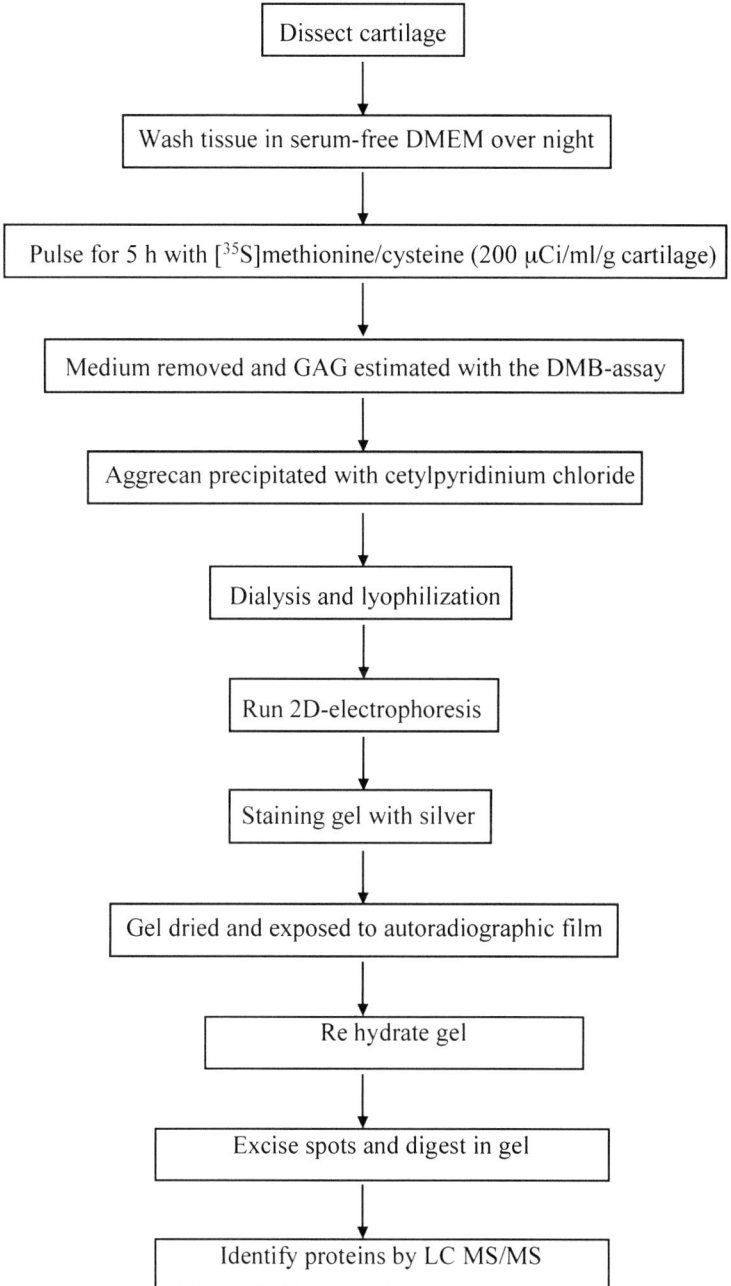

Fig. 1. Overview of the procedure for analysis of secreted proteins from cartilage by 2D-electrophoresis.

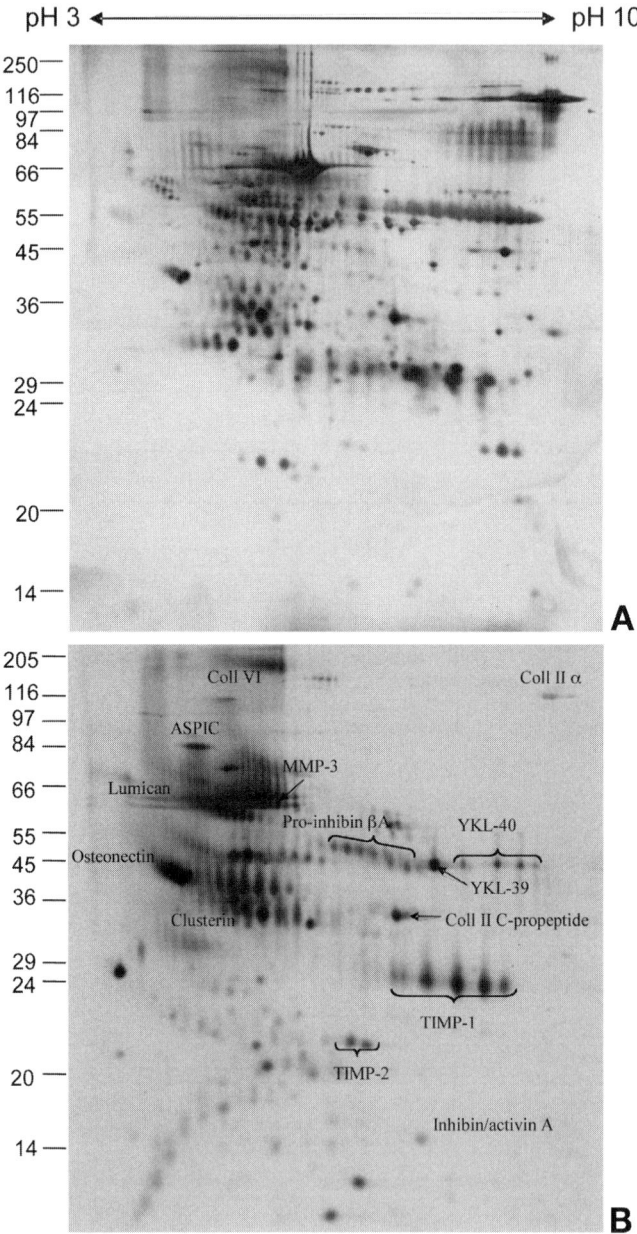

Fig. 2 (A) Silver stained 2D-gel of proteins secreted from osteoarthritic cartilage explants. (B) The corresponding autoradiograph. Three grams of cartilage were removed from a femoral condyle of a 65-yr-old male donor, and processed as described. Proteins were identified by HPLC MS/MS.

human tissue, for example cartilage dissected from porcine metacarpophalangeal joints.
2. Culture medium: Dulbecco's modified Eagle's medium (DMEM) (Cambrex; cat. no. BE12-604F), Penicillin/streptomycin mixture (100 U/mL penicillin, 100 μg/mL streptomycin) (Cambrex; cat. no. DE17-602E), Amphotericin B, 2 μg/mL (diluted from fungizone antimycotic liquid; Invitrogen, 15290-018), HEPES buffered saline, 25 mM (diluted from 1M stock solution) (Cambrex; cat. no. BE17-737E).
3. Culture medium for radiolabeling: Methionine/cysteine/glutamine free DMEM 100 mL (ICN Biomedicals; cat. no.1642454), Glutamine, 1% *(v/v)* (Cambrex; cat. no. DE17-605E), Penicillin/streptomycin mixture (100 U/mL penicillin, 100 μg/mL streptomycin) (Cambrex; cat. no. DE17-602E), HEPES buffered saline, 25 mM (diluted from 1M stock solution) (Cambrex; cat. no. BE17-737E).
4. Radiolabeled methionine and cysteine: Redivue Promix [^{35}S] in vitro cell labeling mix, 200 μCi/mL/g cartilage (GE Healthcare).

2.2. Preparation of Samples for 2D-Electrophoresis

2.2.1. Precipitation of Proteoglycans With Cetyl Pyridinium Chloride (CPC)

1. Shark chondroitin 6-sulfate, 0 to 100 μg/mL in DMEM (Sigma; cat. no. C4384).
2. 1,9-Dimethyl-methylenen blue solution (DMB). Dissolve 16 mg DMB (Aldrich; cat. no. 34,108-8) in 5 mL ethanol, add 2 g sodium formate, and 2 mL formic acid and make up to 1 L with deionized water. If stored in a dark bottle, the solution is stable for at least 1 yr at room temperature.
3. Cetyl pyridinium chloride (CPC) 5% aqueous *(w/v)* (hexadecyl pyridinium chloride) (Sigma; cat. no. C-0732). Dissolve in deionized water with gentle heating or ultrasonication if necessary. The solution should be transparent and colorless. Store at room temperature and discard if it acquires a brownish tinge.

2.2.1. Dialysis and Lyophilization

1. Disposable dialysis cassettes 10,000 MWCO (Pierce; cat. no. Slide-A-Lyser, 66410).
2. Freeze-drier (e.g., Edwards modulayo).

2.3. Two-Dimensional Electrophoresis

2.3.1. Solutions

1. Lysis solution (This may be made in 10 mL batches and 1 mL aliquots stored at –80°C; the protease inhibitor cocktail should be added immediately before use) : Urea 9.5M (GE Healthcare; cat. no. 17-1319-01), 1% *(w/v)* dithiothreitol (DTT) (Alexis Biochemicals; cat. no. 280-001-G025), 2% CHAPS (GE Healthcare; cat. no. 17-1314-01), and 0.5% immobilized pH gradient (IPG) buffer (pH 3.0 –10.0 (GE Healthcare; cat. no. 17-6000-87).

2. Protease inhibitor cocktail: 1 m*M* phenyl methyl sulfonyl fluoride (PMSF), 10 μg/mL pepstatin, 10 μg/mL aprotenin, 10 μ*M* E-64.
3. Equilibration buffer (may be aliquoted and stored at –20°C): 2% SDS, 6*M* urea, 30% Glycerol, 150 m*M* Tris HCl (pH 8.8).
4. 65 m*M* DTT in equilibration buffer.
5. 260 m*M* iodoacetamide in equilibration buffer.
6. 380 m*M* Tris HCl (pH 8.8).
7. Acrylamide 12.5% *(w/v)* (Severn Biotech Ltd; cat. no. 20-2100-10).
8. 0.34% bis-acrylamide, 0.34% *(w/v)* (Severn Biotech Ltd; cat. no. 20-2100-10).
9. 0.1% SDS *(w/v)*.
10. 0.07% ammonium persulfate.
11. 0.07% TEMED.
12. Running buffer: 25 m*M* Tris base, 190 m*M* glycine, 0.01% SDS *(w/v)*.
13. 4X SDS sample buffer: 0.4*M* Tris-HCl (pH 6.8), 40% glycerol *(v/v)*, 0.4% SDS *(w/v)*, 1% mercaptoethanol, and 0.01% bromophenol blue.

2.3.2. Electrophoresis

1. Immobilized polyacrylamide gradient (IPG) strip 13 cm, linear pH 3.0 to 10.0 (GE Healthcare; cat. no. 17-6001-14).
2. IPG strip rehydration tray (GE Healthcare; cat. no. 80-6371-84).
3. Dow-Corning silicone fluid 200/10cs (VWR international; cat. no. 63004 6T)
4. Multiphor II flatbed electrophoresis system (GE Healthcare).
5. ProteanR IIxi cell apparatus (Biorad).
6. NuSieve GTG Agarose, 0.5% (FMC BioProducts; cat. no. 50080).
7. Wide range markers (Sigma; cat. no. M-4038).
8. IEF sample application pieces (GE Healthcare; cat. no. 80-1129-46).

2.4. Silver staining (8)

Solutions should be freshly made with high quality (18.2*M* Ω) deionized or double distilled water.

1. 50% Methanol and 5% acetic acid.
2. 50% Methanol.
3. Sodium thiosulfate, 0.02% (sensitizer).
4. 0.1% Silver nitrate, chilled (Fisher Chemicals; cat. no. S/1280/48)
5. "Formalin" (35% formaldehyde in water), 0.04% in 2% sodium carbonate (developer).
6. 5% Acetic acid.
7. 3% Glycerol, and 30% methanol.

2.5. Autoradiography

1. Medical X-ray film (Fuji Photo Film Ltd.).
2. Film processor (Curix 60) (AGFA).

3. Methods
3.1. Preparation of Cartilage Explant Conditioned Medium

1. Dissect cartilage explants into serum-free DMEM (10 mL/g cartilage), wash once and rest overnight in DMEM at 37°C to deplete plasma and other nonchondrocyte proteins (*see* **Note 1**).
2. Next day, wash explants with methionine/cysteine-free DMEM (10 mL/g cartilage) for 30 min and then incubate with [^{35}S]methionine/[^{35}S]cysteine in methionine/cysteine-free DMEM (200 µCi/mL/g cartilage) for 5 h at 37°C.
3. Remove the supernatant and centrifuge for 10 min to remove cartilage debris. The sample may be frozen and stored at this stage. Alternatively proceed to **Subheading 3.2.**

3.2. Preparation of Sample for 2D-Electrophoresis
3.2.1. Removal of Proteoglycans With Cetyl Pyridinium Chloride (CPC) (see **Note 2**)

Proteoglycan content is estimated as glycosaminoglycan using the dimethylmethylene blue (DMB) assay against a chondroitin sulfate standard *(3)*.

1. Mix 40 µL of the sample sample from **Subheading 3.1.** with 250 µL DMB solution and measure the absorbance 540 nm. Compare the absorbance with a standard curve constructed using shark chondroitin sulfate over the range of 0 to 100 µg/mL (*see* **Note 3**).
2. Add 5% cetyl pyridinium chloride to each sample to give a final concentration of 3 mg CPC/mg GAG.
3. After 30 min at room temperature, centrifuge at 13,000 rpm for 10 min to precipitate the proteoglycan/CPC complex and recover and retain the supernatant (*see* **Notes 4** and **5**).
4. Repeat the DMB assay using 40 µL of supernatant to verify adequate proteoglycan depletion. Less than 10 µg/mL GAG should not interfere with subsequent isoelectric focusing.

3.2.2. Dialysis and Lyophilization

1. Dialyze (10 kDa cut off) the samples against water overnight at 4°C, to remove free CPC and salt. Change the water once.
2. Next day, transfer the samples into Falcon tubes, cover the open end with perforated aluminium foil, and freeze at –80°C. Lyophilize overnight using a freeze-drier, used according to the manufacturer's instructions.

3.3. 2D-Electrophoresis
3.3.1. Sample Loading by in Gel Rehydration

1. Dissolve the lyophilized residues in 245 µL urea lysis solution supplemented with protease inhibitors (*see* **Note 6**). Centrifuge at 13,000 rpm for 5 min to pellet

undissolved material. Remove 5 µL for scintillation counting to estimate incorporation of [^{35}S].
2. Pipet the sample into a lane of the IPG rehydration tray and position a pH 3.0 to 10.0 IPG strip (gel side down) on top.
3. Overlay the strip with 2 mL silicon fluid to minimize evaporation and urea crystallization.
4. Rehydrate overnight (>12 h), at room temperature.

3.3.2. Isoelectric Focusing

Isoelectric focusing may be performed on any commercial flatbed electrophoresis system operated according to the manufactures instructions. The conditions below are appropriate for the multiphor II apparatus (GE Healthcare).

1. Focus at 300 V for 1 min, then increase to 3500 V over1.5 h and maintain at 3500 V for 3.5 h. The temperature should be controlled at 20°C.
2. After isoelectric focusing, either proceed immediately to equilibration and second dimension SDS-PAGE, or alternatively, store the IPG strips between 2 sheets of plastic film at –20°C until required.
3. Prior to the second dimension separation, equilibrate the IPG strips in 65 m*M* DTT-containing equilibration buffer (5 mL/strip in a glass tube) with gentle agitation for 15 min, to cleave disulphide bonds. Alkylate the resulting -SH groups by addition of 260 m*M* iodoacetamide in equilibration buffer (5 mL/strip in a glass tube) for 15 min (*see* **Note 7**).

3.3.3. Second Dimension SDS-PAGE.

1. Medium format 12.5% polyacrylamide slab gels are cast between glass plates using 1-mm spacers, according to the protocol of Laemmli (without stacking gel).
2. Transfer each equilibrated strip to the top of an SDS-PAGE gel and secure with approximately 1 mL of 0.5% low-melting-point agarose.
3. Mix 10 µL sample buffer with 3 µL of wide range molecular weight marker solution and add to a sample application paper positioned at the top left hand corner of the gel.
4. When the agarose is set, transfer the gels to the gel running tank and fill the upper and lower reservoirs with running buffer (*see* **Note 8**).
5. Run the gels overnight at 5 mA/gel. Ensure the temperature is maintained at 15°C by pumping coolant through the coils.
6. Next morning check that the dye front has reached the bottom of the gel; if not, increase the current to 45 mA and continue electrophoresis until the dye front is between 5 and 10 mm from the edge of the gel.

3.4. Silver Staining

Silver staining is performed in glass trays, with gentle agitation (e.g., by means of a rocker platform).

1. Fix the gel in 50% methanol, 5% aqueous acetic acid for a minimum of 20 min. Wash with 50% aqueous methanol for 10 min and then with deionized or double distilled water for at least 10 min more (*see* **Note 9**).
2. Immerse the gel in sodium thiosulfate solution for 1 min, then rinse 2 times with water (1 min each rinse), discarding the washings (*see* **Note 10**).
3. Soak the gel in chilled (4°C) silver nitrate solution for 20 min, with gentle agitation. Discard the silver solution and rinse 2 times with water (1 min each rinse) (*see* **Note 11**).
4. Add formaldehyde to the sodium carbonate solution and immediately pour over the gel to develop the image. If the developer turns brown it should be discarded and replaced (*see* **Note 12**).
5. Terminate development after 5 to 10 min by adding 5% acetic acid. The stained gels can be stored in distilled water or in 1% acetic acid at 4°C.

3.5. Gel Drying and Autoradiography

1. Prior to drying, soak the gels overnight in 3% glycerol and 30% aqueous methanol (*see* **Note 13**).
2. Dry onto filter paper (Whatman 3MM Chr) using a gel dryer (e.g., BioRad model 583) increasing the temperature gently to avoid cracking of gels (*see* **Note 14**).
3. Expose the dried gels to X-ray film and develop after 2 to 7 d, depending on the level of activity.

4. Notes

1. It is important to wash out as much blood and synovial fluid as possible because proteins from these sources (e.g., albumin) complicate the 2D-pattern and can obscure chondrocyte-derived components.
2. In principle glycosaminoglycan chains can be removed by treatment with suitable endoglycosidases such as chondroitinase ABC lyase *(10)*. However, all the commercially available chhondroitinase preparations we have tested are heavily contaminated with extraneous proteins. Thus, although well-resolved gels are obtained, the pattern is dominated by proteins of noncartilage origin.
3. Typically samples contain between 100 and 200 µg/mL GAG and need to be diluted two- to fourfold to obtain a reading within the standard curve.
4. The proteoglycan-CPC precipitate is white but the pellet is usually reddish because the phenol red from the culture medium binds to CPC.
5. Proteoglycans and any proteins binding strongly to them can be recovered from the pellet for subsequent analysis by washing with $0.4M$ sodium acetate in 90% ethanol and then with ethanol alone (to remove sodium acetate). Exchange of cetylpyridium and sodium ions leads to disruption of the complex.
6. The lyophilized residue can be initially dissolved in 100 µL of water for estimation of the protein concentration by Bradford assay. The volume can then be made up to 245 µL with urea lysis solution.
7. The strips should be soaked in iodoacetamide-containing buffer for a minimum of 15 min buffer, but can be left there until ready to load onto the second dimension gel.

8. It is easiest to pour in the agarose first and then carefully position the IPG strip and molecular weight marker paper. Do not wait too long before starting the electrophoresis as the marker solution will start to diffuse away.
9. Longer fixation can improve the staining and the gel can be left in fixing solution for up to 1 wk without noticeable effect. Increasing the time of the initial wash with deionized water to 12 h or overnight also may improve the quality of the stained pattern.
10. The incubation should not exceed 1 min, as longer periods in the presence of sensitising solution can increase the background staining.
11. It is important that this step is performed at 4°C to avoid the reduction of the silver ions to silver prior to development, which otherwise will increase the background staining.
12. The brown color is caused by formation of silver carbonate and could increase the background staining if not removed. On completion of staining it is good practice to transfer the gel to a fresh tray, because residual silver ions can cause darkening. An alternative MS-compatible silver staining protocol also gives good results in our hands *(11)*.
13. Inclusion of glycerol in the drying solution increases the durability of the dried gel. It also allows the gel to be rehydrated without splitting, prior to excision of spots for mass spectrometric identification. The risk of splitting can be further reduced by trimming 1 mm from the four edges of the gel with a scalpel before rehydration.
14. Imaging of the dried gel prior to rehydration facilitates correlation of the silver stained and autoradiographic patterns *(9)*, because the dimensions of both images are identical and can be superimposed.

References

1. Bjellqvist, B., Ek, K., Righetti, P. G., et al. (1982) Isoelectric focusing in immobilized pH gradients: Principle, methodology and some applications. *J. Biochem. Biophys. Methods* **6**, 317–339.
2. Bradford, M. M. (1976) A rapid and sensitive method for the quantitation of microgram quantities of protein utilizing the principle of protein-dye binding. *Anal. Biochem.* **72**, 248–254.
3. Farndale, R. W., Sayers, C. A., and Barrett, A. J. (1982) A direct spectrophotometric microassay for sulfated glycosaminoglycans in cartilage cultures. *Connect. Tissue Res.* **9**, 247–248.
4. Hermansson, M., Sawaji, Y., Bolton, M., et al. (2004) Proteomic analysis of articular cartilage shows increased type ii collagen synthesis in osteoarthritis and expression of inhibin betaa (activin a), a regulatory molecule for chondrocytes. *J. Biol. Chem.* **279**, 43,514–43,521.
5. Klose, J. (1975) Protein mapping by combined isoelectric focusing and electrophoresis of mouse tissues. A novel approach to testing for induced point mutations in mammals. *Humangenetik* **26**, 231–243.

6. Laurent, T. C., and Scott, J. E. (1964) Molecular weight fractionation of polyanions by cetylpyridinium chloride in salt solutions. *Nature* **202**, 661–662.
7. O'Farrell, P. H. (1975) High resolution two-dimensional electrophoresis of proteins. *J. Biol. Chem.* **250**, 4007–4021.
8. Shevchenko, A., Wilm, M., Vorm, O., and Mann, M. (1996) Mass spectrometric sequencing of proteins silver-stained polyacrylamide gels. *Anal. Chem.* **68**, 850–858.
9. Westbrook, J. A., Yan, J. X., Wait, R., and Dunn, M. J. (2001) A combined radio-labelling and silver staining technique for improved visualisation, localisation, and identification of proteins separated by two-dimensional gel electrophoresis. *Proteomics* **1**, 370–376.
10. Yamagata, T., Saito, H., Habuchi, O., and Suzuki, S. (1968) Purification and properties of bacterial chondroitinases and chondrosulfatases. *J. Biol. Chem.* **243**, 1523–1535.
11. Yan, J. X., Wait, R., Berkelman, T., et al. (2000) A modified silver staining protocol for visualization of proteins compatible with matrix-assisted laser desorption/ionization and electrospray ionization-mass spectrometry. *Electrophoresis* **21**, 3666–3672.

25

Mapping Lymphocyte Plasma Membrane Proteins

A Proteomic Approach

Matthew J. Peirce, Jeremy Saklatvala, Andrew P. Cope, and Robin Wait

Abstract

The pathological importance of tumor necrosis factor (TNF)-α in rheumatoid arthritis (RA) is now widely accepted. Ex vivo data from synovial cell cultures suggest that direct cell contact between activated T-cells and macrophages may be an important driver of macrophage TNF-α production in the RA joint. However, the ligand/receptor pairs driving this cell contact signal remain obscure. One reason for this is that plasma membrane (PM) proteins are resistant to systematic analysis using traditional proteomic approaches. In this chapter we present a method for the enrichment and resolution of PM proteins from murine T-cell hybridomas as a prelude to identification by tandem mass spectrometry. We used cell surface biotinylation, differential centrifugation and subsequent streptavidin affinity capture, followed by solution phase iso-electric focussing and tandem mass spectrometry to identify 75 PM proteins and make semiquantitative comparisons of resting and activated cells. The method is applicable to a wide variety of cell types.

Key Words: Plasma membrane; cell surface; biotinylation; proteomics; solution-phase isoelectric focusing; tandem mass spectrometry; 2D electrophoresis; T-cell; murine; cell contact.

1. Introduction

It is now generally accepted that tumor necrosis factor (TNF)-α plays a pivotal role in the pathogenesis of rheumatoid arthritis (RA) and that the macrophage represents the dominant source of TNF-α in the rheumatoid joint. In contrast, the nature of the stimulus driving macrophage TNF-α production remains elusive. However, a growing body of data suggest that cell contact-dependent signals may be important. Monocyte/macrophage cell lines *(1)* or primary cells from peripheral blood *(2)* or dispersed synovial cultures *(3)* have been shown to generate an array of pathogenic inflammatory mediators, including TNF-α *(2,3)* , interleukin (IL)-1b *(4)* and matrix matalloproteinases *(5)*, following direct cell–cell contact with appropriately activated T-cells. The ligand/receptor pairs driving cytokine production are still unknown however, and whereas some candidate molecules (e.g., CD40 *[6]*, CD69 *[2,7]*, and LFA-1 *[2]*) have been implicated, the picture is far from complete. An unbiased, inclusive proteomic approach to profiling T-cell plasma membrane (PM) proteins might help to expand the search for cell-surface molecules involved in contact-dependent macrophage activation beyond "the usual suspects."

Classically, such approaches utilize two-dimensional electrophoresis (2DE) in which hundreds of proteins can be resolved and visualized on a single gel prior to identification by mass spectrometry. However, applying the power of this technique to PM proteins has been beset by difficulties *(8)*. Many PM proteins are of relatively low abundance and have regions of high hydrophobicity which can result in protein precipitation during isoelectric focusing (IEF) in immobilized pH gradient (IPG) gel strips and, consequently, under representation of PM proteins in the final gel *(9)*.

The protocols described below attempt to address these issues. We first describe how to generate a highly enriched PM protein preparation using cell surface biotinylation with a water-soluble reagent followed by differential centrifugation and affinity capture of the labeled proteins using an immobilized streptavidin matrix. We then describe how the recovered proteins can be resolved using solution-phase IEF which, compared with IPG-based IEF, better retains PM proteins in solution and markedly improves protein recovery. The technique allowed identification of 75 PM proteins from murine lymphocytes and, furthermore, enabled semiquantitative comparison of PM protein profiles on resting and activated primary splenocytes *(10)*.

2. Materials

The protocol described here applies to T-cell hybridomas but essentially the same protocol has been used to recover PM proteins from primary murine splenocytes and RAW 264.7 murine monocyte/macrophages (*see* **Note 1**).

2.1. Cell Surface Labeling

1. Labeling buffer: ice cold borate buffered saline (BBS) (pH 8.1), 10 mM sodium orthoborate, 2.3 mM sodium tetraborate (Borax), and 115 mM NaCl (*see* **Note 2**).
2. Quenching buffer: ice cold RPMI 1640 medium containing glutamine (*see* **Note 3**).
3. 10 mg/mL Sulfo-NHS-SS-biotin (Pierce); in H_2O.

2.2. Cell Homogenization and Fractionation

1. Hypotonic lysis buffer: 20 mM Tris-HCl (pH 7.4), 5 mM ethylene diamine tetraacetic acid (EDTA), protease inhibitor cocktail (1 µL/10^7 cells) (Sigma).
2. Dounce homogenizer with tight fitting plunger.
3. Detergent lysis buffer: hypotonic lysis buffer supplemented with; 1% *(v/v)* Triton X-100, 150 mM NaCl.

2.3. Affinity Purification

1. Immobilized streptavidin-agarose (Pierce) (1 µL/10^6 cell equivalents) (*see* **Note 4**).
2. RIPA wash buffer: 50 mM Tris-HCl (pH 7.4), 1% *(v/v)* Triton X-100, 1% *(w/v)* deoxycholate, and 0.1% sodium dodecyl sulfate (SDS).
3. Wash buffer 2: 1% Triton X-100.
4. Elution buffer (1 mL/500 µL of added beads): 9M urea, 1% *(w/v)* dithiothreitol (DTT), 1% *(v/v)* Triton X-100, protease inhibitor cocktail (2 µL/mL).

2.4. Solution-Phase Isoelectric Focusing

The protocol described here is for a pH 3.0 to 10.0 gradient which is adequate for the majority of samples and is advisable in the first instance. Use of other gradients (e.g., pH 3.0–7.0) requires alternative ampholytes and acid/base reservoir buffers.

1. IEF buffer: elution buffer supplemented with 1% *(v/v)* (pH 3.0–10.0) ampholytes (Bio-Rad).
2. Acid reservoir buffer: 0.1M acetic acid.
3. Base reservoir buffer: 0.1M NaOH.
4. Rotofor isoelectric focusing unit (Bio-Rad).

2.5. Sample Concentration and Cleanup

The Rotofor fractions should be approx 2 mL each and require concentration and removal of ampholytes, which interfere with subsequent silver staining, before separation by polyacrylamide gel electrophoresis (PAGE).

1. Centrifugal concentrators (6 mL capacity) (Vivascience).
2. SDS-PAGE clean-up kit (Amersham Bioscience).

3. Methods
3.1. Cell Surface Biotinylation

1. Wash cells 3 times in ice cold BBS and resuspended at 1 to 5 × 10^7/mL.
2. Prepare a 100X solution of sulfo-NHS-SS biotin (10 mg/mL in BBS).
3. Add the biotinylating reagent to the cells to a final concentration of 0.1 mg/mL and agitate gently for 20 min at 4°C.
4. Pellet cells and wash 2 times in ice-cold glutamine-containing serum free medium (e.g., RPMI or Dulbecco's modfied Eagle's medium [DMEM]).

3.2. Cell Homogenization and Fractionation

1. Pellet cells from quenching buffer and resuspend in ice cold hypotonic lysis buffer at 2 to 5×10^7/mL. Allow to swell and lyse for 10 min on ice.
2. Decant cells in to Dounce homogenizer and subject to 30 strokes with a tight fitting pestle (*see* **Note 5**).
3. Decant homogenate in to ultracentrifuge tubes and spin at $4000g$ for 15 min at 4°C.
4. Recover the post nuclear supernatant (PNS) and clarify at $20,000g$ for 30 min at 4°C.
5. Aspirate as much of the supernatant of this spin (s20) as possible and recover the pellet (p20) in a small volume of detergent lysis buffer (approx 2 µL/million cell equivalents). Disrupt the pellet by repeated pipetting and solubilize by agitating gently for 30 min at 4°C.
6. Pellet the insoluble material ($13,000g$ for 5 min at 4°C) and recover the supernatant.

3.3. Affinity Purification

1. To the detergent-soluble p20 fraction add immobilized streptavidin agarose beads (approx 1 µL bead slurry/million starting cells)
2. Incubate overnight at 4°C with end-over-end agitation.
3. Pellet beads and remove supernatant by puncturing eppendorf and centrifuging briefly (*see* **Note 6**).
4. Wash beads 5 times with 1 mL of ice cold RIPA wash buffer, and then 4 times with 1 mL of wash buffer 2 to remove traces of sodium dodecyl sulfate (SDS) from beads. Use punctured eppendorf to remove completely supernatant of each wash.
5. Incubate over night at 4°C elution buffer/µL of bead slurry used. Recover eluate by centrifugation of punctured eppendorf as above (*see* **Note 7**).

3.4. Solution-Phase IEF

1. Retain an aliquot (50 µL) of the eluate from **Subheading 3.3.** and dilute the remainder in IEF buffer and supplement with ampholytes (pH 3.0–10.0) to give a final volume of 55 mL and a final ampholyte concentration of 1%.
2. With a 50-mL syringe and blunt 18-gage needle fill the Rotofor chamber with the sample, minimizing introduction of air bubbles, and connect apparatus to a 4°C water cooler and allow sample 10 min to equilibrate.
3. Begin focussing under fixed power of 12 W monitoring current and voltage readings intermittently.
4. Allow focussing to continue for 3 to 4 h, or until the voltage reading is stable. Turn off the machine and under a strong vacuum harvest the fractions in to prelabeled and rinsed 6-mL centrifuge tubes (*see* **Notes 8–10**).

3.5. Sample Concentration and Cleanup

1. Pool samples as required and transfer carefully to 6 mL 10 kD cut-off centrifugal concentrator tubes (Vivascience). When sample volume has been reduced to between 50 and 100 µL, transfer to Eppendorf tubes prerinsed with d-H_2O.

2. Perform clean up of sample to remove ampholytes using SDS-PAGE clean up kit (Amersham Biosciences). Concentrated and cleaned samples can be used immediately or stored at −20°C.
3. Sample is then ready for SDS-PAGE and visualization and analysis as required (*see* **Note 11**).

4. Notes

1. At every stage of the protocol it is essential that all tubes, instruments and surfaces are kept scrupulously clean by washing with polished water and/or 70% EtOH to avoid contamination of the sample with environmental proteins (e.g., keratin) especially during the final concentration steps.
2. The biotinylation reaction targets primary amine groups (i.e., lysine residues of proteins). It is essential that the buffer in which the reaction is performed is free of primary amine groups. For this reason Tris-based buffers should be avoided. Either borate or phosphate buffered saline (pH 8.1) should be used. It is also important that as far as possible extraneous proteins (e.g., from serum-containing growth media) are removed before biotinylation takes place.
3. The buffer used to quench the biotinylation reaction can be any in which free primary amine groups are abundant. For example, glycine-containing buffers are often used but any glutamine-containing serum-free growth media will also make a good quenching buffer.
4. We found streptavidin-agarose beads from Pierce to yield much less background staining compared to those from Sigma under the same conditions.
5. The optimal amount of homogenization may vary between cell types. Conditions required to give complete cell lysis should be established empirically by monitoring trypan blue exclusion. The extent of cell lysis can be increased by prolonging the period of incubation in the hypotonic lysis buffer or by increasing the number of strokes with the homogenizer. Alternatively, for cells resistant to lysis (e.g., RAW 264.7), intact cells can be separated from broken cells by centrifugation (1000g for 5 min at 4°C). The supernatant is retained, the intact cell pellet subjected to a second round of homogenization in fresh hypotonic lysis buffer then recombined with the supernatant of the first spin.
6. To ensure effective washing of streptavidin-agarose beads and complete recovery of the DTT eluate use a 26-gage needle to puncture the bottom of an Eppendorf tube such that the liquid contents of the tube are recovered by placing the eppendorf inside a 6-mL plastic test tube and centrifuging (10,000g for 1 min). All agarose beads should be retained in the Eppendorf.
7. If an IEF step is not required (i.e., if the sample is to be resolved by 1DE only), the bead bound proteins can be eluted equally well after the washing steps by boiling (5 min) in an appropriate volume of SDS-PAGE sample buffer. Note that should you wish to visualize the biotinylated proteins using a streptavidin-HRP Western blot, reducing agent should be left out of the elution buffer in order to avoid cleavage of the biotin moiety.

8. Before harvesting the Rotofor fractions ensure the tubing is clean by rinsing through with $0.1M$ NaOH followed by 2 washes with Ultrapure water using a plastic transfer pipet.
9. When harvesting the Rotofor fractions, ensure that the metal teeth of the harvesting device penetrate each fraction simultaneously to avoid mixing of adjacent fractions.
10. To ensure a strong vacuum is maintained, thereby optimizing sample recovery, coat the upper edges of the plastic sample box with vacuum grease.
11. Following the clean-up step the sample should be in a volume of approx 100 µL. The gel system chosen for the second dimension should have sufficient capacity to accommodate the entire sample. The 20-cm Bio-Rad gel system using 1.5-mm thick combs worked well for these purposes as well as providing an extra few centimeters for greater resolution. It should also be noted that the use of gradient gels (e.g., 4–18%) gave greatly enhanced protein resolution compared to single percentage gels.

Acknowledgments

This work was funded by the Arthritis Research Campaign, The Wellcome Trust, and The MRC UK.

References

1. Li, J. M., Isler, P., Dayer, J. M., and Burger, D. (1995) Contact-dependent stimulation of monocytic cells and neutrophils by stimulated human T-cell clones. *Immunology* **84**, 571–576.
2. McInnes, I. B., Leung, B. P., Sturrock, R. D., Field, M., and Liew, F. Y. (1997) Interleukin-15 mediates T cell-dependent regulation of tumor necrosis factor-alpha production in rheumatoid arthritis. *Nat. Med.* **3**, 189–195.
3. Brennan, F. M., Hayes, A. L., Ciesielski, C. J., Green, P., Foxwell, B. M., and Feldmann, M. (2002) Evidence that rheumatoid arthritis synovial T cells are similar to cytokine-activated T cells: involvement of phosphatidylinositol 3-kinase and nuclear factor kappaB pathways in tumor necrosis factor alpha production in rheumatoid arthritis. *Arthritis Rheum.* **46**, 31–41.
4. Vey, E., Dayer, J. M., and Burger, D. (1997) Direct contact with stimulated T cells induces the expression of IL-1beta and IL-1 receptor antagonist in human monocytes. Involvement of serine/threonine phosphatases in differential regulation. *Cytokine* **9**, 480–487.
5. Lacraz, S., Isler, P., Vey, E., Welgus, H. G., and Dayer, J. M. (1994) Direct contact between T lymphocytes and monocytes is a major pathway for induction of metalloproteinase expression. *J. Biol. Chem.* **269**, 22,027–22,033.
6. Ribbens, C., Dayer, J. M., and Chizzolini, C. (2000) CD40-CD40 ligand (CD154) engagement is required but may not be sufficient for human T helper 1 cell induction of interleukin-2- or interleukin-15-driven, contact-dependent, interleukin-1beta production by monocytes. *Immunology* **99**, 279–286.

7. Isler, P., Vey, E., Zhang, J. H., and Dayer, J. M. (1993) Cell surface glycoproteins expressed on activated human T cells induce production of interleukin-1 beta by monocytic cells: a possible role of CD69. *Eur. Cytokine Netw.* **4**, 15–23.
8. Santoni, V., Molloy, M., and Rabilloud, T. (2000) Membrane proteins and proteomics: un amour impossible? *Electrophoresis* **21**, 1054–1070.
9. Gygi, S. P., Corthals, G. L., Zhang, Y., Rochon, Y., and Aebersold, R. (2000). Evaluation of two-dimensional gel electrophoresis-based proteome analysis technology. *Proc. Natl. Acad. Sci. USA* **97**, 9390–9395.
10. Peirce, M. J., Wait, R., Begum, S., Saklatvala, J., and Cope, A. P. (2004). Expression profiling of lymphocyte plasma membrane proteins. *Mol. Cell Proteomics* **3**, 56–65.

26

In Vivo Phage Display Selection in the Human/SCID Mouse Chimera Model for Defining Synovial Specific Determinants

Lewis Lee, Toby Garrood, and Costantino Pitzalis

Abstract

Phage display has represented a phenomenal technological advance of the last two decades. This technique is a very effective way of producing large numbers (up to 10^{12}) of diverse peptides and proteins (including antibodies), presented as fusion proteins on the viral capsid, that can be used for isolating specific molecules for therapeutic targeting. The increasing realization of the importance of the vascular endothelium in chronic inflammation as well as in neoplastic growth/spreading has prompted the targeting of blood vessels using phage display. This technique has been very successful in vivo in animals in selecting tissue specific vascular determinants. However, one disadvantage of using "pure" animal models is that the ligands obtained in this way are, by definition, specific for the targeted animal and might not bind to the human homologues. For this reason we have developed a novel approach using in vivo phage display selection against human tissues transplanted into SCID animals. In particular, we have focused on the transplantation of human synovium, although we have also successfully grafted skin, lymphoid, and fetal gut into these animals. The strength of this model is that the human graft blood vessels form functional anastomoses with mouse subdermal vessels that allow the target of lumenally expressed human molecules via the mouse circulation. Here we first describe the technical procedure for the in vivo selection of synovial homing phage using a commercially available peptide phage library in SCID mice transplanted with human synovium. This is followed by the description of the quantification and isolation of putative synovial specific peptide sequences. Finally, we describe the methodology used to confirm peptide-binding specificity including a competitive inhibition assay with synthetic peptide and the parent phage. The information provided should enable the reader to apply this technology in an in vivo setting to target human tissues in order to identify novel organ specific determinants as well as to develop tissue specific drug delivery systems.

Key Words: Phage; phage display; peptide; homing; microvascular endothelium; SCID mice; xenotransplantation; synovium; rheumatoid arthritis.

1. Introduction

Phage display, since its conception two decades ago *(1)*, has almost become a standard laboratory technique for the molecular biologist in the investigation of ligand interactions. The principle of the technique is that from a phage library expressing individual reagents, specific novel molecules can be obtained by affinity selection without prior knowledge of the structure or properties of the target. Selected phage will be then amplified in bacterial host and the specific expressed reagent identified. Of the molecules which can be presented by the phage the simplest are peptides *(2)*. An advantage of a peptide library is that, because of their size, peptides cause minimal disruption to the assembly of the phage in the host thus making these libraries very stable *(3)*. Moreover, peptides form very precise interactions with the binding site which results in the selection of a specific sequence consensus motif to occur *(3)*, and thereby allows the identification of the amino acid sequence involved in binding. However, the affinities of the peptides may be low because of the limited number of interactions with the binding site. Higher affinity reagents can be produced using antibodies display libraries, whereby the same nature of molecular display can produce a greater number of ligand-receptor interactions *(4)*. Another advantage of these libraries is that antibodies can bind to conformational as well as linear epitopes and have a greater retention time in the circulation because of their size; hence, generally speaking, they are more suited to therapeutic application than peptides. However, because of the complexity of binding, it is almost impossible to deduce the nature of interaction with the binding site to allow for any modifications to the molecule. For this reason peptide phage display has been favored when attempting to define ligand-receptor interactions *(3)*. Other display systems have also been described, such as bacterial *(5)*, and yeast *(6)*. These vectors are relatively large compared with the phage and potentially could be cleared quickly by organs such as the liver and spleen in in vivo selection procedures. The recent development of ribosomal display is a promising technology, however, selection can only be performed at temperatures below 4°C *(7)*.

The optimal method of selection for phage display should be chosen depending on the characteristics of target molecule. Simple molecules such as streptavidin can be presented on a solid support *(8)*, whereas complex molecules which have been found to be structurally unstable on solid phase are better and correctly presented on the surface membrane of the whole cell. For example, it has been found that the trimeric form of HLA class I molecules, consisting of the polymorphic α-chain, the associated $β_2$-microglobulin and a bound peptide, does not maintain its conformation during immobilisation onto plastic *(9)* but can be expressed on cells in their native conformation for phage display selection *(10)*. More interestingly phage selection using whole cells may be used to reveal unknown receptors specific for cell types such as neutrophils *(11)* or pathological conditions such as tumor cells *(12)*.

Given the importance of the microvascular endothelium (MVE) in physiology and pathology, several attempts have been made to target by phage display the MVE in vivo in various animal models *(13–15)* including disease states such as RA *(16)* and carotid injury *(17)*. However, though not known in all cases, it is likely that the determinants selected in this way are species-specific; hence, the targeting of human MVE remains a considerable challenge. This results mainly from the technical difficulties and ethical considerations of applying this technology in vivo to humans. The culturing of the MVE in vitro (from humans or animals), on the other hand, leads to the loss of important tissue-specific traits such as tight junctions in brain MVE or loss of MAdCAM-1 expression by intestinal MVE *(18,19)*. Thus, the transplantation of human tissues into SCID mice, which has been shown in various studies to be a successful means of maintaining human tissue in an authentic native and functional state *(20–22)*.

In this respect, we have pioneered the use of in vivo peptide phage display selection in SCID mice grafted with human tissues, by targeting in the first instance human synovium *(23,24)*. In this chapter, we describe the strategy (*see* **Fig. 1**) as well as the technical practical aspects utilized for the isolation and identification of synovial homing phage/peptides. These methodologies could be adapted to target other transplantable human tissues further enhancing the means of identify, in an in vivo setting, novel organ-specific determinants which could be exploited for diagnostic and/or therapeutic purposes.

2. Materials

1. $\alpha_V\beta_3$ (Chemicon)
2. TBS+: 1 mM CaCl$_2$, 1 mM MgCl$_2$, 0.01 mM MnCl$_2$, 150 mM NaCl, and 50 mM Tris-HCl (pH 7.4).
3. TBST: TBS+, 0.1% *(v/v)* Tween (Sigma).
4. Blocking solution: 35 mg/mL bovine serum albumin (BSA) (BDH) in TBS+
5. Peptide phage Kit Ph.D.C7C^{TM} system, (New England Biolabs).
6. *Escherichia coli* ER2738 phage host, glycerol stock (New England Biolabs).
7. Luria broth (LB) media: 10 g/L bacto-tryptone (Sigma), 5 g/L yeast extract (Sigma), and 5 g/L NaCl (BDH)
8. LB + tetracycline (Tet) media: LB media and 20 mg/L Tet (Sigma).
9. LB+Tet plates: LB+ Tet media and 14 g/L agar.
10. Agarose top, LB media, 7 g/L agar and 1 g/L MgCl$_2 \cdot$ 6H$_2$O (BDH).
11. Isopropyl β-D-thiogalactoside (IPTG) (Kramel Biotech).
12. 5-Bromo-4-chloro-3-indolyl-β-D-galactoside (Xgal) (Kramel Biotech).

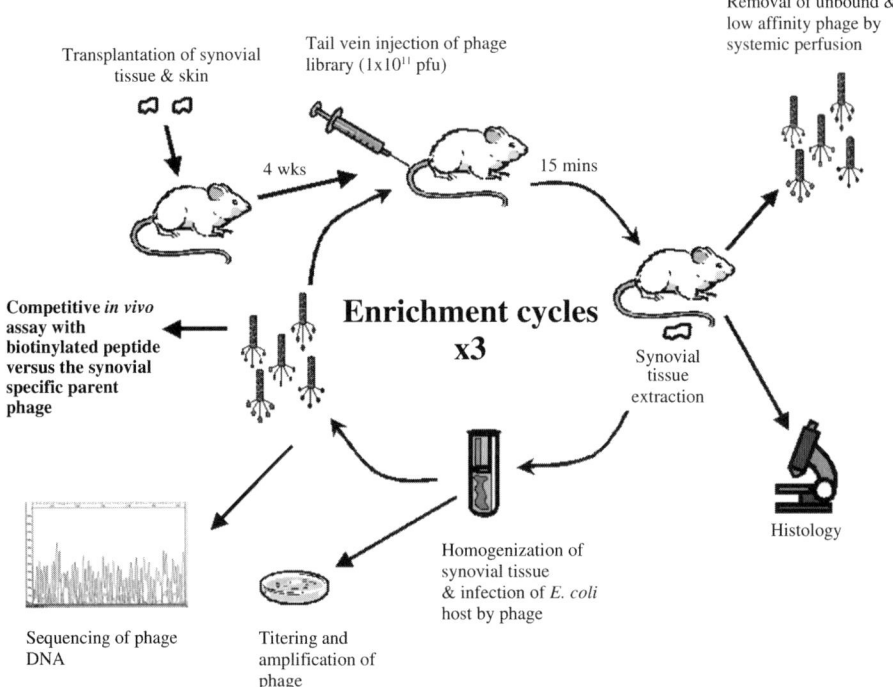

Fig. 1. Overview of in vivo phage display selection cycle in the human/SCID mouse transplantation model. The injection of the library into a transplanted SCID mouse allows the selection of specific synovial binding phage in the graft. Graft homing phage are recovered, amplified and injected into another transplanted animal to further enrich the phage pools homing specificity.

13. Xgal/IPTG agar plates: LB medium, 14 g/L agar, 0.2 *(w/v)* Xgal, and 0.25 *(w/v)* IPTG .
14. Polyethylene glycol (PEG)/NaCl; 200 g/L PEG-8000 (Sigma) and 147 g/L NaCl (autoclaved).
15. NOD/Lt Sz-SCID/SCID mice (beige SCID C.B-17).
16. Hanks balanced salt solution (HBSS) (Gibco).
17. Freezing medium: 20% dimethyl sulfoxide (DMSO) (Sigma) heat inactivated foetal calf serum (FCS) (PAA Labs GmbH).
18. Medetomide (Dormitor, 0.1 mg/mL) (Pfizer).
19. Ketamine (Ketalar, 0.1 mg/mL) (Parke-Davis).
20. Pentobarbitone sodium (Sagatal, 60 mg/mL) (Rhone Merieux Ltd.).
21. 21-gage 3/4" 0.8 × 19mm UTW tube winged infusion set (Terumo).
22. Saline: 9 g/L NaCl, sterile water for irrigation (Baxter's).

23. Protease inhibitor cocktail (Sigma).
24. Elution buffer: 0.1M glycine (pH 2.0) (Sigma).
25. 2M Tris base (Sigma).
26. OCT medium (Miles).
27. Isopentane (BDH).
28. Poly-L-lysine (Sigma).
29. Anti-M13 unconjugated monoclonal antibody (mAb) (Pharmacia).
30. Biotinylated goat anti-mouse secondary antibody (Dako).
31. Avidin-biotin-complex alkaline phosphatase system (AP-ABC) (Dako).
32. Vector Red substrate kit (Novacastra).
33. Fluorescein isothiocyanate (FITC) conjugated anti-human von Willebrand factor (vWf) (Serotec).
34. Anti-murine CD31 (clone MEC13.3) (Pharmingen).
35. Biotinylated rabbit anti-rat (Dako).
36. NaI buffer: 4M NaI, 10 mM Tris-HCl (pH 8.0) (Sigma).
37. Ethanol (BDH).
38. BigDye™ Terminator Cycle sequencing kit (Applied Biosystems, Warrington, UK).
39. Primer -96gIII (5'- ^{HO}CCC TCA TAG TTA GCG TAA CG –3') (New England Biolabs).
40. Dimethyl sulfoxide (DMSO) (BDH).
41. Ammonium acetate (pH 6.0) (Sigma).

3. Methods

The sections below describe the methods related to (1) validation of the phage display selection methodology in vitro, (2) in vivo peptide phage display selection in the human/SCID transplantation model, and (3) identification of candidate peptide sequences and confirmation of tissue specificity.

3.1. Validation of the Phage Selection Methodology In Vitro

As phage panning may cause the loss/reduction of diversity, it is important to validate the library and the techniques used in the selection process. As an example, we describe the methodology in the validation of the library using a defined target $\alpha_V\beta_3$. $\alpha_V\beta_3$ is a well-characterised integrin that typically binds the canonical RGD peptide motif, a common recognition/binding motif for most integrins. This motif has been previously isolated by several other peptide phage display libraries *(25–27)*. Therefore, the selection process of $\alpha_V\beta_3$ specific phage will be used as prototypic example to illustrate the methods for library validation and phage display selection in vitro.

The methods described for this work will be appropriate for the peptide phage kit Ph.D.C7C™ system, which is a commercially available peptide phage display library (New England Biolabs). As mentioned above, the principle of phage display is that each phage contains a single copy of the DNA sequence, which is translated as a specific peptide sequence on the phage capsid protein. Phage clones in the library carrying a specific sequence bind to the target molecule, whereas nonspecific unbound phage can be removed by washing. The bound phage clones can be eluted and replicated many fold by infecting into an appropriate bacterial host to produce progeny with the identical phenotype as the specific parent phage. This procedure increases the number of specific phage in the amplified pool. Reapplying the amplified pool in 2 to 3 other rounds of selection, the number of specific high affinity phage clones can be further enriched.

3.1.1. $\alpha_V\beta_3$ In Vitro Peptide Phage Display Selection

1. Coat a 96-well titer plate with 100 µL/well of 5 µg/mL of highly purified $\alpha_V\beta_3$ in TBS+ and incubate overnight at 4°C. As a control, coat half the plate with blocking solution only.
2. After incubation decant the $\alpha_V\beta_3$ from the plate and add 200 µL of blocking solution (BSA, 35 mg/mL) and incubate for 1 h at 4°C.
3. Dilute an aliquot of the phage display library stock to concentration of 1×10^{10} pfu/mL TBS+ containing 1 mg/mL of BSA.
4. Decant the blocking solution and add 100 µL/well of the diluted phage stock to 10 of the $\alpha_V\beta_3$ coated well and 10 of the BSA control wells and incubate again for 1 h at 4°C.
5. Wash the wells with TBST by adding 200 µL/well, decant and then blot the plate onto absorbent tissue. Repeat this 10 times ensuring to use a clean piece of tissue between each blot.
6. Elute the bound phage in the wells by adding 100 µL of elution buffer to each well and incubate for 10 min at room temperature.
7. Pool the respective wells together and neutralize the eluate with 6 µL of 2*M* Tris base and store at 4°C for titering and amplification.

3.1.2. Determination of Phage Concentration (Titration in EscherichiaColi Host)

To assist the identification of the peptide phage the M13KE vector has been engineered to carry the *lacZ a* gene. Consequently, peptide phage infecting *E. coli* host on an Xgal and IPTG agar plate will appear as blue plaques. Therefore, the concentration of a phage solution may be determined by counting blue plaques from a serial dilution of phage propagated on Xgal/IPTG agar plate.

3.1.2.1. THE PROPAGATION OF COMPETENT *ESCHERICHIA COLI* HOST

The peptide phage requires the presence of F-pilus on the male *E. coli* for infection. To ensure the entire *E. coli* host is competent, the ER2738 strain

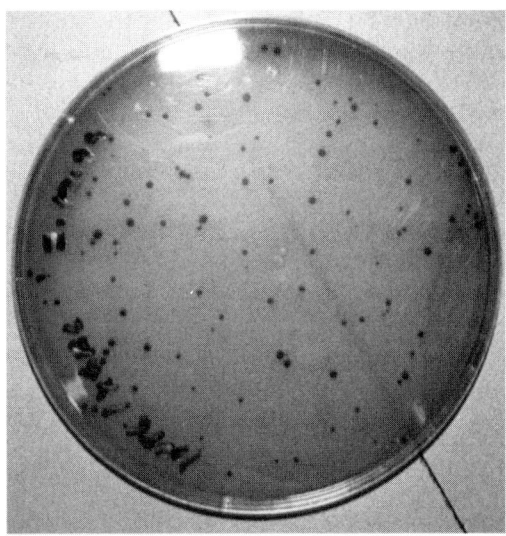

Fig. 2. Xgal/IPTG agar plate with peptide phage and *E. coli* host. The peptide phage titered on an Xgal/IPTG plate on a lawn of E.coli host appear as discrete blue plaques.

supplied by the manufacturer contains a mini-transposon, which confers tetracycline resistance, so that cells harbouring the F-factor can be preferentially propagated on selective media.

1. To propagate F-pilus positive cells, streak the glycerol stock of *E. coli* host, onto LB+Tet media plates and incubate overnight at 37°C.
2. Check the following day for the growth of single colonies and then the plate may be stored inverted at 4°C for up 3 mo.
3. For a working stock culture of F-factor positive, pick a single colony from the plate using sterile disposable inoculation loops and transfer into 20 mL of LB+Tet media.
4. Incubate the culture overnight at 37°C a rotary incubator.
5. After incubation the plate may be stored at 4°C, for up to 1 wk, until required.

3.1.2.2. DETERMINATION OF PHAGE CONCENTRATION BY TITERING ON XGAL/IPTG AGAR PLATES

The number of phage after selection against $\alpha_V\beta_3$ is determined by titering then counting the number of blue plaques formed on Xgal/IPTG agar plate with the *E. coli* host (*see* **Fig. 2**). To ensure consistency in titer determinations, an intra assay reference standard is also prepared from the native phage library stock (with a known original concentration of 2×10^{13} pfu/mL) and compared with the unknown test samples. **Important:** phage which appear white are wild type environmental phage and should be excluded from any titer determinations or clone selection (*see* **Note 1**).

1. Prepare the intra assay standard by diluting 10 µL sample of native library in 10 mL of TBS to give an expected phage concentration of approx 2×10^{10} pfu/mL. Then further diluted to final concentration of 2×10^2 pfu/mL in a series of stepwise 100-fold dilutions.
2. In triplicate, add 1 mL of the 2×10^2 pfu/mL to 200 µL of 1:50 diluted overnight stock culture of *E. coli* host. Mix by vortexing and incubate of 5 min.
3. Add 3 mL of molten agarose top (55°C) to the mixture and then immediately spread evenly onto a Xgal/IPTG agar plate.
4. Once the agar has hardened, invert the plates and then incubate overnight at 37°C. The next day, count the number of blue plaques formed by the phage clone.
5. For accuracy, use plaque densities of approx 200 pfu/plate for plate counting. Calculate the original phage concentration by multiplying the plate plaque count with dilution factor of the sample volume and express as pfu/mL.
6. Determine the eluted phage concentration by diluting and plating in triplicate 10-fold serial dilutions of the sample as described above.
7. Normalize the unknown phage concentration with the mean intra-assay reference standard value.

3.1.3. Amplification of Phage by Plate Culture

To increase the number of specific phage clones the phage is propagated in the *E. coli* host by plate culture (*see* **Notes 1 and 2**).

1. Amplify the eluate phage by first plating at a density of 1×10^4 pfu/plate as described in **Subheading 3.1.2.2.** for phage titering on Xgal/IPTG agar plates.
2. The next day, scrap the top agar layer, containing the blue phage plaques, into a 50-mL centrifuge tube. Add 20 mL of LB media to the tube homogenize the phage-containing agar by aspirating the agar/LB media mixture repeatedly through a 20-mL syringe fitted with a 16-gage needle.
3. Incubate homogenate on rotary mixer at 4°C for 1 h and then centrifuge at 11,000g at 4°C for 30 min.
4. Decant the supernatant containing the phage into a fresh tube.
5. Precipitate the phage by adding 1/6 of the volume with polyetyhlene glycol (PEG)/NaCl then incubate for 1 h on ice with, then centrifuge at 20,000g for 30 min.
6. Reconstitute the phage pellet in 1 mL of TBS, microcentrifuge for 10 min at 14,000 rpm to remove any residual cells.
7. Decant phage-containing supernatant in a fresh sterile tube and determine the concentration of the sample by titering onto Xgal/IPTG plates as described above.
8. Store the phage at 4°C until required.

3.1.4. Enrichment of the Phage Pools by Repetitive Rounds of Selection

Several rounds of selection (3–4) are generally required to obtain target specific phage. The main reason for this is the need of "selecting out" nonspecific

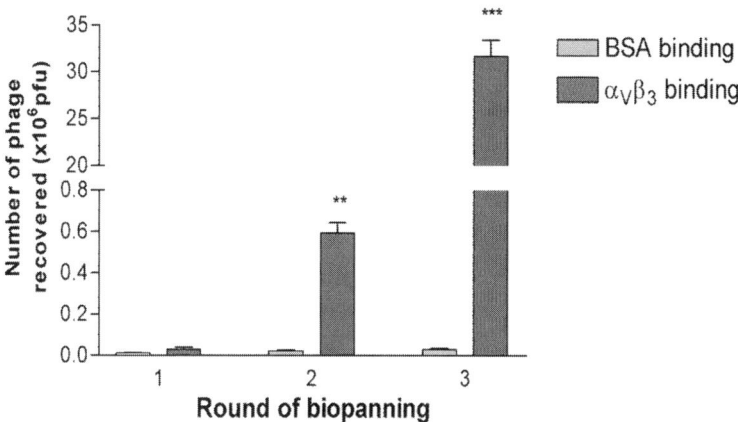

Fig. 3. In vitro selection of peptide-phage specific for $\alpha_V\beta_3$. The number of phage recovered from each round of selection against BSA control and $\alpha_V\beta_3$ coated wells are shown. The number of phage recovered from $\alpha_V\beta_3$ coated wells after each round increase significantly above that of the initial round of selection and BSA control wells. The data are expressed as the mean ± standard deviation of triplicate wells. ** $p = 0.004$, *** $p < 0.0001$ (difference of 2 and 3 compared with 1, unpaired, two tailed *t*-test).

binding phage while, at each round, increasing the specific binders. In addition, a higher density of specific phage competing for a limited number of receptors in each round will result in the selection of phage with the highest affinity.

It is important to ensure that the input number of phage in each round of selection is adjusted to the same concentration (in our conditions of selection 1×10^{10} pfu/mL).

As a result of enrichment the number of selected phage after each round should rise as the number of specific binders in the pool increase as shown in **Fig. 3**. Here it can be seen that there is almost a 1000-fold increase in the third round over first round.

3.1.5. DNA Sequencing of Selected Phage Clones

The precise nature of binding following peptide library selection should produce a characteristic peptide binding sequence for the target. In the case of integrins it has been shown that there is a preferential selection of peptide sequences, which contains the triple peptide motif RGD. To identify and confirm specificity of the phage from the selected pool, the phage DNA can be sequenced by dye terminated polymerase chain reaction (PCR) amplification using specific primers that will allow the precise determination of the peptide insert coding region and ultimately the peptide sequence itself.

3.1.5.1. PREPARATION OF PHAGE DNA TEMPLATE

For sequencing of the phage DNA, the phage clones from the pool must be isolated, amplified and then purified from the viral proteins and media.

1. Titer for each round of selection the pool of phage as described above to produce plates with 200 pfu/plate.
2. From these plates, select at random 10 to 100 clones from each plate. Excise each phage plaque separately from the agar and transfer each into 5 mL of *E. coli* host starter culture (1:200 dilution of stock in LB media) and then incubate in a 37°C rotary incubator 4.5 h for amplification.
3. Once amplified, centrifuge the cultures at 20,000g for 30 min.
4. Decant the phage supernatant and 1 mL of PEG/NaCl and incubate for 30 min on ice to precipitate the phage.
5. Centrifuge the phage precipitate at 20,000g for 30 min decant the supernatant.
6. Into each tube, add 100 µL of iodide buffer and mix thoroughly by aspirating several times with the pipet.
7. Recover the DNA by standard ethanol precipitation. Then dissolve the DNA with 10 mM Tris-HCl (pH 8.0) for sequencing.

3.1.5.2. SPECIFIC AMPLIFICATION OF PEPTIDE ENCODING DNA BY SPECIFIC PRIMERS

The sequence of the peptide encoding DNA insert in the phage vector may be determined by dye-terminated automated sequencing (BigDye^{TM} Terminator Cycle sequencing) using the specific primer -96gIII (5'- ^{HO}CCC TCA TAG TTA GCG TAA CG –3'). The 96gIII primer when annealed to the vector lies (from the 3'-end) 96 nucleotides back from the peptide encoding sequence and is specifically suited to automated sequencing.

In the specific example given, PCR sequence the phage DNA template using 3.2 pmoL of –96gIII primer and 0.1 µg of phage DNA. Then resolve the amplified DNA PCR products by electrophoresis sequencing on an ABI PRISM 377 sequencer, using standard techniques to determine the insert DNA sequence. The resultant insert DNA sequence can be decoded using a website based translation utility (*http:bio.lundberg.gu.se/edu/translat.html*) to determine the displayed peptide sequence.

3.1.6. Identification of Consensus Amino Acid Motifs From Specific Phage Peptide Sequences

As mentioned above, the advantage of peptide based phage display selection is that in some cases the selection process against a binding site is very precise and specific high affinity peptide binding configuration will be pre-

Table 1
Sequence Alignment of $\alpha_V\beta_3$ Selected Clones Show RGD/RLD Consensus Motif Enrichment

Clone No	Round 1	Round 2	Round 3
01	CHFPLTSGC	C-**RGD**—C	C—-**RLD**—C
02	CVSPQAKWC	C——**RGD**C	C—-**RGD**—C
03	CGVVGYQHC	C**RGD**——C	CPATHSLTC
04	CKQGELPFC	CIPLQTVLC	C—-**RGD**—C
05	CDIEPQPFC	CSNTHIPRC	C-**RGD**—C
06	CLVTADSSC	C—-**RLD**—C	C—-**RGD**—C
07	CNTAPLPGC	C-**RGD**—C	C—-**RGD**—C
08	CPTPGREMC	C**RGD**——C	CDETGRSSC
09	CTHHPELTC	C-**RGD**—C	C—-**RGD**—C
10	CQSLANRLC	CTPSQPMLC	C-—-**RGD**—C

Note: Sequence analysis of 10 clones, chosen at random, from each round of in vitro selection. The 7-peptide insert sequences are shown between the two flanking cysteines. RGD and RLD sequences (shown in **bold**), specific for $\alpha_V\beta_3$, occur in rounds 2 and 3 but not in round 1.

dominant in the sequences of selected phage pool. These consensus motifs are a hallmark for specific binding, in the case of integrins such as $\alpha_V\beta_3$ it has been shown that peptide phage libraries select for RGD or RLD in their displayed sequences.

In the example of peptide phage selection against $\alpha_V\beta_3$, it is shown from the random selection and sequencing of 10 clones from each round of in vitro selection that there is indeed enrichment of RGD sequences in many of the clones in the second and third round of selection (**Table 1**). Moreover, this experiment confirms the diversity of the library with no RGD/RLD sequences present in the first round of selection. In the next section these same techniques will be applied to the in vivo phage selection.

3.2. In Vivo Peptide Phage Display Selection in the Human/SCID Mouse Transplantation Model

There is a large body of evidence for the existence of organ specific vascular determinants both from leukocyte homing studies and from in vivo phage display selection in a "pure mouse model." This supports the hypothesis that organ specific MVE determinants are present in most tissues, which may be responsible for the coordination of interactions that regulate the specific trafficking of leukocyte subpopulations to various tissues. The grafting of these tissues into SCID mice has been shown to preserve the MVE in its own microenvironment, which should maintain tissue specific traits that are normally lost during culturing of the MVE in vitro. Furthermore, leukocyte homing studies have shown that the graft MVE is functional in facilitating human cell migration to human transplanted tissues. Therefore, it would be expected that targeting transplanted human synovial tissue with phage display would reveal MVE determinants involved in tissue specific leukocyte trafficking. In this section we will describe the SCID mouse transplantation model for human synovium and skin, the in vivo selection process and the histological analysis of tissue.

3.2.1. The Human/SCID Mouse Transplantation Model

3.2.1.1. PRESERVATION AND STORAGE OF HUMAN TISSUE FROM SURGERY FOR TRANSPLANTATION

Synovial and skin tissue samples for transplantation maybe obtained from surgical specimens of patients requiring joint replacement or synovial biopsies.

1. Tissue samples, placed after surgery into *sterile* specimen jar for collection, are processed within 3 h of removal from the patient (*see* **Note 3**).
2. Tissue samples are cut into approx 5-mm^3 pieces in a class II laminar flow hood under sterile conditions.
3. Tissues for transplantation are washed in sterile HBSS then place into 2 mL freezing vials containing 1 mL of 20% DMSO in heat inactivated FCS for transplantation (*see* **Note 4**) or snap frozen in liquid nitrogen chilled isopentane mounted in OCT media for histology.
4. Cool the tissues for transplantation to –80°C in an isopropanol cell-freezing chamber, then stored under liquid nitrogen until required for implanting into mice (*see* **Note 4**).

3.2.1.2. TRANSPLANTATION OF SCID MICE STRAIN

SCID mice kept under pathogen free conditions are singly or double transplanted with human synovial and skin tissue between 4 and 6 wk of age (*see* **Note 5**). The surgical procedures are performed in an isolator or under sterile conditions.

1. Thaw the tissue samples from liquid nitrogen stock at 37°C of 5 min and then wash the tissue in sterile HBSS.
2. For transportation, keep the tissue on moistened sterile gauze placed in a sterile tube until transplantation.
3. Anaesthetize the mice with an i.p. injection of 0.2 mL Dormitor, (0.1 mg/mL) and 0.1 mL Ketalar (0.1 mg/mL)
4. To transplant an animal, make an incision in the dorsal skin behind the ear and insert the tissue subcutaneously.
5. Close the wound with insoluble suture material and allow the animals 3 to 4 wk to recover. After 3 to 4 wk the transplants will be imbedded in the mouse subdermal tissue and feed by surrounding murine vessels (*see* **Fig. 4**).

Phage Display for Defining Synovial Specific Determinants

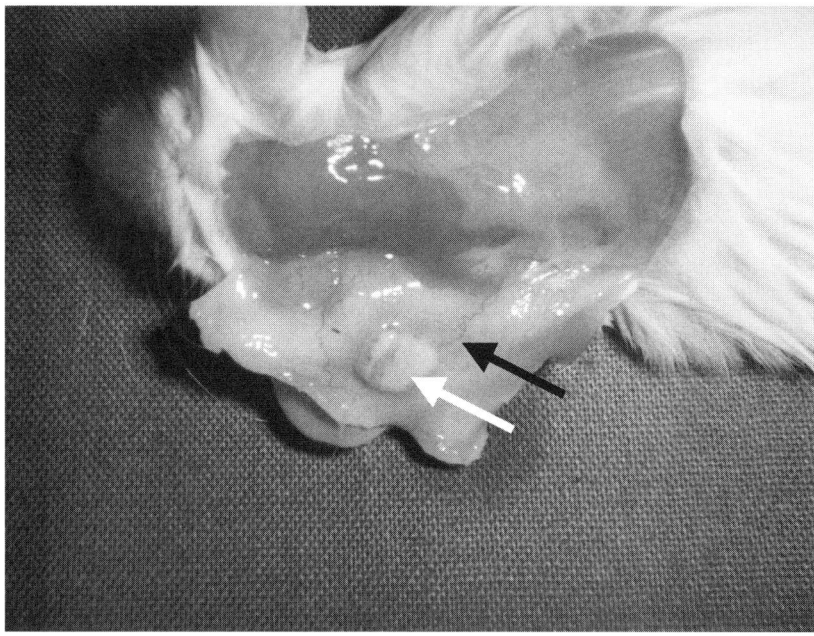

Fig. 4. Engrafted synovial tissue in SCID mouse. The figure shows synovial tissue 4 wk post-transplantation (indicated by the *white arrow*). Subdermal murine vessels can be seen leading into the area of the transplant (indicated by the *black arrow*).

3.2.2. In Vivo Phage Display in Human Synovium and Skin Transplanted SCID Mice

The methods described here include the injection of the library, the removal of the unbound phage from the circulation, recovery of the phage from the tissue and the enrichment of synovial homing phage by two further rounds of in vivo selection.

1. Four weeks post transplantation with human rheumatoid arthirits (RA) synovial and skin tissue; inject 1×10^{11} pfu of the peptide phage library into the mouse (with a maximum volume of 200 µL), via the tail vein.
2. Allow the phage to circulate for 15-min (*see* **Note 6**). Then anaesthetize the animal with 150 µL sodium phenobarbitone.
3. Remove nonspecific phage from the circulation by perfusing animals with normal saline (*see* **Note 7**). Cut at the level of the diaphragm to reveal the thoracic and peritoneal walls, then remove the rib cage to reveal the heart and insert a 21-gage butterfly needle, attached to a 50-mL syringe filled with saline into the left ventricle (*see* **Fig. 5**).

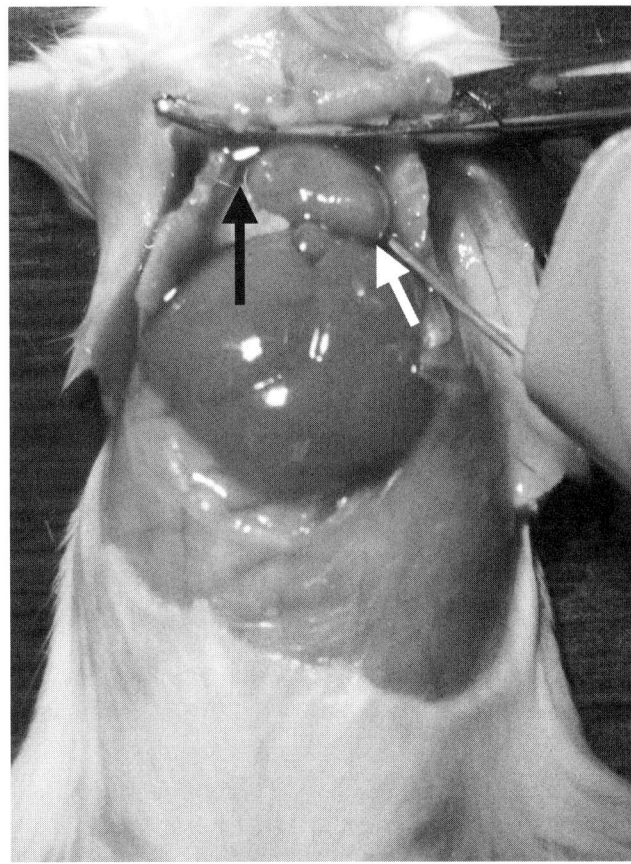

Fig. 5. Cannulation of murine heart for perfusion. In preparation of cardiac perfusion, under anaesthesia, the rib cage is removed to expose the heart. The heart is cannulated by the insertion of a butterfly needle into the left ventricle (indicated by the *white arrow*). To avoid fluid overload, an incision is made in the right atrium (indicated by the *black arrow*). Saline is then injected through the cannula into the left ventricle to clear the blood pool from the systemic and tissue circulation.

4. To avoid fluid overload at the beginning of the perfusion, make a small incision on the right arterial chamber then perfuse with 100 mL of saline (tissue blanching, especially in the liver, is taken as an indicator of successful blood pool clearance from the tissues).
5. Once the blood pool has been cleared remove the transplants from the animal and organs for phage recovery and titering.
6. For immunohistology, prepare frozen sample sections from the tissues collected by embedding a small sample from each in OCT (supported on 1-cm^2 cork tiles)

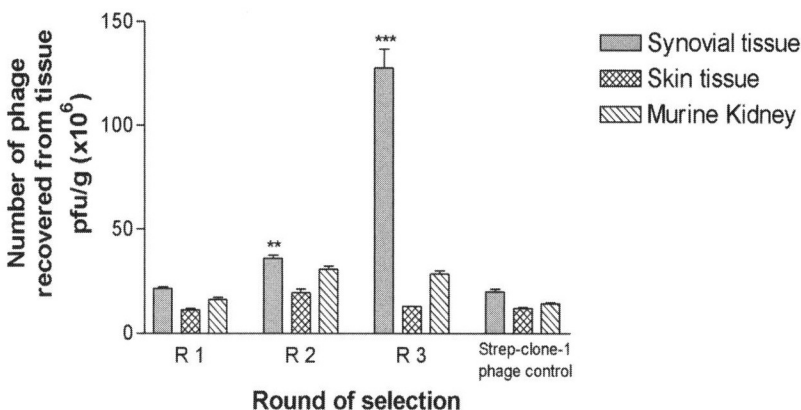

Fig. 6. In vivo selection of peptide phage specific for human synovium in human RA synovium transplanted SCID mice. The numbers of peptide-phage (pfu/g tissue) recovered from 3 consecutive rounds of in vivo selection are shown. Mouse kidney has been included as murine control tissue. The results are expressed as the mean ± standard deviation of triplicate plate readings. It can be seen that in the first round of selection the level of phage recovered from the synovial graft is similar to that of the control phage. After the second and third round there is a stepwise increase in the number of phage recovered from the synovial graft. Whereas in the skin graft and murine kidney's the number of phage remain relatively constant. Differences seen in enrichment rounds, 2 and 3, in the synovial tissue are statistically significant compared to R1 and strep-clone-1 phage control. $**p = 0.0004$, $***p < 0.0001$. There are no significant differences in sequential rounds of enrichment in the skin transplants, the mouse kidney and in the strep-clone-1 study, $p > 0.05$ (unpaired, two tailed t-test).

and then snap freeze in liquid nitrogen chilled isopentane then store at $-20°C$ until required.
7. Wash the synovial transplants three times in TBS then homogenize in 1 mL of TBS containing protease inhibitor cocktail on ice using an electric homogenizer.
8. Elute the phage from the homogenate by adding 2.5 mL of $0.1M$ glycine (pH 2.0) and incubate on ice for 10 min.
9. Neutralize the mixture with 36 µL of $2M$ Tris base and then titer and amplify the phage on plate culture as described above.
10. To further enrich the selected pool repeat two further cycles *of in vivo* selection as above (*see* **Note 8**). An example of 3 rounds of selection of in vivo selection for synovial specific phage is shown in **Fig. 6**.

3.2.3. Histology

In addition to the titers, binding specificity of the phage can also be confirmed by visualizing the phage distribution in the tissue by immunohistochemistry. Tissue sample from the selection process can be specifically stained using an anti-M13 phage specific antibody in conjunction with antibodies to species-specific vessel markers.

3.2.3.1. DOUBLE IMMUNOFLUORESCENCE STAINING FOR PHAGE AND HUMAN VESSELS WITHIN GRAFT TISSUE

The phage in graft tissue sections after in vivo selection of the library may be detected by standard single or double immunohistochemistry/fluorescence as previously described *(23,28)*. Phage can be detected with anti-M13 antibody followed by indirect immunoalkaline phosphatase immunohistochemistry (AP-ABC) visualized by Vector Red substrate under fluorescence microscopy. For detection of human and murine vessels within the grafts, sections may be double stained with either FITC-conjugated sheep anti-human von Willebrand factor or rat anti-murine CD31.

To illustrate the vessel specificity of the selected phage, the double staining of the same section as above shows that the phage staining colocalize with the human vessel but not murine vessel in the graft or in the murine kidney (*see* **Fig. 7**).

3.2.3.2. ASSESSMENT AND WUANTIFICATION OF HUMAN AND MURINE VASCULATURE WITHIN THE GRAFTS

To exclude the possibility that the specific localization of the phage to the synovial tissue results from differences in size of the vascular bed feeding the transplant, it is important that the degree of vascularisation is determined.

The density of human and murine endothelial surface can be determined by immunohistochemistry using the above-mentioned species-specific anti-human and anti-murine vessel markers. The volume fraction (Vv) of immunostained human and murine vessels can be determined microscopically using a point counting method as previously described *(29)*.

The example shown (*see* **Fig. 8**) compares the skin and synovial graft tissue vascularity from the third round of in vivo selection. The data indicates that the skin graft contains a significantly greater number of vessels compared to the syn-

---

Fig. 7. Histological analysis show peptide phage bind to human but not mouse microvasculature. Histological localization of the peptide phage within the synovial grafts and mouse kidney detected by double immunohistology using anti-M13 mAb and species-specific vascular markers and visualized by fluorescence microscopy in the same section. Representative microscopic fields from the final round of selection are shown. Discrete M13 staining can be clearly seen in (A), whereas the isotype matched negative antibody control showed no staining (C). M13 staining typically colocalizes with

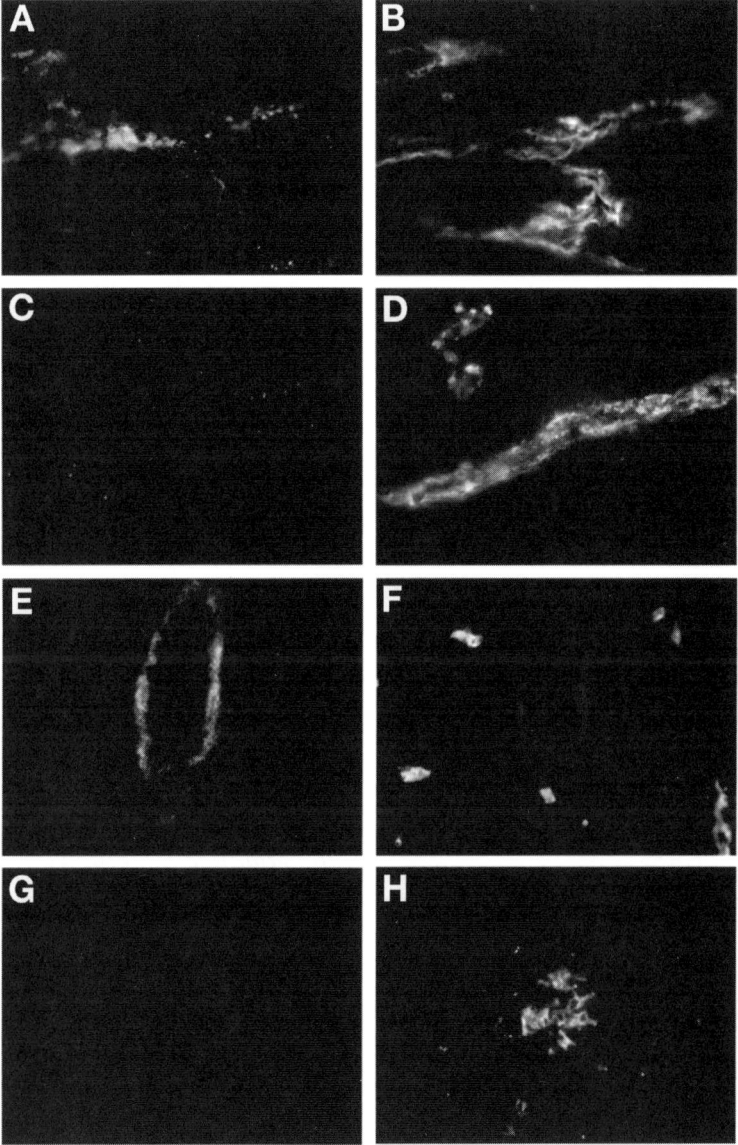

the human vasculature visualized with FITC conjugated anti-human vWf antibody (B, D). However, M13 immunoreactivity (E) shows no colocalization with murine vasculature within the grafts detected with FITC conjugated anti-murine CD31 secondary antibody (F). Likewise, sections of murine kidney, which were taken from the same animal, show no M13 immunoreactivity (G) in the glomerular capillaries clearly positive for murine CD31 (H). Magnification: ×400.

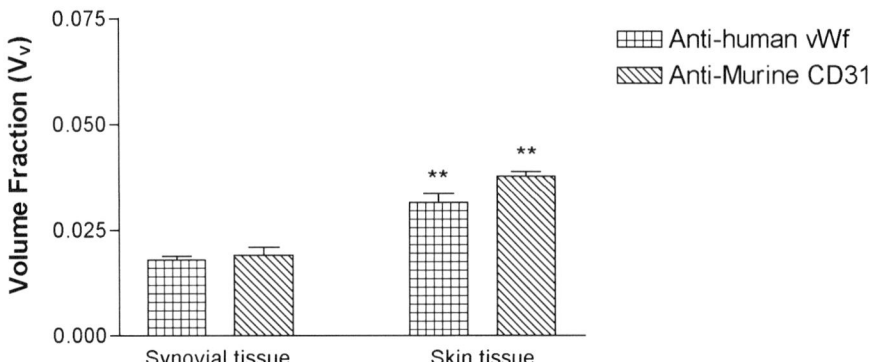

Fig. 8. Vascularity determinations of synovial and skin transplants from histological sections. The degree of vascularization in frozen tissue aliquots of the samples for third circulated phage pools was determined by immunohistochemistry by the staining of the human and mouse vascular endothelium, using species-specific anti-human vWf and murine CD31 antibodies. The volume fraction (Vv) of immunostained human and murine vessels was determined microscopically by point counting. The results shown are expressed as the mean ± standard deviation from three cutting levels. There is a slight but statistically significant lower endothelial volume in the synovial grafts compared with skin grafts in both experiments. This applied to both human (**$p = 0.0043$) and murine (**$p = 0.001$) vessels (unpaired, two tailed t-test).

ovial graft. Therefore, the higher number of phage in the synovial tissue cannot be simply attributed to a difference in tissue vascularity.

3.3. Identification of Candidate Peptide Sequence Specificity From Selected Phage Clones

The in vivo selection process so far has described the selection of the phage pool, however, to identify specific determinants it would be necessary to isolate single phage clones. To achieve this, as described for $\alpha_V\beta_3$ selection, the clones from the final round of selection are picked at random and DNA sequenced to determine the peptide sequence displayed at clonal level. By comparing these sequences, clones that contain recurrent motifs may be investigated further in vitro/in vivo competition assays. These experiments provide invaluable information on which sequences to select for subsequent in vitro peptide synthesis and functional analysis.

3.3.1. Identification of Synovial Specific Candidate Peptide Sequences From In Vivo Selected Phage by DNA Sequencing

To identify possible tissue specific sequences phage clones can be selected from the enriched pools for DNA sequencing and peptide alignment. As mentioned above, an advantage of phage display is that the interaction with a target ligand can be precise and can be characterized by specific peptide motifs. For in vivo phage display it has been shown that recurrent peptide motifs can identify specific peptide interactions with endothelial molecules *(15)*.

To identify specific clones by motif comparisons, the final round of phage selection is titered, a sample of clones is chosen at random and DNA sequenced to determine the displayed peptide sequence as described in **Subheading 3.1.5.** As an example to illustrate identification of a candidate peptide, the clones from an in vivo selection experiment in SCID mice transplanted with human synovium have been aligned and show number of peptide motifs that have occurred (**Table 2**). The binding specificity of phage displaying motif sequences was confirmed at the single clone level in vivo in the same SCID mouse chimera model. The peptide sequence CKSTHDRLC from clone 3.1 containing a DRL motif, was found to have the highest tissue localization and hence was chosen as a candidate peptide for further studies using the synthetic form of the peptide.

3.3.2. Confirmation of Binding Specificity by Selected Peptides

To determine if a candidate peptide can maintain the functional capacity to localize specifically to synovial grafts in the absence of phage, a synthetic biotinylated peptide of the sequence can be used in histology for specific vessel staining after in vivo circulation. In addition, the synthetic peptide may also be used in vivo to compete with the parent phage. As mentioned above, because of the high degree of tissue localization and the occurrence of a triple peptide motif, the 3.1 phage clone displaying the CKSTHDRLC peptide will be used to illustrate the process of synthetic peptide in vivo localization and competition to the parent phage in transplanted SCID mice.

3.3.2.1. SYNTHESIS OF CANDIDATE PEPTIDE SEQUENCE

The candidate peptide was custom synthesized in vitro by Alta Biosciences (Birmingham University) using FMOC chemistry in an automated peptide synthesizer that also allowed the incorporation of a biotin label. The peptide was prepared to a purity of >95% by reverse phase chromatography and freeze dried in 2 mg/vial aliquots. Prior to use, the peptide was solubilized in 10 µL of DMSO and reconstituted to a final concentration of 4 mg/mL in $0.1M$ ammonium acetate (pH 6.0).

3.3.2.2. HISTOLOGICAL DISTRIBUTION OF THE BIOTINYLATED PEPTIDE

As mentioned above, by taking advantage of the biotin tag on the peptide the distribution after in vivo injection of the candidate peptide can be visualized in histological sections of the graft using the AP-ABC detection systems,

Table 2
The Alignment of Peptide Sequences by Amino Acid Properties

Consensus peptide motif	Clone number	Alignment of sequences
NQR	2.27	QTHNQRY
	2.29	TNQRLAI
RLP	1.17	HPRLPFA
	2.14	APNWRLP
SPS	2.2	SPSPFRA
	1.22	SPSRFDQ
	2.15	VSPSRTT
HSS	2.16	PLSSAQR
	1.29	TWSATST
SAT	1.23	THSSATQ
	3.13	HTHSSNL
SSR	2.6	PNHSSPH
	1.30	ADHSSRH
SSA	1.15	SDYSSRS
DRL	3.1	KSTHDRL
	3.4	PFHDRHS
HDR	2.12	HPSDRLS
	2.10	DRLNHQF

Key to shaded coding:
- Non polar (hydrophobic)
- Polar (uncharged)
- Basic
- Acidic

Note: The peptide sequences from in vivo phage display have been divided into 5 main groups with similar triple peptide consensus motifs. The amino acids have been shaded according to their chemical properties to reveal further sequence similarities.

in relation to human vessels, indicated by FITC conjugated anti-human vWf as previously described.

The histological analysis of the distribution biotinylated 3.1 CKSTHDRLC peptide in the graft shows that the peptide co-localizing with the human blood vessels in the graft tissue, whilst the control peptide strep-clone-1 (containing the sequence CGTWSHPQC) at the equivalent concentration shows negative staining (see **Fig. 9**).

3.3.2.3. COMPETITIVE IN VIVO ASSAY WITH CANDIDATE SYNTHETIC PEPTIDE

The specificity and binding capacity of the peptide can be determined by the competitive effect of the peptide to the synovial homing capacity of the parent phage, which can be quantified by titering (see **Note 9**).

For the in vivo competitive assay, dose groups of the candidate 3.1 CKSTHDRLC peptide and the strep-clone-1 peptide control are injected simultaneously with a constant number of 3.1 phage into RA synovial tissue transplanted mice.

1. Prepare dose groups of candidate 3.1 CKSTHDRLC peptide (i.e., 500, 250, and 50 µg/mouse) and premix with a fixed concentration of phage (1×10^{11} pfu/mouse) in dose volume of 200 µL/ mouse.
2. For the control experiment, as above prepare the equivalent doses of the control peptide then mix with 1×10^{11} pfu of 3.1 phage clone.
3. Inject three mice per dose group with the phage/peptide mixture via the tail vein and allow to circulate for 15 min.
4. As described in **Subheading 3.2.2.**, perfuse the animals and then quantify the number of phage in the tissue by titering.

The titers of the 3.1 phage, in this example, show the decrease in binding of the phage when the concentration of the biotinylated 3.1 CKSTHDRLC peptide is increased (see **Fig. 10**). Whereas the control peptide has no effect on the 3.1 phage binding at the equivalent peptide concentrations. This confirms the peptide binding specific to the synovial graft and independent of the phage molecule.

4. Notes

1. Any wild type phage contamination in the amplification culture will result in a final pool of mainly wild type phage. This results from the displayed oligopeptide sequences on the pIII (which mediates the docking to the F-pilus of the *E. coli* host) reducing the reproductive potential of the library. Therefore, to avoid wild-type phage outgrowth and ensure minimal bias in propagation, it is recommended that the phage clones be amplified on plates as individual phage plaques.
2. Phage plaques, which appear white on Xgal/IPTG plates, are wild-type environmental phage and should be excluded from titer determination or clone selection.

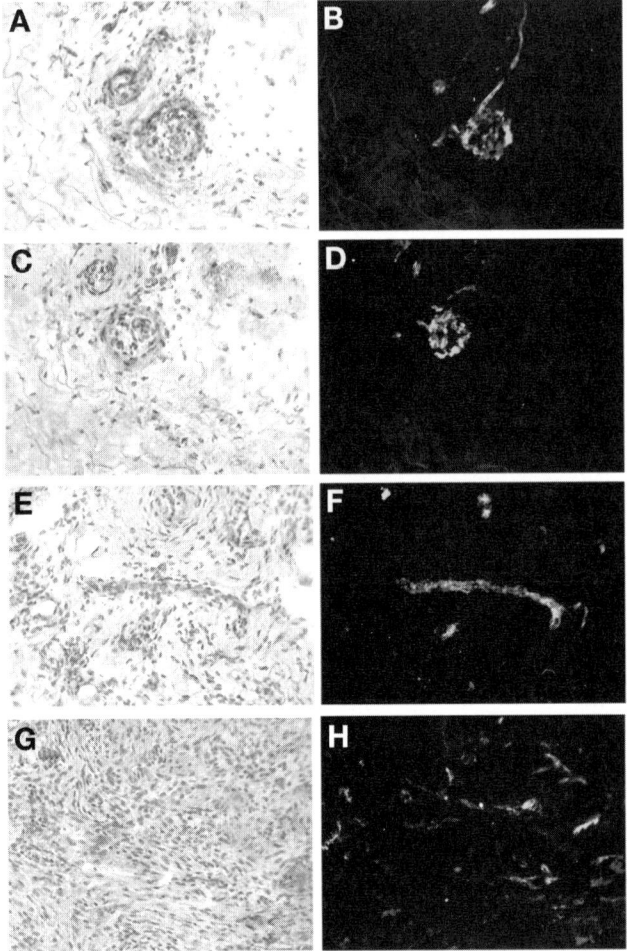

Fig. 9. Histochemical analysis for the biotinylated 3.1 CKSTHDRLC peptide confirms localization in vivo to human vessels in synovial grafts compared to the strep-clone-1 peptide. Frozen tissue sections of synovial samples analyszd by histochemistry by applying the AP-ABC detection system and stained with Vector red substrate. Sections were then double stained with FITC conjugated anti-human vWf antibody. The 3.1 CKSTHDRLC peptide clearly shows specific reactivity colocalizing with human vasculature (A,B). No specific staining was seen when AP-ABC was omitted from the sequential section (C), although human vWf positive vessel are present (D). Grafts from mice injected with strep-clone-1 peptide show no reactivity for the AP-ABC detection system (E) in relation to the human vessel in the graft (F). Omission of AP-ABC detection system in sequential section of strep-clone-1 peptide control tissue again shows no reactivity (G) in relation to human vWf positive blood vessels (H). Magnification: ×400

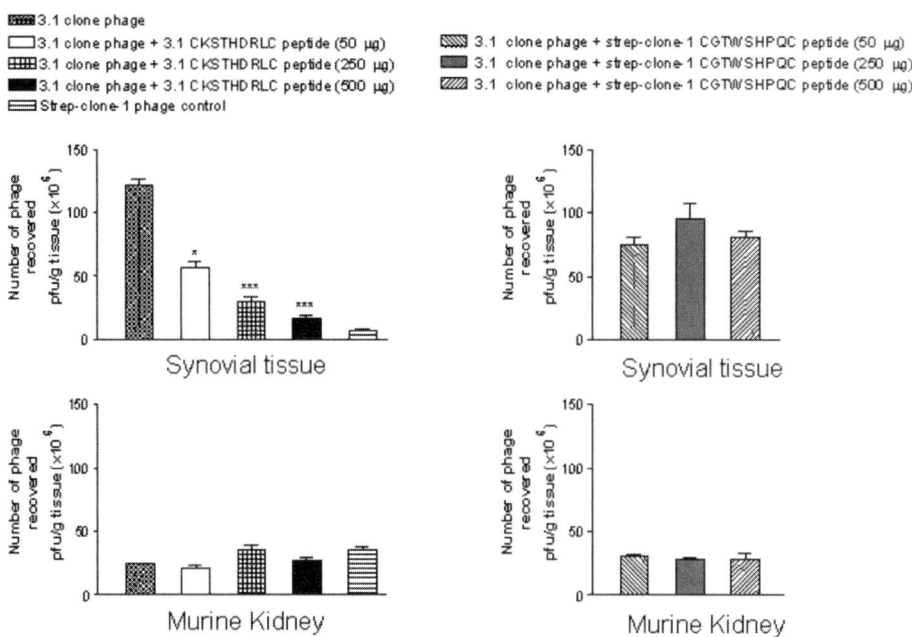

Fig. 10. The synthetic biotinylated 3.1 CKSTHDRLC peptide competes in vivo in RA synovial grafts for a putative synovial vascular ligand(s) against the parent 3.1 phage clone in dose dependent manner. SCID mice transplanted with human synovial tissue (2 grafts/animal) were injected intravenously with 1×10^{11} pfu of 3.1 phage clone with the 3.1 CKSTHDRLC peptide or with the control strep-clone-1 peptide at three increasing doses (50, 250, and 500 μg/mouse in 200 μL dose volume). Equal concentrations of strep-clone-1 phage were used as a negative phage control. It can be seen that there is a highly significant statistical difference in the number of phage recovered from the grafts in the presence and the absence of the biotinylated 3.1 peptide (***$p < 0.0001$, unpaired, two tailed t-test) with an inhibition of over 70% at the highest concentration of the peptide. No difference was observed between the various groups in the number of phage recovered from murine kidneys or in the control peptide treatments. $p > 0.05$ (unpaired, two tailed t-test).

3. SCID mice are unable to mount an effective immune response, therefore it is critical that tissue transplants are sterile, otherwise the mice will quickly die from infection
4. To allow for the coordination of time between tissue collection and transplantation for experimentation, DMSO freezing is used to preserve the tissue.
5. Severe combined immunodeficient (SCID) mice have a recessive mutation on chromosome 16 affecting the mechanisms critical for T-cell receptor and immunoglobulin VDJ segment rearrangements, and as a result of this mutation they

lack functional T- and B-cells *(30)*. Consequently, SCID mice are unable to reject various types of transplanted tissue *(31)*. However, the arrest in lymphocyte development is not absolute and as these mice become older, they generate a few clones of functional B- and T-cells. By 10 to 14 wk of age the majority of these mice will be "leaky," producing antibodies. Therefore, mice used in this study should be between 4 and 10 wk of age.
6. In vivo selection has the advantage of binding specific phage underflow conditions and encourages the selection of phage with high affinity.
7. The cardiac perfusion of the mice following in vivo phage selection allows the removal of nonspecific phage from the circulation. This greatly improves the efficiency of enrichment in each of the rounds of selection.
8. There is also a process of de-selection of phage recognizing common (mouse/man) vascular specificity by the mouse systemic circulation.
9. The competition of the synthetic peptide and phage is an important functional aspect that can confirm binding to the same determinant.

References

1. Smith, G. P. (1985) Filamentous fusion phage: Novel expression vectors that display cloned antigens on the virion surface *Science* 1315–1317.
2. Smith, G. P. and Scott, J. K. (1993) Libraries of peptides and proteins displayed on filamentous phage. *Methods Enzymol.* **217,** 228–257.
3. Parmley, S. F. and Smith, G. P. (1988) Antibody-selectable filamentous fd phage vectors: affinity purification of target genes. *Gene* **73,** 305–318.
4. Winter, G., Griffiths, A., Hawkins, R. E., and Hoogenboom, H. R. (1994) Making antibodies by phage display technology. *Ann. Rev. Immunol.* **12,** 433–455.
5. Samuelson, P., Gunneriusson, E., Nygren, P. A., and Stahl, S. (2002) Display of proteins on bacteria. *J. Biotechnol.* **96,** 129–154.
6. Weaver-Feldhaus, J. M., Lou, J., Coleman, J. R., Siegel, R. W., Marks, J. D., and Feldhaus, M. J. (2004) Yeast mating for combinatorial Fab library generation and surface display. *FEBS Lett.* **564,** 24–34.
7. Hanes, J., Schaffitzel, C., Knappik, A., and Pluckthun, A. (2000) Picomolar affinity antibodies from a fully synthetic naive library selected and evolved by ribosome display. *Nat. Biotechnol.* **18,** 1287–1292.
8. Giebel, L. B., Cass, R. T., Milligan, D. L., Young, D. C., Arze, R., and Johnson, C. R. (1995) Screening of cyclic peptide phage libraries identifies ligands that bind streptavidin with high affinities. *Biochemistry.* **34,** 15,430–15,435.
9. Pistillo, M. P., Hammer, J., Bono, E., and Sinigaglia, F. (1997) A novel approach to human anti-HLA mABs production: Using of phage display library. *Human Immunol.* **57,** 19–27.
10. Marget, M., Sharma, B. B., Tesar, M., et al. (2000) Bypassing hybridoma technology: HLA-C reactive human single-chain antibody fragments (scFv) derived from a synthetic phage display library (HuCAL) and their potential to discriminate HLA class I specificities. *Tissue Antigens.* **56,** 1–9.

11. Mazzucchelli, L., Burritt, J. B., Jesaitis, A. J., et al. (1999) Cell-specific peptide binding by human neutrophils *Blood* **93,** 1738–1748.
12. Roovers, R. C., van der, L. E., de Bruine, A. P., Arends, J. W., and Hoogenboom, H. R. (2001) Identification of colon tumour-associated antigens by phage antibody selections on primary colorectal carcinoma. *Eur. J. Cancer.* **37,** 542–549.
13. Arap, W., Pasqualini, R., and Ruoslahti, E. (1998) Cancer treatment by targeted drug delivery to tumor vasculature in a mouse model. *Science.* **279,** 377–380.
14. Arap, W., Haedicke, W., Bernasconi, M., et al. (2002) Targeting the prostate for destruction through a vascular address *Proc. Natl. Acad. Sci. USA* **99,** 1527–1531.
15. Rajotte, D. and Ruoslahti, E. (1999) Membrane dipeptidase is the receptor for a lung-targeting peptide identified by in vivo phage display *J. Biol. Chem.* **274,** 11,593–11,598.
16. Gerlag, D. M., Borges, E., Tak, P. P., et al. (2001) Suppression of murine collagen-induced arthritis by targeted apoptosis of synovial neovasculature. *Arthritis Res.* **3,** 357–361.
17. Herrmann, A., Pieper, M., and Schrader, J. (1999) Selection of cell specific peptides in a rat carotid injury model using a random peptide-presenting bacterial library. *Biochim. Biophys. Acta.* **1472,** 529–536.
18. Borsum, T., Hagen, I., Henriksen, T., and Carlander, B. (1982) Alterations in the protein composition and surface structure of human endothelial cells during growth in primary culture *Atherosclerosis* **44,** 367–378.
19. de Bono, D. P. and Green, C. (1984) The adhesion of different cell types to cultured vascular endothelium: effects of culture density and age. *Br. J. Exp. Pathol.* **65,** 145–154.
20. Wahid, S., Blades, M. C., De Lord, D., et al. (2000) Tumour necrosis factor-alpha (TNF-alpha) enhances lymphocyte migration into rheumatoid synovial tissue transplanted into severe combined immunodeficient (SCID) mice. *Clin. Exp. Immunol.* **122,** 133–142.
21. Rendt, K. E., Barry, T. S., Jones, D. M., et al. (1993) Engraftment of human synovium into severe combined immune deficient mice. Migration of human peripheral blood T cells to engrafted human synovium and to mouse lymph nodes. *J. Immunol.* **151,** 7324–7336.
22. Jorgensen, C., Couret, I., Canovas, F., et al. (1996) Mononuclear cell retention in rheumatoid synovial tissue engrafted in severe combined immunodeficient (SCID) mice is up-regulated by tumour necrosis factor-alpha (TNF-alpha) and mediated through intercellular adhesion molecule-1 (ICAM-1). *Clin. Exp. Immunol.* **106,** 20–25.
23. Lee, L., Buckley, C., Blades, M. C., Panayi, G., George, A. J., and Pitzalis, C. (2002) Identification of synovium-specific homing peptides by in vivo phage display selection. *Arthritis Rheum.* **46,** 2109–2120.
24. George, A. J., Lee, L., and Pitzalis, C. (2003) Isolating ligands specific for human vasculature using in vivo phage selection. *Trends Biotechnol.* **21,** 199–203.
25. Koivunen, E., Gay, D. A., and Ruoslahti, E. (1993) Selection of peptides binding to the alpha 5 beta 1 integrin from phage display library. *J. Biol. Chem.* **268,** 20,205–20,210

26. Pasqualini, R., Koivunen, E., and Ruoslahti, E. (1995) A peptide isolated from phage display libraries ia a structual and functional mimic of an RGD-binding site on integrins. *J. Cell Biol.* **130,** 1189–1196.
27. Koivunen, E., Wang, B., and Ruoslahti, E. (1995) Phage libraries displaying cyclic peptides with different ring sizes: ligand specificities of the RGD-directed integrins. *Biotechnology (N. Y.).* **13,** 265–270.
28. Pitzalis, C., Cauli, A., Pipitone, N., et al. (1996) Cutaneous lymphocyte antigen-positive T lymphocytes preferentially migrate to the skin but not to the joint in psoratic arthritis. *Arthritis Rheum.* **39,** 137–145.
29. Blades, M. C., Manzo, A., Ingegnoli, F., et al. (2002) Stromal cell-derived factor 1 (CXCL12) induces human cell migration into human lymph nodes transplanted into SCID mice. *J. Immunol.* **168,** 4308–4317.
30. McCune, J. M., Namikawa, R., Kaneshima, H., et al. (1988) The SCID-hu mouse: murine model for the analysis of human hematolymphoid differentiation and function. *Science* **241,** 1632–1639.
31. Barry, T. S., Jones, D. M., Richter, C. B., and Haynes, B. F. (1991) Successful engraftment of human postnatal thymus in severe combined immune deficient (SCID) mice: differential engraftment of thymic components with irradiation versus anti-asialo GM-1 immunosuppressive regimens. *J. Exp. Med.* **173,** 167–180.

27

Adenoviral Targeting of Signal Transduction Pathways in Synovial Cell Cultures

Alison Davis, Corinne Taylor, Kate Willetts, Clive Smith, and Brian M. J. Foxwell

Abstract

Methods for high efficiency gene transfer into primary cells of various lineages and disease states are desirable, as they remove the uncertainties associated with using transformed cell lines. Adenoviruses have evolved to deliver their genes into cells with high efficiency and in recent years have been exploited as a gene transduction system. Prior to the discovery of adenoviruses, efficient expression of transgenes was only possible by cloning stably transfected cells; this was limited to cell lines and was not an option for primary cells. Here we describe a method of transgene expression, which enables previously untransfectable cells, such as primary myeloid cells or diseased synovium, to express protein at extremely high levels with nearly 100% of cells expressing the transgene. This allows us to examine the effect of target genes on signaling pathways in primary cells without the need for cell sorting or the simultaneous transfection of reporter genes. This is very important in studies of tissues such as rheumatoid synovium where sorting of cells will damage the biological value of the system.

Key Words: AdEasy system; homologous recombination; adenovirus; adenovirus grow-up; adenovirus purification; adenovirus infection; cytokine signaling; rheumatoid synovium.

1. Introduction

The dissection of cytokine signal transduction mechanisms has generally relied upon the transfection of immortalized cell lines by techniques such as calcium phosphate precipitation, cationic lipids or electroporation. These methods often proved limiting as many cells and cell lines are resistant to such approaches. This problem was further compounded by the growing understanding that the function of transformed cell lines did not always reflect the function of primary cells. As an alternative to transfection, it was realized that the capability of viruses to transport DNA to cells could be harnessed for delivering transgenes. Over the years, retroviral vaccinia, adenoviral, and most recently lentiviral systems have been developed. All these systems have strengths and weaknesses. Retroviruses are simple to generate but only infect dividing cells. Vaccinia viruses are comparatively large and complex. Lentiviral and retroviral systems integrate DNA with it's host genome and are useful for generating stable cells. Adenoviruses will infect both proliferating and quiescent cells and remain endosomal, making them suitable for transient studies on terminally differentiated cells such as macrophages.

Advances in recombinant adenovirus technology have expanded the range of cell types amenable to genetic modification. Primary cells and diseased tissue, such as inflamed synovium, are notoriously difficult to transfect. By using adenoviruses, high levels of transgene expression and infection of up to one hundred percent of cells are now possible *(1)*.

Adenoviruses are efficient tools for delivering genes to cells both in vitro and in vivo, achieving high levels of expression of the encoding transgene. Adenoviruses attach to cells via the human protein CAR (coxsackie virus and adenovirus receptor) *(2)*. The second stage of infection utilises the RGD (Arg-Gly-Asp) motif of the adenovirus's penton base which interacts with $\alpha v \beta 3$ and $\alpha v \beta 5$ intergrins on the cell surface and triggers receptor-mediated endocytosis *(3)*.

The first generation of replication-deficient adenoviral vectors involved removing the E1 region and replacing it with the "gene of interest" (*see* **Fig. 1**). Viruses were generated in HEK 293 cells, which stably express the E1A protein thereby allowing viral replication. The problem from a DNA manipulation point of view was that at approximate 37 Kb the standard adenoviral genome was too large to be manipulated by conventional molecular biological techniques. Consequently the virus is generated in a two-stage process requiring the use of a "shuttle" vector that includes some of the viral sequences as well as a multiple cloning site for the insertion of genetic material. As the shuttle vector is significantly smaller it is more amenable to manipulation. Originally, the generation of a complete viral genome was achieved by cotransformation of the shuttle vector and a modified viral genome into HEK 293 cells, where the production of virus relied upon the homologous recombination of these two constructs. This was a very inefficient process and purification of virus required several rounds of plaque purification. Therefore, subsequent adenoviral vector systems have sought to improve the efficiency of the recombination. The AdEasy™ System has achieved this by allowing the recombination of the shuttle and genomic plasmids to occur in recombination competent bacteria. There are other adenoviral systems that involve classical ligation of shuttle vector into the viral genome or using recombination in vitro. However our laboratory has found the AdEasy™ System most advantageous to use.

Adenoviral Methods

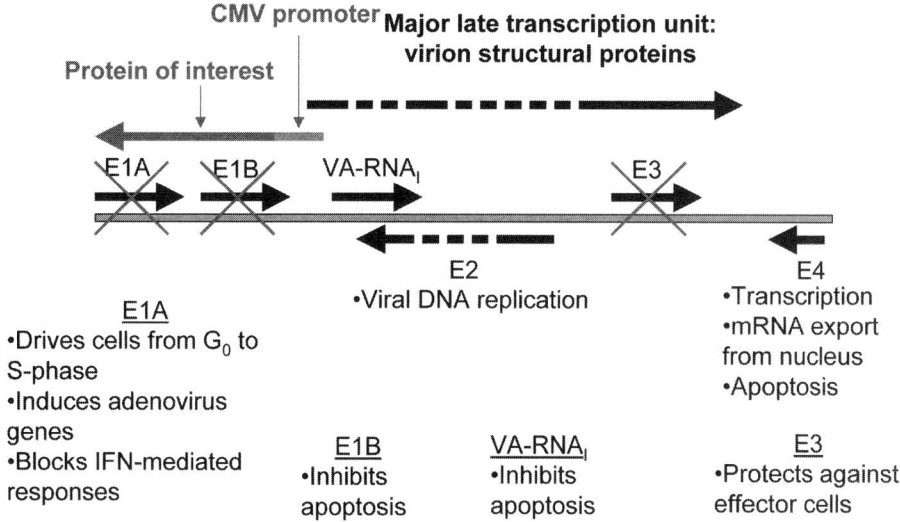

Fig. 1. Modification of the E1/E3 regions of the adenoviral backbone rendering it replication incompetent.

The AdEasy™ System was developed by T.C. He et al. in 1998 *(4)* as a fast and easy alternative to traditional adenoviral systems. Construction of a recombinant adenovirus is a two-step process in which the desired transgene is initially subcloned into a transfer vector and subsequently transferred to the adenoviral genome by means of homologous recombination in *E. coli*. The AdEasy™ vector system (*see* **Fig. 2**) *(5)* employs a backbone vector known as AdEasy-1, which contains most of the adenoviral genome. The cDNA of interest is cloned into a transfer vector (e.g., pAdTrackCMV) (*see* **Fig. 3**) *(5)*. The resulting plasmid is then linearised with *Pme I* and is electroporated into *E. coli* (BJ5183) along with the pAdEasy-1 viral plasmid (*see* **Fig. 4**) *(5)*. Upon recombination, the kanamycin resistance gene is transferred into the viral plasmid along with the gene(s) of interest allowing for the detection of recombinants. The recombined adenoviral DNA is digested with *Pac I*, exposing the inverted terminal repeats (ITR) and is transfected into AD-293 cells (AdEasy-1 is E1 and E3 deleted and its functions are complemented in AD-293 cells) resulting in the formation of viral plaques (*see* **Fig. 5**). Plaque purification is essential for the production of *de novo* clonal viruses and for the maintenance of good viral stocks. Cells are infected with the eluted virus until they reach full cytopathic effect (CPE) and virus overexpression of the transgene is confirmed by Western blotting (successive transfer of viral lysate to AD-293 cell monolayers of increasing surface area achieves viral amplification efficiently).

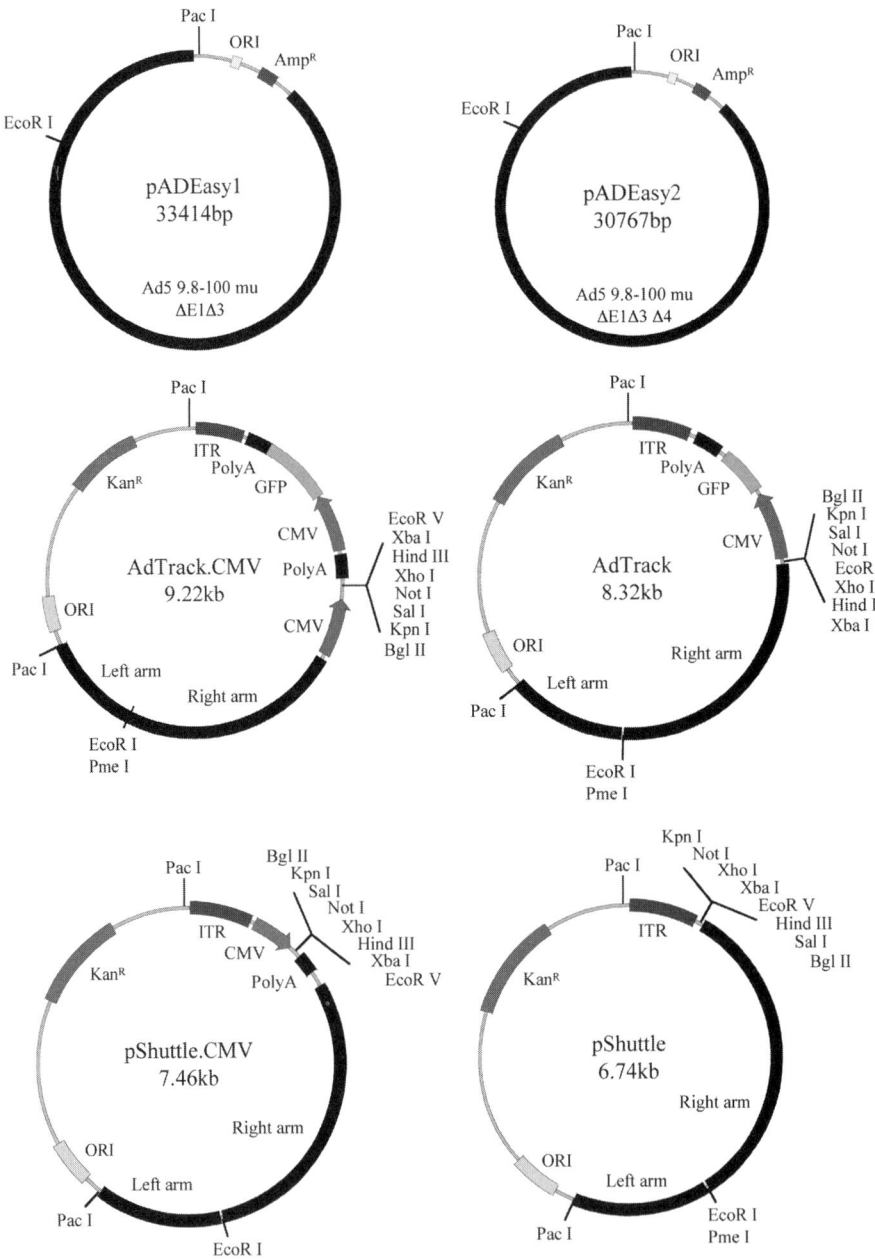

Fig. 2. Pictorial representation of the AdEasy system plasmid vectors. Adenoviral plasmids and shuttle vectors associated with the AdEasy system. For abbreviations refer to text. (Color illustration appears in insert following p. 138.)

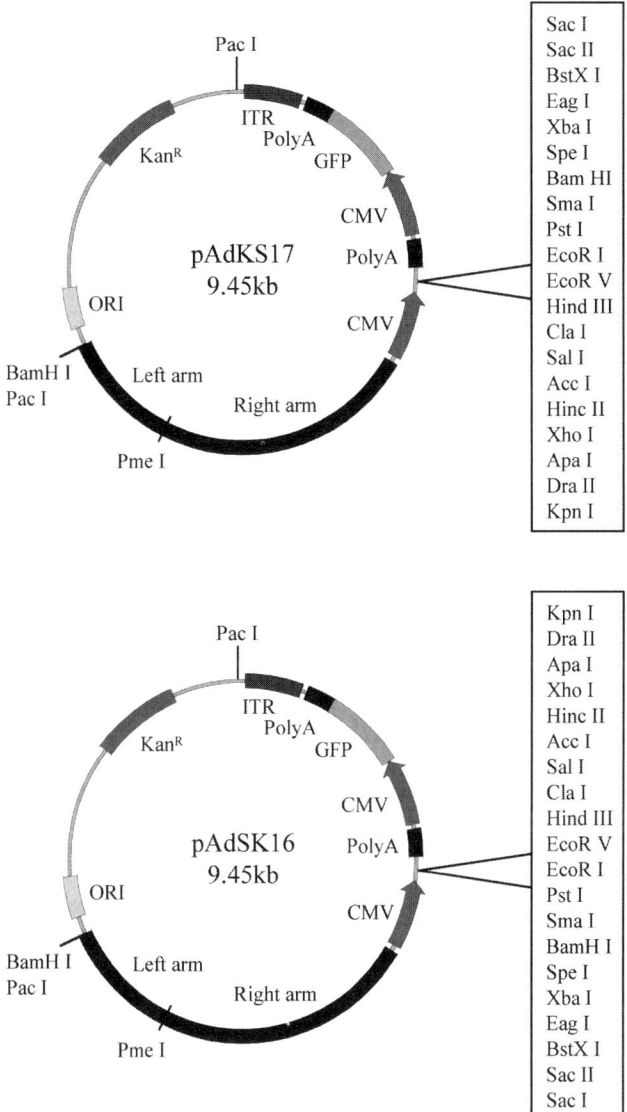

Fig. 3. Pictorial representation of the transfer vectors pAdKS17 and pAdSK16. The vectord pAdKS17 and pAdSK16 were produced by replacing the MCS (*EcoR* V-*Bgl*II) of pAdTrack. CMV.RI⁻ with the *BssH II*-flanked MCS of pBC KS+. For abbreviations refer to text. (N.B. Not all the sites within the MCS are guaranteed unique as the original pAdTrack. CMV transfer vector sequence was ill characterized.) (Color illustration appears in insert following p. 138.)

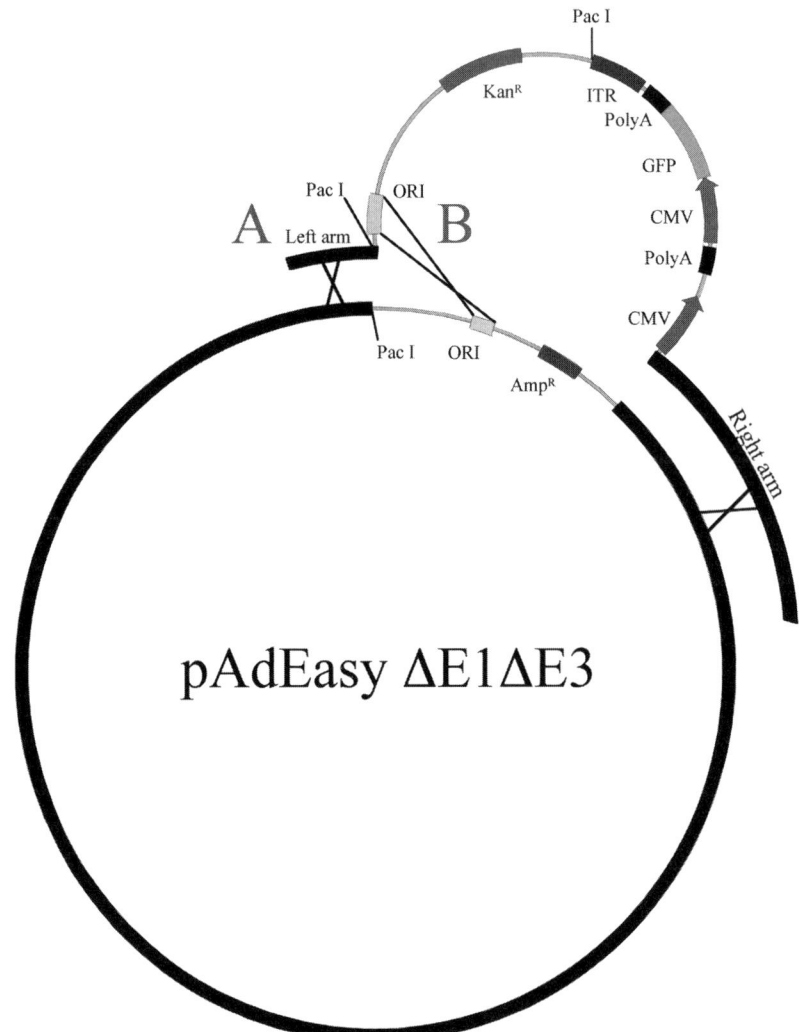

Fig. 4. Modes of in vivo bacterial recombination of the AdEasy vectors. A recombinant transfer vector containing the gene of interest, within the MCS is first linearized with *Pme*I and then gel extracted using Quiagen. Cotransformation with an AdEasy genomic vector into competent BJ 5183 *E. coli* cells follows. Recombination across the right arm, coupled with recombination across the left arm (A) or ORI (B), produces a recombinant adenoviral genome in plasmid form that can be selected for by virtue of kanamycin resistance. Recombination is confirmed by *Pac*I restriction. Recombination type A yields a 3.0 Kb fragment plus a large genomic fragment. Recombination type B yields a 4.5 Kb fragment plus a large genomic fragment. (Color illustration appears in insert following p. 138.)

Adenoviral Methods

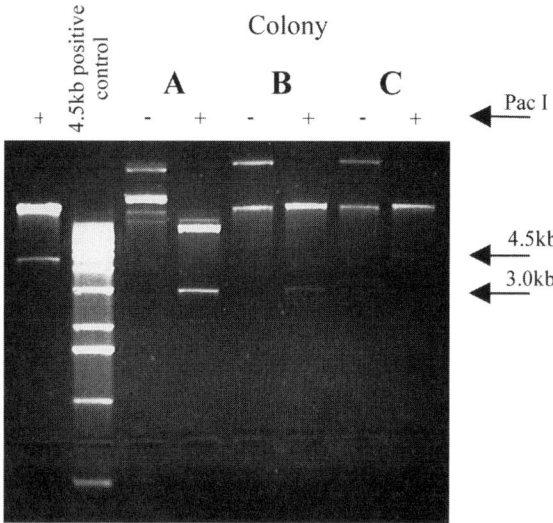

Fig. 5. *Pac*I restriction analysis of putative *AdE*1 recombinants for pAdKS17.flgNIK and pAdKS17.flgNIK.kd. One microgram of *Pme*I linearized, gel purified pAdKS17.flgNIK or pAdKS17.flgNIK.kd was cotransformed with 200 ng of AdEasy1 into BJ5183 cells (*see* Subheading 2.4.1.). Following selection upon LB/agar plates containing 50 µg/mL kanamycin. Plasmid DNA was isolated via standard alkaline lysis and subjected to *Pac*I restriction. Restriction products were resolved upon a 1% agarose/TAE gel. Colonies A and B were produced from AdKS17.flgNIK cotransformation with AdE1. Colony C was produced from pAdKS17.flgNIK.kd cotransformation with *AdE*1.

At the large scale grow up stage cells are harvested before they reach full CPE, in order to contain the virus within the cells. Successive caesium chloride gradients are employed for the purification of adenoviral particles. The virus titre is then assessed utilising AdenoX Rapid Titer Kit (BD Clontech, UK), or alternatively via traditional plaque assay technique.

2. Materials
2.1. Generation of Recombinant Adenoviral Constructs

1. AdEasy Vector System (Q-BIOgene).
2. Restriction enzymes, T4 DNA ligase (NEB).
3. Gel extraction and purification, miniprep and maxiprep kits (Qiagen).
4. *Escherichia coli* strains XL-1 Blue and BJ5183 (Q-BIOgene).
5. Agarose gel electrophoresis equipment.
6. Luria-Bertani (LB) broth and agar.
7. Kanamycin (Invitrogen Life Sciences).

8. Tris-ethylne diamine tetraacetic acid (EDTA) buffer (TAE): 10 mM Tris-HCl Base (pH 8.0), 1 mM EDTA.
9. Lipofectamine 2000 (Invitrogen Life Sciences).
10. OptiMEM media (Invitrogen Life Sciences (Gibco).
11. Sodium dodecyl sulfate polyacrylamide gel electophoresis (SDS-PAGE) equipment.
12. Western Blotting Apparatus.
13. Electroporater (Biorad).
14. ScaPlaque GTG Agarose (BioWhittaker Molecular Applications).
15. 2X Modified Eagles Medium (MEM) (Invitrogen Life Sciences).
16. Dulbecco's modified Eagle's medium (DMEM) supplemented with 4.5 g/L glucose and 584 mg/L L-glutamine (Cambrex).
17. AD-293 Cells (Stratagene).
18. Heat-inactivated fetal calf serum (FCS) (PAA Laboratories).

2.2. Scaling up of Viral Particle Production in AD-293 Cells

1. Dulbecco's Modified Eagle's Medium (DMEM) supplemented with 4.5 g/L glucose and 584 mg/L L-glutamine (Cambrex,).
2. AD-293 Cells (Stratagene, La Jolla).
3. Heat-inactivated FCS (PAA Laboratories).
4. Phosphate buffered saline (PBS) 10X concentrate (BDH).
5. Lysis Buffer: 1% *(v/v)* Triton X-100, 10 mM Tris-HCl (pH 7.6), 150 mM NaCl, 5 mM EDTA. Add fresh 1 mM 4-(2-aminoethyl) benzenesulfonylfluoride, HCl (AEBSF) and 5 µg/mL aprotinin.
6. 4X Gel sample buffer: 8% *(w/v)* SDS, 40% *(v/v)* glycerol, 2% *(v/v)* 2-mercaptoethanol, 250 mM Tris-HCl (pH 6.8).

2.3. Caesium Chloride Purification of Recombinant Adenovirus

1. 0.01M Tris-HCl buffer (pH 8.0).
2. EDTA.
3. PBS MgCl$_2$ and CaCl$_2$ (PBS $^{2+}$): add 100 mL 10X PBS, 10 mL 1% *(w/v)* CaCl$_2$, and 10 mL 1% *(w/v)* MgCl$_2$ to 880 mL distilled water.
4. Saturated caesium chloride (CsCl). Add CsCl (BD Biochemicals, UK) to 35 mL of 0.01 M Tris, 1 mM EDTA until the CsCl falls out of solution. Store at 4°C.
5. 1.34 g/mL density caesium chloride (CsCl). Add 534 g CsCl to 1 L of distilled water (1.4 g/mL density CsCl). Add 212.5 mL 1.4 g/mL CsCl to 37.5 mL PBS $^{2+}$ (1.34 g/mL CsCl).
6. Microsol ^{3} Solution and Wipes (Anachem).
7. 5 mL and 10 mL syringes (Becton Dickinson) and 5 mL Luer Lock syringes (Becton Dickinson)
8. Ultracentrifuge tubes with collars and screws (Sorvall).
9. 150 × 150 × 150 mm small Biohazard Bags (Jencons Scientific Ltd).
10. PD-10 columns (Amersham Biosciences).

11. 10 ml cryogenic vials (Alpha Laboratories).
12. 0.45 µm and 0.2 µm filters that accommodate luer lock syringes (Becton Dickinson).
13. Blunt, 19-gage needles (Kendall).
14. Sharp, 19-gage needles (Terumo).

2.4 Recombinant Adenoviral Titration
2.4.1. Plaque Formation

1. 6-well plate of AD-293 cells (Stratagene) 1 d post confluence.
2. 2X MEM (Invitrogen Life Sciences).
3. Heat-inactivated FCS (PAA Laboratories).
4. 1.5% SeaPlaque GTG Agarose (BioWhittaker Molecular Applications) in sterile distilled water.

2.4.2. AdenoX Rapid Titre Kit

1. AD-293 Cells (Stratagene).
2. 100% Ice-cold methanol.
3. PBS + 1% BSA (Sigma).
4. Mouse anti-hexon antibody diluted 1:1000 in PBS + 1% BSA (BD Clontech). Store at –20°C.
5. Rat anti-mouse antibody diluted 1:500 in PBS + 1% BSA (BD Clontech). Store at 4°C.
6. 10X DAB substrate (store at –20°C) diluted 1:10 with 1X stable peroxidase buffer (BD Clontech). Store at 4°C.

2.5. Replication Competent Test

1. One confluent 25 cm^3 flask of human foreskin fibroblasts (HSF).

3. Methods
3.1. Construction of Vectors for Homologous Recombination in E. coli
3.1.1. AdEasy™ System Vectors

AdEasy-1 (Q-BIOgene) (*see* **Note 1**) is a supercoiled adenoviral plasmid, containing all the adenovirus serotype 5 (Ad5) sequences except the E1 and E3 regions (nucleotides 1-3,533 and 28,130-30,820 respectively), of which the E1 function can be complemented in AD-293 cells (*see* **Note 2**). The viral DNA was derived from pBR322 backbone, therefore allowing for bacterial selection by ampicillin resistance.

The transfer vectors (*see* **Fig. 3**) are used to replace the E1 deleted region and for expression of the foreign transgenes. Homologous recombination occurs across adenoviral sequences known as the right and the left "arms." The left arm is comprised of the left hand inverted terminal repeat (ITR), plus the packing signal sequence required for viral production in mammalian cells. The right arm contains Ad5 sequence, which allows for the homologous recombination with AdEasy vectors to occur. Separating the two arms is a *Pme*I restriction site, responsible for linearizing the plasmid prior to recombination. The vector has been modified with a kanamycin resistance gene to allow for screening of positive recombinants.

These transfer vectors are available with or without green fluorescent protein (GFP) (AdTrack and pShuttle, respectively) that acts as an internal control for infection efficiency, and additionally with or without a cytomegalo virus (CMV) promoter (*see* **Note 3**). Furthermore, the multiple cloning site (MCS) of the transfer vector pAdTrackCMV was modified by inserting the MCS from pBluescript SK+ to allow for more diverse cloning possibilities. The obsolete *Eco*RI restriction site at nucleotide 5282 was also destroyed (*see* **Note 4**).

3.1.2. Subcloning Into Transfer Vectors

DNA manipulation, for the insertion of the gene of interest, is performed by standard molecular biological techniques (**6**). The sequence of the cDNA insert should be confirmed by automated sequencing, whereas transgene expression is determined by transient transfection into AD-293 cells using Lipofectamine 2000 (*see* **Note 5** and **Subheading 3.2.3.**) and detected by Western blotting.

3.2. Generation of Recombinant Adenoviral Plasmids by In Vivo Homologous Recombination in Bacteria

3.2.1. Linearization and Purification of Transfer Vector

1. Linearise 2 µg of transfer vector, containing the gene of interest, with *Pme*I (NEB) restriction enzyme for 2 h.
2. Excise linearized vector from 1% agarose gel, purify and resuspend in 30 µL of sterile distilled water (Qiagen).

3.2.2. Homologous Recombination in Bacteria

Cotransform 3 µL of this purified fragment with 200 ng of AdEasy-1 genomic vector in electrocompetent BJ5183 *E. coli* (Q-BIOgene). This particular strain of *E. coli* is used because it possesses the *recombinase A* enzyme, responsible for catalysing recombination between homologous DNA sequences.

3.2.3. Screening of Clones for Positive Recombinants

Select for positive recombinants by virtue of kanamycin resistance and restriction analysis using *Pac*I (Q-BIOgene), to confirm their molecular weight and restriction digest banding pattern (*see* **Note 6**). *Pac*I restriction digest of recombinants usually results in a large fragment of ~30 Kb and a smaller fragment of either 3 or 4.5 Kb, dependant on whether recombination occurred across the left arm or the point of origin (ORI) respectively (see **Fig. 5** and **Note 7**).

3.2.4. Transformation of Positive Recombinants into XL-1 Blue E. coli

To prevent further recombination occurring and to allow for large-scale preparation of the recombined viral genome (Qiagen), extract 2 µL of the DNA from two of the positive recombinants and transform into electrocompetent XL-1 Blue *E. coli* (*see* **Note 8**).

3.3. Generation of Recombinant Adenoviral Particles via Lipofectamine Mediated Transfection of AD-293 Cells

3.3.1. Cell Preparation

Plate AD-293 cells at 1×10^6 cells/well in a 6-well plate, to obtain approx 90% confluency by the following day.

3.3.2. PacI Linearization of the Recombinant Adenoviral Construct

Digest 10 µg of the recombinant adenoviral construct (maxiprep) with *Pac*I restriction enzyme for 2 to 4 h to expose its ITR. Then carefully ethanol precipitate the linearized vector and resuspend it in 500 µL of OptiMEM media, (preincubated to 37°C) to prevent air-drying. In order to reduce damage to the DNA by mechanical shearing, great care needs to be taken when manipulating these large vectors.

3.3.3. Lipofectamine Mediated Transfection

1. Incubate the DNA precipitate at 37°C for 30 min to aid resuspension.
2. Add 20 µL of Lipofectamine 2000 transfection reagent to the adenoviral DNA resuspension, mix gently and incubate for a further 20 min at room temperature to allow for the formation of DNA/lipid complexes.
3. While the complexes are forming, remove the growth media from the AD-293 cells and replace with 1 mL of OptiMEM (preincubated to 37°C) and incubate for 30 min to allow the cells to equilibrate.
4. Transfer the lipid/DNA complexes to the cells and incubate for 2 h (*see* **Note 9**), before aspirating and replacing with growth media (DMEM + 10 % FCS) overnight.

3.4. Agarose/MEM Overlay of Adenoviral Transfected AD-293 Cells

1. After 24 h, monitor the transfected cells for GFP expression using blue light microscopy. The transfection efficiency with Lipofectamine is approx 30%.
2. Melt a sterile solution of 1.5% *(w/v)* SeaPlaque GTG-agarose in water and then cool to approx 45°C.
3. Mix 2X MEM media, containing 4% *(v/v)* FCS and 2% *(v/v)* Pen/Strep (preincubated at 37 °C), with the 1.5% *(w/v)* SeaPlaque agarose at a ratio of 1:1.

4. Gently add 3 mL to the AD-293 cells taking care not to disturb the monolayer or introduce any bubbles.
5. Allow the overlayed agarose/MEM solution to solidify at room temperature for 5 min and then incubate at 37°C.
6. Recombinant adenoviral plaques will be visible after 7 to 10 d (*see* **Fig. 6**) *(7)*.

3.5. Plaque Purification and Virus Recovery

Plaques become well-defined 10 to 14 d after overlay and can be visualized as regions of clearing in the cell monolayer. This is known as the cytopathic effect (CPE). GFP plaques can be picked using a fluorescent microscope after approximately 7 d. Non-GFP plaques are visible by eye as regions of clearing on the bottom of a 6-well plate, or with the aid of a microscope at 10 to 14 d.

All adenoviral work must be performed in an authorized Biohazard Class 2 Laboratory. Although the empty, replication deficient adenoviral vector requires only category 1 containment, the encoding transgene can increase this to a category 2. Oncogenes, soluble factors such as cytokines, and allergens, are all considered to be Class 2 molecules, therefore this laboratory routinely conducts all viral work at category 2 containment level.

Perform in Class 1 Hood:

1. Puncture the agar overlay with the tip of a 10 μL Gilson pipette and remove 1 to 5 μL of media from the centre of an isolated plaque. Dilute the virally infected media in 500 μL of serum free DMEM in 1 well of a 48-well plate. Pick between 6 and 8 plaques in total. Leave overnight at 37°C to allow virus to elute into the media (*see* **Note 10**).
2. Aspirate media from 80 to 100 % confluent 12-well plate of AD-293 cells. Add 500 μL of viral lysate to 1 well of the 12-well plate. Repeat with all the wells of viral lysate. Incubate cells for 1 to 2 h at 37°C. Add 500 μL of DMEM supplemented with 4% FCS, giving a final concentration of 2% FCS per well (*see* **Note 11**). Incubate at 37°C until full CPE is seen in the majority of wells. Full CPE can usually be observed 3 to 7 d post infection.

3.6. Small Scale Adenoviral Amplification

Amplification is achieved by successive increases in the surface area of the AD-293 monolayer infected with adenovirus. Cells are left until full CPE is observed, this is characterized by the cells rounding up and no longer adhering to the bottom of the plate. At the 6-well plate stage, clones are screened for recombinant gene expression.

1. Transfer each viral lysate from the 12-well plate to a 2 mL cryovial and number sequentially.
2. Snap freeze the vial in liquid nitrogen and thaw in 37°C water bath, this ensures that all virions have been released from any unlysed cells. Freeze thaw a total of three times (*see* **Notes 12** and **13**).

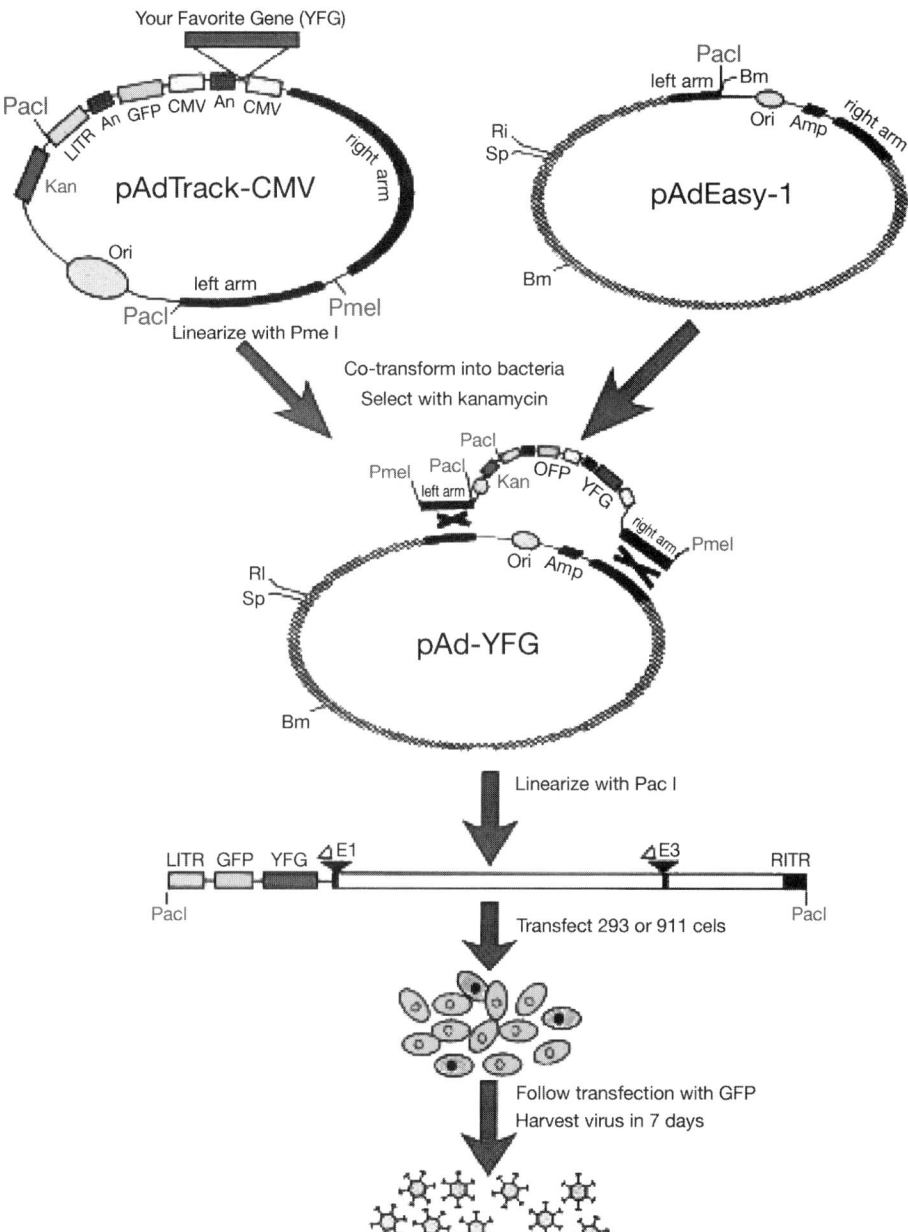

Fig. 6. Schematic representation of the AdEasy™ Vector System for the generation of recombinant adenoviruses. Reproduced with kind permission from Qbiogene (as published in AdEasy Vector System Application Manual). (Color illustration appears in insert following p. 138.)

3. Aspirate media from two 80 to 100% confluent 6-well plates of AD-293 cells. Label the wells on both plates 1 through to 6.
4. Add 500 µL of serum free DMEM to each well. Add 500 µL of lysate from cryovial 1 to each of the wells labeled 1 on the two 6-well plates. Repeat until five wells on both plates are infected, leave 1 well uninfected.
5. Incubate cells for 1 to 2 h at 37°C. Add 1 mL of DMEM supplemented with 4% FCS, giving a final concentration of 2% FCS/well.
6. After 24 h take one of the plates and harvest the cells in preparation for analysis by western blotting (*see* **Note 14**).
7. Aspirate media and wash cells gently in sterile 1X PBS. Dislodge the cells from the bottom of the plate by pipetting vigorously with 1 mL of 1X PBS per well and transfer to 1.5 mL screw top Eppendorfs. Spin cells down in a microfuge for 5 min at top speed at 4°C.
8. Remove the supernatant and resuspend the cells in 80 µL of lysis buffer. Leave on ice for 10 min.
9. Spin cells down in a microfuge for 10 min at top speed at 4°C.
10. Remove supernatant and transfer to fresh Eppendorfs containing 20 µL of 5X gel sample buffer.
11. Run samples on Tris-HEPES-SDS mini gel, transfer onto PVDF membrane and probe with the appropriate antibody.
12. Leave second plate at 37°C until cells display full CPE.
13. Once the clone has been shown to express protein, use some of the viral lysate for large-scale amplification and freeze the remainder at –70°C as archive stocks. Also freeze all other positive clones as archive stocks (*see* **Note 15**).

3.7. Large Scale Adenoviral Amplification

1. Transfer 1 mL of positive viral lysate to a 2 mL cryovial and freeze/thaw as with the 12-well plate (*see* **Notes 12** and **13**).
2. Make up viral lysate to 5 mL with serum free DMEM and mix thoroughly.
3. Remove media from 80 to 100% confluent 75cm^3 flask. Add the viral lysate to the flask on the opposite side to the cells.
4. Incubate cells for 1 to 2 h at 37°C. Ensure flask is level in incubator for even infection of cells. Add 5 mL of DMEM supplemented with 4% FCS, giving a final concentration of 2% FCS in the flask.
5. Incubate at 37°C until full CPE is seen. This should take between 1 and 2 d.
6. Gently tap the side of the 75 cm^3 flask to dislodge any adherent cells. Divide all the viral lysate from the 75 cm^3 flask between two 10 mL cryovials and freeze/thaw as with the 12-well plate (*see* **Note 12**).
7. Make up viral lysate to 100 mL with serum free DMEM and mix thoroughly.
8. Remove media from ten 80 to 90% confluent 162 cm^3 flasks. Add 10 mL of the viral lysate to each of the flasks on the opposite side to the cells.
9. Incubate cells for 1 h at 37°C. Ensure flasks are level in incubator for even infection of cells. To each flask add 10 mL of DMEM supplemented with 4% FCS, giving a final concentration of 2% FCS.
10. Incubate at 37°C for approx 40 h.

3.8. Caesium Chloride Purification of Recombinant Adenovirus

Harvest the ten virally infected 162 cm^3 flasks forty hours post-infection (*see* **Notes 16** and **17**); 80 to 90% of the cells will be nonadherent at this stage. The rest will be easily removed by gently tapping the flask. The adenovirus is purified following a standard double caesium chloride purification procedure over a 3 d centrifugation program. The first stage of the purification removes cellular contaminants and any unpackaged viral particles. The second stage achieves complete separation of the recombinant adenoviral particle and the final stage removes contaminating CsCl salts via column chromatography (*see* **Figs. 7** and **8**)

Day One—Perform in Class 1 Hood.

1. Harvest the cells and media from the ten, 162 cm^3 flasks into four, 50 mL Falcons and centrifuge for 5 min at 250*g*.
2. Gently pour off the supernatants into a waste pot of 10% microsol (*see* **Note 18**) and resuspend all four pellets in 10 mL of serum free DMEM. Centrifuge again for 5 min at 250*g*.
3. Pour off the supernatant and resuspend the pellet in 10 mL of Tris-HCL (pH 8.0) (*see* **Note 19**).
4. Divide the 10 mL solution of cell lysates into two 10-mL cryogenic vials and freeze/thaw 3 times.
5. Place the lysates in a clean, 50 mL falcon and pass through a blunt, 19 gage needle 3 to 4 times to shear the chromatin.
6. Spin at 1000*g* for 5 min to pellet the cell debris.
7. Save the supernatant and discard the cells debris. Note the volume of the supernatant and make it up to 11.4 mL with 0.1*M* Tris-HCL (pH 8.0) in a clean falcon. Add 6.6 mL of saturated CsCl and mix thoroughly.
8. Using a blunt, 19 gage needle and 10 mL syringe, divide the solution into two ultracentrifuge tubes. Fill the tube up to the neck with no air bubbles (*see* **Note 20**). Place the collar on the tube and screw in place. Wipe up any residual solution that leaks out while the screw is being inserted with a microsol wipe.
9. Transfer the tubes to the rotor and spin at 180,000*g* overnight (minimum of eight hours) at 4°C. Acceleration 9, deceleration 1.
10. Place all solid waste in a small biohazard bag and dispose of by autoclaving. All needles and accompanying syringes should be disposed of in an appropriate sharps bin.

Day Two—Perform in Class 1 Hood with double gloves and safety goggles.

1. Remove rotor from the ultracentrifuge, check for leaks and then transfer to the class 1 hood. Remove the ultracentrifuge tubes using the collar gadget and check again for any leaks.
2. Place tube in clamp and release the airtight seal by removing the screw. Mop up

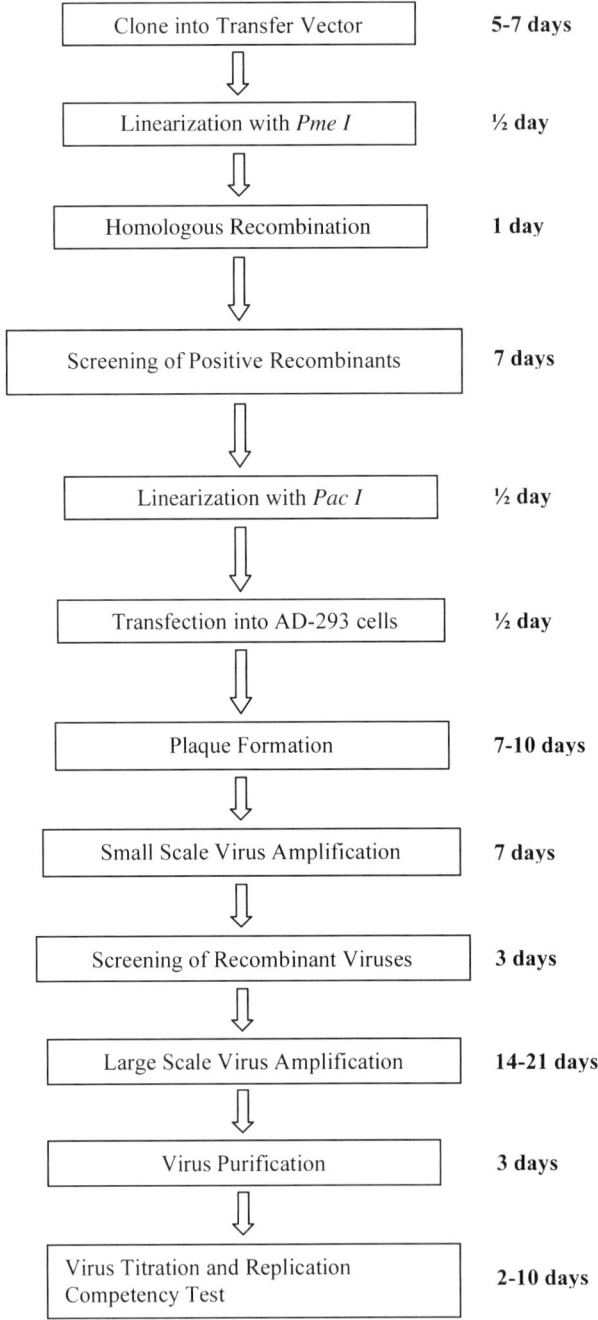

Fig. 7. Schematic representation of the time scale involved in the production of recombinant adenoviruses.

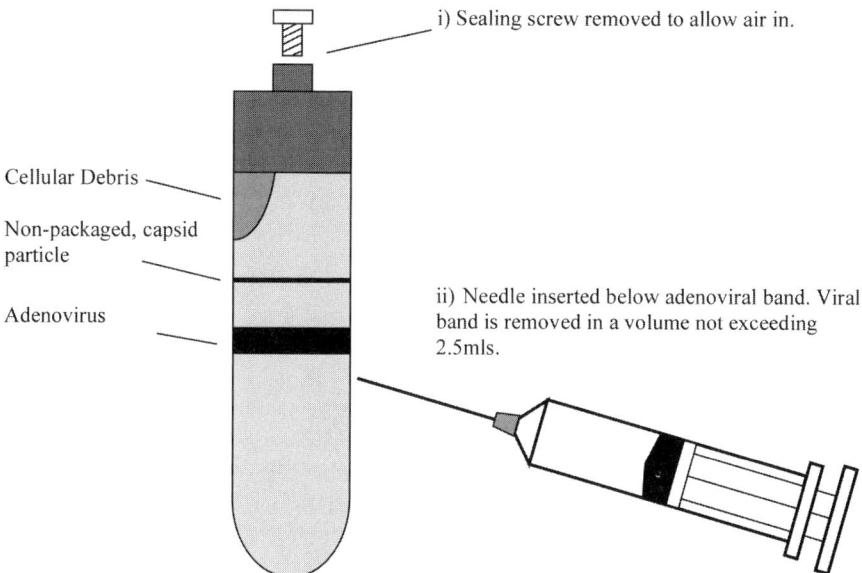

Fig. 8. Isolation of the recombinant adenoviral band after caesium chloride density-gradient centrifugation *(9)*. Ultracentrifugation results in the recombinant adenoviral particles banding at a density of 1.34 g/mL thereby separating them from any cellular debris and nonpackaged, capsid particles. The airtight seal is released by removing the sealing screw. The viral band is then isolated using a 5 mL syringe and a 19 gage needle by puncturing the ultracentrifuge tube.

any solution that is released with a microsol wipe. Place the screw in diluted, 10% microsol (*see* **Note 18**). Raise the tube to eye level.
3. Using a sharp, 19-gage needle and 5 mL syringe, pierce the tube a few millimetres below the viral band (*see* **Fig. 8**) *(10)* (*see* **Note 21**). Carefully draw the viral band into the syringe (*see* **Note 22**). Place your finger on top of the tube and then withdraw the needle. Withdraw the syringe further at this stage to ensure that all the virus solution is in the body of the syringe and not in the needle. Do not recap the needle, but place the syringe/needle safely on one side.
4. Using a 10 mL syringe with a blunt, 19 gage needle, remove the contaminated CsCl till below the puncture mark in the tube. Rinse the needle in 10% microsol solution (*see* **Note 18**) and place in sharps bin.
5. Change outer layer of gloves.
6. Using the collar gadget, remove the tube from the clamp and take off the collar. Place the collar in dilute, 10% microsol (*see* **Note 18**). Any liquid remaining in the tube should be poured gently into a 10% microsol solution (*see* **Note 18**).
7. Replace outer layer of gloves and repeat the procedure with the second tube.

8. Carefully inject the virus material from both 5 mL syringes into a new ultracentrifuge tube. Using a 10 mL syringe with a blunt, 19 gage needle, top up the tube with 1.34 g per/mL density CsCl. Fill the tube up to the neck with no air bubbles (see **Note 20**). Place the collar on the tube and screw in place. Wipe up any residual solution that leaks out while the screw is being inserted with a microsol wipe.
9. Transfer the tube to the rotor (see **Note 23**). Spin at 180,000g overnight (minimum of 8 h) at 4°C. Acceleration 9, deceleration 1.
10. Place all solid waste in a small biohazard bag and dispose of by autoclaving.

Day Three—Perform in Class 1 Hood with double gloves and safety goggles.

1. Equilibrate a PD-10 column with 25 mL PBS^{2+}. Set up the column over a waste tray containing dilute microsol (see **Note 18**) and uncap from the bottom and then the top and allow liquid to drain. Do not let the column dry out. Place 5 mL PBS^{2+} in the column at a time until 25 mL PBS^{2+} has passed through. Before the last 1 mL has drained from the column, cap again at the bottom and set aside.
2. Prepare a sterile bijou tube containing approximately 10 drops of sterile glycerol. Carefully drop the glycerol to the bottom of the tube. The final concentration of the eluted virus and glycerol is approx 10%.
3. Remove the ultracentrifuge tubes from the rotor and take off the virus band following the above outlined procedure. Do not remove the virus band in a volume exceeding 2.5 mL (see **Note 22**).
4. Allow the last 1 mL of PBS^{2+} to drain from the PD-10 column and then carefully inject the virus onto the top of the column. Do not pierce the column with the needle. Allow the virus to drain into the column. If the virus band was taken off in less then 2.5 mL, add PBS^{2+} up to 2.5 mL (see **Note 24**). Collect the eluted PBS^{2+} in the waste tray.
5. To elute the virus from the column add 3.5 mL or less of PBS^{2+}. This will depend on the volume the virus band was removed in (see **Note 24**). Collect the eluate in the bijou containing the glycerol.

Perform in Class II Hood with double gloves and safety goggles.

1. Mix the virus and the glycerol thoroughly using a blunt, 19 gage needle and 5 mL luer lock syringe (see **Note 25**). Take up the total volume of virus/glycerol into the body of the 5 mL syringe and then carefully remove the blunt needle and place it in a sharps bin. Then attach a 0.45 µm luer lock filter followed by a 0.22 µm luer lock filter onto the 5 mL syringe and gently push the virus/glycerol through the two filters (see **Note 26**) into a new sterile bijou tube.
2. Aliquot the virus into sterile 0.5 mL screw cap Eppendorfs and freeze these at –70°C.

3.8. Titration of the Adenovirus

In order to maintain uniformity between different experiments and an optimum level of gene expression, it is important to quantify the number of infectious adenoviral particles present in your purified stock. The following two procedures are examples of the many ways in which a virus titer can be obtained. It must be noted that a certain amount of inconsistency exists between all these methods and reproducibility of titers between adenoviral stocks is variable.

3.8.1. Plaque Formation

Titration by plaque assay makes use of the support of agar over a monolayer of 293 cells to help generate plaques/holes caused by sequential lytic infections. These plaques can then be counted to determine the plaque forming units/mL (pfu/mL).

1. Generate a series of viral dilutions (10^{-2} to 10^{-6}) in serum free DMEM.
2. Remove media from the 6-well plate of AD-293 cells and add 1 mL of dilution 10^{-6}, 10^{-7}, 10^{-8}, 10^{-9}, 10^{-10} and one control well (*see* **Note 27**).
3. Incubate the plate for 2 h in a 37°C tissue culture incubator.
4. Microwave the sterile solution of 1.5% SeaPlaque agarose and then cool it in a 37°C tissue culture incubator for one hour prior to use. At the same time, prepare 10 mL 2X MEM, 4% FCS, 2% penicillin/streptomycin, and warm up to 37°C in a water bath.
5. Remove media from the cells and add 3 mL of 2X MEM, 4% FCS, 2% penicillin/streptomycin to 3 mL of 1.5% SeaPlaque agarose. Mix carefully and quickly avoiding bubbles (*see* **Note 27**).
6. Take up the whole 6 mL and carefully add drop wise to two wells of the 6-well plate (3 mL/well).
7. Cover the plate and leave level in the hood until set (approx 5 min).
8. Place carefully in a 37°C tissue culture incubator and leave for 7 to 8 d for GFP adenoviral plaques to develop or 10 d for non-GFP adenoviral plaques to develop.
9. Count the number of plaques in a dilution that gives well-isolated plaques. Multiply the number of plaques by the dilution factor to obtain the plaque forming units/mL (pfu/mL)

3.8.2. Adeno X Rapid Titer Kit

AD-293 cells are infected with a succession of serial dilutions of purified adenoviral stock. After an incubation period, which allows the virally infected cells to produce hexon proteins, the cells are fixed and treated with an anti-hexon protein specific antibody. A secondary, HRP-conjugate antibody is then added and allowed to develop with DAB substrate. Any positive cells will turn brown/black and can be counted using a ×20 objective. The infectious units/mL (ifu/mL) can then be calculated as the average number of positive cells/dilution.

Day One:

1. Seed 1 mL of AD-293 cells (5×10^5 cells/mL) in each well of a 12-well plate (*see* **Note 28**). Using either PBS or serum free DMEM, dilute the virus into 10-fold serial dilutions (*see* **Note 29**). Include a negative control.

2. Add 100 μL of each dilution drop wise to a well.
3. Incubate the plate at 37°C for 28 h.

Day Two:
1. Remove media from the cells into a microsol waste pot (*see* **Note 18**). Some cells in the higher dilution wells may be non-adherent. Allow the plate to dry for 5 min in class 1 hood.
2. Gently add 1 mL of ice-cold methanol to each well so as not to disrupt the cell layer.
3. Incubate the plate at –20°C for 10 min. Once the cells have been fixed, sterile conditions are not required.
4. Aspirate the methanol and gently wash the wells three times with PBS + 1% BSA.
5. Once the final wash has been removed, add 0.5 mL of mouse anti-hexon antibody dilution (1:1000) to each well.
6. Incubate the plate at 37°C for 1 h.
7. Aspirate the mouse anti-hexon antibody and then gently wash the wells three times with PBS + 1% BSA.
8. Add 0.5 mL rat anti-mouse antibody dilution (1:500) to each well and incubate the plate at 37°C for 1 h.
9. Prepare the DAB working solution (10X DAB substrate diluted 1:10 with 1X stable peroxidase buffer) and allow it to warm to room temperature. Do not allow the 10X DAB substrate to warm to room temperature.
10. Aspirate the rat anti-mouse antibody and gently rinse the wells three times with PBS + 1% BSA.
11. Add 500 μL of the DAB working solution to each well and incubate the plate at room temperature for 10 min.
12. Aspirate the DAB and rinse each well once with 1X PBS.
13. Using a microscope with ×20 objective, count a minimum of 3 fields of brown/black positive cells in a dilution that has 10% or fewer positive cells. A field should have no less then 5 positive cells and a maximum of 50 positive cells. Ideally between 20 and 40 positive cells.
14. Calculate the ifu/mL as follows:

$$\frac{(\text{infected cells/field}) \times (\text{fields/well})}{\text{volume virus (mL)} \times \text{dilution factor}}$$

For example, infected cells/field = 25 at dilution 5×10^{-5}
fields/well = 573 (at ×20 objective)
volume of virus = 0.1 mL

3.9. Replication Competence Test

To ensure that the recombinant adenoviruses that are produced adhere to all safety guidelines, each purified, titered virus needs to be checked for replica-

Table 1

Cell type	Multiplicity of infection (moi)	Incubation time with virus (hours)	Infection conditions	Optimal cell confluence for infection
HeLa	10–50	1	Serum Free	80%
Macrophages *(6)*	50–200	1–2	Serum Free	90%
Dendritic Cells *(8)*	100–300	1–2	Serum Free	90%
Rheumatoid Synovium *(8)*	50–200	1–2	Serum free	90%
Synovial Fibroblasts *(9)*	100–500	1–4	Serum free – 1%	80–90%
HUVECS *(9)*	50–100	1	Serum Free	60 %

tion competency in human cells. As the E1 and E3 regions have been deleted from the virus, all recombinant adenoviruses should be replication incompetent in human cells. However on rare occasions, low frequency homologous recombination between the human chromosome and the recombinant E1 deleted adenovirus has led to a reacquisition of replication competency.

1. Remove media from confluent, 25 cm^3 flask of HSF cells.
2. Add 2 mL of serum free DMEM containing 10 µL of purified adenovirus stock.
3. Leave the flask in a 37°C tissue culture incubator for 1 to 2 h.
4. Remove the virus and replace with 5 mL of 10% FCS, DMEM.
5. Observe the HSF cells over 7 to 10 d. The virus will infect the cells (i.e., the cells will appear green if infected with GFP adenovirus), but not be able to replicate. Thus no discernable effects of the viral infection should be observed. If the virus is replication competent a cytopathic effect will be observed and this batch of virus must be destroyed.

3.10. Infection of Primary Cells and Cell Lines with Adenovirus

Adenoviral vectors are able to infect both quiescent and dividing cells with high infection efficiency. They can also be used to infect diseased tissue cultures like rheumatoid arthritis synovial cells (**Table 1**).

4. Notes

1. In addition to the pAdEasy-1 adenoviral vector, there is an alternative version available known as pAdEasy-2. This is identical to the pAdEasy-1 vector except that it contains an additional deletion of Ad5 (nucleotides 32,816-35,462), which encompasses the E4 genes (*see* **Fig. 2**).

2. Other mammalian cell lines also suitable for the production of adenovirus are different HEK293 lines (such as HEK293A), 911 or 911E4. We use AD-293 cells due to their improved cell adherence and plaque forming properties.
3. We advise where possible to employ a transfer vector that expresses GFP, as this is a useful marker for tracking efficiency of adenoviral infection and consequently acts as a guide to the success of transgene expression as both are driven by identical CMV promoters. GFP expression is also advantageous when picking recombinant adenoviral plaques as they are clearly visible by fluorescent microscopy and the quality of plaque extractions can be closely monitored. Adenoviral plaques not expressing GFP can be difficult to observe with the naked eye, or under normal light microscopy conditions.
4. Our modified version of the pAdTCMV transfer vector was used in preference to the alternatives as it allowed us more scope for subcloning of transgenes and the removal of the obsolete *EcoR*I site has proved essential for many of our cloning strategies.
5. Other brands of transfection reagent (Superfect, Qiagen UK and Fugene, Roche UK) have also been used at this stage to test transgene expression, but we have found Lipofectamine 2000 (optimised for use of large plasmids) to be the most effective with a transfection efficiency of about 30 %.
6. To minimize the amount of transfer vector contamination on the recombination plates, we recommend digesting the 2 µg of transfer vector with at least 3 µL of *Pme*I (10,000 U/mL) restriction enzyme for at least 2 h. However it should be noted that there will still be contaminating circularized transfer vector within the digested sample that will therefore be able to grow on the kanamycin agar plates. Colony size serves as a means of distinguishing between transfer vector contamination and positive recombinants. Your plate should contain an even number of both large and small colonies. Generally, the small colonies are recombinants and the larger colonies contain only transfer vector. This difference in size is caused by the delay from the recombination process and the fact that recombinants are large, low-copy plasmids. If your recombination plate only produces large colonies, it usually indicates the *Pme*I digest was insufficient and should be repeated under more stringent conditions, such as longer incubation or a higher concentration of enzyme. As a fast way of screening a large number of colonies on a particularly bad plate, we often employ PCR screening techniques using primers, which bind to a selected region of the pAdEasy-1 vector.
7. Recombinants yielding a 4.5 Kb fragment upon *Pac*I restriction digest, are selected in preference to those with a 3 Kb fragment as the former can only be the product of a recombination event. The 3 Kb fragment, as a consequence of incomplete *Pme*I linearization, can also be produced from non-recombined transfer vector. The size and intensity of the remaining vector fragment further distinguishes recombinants from transfer vector only (*see* **Fig. 5**).
8. Visualization of positive recombinants on an ethidium bromide agarose gel at the *Pac*I screening stage in the BJ 5183 cells is very poor. Often only the top 30 Kb band can be seen faintly. It is therefore recommended that unless a clear positive is

observed, two "possible" positive clones should be transformed into XL-1 blue cells to obtain higher DNA yields allowing for easier detection of recombinants.
9. OptiMEM media itself is toxic to the cells. Therefore it is not recommended that complexes be left on the cells for more than 2 h. Incubations of 4 h or more can induce cell death.
10. At this stage viral media diluted in serum free DMEM can be frozen and stored at –20°C. Viral elution will work if infected media is added straight onto the AD-293 cells although it may take longer for cells to reach full CPE.
11. Different batches and suppliers of FCS may need to be tested before use, as they can have differing effects on the adenoviruses ability to infect AD-293 cells.
12. At this stage lysates can be frozen and stored at –20°C, only two further freeze/thaw cycles are then required.
13. If cells have reached full CPE, three freeze/thaw cycles are not vital at this stage.
14. If cells have not detached from the plate at this stage, wash gently with 1X PBS, add lysis buffer directly to each well and leave plate on ice for 10 min.
15. Different clones are often found to exhibit different properties (e.g., their ability to express proteins or their efficiency of infection, so it is best to check more than one plaque).
16. The saturated caesium chloride should be stored at 4°C overnight prior to use to ensure its stability.
17. The timing of harvesting the ten 162 cm^3 flasks needs to be monitored carefully as different adenoviruses show varying rates of infection. The optimum time for harvesting is between the stages where the cells become non-adherent and the cells lysing. This is usually a time frame of about eight hours. The purification procedure for the adenovirus is such that any viruses released into the media following cell lysis will be lost.
18. All liquid waste needs to be disposed of in a final volume of 10% microsol, overnight. All pipets and filter tips need to be rinsed with 10% microsol before they are disposed of. Collars and screws must be decontaminated for one hour only to minimize any damage caused by soaking in microsol.
19. The procedure may be stopped at this point by freezing the 10 mL cell lysate in two cryogenic vials until required.
20. To ensure that no bubbles occur in the ultracentrifuge tube, gently place the eye of the needle against the side of the tube while you fill it until the eye of the needle is covered with liquid. If any bubbles appear, fill the tube up to the top of the neck and gently tap the tube. This should release the bubbles into the neck where they can be removed easily.
21. The eye of the needle should be facing down as you pierce the tube, but once the tube has been punctured the eye of the needle is rotated so that it is facing upwards, towards the viral band.
22. At this stage, the volume that the viral band is extracted from is unimportant. On the third day, however the viral band cannot be taken off in more than 2.5 mL. This limitation is a result of the capacity of the PD-10 columns used to desalt the recombinant adenovirus.

23. When purifying odd numbers of recombinant adenoviruses, a balance will be required. This can be made using the 1.34 g/mL density CsCl.
24. The PD-10 column capacity is 2.5 mL. To ensure that the virus is the first thing to be eluted from the column, the total volume of virus to be added to the column must be 2.5 mL. To ensure that the eluted virus is not too dilute, never elute with a volume of PBS^{2+} that exceeds the viral band volume by more than 1 mL. For example if the viral band was taken off in a volume of 1.5 mL then only elute the virus with 2.5 mL of PBS.
25. The use of luer lock syringes at this stage is to prevent the filters from coming off the end of the syringe as a result of the pressure caused by pushing the virus/glycerol through the filters.
26. Firm pressure will be required to push the virus/glycerol through both syringes; some of the total volume of the virus will be lost in the filters.
27. Tip the plate and remove the media from all of the 6 wells as you will not be able to tip the plate again once you have begun adding the agarose. It is important to work relatively rapidly at this point to prevent the agarose from solidifying in the pipette.
28. The AD-293 cells do not need to be left to adhere before addition of the diluted adenovirus.
29. Serial dilutions should range between 10^{-2} and 10^{-6}, although the number of dilutions used can be reduced according to the expected titer of the virus. To improve the accuracy of the titer, dilutions may also be adjusted to half log increments.

References

1. Horwood, N. J., Smith, C., Andreakos, E., et al. (2002) High efficiency gene transfer into non-transformed cells:utility for studying gene regulation and analysis of potential therapeutic targets. *Arthritis Research* 4 Suppl. 3, S215–S225
2. Bergelson, J. M., Cunningham, J. A., Droguett, G., et al. (1997) Isolation of a common receptor for Coxsackie B viruses and adenoviruses 2 and 5. *Science* 28(275), 1320–1323.
3. Wickham, T. J., Carrion, M. E., and Kovesdi, I. (1995) Targeting of adenovirus penton base to new receptors through replacement of its RGD motif with other receptor-specific peptide motifs. *Gene Therapy* 2(10), 750–756.
4. He, T. C., Zhou, S., Da Costa, L. T., Yu, J., Kinzler, K. W., and Vogelstein, B. (1998) A simplified system for generating recombinant adenoviruses. *Proc Natl Acad Sci USA* 95(5), 2509–2514.
5. Smith, C. (1999) An investigation of intracellular TNFα signaling mechanisms.
6. Foxwell, B., Browne, K., Bondeson, J., et al. (1998) Efficient adenoviral infection with IkappaB alpha reveals that macrophage tumour necrosis factor alpha production in rheumatoid arthritis is NF-kappaB dependent. *Proc Natl Acad Sci USA* 95, 8211–8215.
7. Sambrook, J., Fritsch, E. F., and Maniatis, T.M. (ed.) (1989) *Molecular Cloning: A Laboratory Manual*. Cold Spring Harbor, NY: Cold Spring Harbor Laboratory Press.

8. Andreakos, E., Smith, C., Monaco, C., Brennan, F. M., and Foxwell, B. M., Feldmann, M. (2003) IκB kinase 2 but not NF-κB–inducing kinase is essential for effective DC antigen presentation in the allogeneic mixed lymphocyte reaction. *Blood* 101, 983–991.
9. Andreakos, E., Sacre, S., Smith, C., et al. (2004) Distinct pathways of LPS-induced NF-κB activation and cytokine production in human myeloid and nonmyeloid cells defined by selective utilization of MyD88 and Mal/TIRAP. *Blood*. 103, 2229–2237.
10. Mc Daid, J.P.(2004). Role of Protein Tyrosine Kinase in Macrophage Activation and Cytokine Production (PhD thesis).

Index

A

α_1-Acid glycoprotein, acute phase response analysis in pristane-induced arthritis, 26, 62, 264

Adenovirus vectors
  AdEasy™ system, 397, 401
  advantages, 396
  cytokine signal transduction pathway analysis in synovial cells
    adenoviral amplification
      large scale, 408, 417
      small scale, 406, 408, 417
    agarose overlay of adenoviral-transfected cells, 405, 406
    cesium chloride purification of virus, 409, 411, 412, 417, 418
    infection of synovial cells, 415
    lipofection of AD-293 cells, 405, 417
    materials, 40, 03
    plaque purification and virus recovery, 406, 417
    plasmid generation by homologous recombination in *Escherichia coli*, 404, 405, 416, 417
    replication competence test, 414, 415
    titration of adenovirus
      Adeno C Rapid Titer Kit, 413, 414, 417
      plaque formation, 413, 418
    vector construction for homologous recombination in *Escherichia coli*, 403, 404, 415, 416

  first-generation vectors, 396
  receptor, 396

Animal models
  antigen-induced arthritis, *see* Antigen-induced arthritis
  cartilage oligomeric matrix protein induction, *see* Cartilage oligomeric matrix protein
  collagen-induced arthritis models, *see* Collagen antibody-induced arthritis; Collagen-induced arthritis
  ethics, 187
  pristane-induced arthritis, *see* Pristane-induced arthritis
  rheumatoid arthritis models, overview, 18, 87
  transgenic mouse models, *see* Del1 mouse; K/BxN mouse arthritis model

Antigen-induced arthritis
  murine antigen-induced arthritis with methylated bovine serum albumin
    arthritis flare induction, 248, 251
    cartilage metabolism measurement, 248, 251
    emulsion preparation, 245
    immunization, 245, 250
    immunohistological grading, 247, 248, 250, 251
    intra-articular injection, 245, 247
    joint swelling scoring, 247
    materials, 244
    methylated bovine serum albumin cationicity control, 244, 245, 250

overview, 243, 244
Apoptosis
  execution, 122
  initiation, 11, 22
  regulation
    Bcl-2, 121, 122
    caspases, 119
    cytochrome *c* release, 121
    initiation receptors, 119, 121
    survival signals, 117, 121
  T-cell assessment in synovial fluid
    arthritis T-cells, 118, 119
    caspase-3 activation
      flow cytometry, 128, 129, 134
      immunofluorescence microscopy, 128, 129, 134
    DNA fragmentation, 129
    materials, 12, 24
    mitochondrial depolarization assays, 12, 27, 132, 133
    phosphatidylserine externalization, 127
    positive selection of T-cells, 124, 125, 132
    synovial fluid processing, 124, 132
    TUNEL assay, 12, 31
Artificial antigen-presenting cell
  antigen recognition overview, 71, 72
  antigen specific T-cell studies
    artificial antigen-presenting cell preparation, 80, 84
    expansion of cells, 70, 71
    identification strategies, 70
    lipid lamellar sheet preparation, 76, 83
    liposome preparation and sizing, 76, 77
    major histocompatibility complex molecule purification
      biotinylation, 79, 80
      cell pelleting and lysis, 77, 78
      immunoaffinity chromatography, 78, 79
      peptide complex preparation, 80, 84
      quality analysis, 79, 83, 84
    materials, 7, 6, 83
    principles, 72, 73
    T-cell capture assay
      flow cytometry, 8, 4
      peripheral blood mononuclear cell immunostaining, 80, 81

B

B-cell
  antigen specific B-cell detection in tissues
    biotinylation of antigen, 2, 3
    data analysis, 23
    double staining for intracellular immunoglobulins, 23, 24
    immunohistochemistry, 2, 4
    materials, 20, 21
    overview, 19, 20
  autoimmune pathogenesis role, 3, 19, 25, 26
  differentiation, 26
  history of study, 3, 4
  phenotypic analysis
    density gradient centrifugation of peripheral blood mononuclear cells, 8, 12, 13
    enzyme-linked immunospot assay
      enumeration of antibody-secreting cells, 11, 14, 15
      enumeration of antigen-specific antibody-secreting cells, 12
      guidelines, 14, 15
    flow cytometry
      antibody secretion assay, 10, 11
      guidelines, 13, 14
      intracellular epitope staining, 10
      surface staining of living cells, 8, 9
  single cell analysis
    cloning and DNA sequencing, 33

Index

immunohistochemistry, 30, 31, 34
materials, 20, 35
microdissection, 31, 32, 34
overview, 26, 27
polymerase chain reaction of
  rearranged V-region genes
  first-round reactions, 32
  preamplification, 32
  primers, 29
  second-round reactions, 33, 35
synovial tissue preparation, 30
Bcl-2, apoptosis regulation, 121, 122
Bovine serum albumin, *see* Antigen-induced arthritis

C

CAIA, *see* Collagen antibody-induced arthritis
Cartilage oligomeric matrix protein (COMP)
  arthritis induction model
    antibody quantification in serum, 237, 238
    injection, 236, 237
    materials, 229, 230, 237
    overview, 227
    protein purification, 23, 34, 227, 229, 237
    scoring, 23, 38
    tissue preparation for protein extraction, 231, 232, 237
  cartilage composition, 225, 226
  cartilage erosion evaluation in pristane-induced arthritis, 26, 64
  structure, 226
Caspases
  apoptosis regulation, 119
  caspase-3 activation assay in T-cells
    flow cytometry, 128, 129, 134
    immunofluorescence microscopy, 128, 129, 134
CFA, *see* Complete Freund's adjuvant
CIA, *see* Collagen-induced arthritis
Collagen antibody-induced arthritis (CAIA)
  clinical evaluation, 220, 221
  materials, 218, 219
  monoclonal antibody purification, 220, 221
  overview, 21, 18
  passive transfer of antibodies, 220, 221
  statistical analysis, 220, 221
Collagen-induced arthritis (CIA)
  mouse model
    anti-collagen immunoglobulin G assay, 19, 97
    arthritis induction and assessment, 194, 195, 197
    collagen type II purification, 194
    materials, 192, 193, 196, 197
    overview, 192
    T-cell response characterization, 196, 197
  rat model
    animal husbandry, 206, 207
    autoantibody assay, 209, 213
    clinical evaluation, 208
    collagen type II purification, 206, 211, 212
    laboratory evaluation, 208, 209
    materials, 204, 210, 211
    overview, 201, 202
    timeline, 207, 212
  type XI collagen-induced arthritis
    antibody quantification in serum, 237, 238
    cartilage function, 225, 226
    injection, 236, 237
    materials, 229, 230, 237
    overview, 227
    protein purification, 227, 229, 236
    scoring, 23, 38
    tissue preparation for protein extraction, 231, 232, 237
COMP, *see* Cartilage oligomeric matrix protein

Complete Freund's adjuvant (CFA), arthritis induction, 255
Cytochrome c, release in apoptosis, 121
Cytokine signaling, *see* Adenovirus vectors

D

DC, *see* Dendritic cell
Dc11 mouse
  grading of degenerative lesions, 288, 289, 291
  histology, 287, 288, 299
  loading effects on osteoarthritis development, 296
  materials, 28, 87, 299
  overview, 28, 85
  polarized light microscopy of matrix collagen, 29, 96, 300
  radiology of hind limbs, 287
  RNA isolation from skeletal tissues, 290
  safranin O staining of glycosaminoglycans, 291, 292, 299
Dendritic cell (DC)
  artificial antigen-presenting cell, *see* Artificial antigen-presenting cell
  preparation for CD8$^+$ T-cell response to bacterial pathogens, 63, 64
  rheumatoid arthritis cell characterization
    flow cytometry
      alternative markers, 171, 172
      cell suspension preparation, 167
      fluorescence-activated cell sorting, 170, 177, 178
      four-color analysis, 170
      three-color analysis, 170
      two-color analysis, 170
    immunohistochemistry
      cytospin staining, 175, 176, 180
      dewaxing and antigen retrieval, 173, 179
      frozen sections, 175, 180
      staining, 17, 75, 179, 180
    magnetic-activated cell sorting, 169, 172, 173, 177
    materials, 166, 167
    mononuclear cell and erythrocyte rosette fraction preparation, 168, 169, 176
    overview, 165, 166
Differential display reverse transcription-polymerase chain reaction
  confirmation with reverse-Northern analysis
    complementary DNA grid preparation, 340, 343, 344
    hybridization and imaging, 340, 341
    labeling, 340
  first-strand complementary DNA synthesis, 336, 337, 342
  gel electrophoresis and autoradiography, 337, 343
  gene identification
    cloning and transformation, 339
    DNA sequencing and database searching, 339, 343
    gel band elution, 33, 39
    polymerase chain reaction to screen for recombinants, 339, 343
  materials, 33, 34
  principles, 33, 32
  radioactive polymerase chain reaction, 337, 342, 343
  rationale and advantages, 329, 330
  RNA isolation
    DNase treatment, 335, 342
    extraction, 335, 342
    quality control, 335, 336, 342
  RNA standardization between samples, 336
  time requirements, 341, 342
DNA microarray
  advantages and limitations, 330

data analysis
　cluster analysis, 309, 312, 324
　materials, 312, 313
　processing of data, 324
　software, 309, 323
　statistical analysis, 324
experimental design, 307
hybridization, 322, 325
labeling reaction
　cleanup, 321
　complementary DNA synthesis, 320
　coupling to label, 321, 325
　quenching and cleanup, 321, 322
　RNA hydrolysis, 320
principles, 306
RNA amplification
　antisense RNA synthesis and purification, 318, 319
　complementary DNA purification, 318
　first strand synthesis, 316, 317
　kit, 316
　second strand synthesis, 318
RNA isolation
　peripheral blood mononuclear cells, 314, 325
　tissue samples, 313, 314
　whole blood, 314, 315, 325
scanning, 323
washing, 322, 323, 325

E

ELISA, *see* Enzyme-linked immunosorbent assay
ELISPOT, *see* Enzyme-linked immunospot assay
Enzyme-linked immunosorbent assay (ELISA)
　anti-collagen immunoglobulin G assay in collagen-induced arthritis, 19, 97, 209, 213
　cytokine production by antigen-specific T-cells, 56
　cytokine secretion in T-helper subsets, 89
Enzyme-linked immunospot assay (ELISPOT)
　B-cell phenotyping
　　enumeration of antibody-secreting cells, 11, 14, 15
　　enumeration of antigen-specific antibody-secreting cells, 12
　　guidelines, 14, 15
　cytokine production by T-cells
　　antigen-specific T-cells, 56, 57, 67
　　cytokine secretion in T-helper subsets, 89

F

Flow cytometry
　antigen-specific T-cell phenotyping
　　antigen stimulation, 58
　　capture antibodies, 59
　　intracellular staining, 58
　artificial antigen-presenting cell T-cell capture assay, 4, 8
　B-cell phenotyping
　　antibody secretion assay, 10, 11
　　guidelines, 13, 14
　　intracellular epitope staining, 10
　　surface staining of living cells, 8, 9
　cytokine analysis in T-helper subsets
　　cell harvesting, fixation, and permeabilization, 92, 94
　　cytoplasmic cytokine staining, 95
　　materials, 91
　　overview, 81
　　T-cell activation, 94
　dendritic cells in rheumatoid arthritis
　　alternative markers, 171, 172
　　cell suspension preparation, 167
　　fluorescence-activated cell sorting, 170, 177, 178
　　four-color analysis, 170
　　three-color analysis, 170
　　two-color analysis, 170

426   Index

natural killer cells
   cytokine analysis, 15, 62
   intracellular granules, 156, 161
   isolation, 158, 159
   surface markers, 154, 155, 161
T-cell apoptosis assays
   annexin V staining, 127
   caspase-3 activation, 128, 129, 134
   TUNEL assay, 131

G

Gel electrophoresis, *see* Differential display reverse transcription-polymerase chain reaction; Two-dimensional gel electrophoresis

H

Heteroduplex analysis, T-cell repertoire analysis
   carrier plasmid stock preparation, 110
   heteroduplex reactions, 111
   polymerase chain reaction, 110, 111
Histology
   ankles in K/BxN mouse arthritis model, 280
   Del1 mouse, 287, 288, 299
   phage display in transplanted synovial-transplanted severe combined immunodeficient mouse, 383, 384, 386

I

Immunohistochemistry
   antigen specific B-cell detection in tissues
      biotinylation of antigen, 2, 3
      data analysis, 23
      double staining for intracellular immunoglobulins, 23, 24
      immunohistochemistry, 2, 4
      materials, 20, 21
      overview, 19, 20

B-cell single cell analysis, 30, 31, 34
dendritic cells
   cytospin staining, 175, 176, 180
   dewaxing and antigen retrieval, 173, 179
   frozen sections, 175, 180
   staining, 175, 179, 180
natural killer cells
   antigen retrieval by microwave, 156
   dewaxing and rehydration of sections, 156
   staining, 156, 158
Isoelectric focusing, *see* Plasma membrane protein mapping; Two-dimensional gel electrophoresis

K

K/BxN mouse arthritis model
   anti-GPI antibody screening, 279, 280
   breeding, 271, 272
   clinical evaluation, 273, 275
   crossing against other backgrounds, 277
   genotyping, 272, 273
   histology of ankles, 280
   overview, 269, 270
   serum transfer and arthritis induction, 27, 77
   transgenic T-cell analysis, 27, 79

M

MACS, *see* Magnetic-activated cell sorting
Magnetic-activated cell sorting (MACS)
   dendritic cells, 169, 172, 173, 177
   natural killer cells, 158, 161
Major histocompatibility complex (MHC)
   antigen-specific CD4$^+$ T-cell tracking with soluble major histocompatibility complex molecules

applications, 48
materials, 42
peptide loading and chromophore-
  labeled tetramer preparation,
  45, 46, 48
principles, 39, 40
recombinant class II molecule
  preparation
    biotinylation, 45
    cell growth, 43
    purification, 45, 47, 48
recombinant S-2 cell preparation
    pRmHa-3 construct, 42, 47
    single cell cloning, 43
    transfection, selection, and
      cloning, 42, 47
staining of cells
    clones, 46
    peripheral blood mononuclear
      cells, 48
molecule purification for T-cell
  capture assay
    biotinylation, 79, 80
    cell pelleting and lysis, 77, 78
    immunoaffinity chromatography,
      78, 79
    peptide complex preparation, 80, 84
    quality analysis, 79, 83, 84
MHC, see Major histocompatibility
  complex
Microarray, see DNA microarray
Monocyte/macrophage, T-cell
  interaction analysis
    activation of T-cells, 142
    contact activation assays of
      monocytes, 14, 47
    fixation of T-cells, 142, 143
    materials, 140, 146, 147
    overview, 139, 140
    peripheral blood monocyte
      preparation, 144, 145
    T-cell membrane preparation and
      detergent extraction, 143
    T-cell preparation
      cell lines, 142
      nylon wool chromatography, 141,
        142, 146, 147
      peripheral blood cells, 141
Murine antigen-induced arthritis, see
  Antigen-induced arthritis

N

Natural killer (NK) cell
  markers, 149
  phenotyping in synovial fluid and
    tissue
      cytokine analysis with flow
        cytometry, 15, 62
      cytotoxicity assays, 160, 162
      flow cytometry
        intracellular granules, 156, 161
        surface markers, 154, 155, 161
      immunohistochemistry
        antigen retrieval by microwave,
          156
        dewaxing and rehydration of
          sections, 156
        staining, 156, 158
      isolation
        fluorescence-activated cell
          sorting, 158, 159
        magnetic selection, 158, 161
      materials, 15, 53, 160
      mononuclear cell isolation, 154,
        160, 161
  subsets, 150
NK cell, see Natural killer cell
Northern blot, see Differential display
  reverse transcription-
  polymerase chain reaction

O, P

Oil-induced arthritis, see Pristane-induced
  arthritis
PCR, see Polymerase chain reaction
Phage display, synovial-transplanted severe
  combined immunodeficient
  mouse

candidate peptide sequence
specificity evaluation
confirmation of binding specificity
competition assay in vivo, 389, 392
histological distribution of biotinylated peptide, 387, 389
peptide synthesis, 387
DNA sequencing, 386, 387
histology, 383, 384, 386
human tissue preparation, 380, 391
materials, 37, 73
microvascular endothelium analysis, 371, 379, 380
peptide phage library injection, 381, 392
phage recovery and enrichment, 382, 383, 392
principles, 370, 371
removal of nonbound phage from circulation, 381, 392
validation in vitro
DNA sequencing of phage clones, 37, 79
enrichment of phage pools, 376, 377
integrin selection, 374
overview, 373, 374
phage amplification, 376, 389, 391
phage titration, 37, 76
xenotransplantation, 380, 391
PIA, see Pristane-induced arthritis
Plasma cell, differentiation, 26
Plasma membrane protein mapping
affinity purification, 364, 365
biotinylation of cell surface, 363
cell homogenization and fractionation, 364, 365
immune cell applications, 362
materials, 362, 363, 365
sample concentration and cleanup, 36, 66
solution-phase isoelectric focusing, 364, 366

two-dimensional gel electrophoresis limitations, 362
Polymerase chain reaction (PCR)
B-cell rearranged V-region genes
first-round reactions, 32
preamplification, 32
primers, 29
second-round reactions, 33, 35
cytokine secretion in T-helper subsets, 88, 89
differential display, see Differential display reverse transcription-polymerase chain reaction
T-cell repertoire analysis
heteroduplex analysis, 110, 111
spectratyping analysis, 111, 112
Pristane-induced arthritis (PIA)
acute phase response analysis with α_1-acid glycoprotein assay, 262, 264
adoptive T-cell transfer, 26, 65
cartilage erosion evaluation with cartilage oligomeric matrix protein, 26, 64
induction and evaluation, 259, 260, 264
materials, 25, 59
overview, 25, 57
systemic inflammation reaction evaluation with interleukin-6, 263, 265
T-cell response, 256, 257
Proteomics, see Plasma membrane protein mapping; Two-dimensional gel electrophoresis

R

Rheumatoid arthritis
animal models, see Animal models
dendritic cell characterization
flow cytometry
alternative markers, 171, 172
cell suspension preparation, 167

fluorescence-activated cell
 sorting, 170, 177, 178
four-color analysis, 170
three-color analysis, 170
two-color analysis, 170
immunohistochemistry
cytospin staining, 175, 176, 180
dewaxing and antigen retrieval, 173, 179
frozen sections, 175, 180
staining, 17, 75, 179, 180
magnetic-activated cell sorting, 169, 172, 173, 177
materials, 166, 167
mononuclear cell and erythrocyte rosette fraction preparation, 168, 169, 176
overview, 165, 166
T-cell phenotype, 118
tumor necrosis factor-a role, 191, 361, 362
RNA isolation, see Del1 mouse; Differential display reverse transcription-polymerase chain reaction; DNA microarray

S

Safranin O, staining of glycosaminoglycans, 291, 292, 299
Severe combined immunodeficient mouse, see Phage display, synovial-transplanted severe combined immunodeficient mouse
Spectratyping, T-cell repertoire analysis
 first-round polymerase chain reaction, 111, 112
 run-off labeling reactions, 112, 113
 visualization, 113

T

T-cell

antigen recognition overview, 71, 72
antigen-specific $CD4^+$ T-cell tracking with soluble major histocompatibility complex molecules
applications, 48
materials, 4, 2
peptide loading and chromophore-labeled tetramer preparation, 45, 46, 48
principles, 39, 40
recombinant class II molecule preparation
biotinylation, 45
cell growth, 43
purification, 4, 5, 47, 48
recombinant S-2 cell preparation
pRmHa-3 construct, 42, 47
single cell cloning, 43
transfection, selection, and cloning, 42, 47
staining of cells
clones, 46
peripheral blood mononuclear cells, 4, 8
antigen-specific cell characterization
antigen preparation, 54, 6, 7
$CD8^+$ T-cell response to bacterial pathogens
dendritic cell preparation, 63, 64
pathogen-specific T-cell line generation, 65
T-cell preparation, 64, 65
cloning and characterization of clones, 60, 61
cytokine production assays
enzyme-linked immunosorbent assay, 56
enzyme-linked immunospot assay, 56, 57, 67
expression library screening for antigen identification
library preparation, 62

screening protocol, 62, 63
statistics, 61, 62
flow cytometry phenotyping of
cytokine producing cells
antigen stimulation, 58
capture antibodies, 59
intracellular staining, 58
materials, 52, 53
overview, 51, 52
proliferation assay, 5, 6
stimulation, 59, 60
synovial fluid mononuclear cell
preparation, 53, 54, 59, 65
apoptosis assessment in synovial
fluid
arthritis T-cells, 118, 119
caspase-3 activation
flow cytometry, 128, 129, 134
immunofluorescence
microscopy, 128, 129, 134
DNA fragmentation, 12, 31
materials, 12, 24
mitochondrial depolarization
assays, 12, 27, 132, 133
phosphatidylserine
externalization, 127
positive selection of T-cells, 124,
125, 132
synovial fluid processing, 124, 132
TUNEL assay, 12, 31
artificial antigen-presenting cell studies
of antigen specific T-cells
artificial antigen-presenting cell
preparation, 80, 84
expansion of cells, 70, 71
identification strategies, 70
lipid lamellar sheet preparation,
76, 83
liposome preparation and sizing,
76, 77
major histocompatibility complex
molecule purification
biotinylation, 79, 80
cell pelleting and lysis, 77, 78

immunoaffinity
chromatography, 78, 79
peptide complex preparation,
80, 84
quality analysis, 79, 83, 84
materials, 7, 6, 83
principles, 72, 73
T-cell capture assay
flow cytometry, 8, 4)
peripheral blood mononuclear
cell immunostaining, 80, 81
collagen-induced arthritis model and
response characterization, 196,
197
monocyte/macrophage interaction
analysis
activation of T-cells, 142
contact activation assays of
monocytes, 14, 47
fixation of T-cells, 142, 143
materials, 140, 146, 147
membrane preparation and
detergent extraction, 143
overview, 139, 140
peripheral blood monocyte
preparation, 144, 145
T-cell preparation
cell lines, 142
nylon wool chromatography,
141, 142, 146, 147
peripheral blood cells, 141
pristane-induced arthritis response,
256, 257
repertoire analysis in synovial fluid
cell and cDNA preparation, 109, 110
heteroduplex analysis
carrier plasmid stock
preparation, 110
heteroduplex reactions, 111
polymerase chain reaction,
110, 111
materials, 99, 100, 10, 06
overview, 99, 107, 109, 113
spectratyping analysis

first-round polymerase chain
        reaction, 111, 112
    run-off labeling reactions, 112,
        113
    visualization, 113
  synovial fluid processing, 109
  rheumatoid arthritis phenotype, 118
  T-helper subsets
    cytokine secretion patterns and
        assays, 89
    flow cytometry analysis of
        cytokine secretion
      cell harvesting, fixation, and
          permeabilization, 92, 94
      cytoplasmic cytokine staining,
          95
      materials, 91
      overview, 81
      T-cell activation, 94
TNF-α, *see* Tumor necrosis factor-α
Tumor necrosis factor-α (TNF-α)
  plasma membrane protein mapping,
      see Plasma membrane protein
      mapping
  rheumatoid arthritis role, 191, 361,
      362
TUNEL assay, T-cell apoptosis
    assessment in synovial fluid,
    12, 31

Two-dimensional gel electrophoresis
  proteomics overview, 349, 350
  secreted proteins from articular
      cartilage
    cartilage explant conditioned
        medium preparation, 355, 357
    gel drying and autoradiography,
        357, 358
    isoelectric focusing, 356, 357
    materials, 350, 353, 354
    principle, 350
    sample loading by in-gel
        rehydration, 35, 57
    sample preparation
      dialysis and lyophilization, 355
      proteoglycan removal, 355, 357
    second dimension electrophoresis,
        356, 358
    silver staining, 35, 58

X

Xenotransplantation, *see* Phage display,
    synovial-transplanted severe
    combined immunodeficient
    mouse